Persistence Strategies of Weeds

Persistence Strategies of Weeds

Edited by

Mahesh K. Upadhyaya

Professor Emeritus, Applied Biology, Faculty of Land and Food Systems, University of British Columbia, Vancouver, BC, Canada

David R. Clements

Professor, Biology and Assistant Dean, Faculty of Natural and Applied Sciences, Trinity Western University, Langley, BC, Canada

Anil Shrestha

Professor, Weed Science and Chair, Dept. of Viticulture and Enology, California State University, Fresno, CA, USA

WILEY Blackwell

This edition first published 2022

Registered Office(s)
John Wiley & Sons, Inc., 111 River Street, Hoboken, NJ 07030, USA
John Wiley & Sons Ltd, The Atrium, Southern Gate, Chichester, West Sussex, PO19 8SQ, UK

Editorial Office
The Atrium, Southern Gate, Chichester, West Sussex, PO19 8SQ, UK

For details of our global editorial offices, customer services, and more information about Wiley products visit us at www.wiley.com.

Wiley also publishes its books in a variety of electronic formats and by print-on-demand. Some content that appears in standard print versions of this book may not be available in other formats.

Library of Congress Cataloging-in-Publication Data applied for

9781119525608

Cover Design: Wiley
Cover Images: Courtesy of Mahesh K. Upadhyaya

Set in 9.5/12.5pt STIXTwoText by Straive, Pondicherry, India
Printed and bound by CPI Group (UK) Ltd, Croydon, CR0 4YY

C9781119525608_280122

Contents

Biography

David R. Clements

David R. Clements (PhD, Queen's University) is a Professor of Biology and Assistant Dean of Science at Trinity Western University (TWU) in Canada. He researches invasive weed biology in British Columbia and other parts of the world such as China and Australia and has published extensively. He has served as an Associate Editor for the *Canadian Journal of Plant Science, Weed Research*, *Agronomy*, *Pacific Science*, and *Invasive Plant Science and Management* and two series on the biology of invasive species that he helped to create: one in *Pacific Science* and the other in *Invasive Plant Science and Management*. He received the Excellence in Weed Science Award from the Canadian Weed Science Society. He manages TWU's field research sites and teaches courses in botany and ecology, including field courses on Salt Spring Island and Hawaii. He is actively involved in local environmental advocacy and writes *The Green Beat*, a monthly column in the local newspaper.

Mahesh K. Upadhyaya

Mahesh K. Upadhyaya (PhD, University of Michigan) has served as a Professor of Plant Science for nearly four decades and as an Associate Dean of graduate studies in the Faculty of Land and Food System at the University of British Columbia (UBC), where he is currently a Professor Emeritus of Applied Biology. His interests include weed biology and ecology, nonchemical weed management, and crop physiology. He has served as an Associate editor of *Weed Science* journal and the *Canadian Journal of Plant Science* and has co-edited a book (with R.E. Blackshaw), the *Non-chemical Weed Management; Principles, Concepts and Technology*. He has taught several courses in biology, crop production and protection, different areas of weed science, and postharvest physiology at the UBC. He has received the Killam Teaching Award for outstanding teaching and J.F. Richards Service award at UBC and the Excellence in Weed Science award of the Canadian Weed Science Society. He is a fellow of the Weed Science Society of America, Canadian Weed Science Society, Indian Society of Weed Science, and the Canadian Society of Agronomy.

Anil Shrestha

Anil Shrestha (PhD, Michigan State University) is a Professor of Weed Science and the current chair of the Department of Viticulture and Enology at California State University, Fresno, CA, USA. He works on weed biology, ecology, and management in annual and perennial cropping systems and has published extensively in these areas. He is an Academic

Editor for PLOS ONE, an Associate Editor for *Agronomy Journal* and *Agricultural and Environmental Letters*, and an editorial board member for the *Journal of Crop Production* and the *Journal of Agroecology and Sustainable Food Systems*. He is a fellow of the American Society of Agronomy and has received the Weed Science Society of America's teaching excellence award, California Weed Science Society's award of excellence, and the California State University, Fresno Provost's excellence in teaching award.

List of Contributors

Steve W. Adkins
School of Agriculture and Food Sciences
The University of Queensland
Brisbane, QLD
Australia

Michael B. Ashworth
Australian Herbicide Resistance Initiative
School of Agriculture and Environment
The University of Western Australia
Perth, WA
Australia

Ali A. Bajwa
Weed Research Unit
New South Wales Department of Primary Industries
Wagga Wagga, NSW
Australia

Diego Batlla
Facultad de Agronomía
Cátedra de Cultivos Industriales
and Cátedra de Cerealicultura
IFEVA (UBA/CONICET)
Buenos Aires
Argentina

Hugh J. Beckie
Australian Herbicide Resistance Initiative
School of Agriculture and Environment
The University of Western Australia
Perth, WA
Australia

Roberto L. Benech-Arnold
Facultad de Agronomía
Cátedra de Cultivos Industriales
and Cátedra de Cerealicultura
IFEVA (UBA/CONICET)
Buenos Aires
Argentina

Fernanda C. Beveridge
School of Agriculture and Food Sciences
The University of Queensland
Brisbane, QLD
Australia

Caio Brunharo
Department of Crop and Soil Science
Oregon State University
Corvallis, OR
USA

Roberto Busi
Australian Herbicide Resistance Initiative
School of Agriculture and Environment
The University of Western Australia
Perth, WA
Australia

Ana L. Caicedo
Biology Department
University of Massachusetts
Amherst, MA
USA

Saul J.P. Carvalho
Department of Agronomy
Federal Institute of Education
Science and Technology of the South
Minas Gerais
Machado, Minas Gerais
Brazil

Pedro Christoffoleti
Crop Science Department
Luiz de Queiroz College of
Agriculture
University of São Paulo, São Paulo
Brazil

David R. Clements
Department of Biology
Trinity Western University
Langley, BC
Canada

Nathalie Colbach
Agroécologie, AgroSup Dijon
INRAE, Univ. Bourgogne
Univ. Bourgogne Franche-Comté
F-21000, Dijon
France

Antonio DiTommaso
Soil and Crop Sciences
School of Integrative Plant Science
Cornell University
Ithaca, NY
USA

Antoine Gardarin
UMR Agronomie
INRAE, AgroParisTech
Thiverval-Grignon
France

Xing-You Gu
Agronomy, Horticulture, and Plant Science
Department
Seed Molecular Biology Laboratory
South Dakota State University
Brookings, SD
USA

Saliya Gurusinghe
School of Agricultural
Environmental and Veterinary Sciences
Charles Sturt University
Wagga Wagga, NSW, 2678
Australia
and
Graham Centre for Agricultural Innovation
Charles Sturt University
Wagga Wagga, NSW, 2678
Australia

Linda M. Hall
Department of Agricultural
Food and Nutritional Science
University of Alberta
Edmonton, Alberta
Canada

Alois Honěk
Functional Biodiversity Group
Crop Research Institute
Praha
Czech Republic

Jitka Klimešová
Department of Experimental and
Functional Morphology
Institute of Botany of the Czech Academy
of Sciences
Třeboň
Czech Republic
and
Department of Botany
Faculty of Sciences
Charles University, Praha
Czech Republic

Sajid Latif
School of Agricultural, Environmental and
Veterinary Sciences
Charles Sturt University
Wagga Wagga, NSW, 2678
Australia
and
Graham Centre for Agricultural Innovation
Charles Sturt University
Wagga Wagga, NSW, 2678
Australia

Li Ma
Institute for Sustainable Horticulture
Kwantlen Polytechnic University
Surrey, BC
Canada

Jana Martínková
Department of Experimental and Functional Morphology
Institute of Botany of the Czech Academy of Sciences
Třeboň
Czech Republic

Helen Metcalfe
Rothamsted Research
Harpenden, Hertfordshire
UK

Alice Milne
Rothamsted Research
Harpenden, Hertfordshire
UK

Nadine Mitschunas
UK Centre for Ecology & Hydrology
Wallingford
UK

Acacio Goncalves Netto
Crop Science Department
Luiz de Queiroz College of Agriculture
University of São Paulo
São Paulo
Brazil

Paul Neve
University of Copenhagen
Department of Plant & Environmental Sciences
Højbakkegård Allé 13
Tåstrup
Denmark

Marcelo Nicolai
Agrocon Agronomic Consulting
Santa Bárbara d'Oeste
São Paulo
Brazil

Pavel Saska
Functional Biodiversity Group
Crop Research Institute
Praha
Czech Republic

Anil Shrestha
Department of Viticulture and Enology
California State University
Fresno, CA
USA

Jonathan Storkey
Rothamsted Research
Harpenden, Hertfordshire
UK

Clarence Swanton
Department of Plant Agriculture
University of Guelph
Guelph, ON
Canada

Mahesh K. Upadhyaya
Faculty of Land and Food Systems
University of British Columbia
Vancouver, BC
Canada

Mark VanGessel
Department of Plant and Soil Science
Research and Education Center
University of Delaware
Georgetown, DE
USA

Saeed Vazan
Department of Plant Agriculture
University of Guelph
Guelph, ON
Canada

Markus Wagner
UK Centre for Ecology & Hydrology
Wallingford
UK

Leslie A. Weston
Graham Center for Agricultural Innovation
Charles Sturt University
Wagga Wagga, NSW, 2678
Australia

Preface

Weeds reduce crop yields, lower the quality of agricultural produce, affect livestock health, and interfere with human life in a variety of ways. A considerable amount of time, money, and other resources are spent in controlling weeds by producers as well as by the general public, yet weeds persist. Unfortunately, we have relied very heavily on the use of synthetic herbicides to control weeds for the past several decades. This has hampered research on other nonchemical options for weed management. It is now well recognized that excessive use of synthetic herbicides carries risks to both the environment and the sustainability of herbicides as a tool. Widespread resistance to herbicides has developed among weed species and for almost all major classes of herbicide chemistry.

Stemming from concerns for human health and the environment, public opposition to the use of synthetic herbicides is progressively increasing. The sustainability of our food production systems is rapidly becoming an important issue globally, and pesticide-free, organic produce is becoming increasingly popular. In order to develop novel and more holistic methods for weed management, a sound understanding of persistence strategies of weeds is necessary. This understanding will help us identify the vulnerabilities of different weeds and could lead to development of novel, safe, and effective weed management strategies by making modifications to our production systems and reduce our dependence on synthetic herbicides. However, because weeds are masters of persistence, have co-evolved with humanity, and are very much the product of how we manipulate agroecosystems and other environments, we need to look beyond short-term simplistic remedies and understand our complex relationship with weeds.

This book takes a comprehensive approach to understand the persistence of weeds and strives to fill the gap in our understanding of the underlying issues behind the problem of weed persistence and serves as a comprehensive source of information for students, researchers, and weed managers. The various topics covered in this book include an overview of weed persistence, the role of seed production, dissemination and seedbanks, variability in seed dormancy, physiology and genetics of seed dormancy, seed longevity, vegetative propagation and propagule banks, the influence of agronomic practices, allelopathy, predation, soil microbes, climate change, weed evolution, and the development of herbicide resistance. Because weeds and their management are global concerns, specialists from around the world have been selected to write chapters on these topics.

The key learning objective of this book for students and other readers is to enhance understanding of what underpins the persistence strategies of weeds. While this book is aimed to serve upper-level undergraduates and/or graduate students, it can also be used as a reference or text for courses in agroecology and organic agriculture. Weed scientists and weed management professionals working for universities and government agencies, agribusiness consultants, organic farmers, and other environmentally conscious producers will find this book a valuable source of information on persistence strategies of weeds. We also expect this book to stimulate research on development of environmentally friendly weed management options.

We thank all the chapter authors of this book for contributing informative chapters in their areas of specialization, external reviewers for their critical and constructive reviews of chapter manuscripts, and our families for their cooperation, patience, and encouragement.

Mahesh K. Upadhyaya,
Professor Emeritus, Applied Biology
University of British Columbia
Canada

David R. Clements, and
Professor, Biology
Trinity Western University
Canada

Anil Shrestha
Professor, Weed Science
California State University
USA

Foreword

Weeds have been a costly problem for farmers since the dawn of agriculture. They have evolved with the crops, the farming methods, and the new environments as crops spread across the world. To be successful, weeds had to be genetically flexible in order to adapt to changing agricultural practices. Many thousands of wild plant species have not become successful weeds in agriculture. Successful weed species have many similar traits, but changing agricultural practices impact the relative importance of different traits. In short, successful weed species prosper in the disturbed and changing environments found in agriculture. Understanding weed biology and genetics is providing insight into the traits that contribute to the persistence and spread of weeds.

This book explores the many traits that make weeds successful. To persist, reproduction and propagation are paramount. A major factor in propagation is prolific production of long-lived, dormant seeds, with properties contributing to dispersal. This is especially true in weed management systems that rely heavily on tillage. The importance of seed biology in the persistence of weeds is illustrated by the fact that nine of the sixteen chapters in this book deal with seed biology. Weed seed dormancy is a complicated and fascinating trait influenced by many genetic, environmental, and even seed morphology factors. Seed biology is also integral to topics covered in other chapters, such as that on herbicide resistance.

Vegetative propagation of weeds is dealt with in Chapter 11. This persistence mechanism is found in many weed species that also produce seeds but is the predominant or even the only means of propagation in some species. Vegetative propagation is highly effective for many highly virulent weed species (e.g. Johnsongrass and purple nutsedge). The evolution of herbicide resistance can occur without seed production (e.g. in the aquatic weed *Hydrilla*).

Weeds have been with us for thousands of years. This has given weeds time to evolve traits allowing them to persist in farming systems practiced prior to introduction of herbicides. For example, weeds that have a similar visual appearance to the crop could avoid hand weeding. The advent of synthetic herbicides in the mid-twentieth century subjected weeds to a novel and intense selection pressure. However, this relatively efficient technology has not put any significant weed species on an endangered species list to my knowledge. The interplay between changing herbicides and evolved resistance continually shifts the relative importance of different weed species in time and space. The genetic flexibility of weeds in resisting herbicides has been remarkable. I think that Darwin would have been gratified to see such rapid, profound, diverse, and widespread responses to selection pressures. This amazing display of evolution in action is covered in Chapter 7.

Plants being sessile, their evolutionary response to pathogens and herbivores (including insects) has largely been chemical, producing secondary compounds that are toxic or fight these enemies in other ways, such as attracting beneficial insects that feed on herbivore insects. Some plants clearly release secondary products (allelochemicals) that inhibit the growth of competing plant species. Chapter 14 deals with this phenomenon in weeds and provides some enlightening case studies.

The influence of climate change on weed persistence is the topic of Chapter 12. Increased atmospheric CO_2 and temperature, as well as precipitation changes, will influence all plants, including crops and weeds. The relative efficacy of different herbicides on different weed species will change, especially between C_3 and C_4 species. The robust adaptability of weeds (as covered in Chapter 15) will ensure that weeds cope with changing climate.

These are but a few of the topics considered in the broad scope of this book that provides a valuable update by international experts on the many ways in which weeds have evolved to persist in agriculture and other disturbed ecosystems. It describes what we understand about weed persistence mechanisms and identifies the questions for which answers are still incomplete. I hope that it will inspire future research to fill these knowledge gaps so that we can devise more successful means of protecting crops and the environment from weeds. I expect that the bookshelves of serious students of weed and pest biology will contain this volume, as it provides the most current and update on how weeds persist, despite intense and costly efforts to manage them.

Stephen O. Duke
National Center for National Product Research, School of Pharmacy,
University of Mississippi, Oxford, MS, USA

1

Persistence Strategies of Weeds: Introduction

Anil Shrestha[1], *David R. Clements*[2], *and Mahesh K. Upadhyaya*[3]

[1] *Department of Viticulture and Enology, California State University, Fresno, CA, USA*
[2] *Department of Biology, Trinity Western University, Langley, BC, Canada*
[3] *Faculty of Land and Food Systems, University of British Columbia, Vancouver, BC, Canada*

1.1 Introduction

Since time immemorial, humanity has been plagued by weeds. Today weeds infest almost every environment on earth that is managed. These environments include wilderness areas where the ideal is a "pristine habitat" free of non-native weeds, expansive rangeland habitats where weeds threaten forage quality for livestock, more intensively managed cropping systems where weeds threaten yields, and even our urban yards where weeds negatively impact "the perfect lawn," which is the goal of many home owners. In any case, by now the reader is probably asking the question every weed scientist gets asked repeatedly – "what is a weed?"

In 1912, Blatchley defined a weed as "a plant out of place or growing where it is not wanted." Although this seems like a pretty all-encompassing definition, there is room for many other perspectives, such that of renowned American transcendentalist and writer Ralph Waldo Emmerson, also in 1912, who referred to a weed as a "plant whose virtues have not yet been discovered" (as quoted in Zimdahl 1999). If a weed is determined to be "out of place" (regardless of any potential virtues), in practical terms the weed is often very difficult to "weed out," especially because biologically weeds have evolved tremendous persistence strategies, which greatly affect human endeavors, especially agriculture.

Agriculture is an important component of the world's economy and an important means of livelihood, especially in developing countries (Alston and Pardey 2014). However, there is a lot of risk involved in crop production, much of which comes from insect pests, weeds, and pathogens. Weeds compete with crops and cause huge economic losses, estimated at an annual yield loss of 9% to global agriculture (Oerke 2006) and in economic terms, about US$ 27 billion to US agriculture alone (Pimentel et al. 2005). Similarly, weeds are also a

Persistence Strategies of Weeds, First Edition. Edited by Mahesh K. Upadhyaya, David R. Clements, and Anil Shrestha.

problem in non-crop systems (e.g. roadsides, waterways, etc.) and natural ecosystems (e.g. forests, landscapes) where invasions of undesirable plants can cause aesthetic and economic losses and reduce biodiversity (Neve et al. 2018). They have continued to evolve and persist, despite humankind's efforts to control them.

1.2 Persistence of Weeds

Merriam-Webster's dictionary defines the word "persist" as "to continue to exist especially past a usual, expected, or normal time" or "to go on resolutely or stubbornly in spite of opposition" ("Persist" Merriam–Webster.com Dictionary n.d.). In this book we refer to "weed persistence" as "the ability of undesirable plant species to continually evolve, survive, thrive, and reproduce under a variety of natural and anthropogenic selection pressures." Weed scientists, farmers, and land managers continue to be baffled by the persistence of these species despite a wide variety of management techniques developed to eliminate them. In this context, some questions that can be raised include the following: (i) What makes a weed so persistent? (ii) Why are the natural and human-developed selection pressures failing to eliminate them? (iii) What lessons can we learn from the strategies that they have developed to persist in the various ecosystems? (iv) How can we use this knowledge of weed persistence to minimize the damage they cause to agroecosystems and other human-managed ecosystems?

Although Baker (1965) listed the factors that enable a species to become a weed, the plants continue to persist despite our knowledge of these factors. For decades, we have tried to manage weeds with the so-called four corners of weed management, i.e. biological, chemical, cultural, and mechanical weed control, and yet weeds have persisted. Therefore, it is essential to revisit this list to discuss weed persistence.

Baker (1965) listed the following characteristics that make a plant successful as a weed: no special environmental requirements for germination, discontinuous germination, longevity of seeds, rapid seedling growth, short vegetative periods, ability to maintain seed production as long as growing conditions permit, self-compatible but not necessarily self-pollinated or apomictic, cross-pollinated types that can be pollinated by various methods, ability to produce numerous seeds even under unfavorable environments, good mechanisms for short- and long-distance dispersal, vigorous and multiple methods of vegetative reproduction in asexually reproducing species, and having good competitive mechanisms. Some very persistent weeds may have only one or two of these characteristics, whereas others may have nearly all; the point is that this set of characteristics represent key characteristics that make weeds so successful. In summary, we are dealing with plants that have very successful persistence mechanisms despite of above- and belowground natural and human-induced selection pressures and management methods (Figure 1.1). In the following text, some of these mechanisms will be discussed briefly to give an overview of the persistence strategies of weeds.

In this chapter, we discuss persistence of weeds in terms of (i) seed dormancy, germination, seedling emergence, seed production, and seed return, (ii) soil seedbank and longevity of seeds in the soil seedbank, and (iii) ability to persist against natural and human selection pressures and management methods. The other chapters in the book will describe these phenomena in detail and provide valuable insights to understand the persistence strategies of weeds.

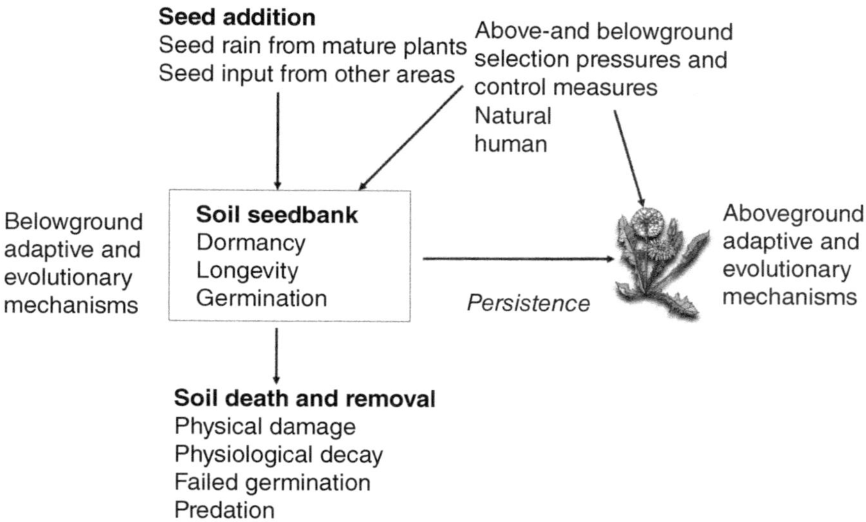

Figure 1.1 Cycle of above- and below-ground natural and human selection and management pressures, continued adaptation, evolution, and persistence of weeds that reproduce by seeds in agroecosystems and other human-managed ecosystems.

1.2.1 Seed Dormancy, Germination, Seedling Emergence, Seed Production, and Seed Return

Seed dormancy has been defined generally as the failure of a seed to germinate despite favorable conditions (Bewley and Black 1994; Benech-Arnold et al. 2000). There are two main types of seed dormancy – primary and secondary (induced) (e.g. Karssen 1982; Bewley and Black 1994). Many weed seeds are known to possess the ability to remain dormant, but viable, in the soil for extended periods of time. For example, the classic, pioneer experiment of W.J. Beal who studied seed dormancy and viability of buried weed seeds in Michigan has been cited in numerous publications (e.g. Kivilaan and Bandurski 1981; Baskin and Baskin 1985; Burnside et al. 1996; Telewski and Zeevart 2002). These and other studies have demonstrated that seeds of some weeds can retain viability and remain dormant for up to more than 100 years. Therefore, seed dormancy can be considered as a primary factor contributing to weed persistence, and knowledge of and a sound understanding of this phenomenon is of prime importance for weed management.

Once dormancy is broken, seeds germinate (emergence of radicle) when the necessary environmental conditions for water, temperature, oxygen, and, in some species, light are met. The range of favorable environmental conditions to stimulate germination depends on the plant genotype (Baskin and Baskin 2001). Seed germination occurs when these factors are present at or above the minimum threshold but below the species-specific maximum levels (Bewley and Black 1994; Baskin and Baskin 2001).

Unlike crop seeds that have a narrower range of optimum conditions for germination, studies have shown that weed seeds can adapt and germinate under a wide range of environmental conditions (e.g. Grundy et al. 2000; Steinmaus et al. 2000; Leon et al. 2004). Whitney and Gabler (2008) suggested that many invasive plant species do not have specialized germination requirements and are highly flexible, making it possible for them to adapt to a wide range of environments. Even under projected climate change scenarios, weed seed persistence is an area of concern. Walck et al. (2011) emphasized the importance of conducting research on seed ecology because they believed that seeds of weedy species could evolve relatively quickly to keep pace with climate change. Furthermore, germination of a few weed seeds and seed production from the resulting plants can add thousands of seeds to the soil seedbank, ensuring the persistence of the weeds.

Since germination is a process primarily driven by a combination of temperature and moisture, researchers have modeled weed seed germination based on the combination of these two factors, i.e. hydrothermal time (e.g. Gummerson 1986; Roman et al. 2000; Bradford 2002). Although these models have enhanced the knowledge of the parameters of weed seed germination, how to use this information from a management standpoint and reduce persistence of weeds in our ecosystems remains a challenge. Nevertheless, modeling is an important component of understanding and predicting population dynamics of weeds and therefore for our understanding of weed persistence. Borgy et al. (2015) emphasized that prediction of population dynamics of annual plants is challenged by soil seedbanks that are difficult to see and presented a time series model to assess weed population dynamics based on data from growing plants and the soil seedbank.

Seed vigor, good seed bed preparation, and proper soil conditions for seedling emergence have been cited as primary needs of seedling establishment of agricultural crops for optimum yields (Finch-Savage and Bassel 2016). However, in the case of weeds, they can emerge and complete their life cycle in less than optimum soil conditions compared with crops. A few weed seedlings that emerge and establish can produce thousands of seeds that go back to the seedbank. Weed seedlings can emerge from the harshest of soil or other substrates, complete their life cycle, and produce seeds (Figure 1.2).

Weeds are capable of producing huge number of seeds. For example, species such as Palmer amaranth (*Amaranthus palmeri*) and horseweed (*Erigeron canadensis*), which are considered to be among the worst agricultural weeds in North America (Weed Science Society of America [WSSA] 2017), can produce up to 600 000 seeds (Keeley et al. 1987) and 200 000 (Weaver 2001) to 800 000 (Shrestha et al. 2010) seeds per plant, respectively. Often, late-season weed control is ignored as the weeds do not cause crop yield loss. However, these late-emerging weeds could deposit thousands of seeds to the soil seedbank. Bagavathiannan and Norsworthy (2012) emphasized that these late-season weeds should not be ignored and suggested the development of thresholds for these weeds using bioeconomic models. Long et al. (2015) emphasized the importance of information on environmental effects on the loss of weed seeds from the soil seedbank, either by germination or death, and proposed a model to predict the length of seed persistence in a range of environments. The ability to germinate, emerge, grow, and produce viable reproductive structures under a wide range of environmental conditions is important for weed persistence. The use of simulation models to predict seedbank and weed population dynamics are therefore needed.

1.2.2 Seed Longevity, Emergence Patterns, and Soil Seedbanks

Since weed seeds and vegetative propagules in soil banks are important for persistence of weeds, they should be the focus of any management strategy for effective weed control

Figure 1.2 Various species of weeds growing, flowering, and producing seeds under soil-less and adverse plant-growing conditions: (a) an *Amaranthus* sp. plant growing through a crack on a sidewalk, (b) an annual sowthistle (*Sonchus oleraceus*) plant growing on the roof of a greenhouse, (c) an annual sowthistle (*Sonchus oleraceus*) plant growing on a gravel road, and (d) a horseweed (*Erigeron canadensis*) plant growing on the roof of a metal lathe house.

(Kremer 1993; Dekker 1999; Walsh et al. 2018). As mentioned earlier, not all the weed seeds that are present in the soil seedbank germinate and emerge immediately or at the same time. Diverse dormancy mechanisms and modification of seed dormancy and germination by environmental cues provide weeds the ability to distribute their germination over time, which is an important factor in persistence of weeds (Baskin and Baskin 2001; Clements et al. 2004). Genotype, environmental factors during seed development and post-dispersal, and soil-borne pests and pathogens influence seed dormancy, longevity, and germination and therefore are important factors in the persistence of weed seeds in soil seedbanks (Egley and Duke 1985). Forcella et al. (1997) collected soil cores from several sites in the US corn belt and monitored weed seed germination and emergence. They found significant variability in the percentage of seedlings that emerged with time and attributed this to effects of microclimate variables. Saatkamp et al. (2011) concluded that the functional role of delayed germination and light and diurnally fluctuating temperatures are key contributors to higher soil seed persistence. Saska et al. (2020) suggested that seeds undergo changes in morphology and viability while in the seedbank. They buried seeds of several weed species in the soil, excavated them six to eight years later and assessed the changes in seed morphology and proportion of fresh seeds to determine seed persistence. They found that seed mass, volume, and proportion of persistent seeds in most species declined with increase in burial periods.

Knowledge of genetic and epigenetic factors, which regulate seed dormancy (Gu et al. 2006), is necessary to comprehend persistence strategies of weeds. A study by Pipatpongpinyo et al. (2019) reported that regulation of soil seedbank longevity was due to genes controlling seed dormancy. Davis et al. (2016) highlighted the importance of the link between inherited weed seed traits in their persistence and suggested that management practices could be tailored to specific weed species based on particular chemical or physical defenses.

Soil microbes have been reported to increase weed seed mortality in soil seedbanks (Gallandt et al. 1999; Kennedy 1999). Ullrich et al. (2011) suggested that this increased mortality could be because of seed coat being damage by soil microbes. However, their study on the effect of microbial abundance on weed seed persistence did not show a strong correlation between seed mortality and seed coat damage, suggesting that soil microorganisms may not play a major role in determining seed mortality. More research is needed to understand the magnitude of effects of soil microbes on seed longevity, viability, and germination and to exploit the potential of reducing the size of seedbanks by enhancing microbial effects.

Since seeds and vegetative propagules can survive and maintain their viability for extended periods, understanding of factors that influence their longevity is essential for understanding weed persistence and for development of effective weed management strategies. Information on seedbank persistence can help in long-term weed management and possibly weed eradication (Panetta and Timmins 2004). The persistence of weed seedbanks, therefore, is an important area of investigation for weed scientists who are working on this and various other aspects of persistence strategies of weeds, including genetics and physiology of seed dormancy and germination, seed longevity, environmental influences, ecophysiology of persistence, and effects of soil biota and allelochemical interactions, to name a few. Auld (2004) looked at the social aspects of weed seed survival in the seedbank and concluded that due to the inevitability of persistent weed populations due to seedbanks, agricultural communities must live with the inevitable presence of weeds and their effects.

Another factor that has gained some attention in terms of weed seed and seedling death is allelopathy (Arroyo et al. 2018; Weston and Duke 2003). Several studies have explored the effects of extracts from several plants, suspected to possess allelochemicals, on seeds, and seedlings of various weed species (e.g. Panasuik et al. 1986; Hoffman et al. 1996; Marles et al. 2010). Scavo et al. (2019a) reported that the crop globe artichoke (*Cynara cardunculus*) produced allelochemicals that reduced the soil seedbank of some weed species. Similarly, Vitalini et al. (2020) reported allelopathic activity of Italian ryegrass (*Lolium multiflorum*) on some weed species of rice (*Oryza sativa*) fields. However, we still need to know much more about allelopathic interactions (Scavo et al. 2019b) to be able to exploit these to manage weed problems. Reasons for the broad variation reported for the role of allelopathy in reducing weed seed longevity, germination, and seedling emergence are still not well understood.

1.2.3 Persistence of Weeds Through Perennial Growth

Although, as described in Section 1.2.2, seed production, dissemination, and seedbanks constitute key life history strategies for the persistence of many weed species, perennial growth is a life history strategy employed by other weed species, or occasionally species may utilize both strategies. Just as seed production results in seedbanks, perennials plants may produce bud banks consisting of underground stems, rhizomes, or roots with buds that are poised to produce new plant ramets under favorable conditions (Ott et al. 2019). Although mechanical management in agroecosystems via use of plows and other implements may kill both annual and perennial weeds, new shoots may emerge from bud banks as a result of plowing (Leakey et al. 1978), thus representing an effective persistence strategy for such perennial weeds. Two key factors determining the function of the bud bank in this way are their vertical distribution and their number (Klimešová et al. 2019). Obviously, a greater number of buds increases persistence, whereas in terms of depth, more deeply placed buds tend to have the highest persistence (Pausas and Paula 2020).

In addition to bud banks, there are numerous other means by which weeds can persist via perennation. However, the ultimate measure of persistence by perennation, and weed persistence in general, is genetic fitness. Most perennial plants have multiple aspects to their fitness, as many produce seeds, and some even have multiple avenues for vegetative propagation. For example, knotweed species such as *Reynoutria sachalinensis* (giant knotweed) or *Reynoutria × bohemica* (Bohemian knotweed) are capable of propagating via seeds, rhizomes, and stem fragments (Gillies et al. 2016). Therefore, for knotweed species, fitness is a measure of how well the combination of these three means of propagation are likely to produce offspring.

Computing fitness for the combination of resprouting (vegetative propagation) or seeding (producing viable seeds) is a complex matter, especially since environmental factors influence variable allocation to different modes of reproduction by a particular plant (Martínková and Klimešová 2016). Allocation to different modes of reproduction even varies over different phases of a perennial plant's life span (Sosnová et al. 2014). Whether or not the fitness of plants with perennial growth can easily be assessed, in many settings perennial growth results in high levels of persistence. Weeds with substantial underground structures such as knotweeds may require repeated herbicide applications over several years to control, and for such weeds alternative control methods are often not recommended, because excessive disturbance could make the situation worse (Clements et al. 2016).

1.2.4 Persistence Under a Wide Range of Environmental and Climatic Conditions

Weed scientists have developed models to describe the phenology of various weed species (Ghersa and Holt 1995; Swanton et al. 2000; Huang et al. 2000; Huang et al. 2001a, 2001b; Shrestha and Swanton 2007). Similarly, studies have shown that several problematic agricultural weed species can produce viable seeds under very low light environments or shade (Bello et al. 1995; Begna et al. 2002; Steckel et al. 2003; Brainard et al. 2005; Chauhan 2013). Furthermore, the abilities of weed species to grow and produce viable seeds under a wide range of stress situations such as drought (Webster and Grey 2008; Chauhan and Johnson 2010; Sarangi et al. 2015; Kaur et al. 2016; Chahal et al. 2018; Mahajan et al. 2018), soil salinity (Chauhan et al. 2013; Shrestha et al. 2018), soil pH (Buchanan et al. 1975), and future predicted environments have also been documented (Ziska and George 2004; Grantz et al. 2010).

Weeds can persist under projected scenarios of climate change. For example, carbon dioxide (CO_2) and ozone (O_3) are projected to increase under future climatic scenarios (IPCC 2013). The effects of these projected elevations are being studied on weed species (e.g. see Patterson 1995; Ziska and George 2004; Clements and DiTommaso 2011 for comprehensive reviews). Clements and DiTommaso (2011) have suggested that under the future scenarios of climate change, many invasive plants may develop adaptations leading to range expansions. One essential prerequisite for this projected scenario to happen is the successful production of reproductive propagules. As discussed earlier, assessment of these factors has focused more on vegetative growth than on seed production in annual (Ziska and Bunce 1997; Ziska and Goins 2006; Naidu and Paroha 2008; Paudel et al. 2016) and perennial (Saebø and Mortensen 1998) weeds. However, some studies have focused on seed and tuber production as well (O'Donnell and Adkins 2001; Ziska et al. 2004; McPeek and Wang 2007; Rogers et al. 2008; Nguyen et al. 2017). Several studies have concluded that the growth of some weed species such as horseweed, Palmer amaranth, and common waterhemp (*Amaranthus tuberculatus*) was not affected very much by elevated O_3 (see Grantz et al. 2008, 2019; Paudel et al. 2016). Grantz et al. (2010) reported that elevated O_3 resulted in more dry matter allocation to reproductive tubers in the perennial weed, yellow nutsedge (*Cyperus esculentus*), while Landesmann et al. (2013) concluded that elevated O_3 could increase persistence of corn spurry (*Spergula arvensis*) in fields by increasing seed dormancy and longevity in the soil seedbank.

In summary, the biological and ecological factors described previously have a major role in improving our understanding of how weeds persist via seed production and other means such as perennial growth, which in turn could provide insights for developing management options. In the face of these inherent persistence characteristics of weeds, it is important to discuss the success and failures of weed management strategies.

1.3 Current Approaches to Manage Weeds and Persistence

Although there are many different methods of controlling weeds including physical, cultural, mechanical, and biological methods, chemical control has become the major weed management tool globally, especially in the developed countries. Synthetic herbicides were developed in the 1940s, and until the past few decades, these chemicals were highly successful in controlling weeds. However, in recent years, the use of synthetic herbicides is being challenged due to environmental and human health concerns. Even the most successful and widely used herbicide globally, glyphosate, long touted for its effectiveness and specificity, is being questioned regarding food safety (Sfiligoj 2019). While issues such as these create awareness and fear of chemical pesticide use, there is another major issue that growers and weed scientists are facing but is probably less known to people outside these professions. This issue is the evolution of herbicide-resistant weed populations worldwide. There are more than 500 unique cases of herbicide-resistant weeds with more than 260 weed species showing resistance to one or more herbicides (Heap 2020). Many farmers in the United States are going back to hand weeding as a last resort because of herbicide-resistant weeds, particularly glyphosate (Roundup®)-resistant weeds (e.g. Sosnoskie and Culpepper 2014). This has an obvious implication for the economics of crop production and weed persistence. Most of the scientific journals in weed science and trade magazines are rife with reports of documented herbicide-resistant weed populations, molecular and physiological mechanisms of evolution of herbicide-resistant weed biotypes, management of herbicide-resistant weed biotypes, economic losses by herbicide-resistant weeds, and social aspects of herbicide resistance. Therefore, most of the efforts seem to be directed toward understanding herbicide resistance and the management of herbicide-resistant weeds. However, if the gist of the problem of weed management and the issues surrounding it were to be summarized, the overarching theme is that these so-called "undesirable species" have very capable strategies to persist adverse situations and conditions. In other words, weeds have a complex of strategies for persistence, and, if they are to be managed, it is important to understand underpinnings of their persistence.

1.3.1 Ability of Weeds to Persist Against Biological, Cultural, Mechanical/Physical, and Chemical Control Measures

Integrated weed management (IWM) generally advocates the use of combinations of biological, cultural, physical, and chemical control methods (Harker and O'Donovan 2013). It is well known that particular control measures have not been able to eliminate weeds completely. While species that are susceptible to a particular control method may be eliminated, other species that are tolerant adapt to the selection pressure and become the dominant species. Such species shifts often change the weed composition from more susceptible to more tolerant species to the method of control used (Aldrich and Kremer 1997). Further, repeated use of the same management technique, chemical or non-chemical, will select for species that will likely resist that technique. Therefore, the use of diverse weed management methods in rotation is important (Harker and O'Donovan 2013).

Wheeler and Schaffner (2013) have summarized examples of weeds that have become tolerant to biocontrol agents and explained the underlying mechanisms. Burdon and Marshall (1981) reported that the degree of control achieved via biocontrol depended on the mode of reproduction of the target weed species as asexually reproducing species were more effectively controlled than sexually reproducing ones. Biocontrol agents are selected to be very host specific (Schwarzländer et al. 2018), and elimination of the targeted species can cause weed species shifts and promote the persistence of tolerant species.

Major strategies of cultural control of weeds generally include tillage, crop rotation, cover crops, manipulation in crop planting dates, crop seeding rates, crop row spacings, stale seedbank techniques, mulching, intercropping, companion cropping, and managing fertilizer and irrigation (Melander et al. 2005). Although there are several examples of weed suppressive ability of these methods, in many cases, these systems can change weed community composition and cause species shift rather than completely eliminate weeds. For example, Streit et al. (2003) reported broadleaf annual weeds to be more abundant in conventionally tilled wheat fields. Shrestha et al. (2002) observed that success of reducing weed densities by tillage was affected by the crop type; while tillage had no effect on weed densities in wheat, weed densities were greater in the conventional tillage than in the no-till systems in three types of bean crops. Tillage and crop rotations tend to alter species composition rather than totally eliminating weeds (Derksen et al. 1993; Clements et al. 1996; Cardina et al. 2002; Murphy et al. 2006). Similarly, there are several examples of sole use of cover crops having inconsistent effects on weed densities (Swanton et al. 1999; Fisk et al. 2001), in some cases suppressing weed biomass but not densities (Hayden et al. 2012). Teasdale (1996) concluded that cover crops provide early but not full season weed control. There are several examples of the success and problems associated with other cultural control strategies listed earlier (see Travlos et al. 2018).

Mechanical weed control methods, although highly successful in reducing weed densities, also have some inherent weaknesses that enable certain species to persist especially when used repeatedly. For example, repeated mechanical cultivation caused a shift in weed species composition to those that mimicked the germination and emergence characteristics of the crop (Radosevich and Holt 1984). Repeated hand weeding in rice favored barnyard grass that showed phenotypic similarities with rice (Barrett 1983). Also, each mechanical tool has different levels of success and problems. Gunsolus (1990) highlighted some of these issues with rotary hoeing and inter-row cultivation where rotary hoeing was effective on weeds that emerged from shallower depth but not on weeds that germinated from deeper than 5 cm and inter-row cultivation was most effective on weeds up to 10–15 cm tall. In an on-farm study conducted at 64 locations in Iowa, Hartzler et al. (1993) reported that using cultivation solely resulted in greater weed densities than with the use of herbicides in 41% of the locations. Furthermore, some perennial weeds that can reproduce vegetatively, such as johnsongrass (*Sorghum halepense*), field bindweed (*Convolvulus arvensis*), or bermudagrass (*Cynodon dactylon*), may spread throughout the field following mechanical cultivation (Colquhoun 2001).

Globally, shortage of labor in agriculture in many parts of the world has resulted in an increase in the use of chemical control and is projected to further increase (Gianessi 2013). Even though there are approximately 29 different modes of action of herbicides (WSSA 2019) and synthetic chemicals have been used since the 1940s, weed populations and communities have persisted. There could be several reasons for this phenomenon including herbicide failure due to environmental reasons, misapplication of herbicides, seed return from weed escapes after the crop growing season, selectivity of herbicides on target weed species only, and restrictions on herbicide use. However, the issue that has drawn the greatest interest of weed scientists in recent years, globally, has been herbicide resistance (Heap 2014). In a perspective published in 2012, S.O. Duke wrote "No new major herbicide mode of action has been introduced in a commercial herbicide active ingredient in the last 20 years" (Duke 2012). This statement still holds true in early 2021. The importance of using alternatives to chemicals was recognized as early as 1929 (Harker and O'Donovan 2013), and the historical sequences and documentation of herbicide-resistant weeds (Heap 2014) strengthen this importance. However, Harker and O'Donovan (2013) highlight that despite the increase in IWM, publications on chemical weed management still dominate the weed science literature. Hence, it can be argued that even with new herbicide modes of action, weeds will continue to persist for the reasons cited earlier for herbicide failures. Weeds will always evolve to the management strategies and will continue to persist. Much of this relates to evolutionary strategies of weeds. Vigueira et al. (2013) discussed the evolutionary process in weeds, explained the mechanisms of their adaptation, and concluded that a better understanding of genetic mechanisms underlying the adaptive process of weedy species is necessary as selective pressures on weeds continue to be applied around the world.

1.4 Conclusions

The study of persistence strategies of weeds is fundamental to the development of IWM strategies. In this respect, it is important to examine how superior seed production, dissemination, maintenance of soil seedbanks, and in many cases perennial growth of weeds facilitate their persistence. The role of dormancy in amplifying the persistence of weed seeds, the genetic basis for seed dormancy and related mechanisms, and environmental interactions need to be understood. The question of how long weed seeds persist in the soil needs to be properly investigated. The evolution of herbicide-resistant weeds and its implication for persistence need to be evaluated. The persistence characteristics of weeds are no secret, and many of these have been well known since humans began tilling the soil and have only multiplied as we have tried to control their proliferation and spread. Still there is merit in going back to the basics and naming the critical factors as Baker (1965) did and revisiting Baker's list of ideal weed characteristics. Furthermore, new weed science techniques have greatly added to our ability to quantify the impacts of weed traits through genetic analysis and weed population modeling.

Weed reproduction does not take place in a vacuum, but rather other aspects of the ecosystem such as climate change, microbial communities, and allelochemicals need to be considered. Furthermore, aspects such as the role of weed seed predation and effects of agronomic practices on weed persistence are important to consider. It is important to note that compared with plants that rely on seedbanks, perennial weeds possess an entirely different strategy for gaining persistence. Some weeds even have both strategies at their disposal. Ultimately, at the root of weed persistence is the ability of weeds to evolve in response to various selection pressures, including the conditions imposed by our attempts to manage them, which may inadvertently apply selection pressure favoring weedy traits. We anticipate that an understanding of the effects of these factors will lead to more knowledge of weed persistence and provide some recommendations on the way forward in this age-old battle between humanity and the weeds.

References

Aldrich, R.J. and Kremer, R.J. (1997). *Principles in Weed Management*, 2e, 35–63. Ames, IA: Iowa State University Press.

Alston, J.M. and Pardey, P.G. (2014). Agriculture in the global economy. *Journal of Economic Perspectives* 28: 121–146.

Arroyo, A.I., Pueyo, Y., Giner, M.L. et al. (2018). Evidence for chemical interference effect of an allelopathic plant on neighboring plant species: a field study. *PLoS One* 13 (2): e0193421.

Auld, B.A. (2004). The persistence of weeds and their social impact. *International Journal of Social Economics* 31: 879–886.
Bagavathiannan, M.V. and Norsworthy, J.K. (2012). Late-season seed production in arable weed communities: management implications. *Weed Science* 60: 325–334.
Baker, H.G. (1965). Characteristics and mode of origin of weeds. In: *The Genetics of Colonizing Species* (eds. H.G. Baker and G.L. Stebbins), 147–172. New York, NY, USA: Academic Press.
Barrett, S.C.H. (1983). Crop mimicry in weeds. *Economic Botany* 37: 255–282.
Baskin, C.C. and Baskin, J.M. (1985). The annual dormancy cycle in buried weed seeds: a continuum. *BioScience* 35: 492–498.
Baskin, C.C. and Baskin, J.M. (2001). *Seeds: Ecology, Biogeography, and Evolution of Dormancy and Germination*. San Diego, CA: Academic Press.
Begna, S.H., Dwyer, L.M., Cloutier, D. et al. (2002). Decoupling of light intensity effects on the growth and development of C_3 and C_4 weed species through sucrose supplementation. *Journal of Experimental Botany* 53: 1935–1940.
Bello, I.A., Owen, M.D.K., and Hatterman-Valenti, H.M. (1995). Effect of shade on velvetleaf (*Abutilon theophrasti*) growth, seed production, and dormancy. *Weed Technology* 9: 452–455.
Benech-Arnold, R.L., Sánchez, R.A., Forcella, F. et al. (2000). Environmental control of dormancy in weed seed soil banks. *Field Crops Research* 67: 105–122.
Bewley, J.D. and Black, M. (1994). *Seeds: Physiology of Development and Germination*. New York, NY: Plenum Press.
Blatchley, W.S. (1912). *The Indiana Weed Book*, 191 pp. Indianapolis, IN: Nature Publishing Co.
Borgy, B., Reboud, X., Peyrard, N. et al. (2015). Model to estimate life history traits from standing plant time series. *PLoS One* 10 (10): e0139278. https://doi.org/10.1371/journal.pone.0139278.
Bradford, K. (2002). Applications of hydrothermal time to quantifying and modeling seed germination and dormancy. *Weed Science* 50: 248–260.
Brainard, D.C., Bellinder, R.R., and DiTommaso, A. (2005). Effects of canopy shade on the morphology, phenology, and seed characteristics of Powell amaranth (*Amaranthus powellii* S. Wats.). *Weed Science* 53: 175–186.
Buchanan, G.A., Hoveland, C.S., and Harris, M.C. (1975). Response of weeds to soil pH. *Weed Science* 23: 473–477.
Burdon, J.J. and Marshall, D.R. (1981). Biological control and the reproductive mode of weeds. *Journal of Applied Ecology* 18: 649–658.
Burnside, O.C., Wilson, R.G., Weisberg, S., and Hubbard, K.G. (1996). Seed longevity of 41 weed species buried 17 years in eastern and western Nebraska. *Weed Science* 44: 74–86.
Cardina, J., Herms, C.P., and Doohan, D.J. (2002). Crop rotation and tillage system effects on weed seedbanks. *Weed Science* 50: 448–460.
Chahal, P., Irmak, S., Jugulam, M., and Jhala, A.J. (2018). Evaluating effect of degree of water stress on growth and fecundity of palmer amaranth (*Amaranthus palmeri*) using soil moisture sensors. *Weed Science* 66: 1–8.
Chauhan, B.S. (2013). Shade reduces growth and seed production of *Echinochloa colona*, *Echinochloa crus-galli*, and *Echinochloa glabrescens*. *Crop Protection* 43: 241–245.
Chauhan, B.S. and Johnson, D.E. (2010). Growth and reproduction of junglerice (*Echinochloa colona*) in response to water stress. *Weed Science* 58: 132–135.

Chauhan, B.S., Abugho, S.B., Amas, J.C., and Gregorio, G.B. (2013). Effect of salinity on growth of barnyardgrass (*Echinochloa crus-galli*), horse purslane (*Trianthema portulacastrum*), junglerice (*Echinochloa colona*), and rice. *Weed Science* 61: 244–248.

Clements, D.R. and DiTommaso, A. (2011). Climate change and weed adaptation: can evolution of invasive plants lead to a greater range expansion than forecasted? *Weed Research* 51: 227–240.

Clements, D.R., Benoit, D.L., Murphy, S.D., and Swanton, C.J. (1996). Tillage effects on weed seed return and seedbank composition. *Weed Science* 44: 314–322.

Clements, D.R., DiTommaso, A., Jordan, N. et al. (2004). Adaptability of plants invading North American cropland. *Agriculture, Ecosystems & Environment* 104: 379–398.

Clements, D.R., Larsen, T., and Grenz, J. (2016). Knotweed management strategies in North America with the advent of widespread hybrid Bohemian knotweed, regional differences, and the potential for biocontrol via the psyllid *Aphalara itadori* Shinji. *Invasive Plant Science and Management* 9: 60–70.

Colquhoun, J. (2001). Perennial weed biology and management. Publication No. EM 8776. Oregon State University Extension Service, Corvallis, OR.

Davis, A.S., Fu, X., Schutte, B.J. et al. (2016). Interspecific variation in persistence of buried weed seeds follows trade-offs among physiological, chemical, and physical seed defenses. *Ecology and Evolution* 6: 6836–6845.

Dekker, J. (1999). Soil weed seed banks and weed management. *Journal of Crop Production* 2: 139–166.

Derksen, D.G., Lafond, G.P., Thomas, A.G. et al. (1993). Impact of agronomic practices on weed communities: tillage systems. *Weed Science* 41: 409–417.

Duke, S.O. (2012). Why have no new herbicide modes of action appeared in recent years? *Pest Management Science* 68: 505–512.

Egley, G.H. and Duke, S.O. (1985). Physiology of weed seed dormancy and germination. In: *Weed Physiology: Reproduction and Ecophysiology*, vol. 1 (ed. S.O. Duke), 27–64. Boca Raton, FL, USA: CRC.

Finch-Savage, W.E. and Bassel, G.W. (2016). Seed vigour and crop establishment: extending performance beyond adaptation. *Journal of Experimental Botany* 67: 567–591.

Fisk, J.W., Hesterman, O.B., Shrestha, A. et al. (2001). Weed suppression by annual legumes cover crops in no-tillage corn. *Agronomy Journal* 93: 319–325.

Forcella, F., Wilson, R.G., Dekker, J. et al. (1997). Weed seedbank emergence across the corn belt. *Weed Science* 45: 67–76.

Gallandt, E.R., Liebman, M., and Huggins, D.R. (1999). Improving soil quality: implications for weed management. *Journal of Crop Production* 2: 95–121.

Ghersa, C.M. and Holt, J.S. (1995). Using phenology prediction in weed management: a review. *Weed Research* 35: 461–470.

Gianessi, L. (2013). The increasing importance of herbicides in worldwide crop production. *Pest Management Science* 10: 1099–1105.

Gillies, S., Clements, D.R., and Grenz, J. (2016). Knotweed (*Fallopia* sp.) invasion of North America utilizes hybridization, epigenetics, seed dispersal (unexpectedly) and an arsenal of physiological tactics. *Invasive Plant Science and Management* 9: 71–80.

Grantz, D.A., Shrestha, A., and Vu, H. (2008). Ozone enhances adaptive benefit of glyphosate resistance in horseweed (*Conyza canadensis*). *Weed Science* 56: 549–554.

Grantz, D.A., Shrestha, A., and Vu, H. (2010). Ozone impacts on assimilation and allocation to reproductive sinks in the vegetatively propagated C4 weed, yellow nutsedge. *Crop Science* 50: 246–252.

Grantz, D., Paudel, R., Vu, H., and Shrestha, A. (2019). Tolerance of ozone and drought in common waterhemp (*Amaranthus tuberculatus*). *Journal of Crop Improvement* 33: 236–253.

Grundy, A.C., Phelps, K., Reader, R.J., and Burston, S. (2000). Modelling the germination of *Stellaria media* using the concept of hydrothermal time. *New Phytologist* 148: 433–444.

Gu, X.Y., Kianian, S.F., and Foley, M.E. (2006). Dormancy genes from weedy rice respond divergently to seed development environments. *Genetics* 172: 1199–1211.

Gummerson, R.J. (1986). The effect of constant temperature and osmotic potentials on the germination of sugar beet. *Journal of Experimental Botany* 37: 729–741.

Gunsolus, J.L. (1990). Mechanical and cultural weed control in corn and soybeans. *American Journal of Alternative Agriculture* 5: 114–119.

Harker, K.N. and O'Donovan, J. (2013). Recent weed control, weed management, and integrated weed management. *Weed Technology* 27: 1–11.

Hartzler, R.G., Van Kooten, B.D., Stoltenberg, D.E. et al. (1993). On-farm evaluation of mechanical and chemical weed management practices in corn (*Zea mays*). *Weed Technology* 7: 1001–1004.

Hayden, Z.D., Brainard, D.C., Henshaw, B., and Ngouajio, M. (2012). Winter annual weed suppression in rye-vetch cover crop mixtures. *Weed Technology* 26: 818–825.

Heap, I. (2014). Global perspective of herbicide-resistant weeds. *Pest Management Science* 70: 1306–1315.

Heap, I. (2020). International survey of herbicide-resistant weeds. www.weedscience.org (accessed 2 June 2021).

Hoffman, M.L., Weston, L.A., Snyder, J.C., and Regnier, E.E. (1996). Allelopathic influence of germinating seeds and seedlings of cover crops on weed species. *Weed Science* 44: 579–584.

Huang, J.Z., Shrestha, A., Tollenaar, M. et al. (2000). Effects of photoperiod on the phenological development of redroot pigweed (*Amaranthus retroflexus* L.). *Canadian Journal of Plant Science* 80: 929–938.

Huang, J.Z., Shrestha, A., Tollenaar, M. et al. (2001a). Effects of photoperiod and temperature on the phenological development of common lambsquarters. *Weed Science* 49: 500–508.

Huang, J.Z., Shrestha, A., Tollenaar, M. et al. (2001b). Effects of photoperiod and temperature on the phenological development of wild mustard (*Sinapis arvensis* L.). *Field Crops Research* 70: 75–86.

IPCC. (2013). Climate Change 2013: The Physical Science Basis. Working Group I Contribution to the Fifth Assessment Report of the Intergovernmental Panel on Climate Change. Cambridge, UK: Cambridge University Press.

Karssen, C.M. (1982). Seasonal patterns of dormancy in weed seeds. In: *The Physiology and Biochemistry of Seed Development, Dormancy and Germination* (ed. A.A. Khan), 243–270. Amsterdam, The Netherlands: Elsevier.

Kaur, S., Aulakh, J., and Jhala, A.S. (2016). Growth and seed production of glyphosate-resistant giant ragweed (*Ambrosia trifida* L.) in response to water stress. *Canadian Journal of Plant Science* 96: 828–836.

Keeley, P.E., Carter, C.H., and Thulen, R.J. (1987). Influence of planting date on growth of palmer amaranth (*Amaranthus palmeri*). *Weed Science* 35: 199–204.

Kennedy, A.C. (1999). Soil microorganisms for weed management. *Journal of Crop Production* 2: 123–138.

Kivilaan, A. and Bandurski, R.S. (1981). The one hundred-year period for Dr. Beal's seed viability experiment. *American Journal of Botany* 68: 1290–1292.

Klimešová, J., Martínková, J., Pausas, J.G. et al. (2019). Handbook of standardized protocols for collecting plant modularity traits. *Perspectives in Ecology, Evolution, and Systematics* 40: 125485.

Kremer, R.J. (1993). Management of weed seed banks and microorganisms. *Ecological Applications* 3: 42–52.

Landesmann, J.B., Gundel, P.E., Martinez-Ghersa, M.S., and Ghersa, C.M. (2013). Ozone exposure of a weed community produces adaptive changes in seed populations of *Spergula arvensis*. *PLoS One* 8 (9): e75820.

Leakey, R.R.B., Chancellor, R.J., and Vince-Prue, D. (1978). Regeneration from rhizome fragments of *Agropyron repens* (L.) Beauv. IV. Effects of light on bud dormancy and the development of dominance among shoots on multi-node fragments. *Annals of Botany* 42: 205–212.

Leon, R.G., Knapp, A.D., and Owen, M.D.K. (2004). Effect of temperature on the germination of common waterhemp (*Amaranthus tuberculatus*), giant foxtail (*Setaria faberi*), and velvetleaf (*Abutilon theophrasti*). *Weed Science* 52: 67–73.

Long, R.L., Gorecki, M.J., Renton, M. et al. (2015). The ecophysiology of seed persistence: a mechanistic view of the journey to germination or demise. *Biological Reviews of the Cambridge Philosophical Society* 90: 31–59.

Mahajan, G., George-Jaeggli, B., Walsh, M., and Chauhan, B.S. (2018). Effect of soil moisture regimes on growth and seed production of two Australian biotypes of *Sisymbrium thellungii* O. E. Schulz. *Frontiers in Plant Science* 9: 1241. https://doi.org/10.3389/fpls.2018.01241.

Marles, S.M., Warkentin, T.D., and Holm, F.A. (2010). Field pea seed residue: a potential alternative weed control agent. *Weed Science* 58: 433–441.

Martínková, J. and Klimešová, J. (2016). Enforced clonality confers a fitness advantage. *Frontiers in Plant Sciences* 7: 1–10.

McPeek, T.M. and Wang, X. (2007). Reproduction of dandelion (*Taraxacum officinale*) in a higher CO_2 environment. *Weed Science* 55: 334–340.

Melander, B., Rasmussen, I.A., and Bàrberi, P. (2005). Integrating physical and cultural methods of weed control: examples from European research. *Weed Science* 53: 369–381.

Murphy, S.D., Clements, D.R., Belaoussoff, S. et al. (2006). Promotion of weed species diversity and reduction of weed seedbanks with conservation tillage and crop rotation. *Weed Science* 54: 69–77.

Naidu, V.S.G.R. and Paroha, S. (2008). Growth and biomass partitioning in two weed species *Parthenium hysterophorus* (C3) and *Amaranthus viridis* (C4) under elevated CO_2. *Ecology, Environment and Conservation* 14: 9–12.

Neve, P., Barney, J., Buckley, Y. et al. (2018). Reviewing research priorities in weed ecology, evolution and management: a horizon scan. *Weed Research* 58: 250–258.

Nguyen, T., Bajwa, A.A., Navie, S. et al. (2017). Parthenium weed (*Parthenium hysterophorus* L.) and climate change: the effect of CO_2 concentration, temperature, and water deficit on growth and reproduction of two biotypes. *Environmental Science and Pollution Research* 24: 10727–10739.

O'Donnell, C.C. and Adkins, S.W. (2001). Wild oat and climate change: the effect of CO_2 concentration, temperature, and water deficit on the growth and development of wild oat in monoculture. *Weed Science* 49: 694–702.

Oerke, E.C. (2006). Crop losses to pests. *Journal of Agricultural Science* 144: 31–43.

Ott, J.P., Klimešová, J., and Hartnett, D.C. (2019). The ecology and significance of below-ground bud banks in plants. *Annals of Botany* 123: 1099–1118.

Panasuik, O., Bills, D.D., and Leather, G.R. (1986). Allelopathic influence of *Sorghum bicolor* on weeds during germination and early development of seedlings. *Journal of Chemical Ecology* 12: 1533–1543.

Panetta, F.D. and Timmins, S.M. (2004). Evaluating the feasibility of eradication for terrestrial weed incursions. *Plant Protection Quarterly* 19: 5–11.

Patterson, D.T. (1995). Weeds in a changing climate. *Weed Science* 43: 685–701.

Paudel, R., Grantz, D.A., Vu, H., and Shrestha, A. (2016). Tolerance of elevated ozone and water stress in a California population of Palmer amaranth (*Amaranthus palmeri* S. Wats.). *Weed Science* 64: 276–284.

Pausas, J.G. and Paula, S. (2020). Grasses and fire: the importance of hiding buds. A response to Moore et al. (2019) 'Effects of drought and fire on resprouting capacity of 52 temperate Australian perennial native grasses.'. *New Phytologist* 226: 957–959.

Persist (n.d.) Merriam–Webster.com Dictionary, Merriam–Webster. https://www.merriam-webster.com/dictionary/persist (accessed 19 December 2020).

Pimentel, D., Zuniga, R., and Morrison, D. (2005). Update on the environmental and economic costs associated with alien-invasive species in the United States. *Ecological Economics* 52: 273–288.

Pipatpongpinyo, W., Korkmaz, U., Hao, W. et al. (2019). Assembling seed dormancy genes into a system identified their effects on seedbank longevity in weedy rice. *Heredity* 124: 135–145.

Radosevich, S.R. and Holt, J.S. (1984). *Weed Ecology. Implications for Vegetation Management*, 265 pp. New York, NY: Wiley.

Rogers, H.H., Runion, G.B., Prior, S.A. et al. (2008). Effects of elevated atmospheric CO_2 on invasive plants: comparison of purple and yellow nutsedge (*Cyperus rotundus* L. and *Cyperus esculentus* L.). *Journal of Environmental Quality* 37: 395–400.

Roman, E.S., Murphy, S.D., and Swanton, C.J. (2000). Simulation of *Chenopodium album* seedling emergence. *Weed Science* 48: 217–224.

Saatkamp, A., Affre, L., Dutoit, T., and Poschlod, P. (2011). Germination traits explain soil seed persistence across species: the case of Mediterranean annual plants in cereal fields. *Annals of Botany* 107: 415–426.

Saebø, A. and Mortensen, L.M. (1998). Influence of elevated atmospheric CO_2 concentration on common weeds in Scandinavian agriculture. *Acta Agriculturae Scandinavica, Section B – Soil and Plant Science* 48: 138–143. https://doi.org/10.1080/09064719809362491.

Sarangi, D., Irmak, S., Lindquist, J.L. et al. (2015). Effect of water stress on growth and fecundity of common waterhemp (*Amaranthus rudis*). *Weed Science* 64: 42–52.

Saska, P., Foffová, H., Martinkova, Z., and Honěk, A. (2020). Persistence and changes in morphological traits of herbaceous seeds due to burial in soil. *Agronomy* 10: 448. https://doi.org/10.3390/agronomy10030448.

Scavo, A., Restuccia, A., Abbate, C., and Mauromicale, G. (2019a). Seeming field allelopathic activity of *Cynara cardunculus* L. reduces the soil weed seed bank. *Agronomy for Sustainable Development* 39, article 41. https://doi.org/10.1007/s13593-019-0580-4.

Scavo, A., Abbate, C., and Mauromicale, G. (2019b). Plant allelochemicals: agronomic, nutritional, and ecological relevance in the soil system. *Plant and Soil* 442: 23–48.

Schwarzländer, M., Hinz, H.L., Winston, R.L., and Day, M.D. (2018). Biological control of weeds: an analysis of introductions, rates of establishment and estimates of success, worldwide. *BioControl* 63: 319–331.

Sfiligoj, E. (2019). The war on glyphosate spreads. *Outlooks on Pest Management* 30 (2): 64–66.

Shrestha, A. and Swanton, C.J. (2007). Parameterization of the phenological development of select annual weeds under noncropped field conditions. *Weed Science* 55: 446–454.

Shrestha, A., Knezevic, S.Z., Roy, R.C. et al. (2002). Effect of tillage, cover crop and crop rotation on the composition of weed flora in a sandy soil. *Weed Research* 42: 76–87.

Shrestha, A., Fidelibus, M.W., Alcorta, M., and Hanson, B.D. (2010). Growth, phenology, and intra-specific competition between glyphosate-resistant and glyphosate-susceptible horseweed (*Conyza canadensis*) in the San Joaquin Valley of California. *Weed Science* 58: 147–153.

Shrestha, A., deSouza, L., Yang, P. et al. (2018). Differential tolerance of glyphosate-susceptible and –resistant biotypes of junglerice (*Echinochloa colona*) to environments during germination, growth, and intra-specific competition. *Weed Science* 66: 340–346.

Sosnoskie, L. and Culpepper, A.S. (2014). Glyphosate-resistant palmer amaranth (*Amaranthus palmeri*) increases herbicide use, tillage, and hand-weeding in Georgia cotton. *Weed Science* 62: 393–402.

Sosnová, M., Herben, T., Martínková, J. et al. (2014). To resprout or not to resprout? Modeling population dynamics of a root-sprouting monocarpic plant under various disturbance regimes. *Plant Ecology* 215: 1245–1254.

Steckel, L.E., Sprague, C.L., Hager, A.G. et al. (2003). Effects of shading on common waterhemp (*Amaranthus rudis*) growth and development. *Weed Science* 51: 898–903.

Steinmaus, S.J., Prather, T.S., and Holt, J.S. (2000). Estimation of base temperatures for nine weed species. *Journal of Experimental Botany* 51: 275–286.

Streit, B., Rieger, S., Stamp, P., and Richner, W. (2003). Weed populations in winter wheat as affected by crop sequence, intensity of tillage and time of herbicide application in a cool and humid climate. *Weed Research* 43: 20–32.

Swanton, C.J., Shrestha, A., Roy, R.C. et al. (1999). Effect of tillage systems, cover crop, and N on the composition of weed flora. *Weed Science* 47: 454–461.

Swanton, C.J., Huang, J.Z., Shrestha, A. et al. (2000). Effects of temperature and photoperiod on the phenological development of barnyardgrass. *Agronomy Journal* 92: 1125–1134.

Teasdale, J.R. (1996). Contribution of cover crops to weed management in sustainable agricultural systems. *Journal of Production Agriculture* 9: 475–479.

Telewski, F.W. and Zeevaart, J.A.D. (2002). The 120-yr period for Dr. Beal's seed viability experiment. *American Journal of Botany* 89: 1285–1288.

Travlos, I.S., Cheimona, N., Roussis, I., and Bilalis, D.J. (2018). Weed-species abundance and diversity indices in relation to tillage systems and fertilization. *Frontiers in Environmental Science* 6: 11. https://doi.org/10.3389/fenvs.2018.00011.

Ullrich, S.D., Buyer, J., Cavigelli, M.A. et al. (2011). Weed seed persistence and microbial abundance in long-term organic and conventional cropping systems. *Weed Science* 59: 202–209.

Vigueira, C.C., Olsen, K.M., and Caicedo, A.L. (2013). The red queen in the corn: agricultural weeds as models of rapid adaptive evolution. *Heredity* 110: 303–311.
Vitalini, S., Orlando, F., Vaglia, V. et al. (2020). Potential role of *Lolium multiflorum* Lam. in the management of rice weeds. *Plants (Basel)* 9: 324. https://doi.org/10.3390/plants9030324.
Walck, J.L., Hidayati, S.N., Dixon, K.W. et al. (2011). Climate change and plant regeneration from seed. *Global Change Biology* 17: 2145–2161.
Walsh, M.J., Broster, J.C., Schwartz-Lazaro, L.M. et al. (2018). Opportunities and challenges for harvest weed seed control in global cropping systems. *Pest Management Science* 74: 2235–2245.
Weaver, S.E. (2001). The biology of Canadian weeds. 115. *Conyza canadensis*. *Canadian Journal of Plant Science* 81: 867–875.
Webster, T.M. and Grey, T.L. (2008). Growth and reproduction of Benghal dayflower (*Commelina benghalenis*) in response to drought stress. *Weed Science* 56: 561–566.
Weed Science Society of America (WSSA). (2017). WSSA survey ranks most common and most troublesome weeds in broadleaf crops, fruits and vegetables. https://wssa.net/2017/05/wssa-survey-ranks-most-common-and-most-troublesome-weeds-in-broadleaf-crops-fruits-and-vegetables (accessed 2 June 2021).
Weed Science Society of America (WSSA). (2019). WSSA-herbicide site of action (SOA) classification list. http://wssa.net/wp-content/uploads/WSSA-Herbicide-SOA-20181205.xlsx (accessed 2 June 2021).
Weston, S. and Duke, S. (2003). Weed and crop allelopathy. *Critical Reviews in Plant Sciences* 22: 367–389.
Wheeler, G.S. and Schaffner, U. (2013). Improved understanding of weed biological controls safety and impact with chemical ecology: a review. *Invasive Plant Science and Management* 6: 16–29.
Whitney, K.D. and Gabler, C.A. (2008). Rapid evolution in introduced species, 'invasive traits' and recipient communities: challenges for predicting invasive potential. *Diversity and Distributions* 14: 569–580.
Zimdahl, R. (1999). *Fundamentals of Weed Science*. San Diego, CA: Academic Press.
Ziska, L.H. and Bunce, J.A. (1997). Influence of increasing carbon dioxide concentration on the photosynthetic and growth stimulation of selected C_4 crops and weeds. *Photosynthesis Research* 54: 199–208.
Ziska, L.H. and George, K. (2004). Rising carbon dioxide and invasive, noxious plants: potential threats and consequences. *World Resources Review* 16: 427–447.
Ziska, L.H. and Goins, E.W. (2006). Elevated atmospheric carbon dioxide and weed populations in glyphosate treated soybean. *Crop Science* 46: 1354–1359.
Ziska, L.H., Faulkner, S., and Lyndon, J. (2004). Changes in biomass and root:shoot ratio of field-grown Canada thistle (*Cirsium arvense*), a noxious, invasive weed, with elevated CO_2: implications for control with glyphosate. *Weed Science* 52: 584–588.

2

Seed Production, Dissemination, and Weed Seedbanks

Acacio Goncalves Netto[1], Pedro Christoffoleti[1], Mark VanGessel[2], Saul J.P. Carvalho[3], Marcelo Nicolai[4], and Caio Brunharo[5]

[1] *Crop Science Department, Luiz de Queiroz College of Agriculture, University of São Paulo, São Paulo, Brazil*
[2] *Department of Plant and Soil Science, Research and Education Center, University of Delaware, Georgetown, DE, USA*
[3] *Department of Agronomy, Federal Institute of Education, Science and Technology of the South Minas Gerais, Machado, Minas Gerais, Brazil*
[4] *Agrocon Agronomic Consulting, Santa Bárbara d'Oeste, São Paulo, Brazil*
[5] *Department of Crop and Soil Science, Oregon State University, Corvallis, OR, USA*

2.1 Introduction

High levels of seed production, long seed dormancy periods, and seed longevity are important traits of weed persistence. Most annual and some perennial weed species reproduce via seeds, and the ability of those seeds to remain viable for extended periods is of great ecological importance in the persistence and regeneration of weed communities in agroecosystems.

A better understanding of weed seeds in the soil (referred to as the seedbank) including the species present, density, survival, and germination can directly aid in the decision-making for weed control and enhancing integrated weed management practices. Each generation of annual weeds originates from seeds, while perennial weed species can reproduce by seeds as well as vegetative structures. Examples of this bimodal feature for perennial weeds include yellow nutsedge (*Cyperus esculentus*) (tubers and basal bulbs), johnsongrass (*Sorghum halepense*) (rhizomes), Bermudagrass (*Cynodon dactylon*) (rhizomes and stolons), and Canada thistle (*Cirsium arvense*) (buds along horizontal roots). These perennial species are also capable of producing a significant number of viable seeds. Other species such as Benghal dayflower (*Commelina benghalensis*) can reproduce through aerial and subterranean seeds, as well as stem fragments (Kaul et al. 2002). Since vegetative propagation is discussed in Chapter 11 of this book, this chapter will focus only on seeds of agriculturally important weeds.

Persistence Strategies of Weeds, First Edition. Edited by Mahesh K. Upadhyaya, David R. Clements, and Anil Shrestha.

Table 2.1 Traits associated with survival of plants and how these traits influence weed competitiveness.

Survival traits	Mode of competitive advantage
High levels of seed production	Increased likelihood of dispersal
	Increased density (plants m^{-2})
Long-distance dispersal	Increased likelihood of colonizing new environments
Seed dormancy	Maintenance of viability even under unfavorable conditions
	Allows for emergence over an extended time period – within a season and over multiple seasons
	Maintain genetic variability in the population
Nonuniform seed germination	Germination dynamics is variable
	Ability to germinate and emerge from greater depths
Asexual reproduction	Alternative reproduction mechanisms

Source: Zimdahl (2018).

2.2 Seed Production

Seeds allow plants to reproduce and persist in their environments for extended periods under a range of conditions. The characteristics that enhance the survival of weed species also contribute to greater competitive ability. These traits, which allow weeds to utilize resources in more efficient ways than crops, include rapid canopy development, rapid vegetative development in response to shading, and rapid initial root development (Zimdahl 2018). Table 2.1 conceptually considers the linkage of traits associated with weed survival and persistence over generations with characteristics associated with the competitiveness of plants.

High seed production per plant maximizes the chances of the species' persistence in agroecosystems. Seed production by the plants present in the field is the biggest contributor to seeds in the soil (Norris 2007). Quantifying seed inputs to the seedbank is critical. Weed seed production values reported in the literature can vary greatly for a given species. This can be due to climate, weed biotypes, or agronomic variables. Without identifying the source of the variation, seed production data available in the literature is limited in its usefulness (Norris 2007).

In the absence of crop competition, Palmer amaranth (*Amaranthus palmeri*) can produce up to 600 000 seeds (Keeley et al. 1987), barnyardgrass (*Echinochloa crus-galli*) over 400 000 seeds (Norris et al. 2001), and horseweed (*Erigeron canadensis*) up to 200 000 seeds per plant (Concenço et al. 2014). There are many other examples of species with high seed production, such as smallflower galinsoga (*Galinsoga parviflora*) (30 000 seeds per plant) and common purslane (*Portulaca oleracea*) (>50 000 seeds per plant) (Zimdahl 2018). High seed production does not mean that all seeds that are produced will be able to germinate; some of the seeds could be immature or nonviable. Seed production alone is not the only indicator of species success and high competitive ability. Wild oat (*Avena fatua*) is considered one of the worst weeds in the world (Holm et al. 1977), yet it produces a modest number of seeds per plant (20–1070 per plant) (Beckie et al. 2012).

The number of seeds produced by a plant is influenced by a number of factors, e.g. time of seedling emergence, inter- and intraspecies competition, and environmental conditions. In general, seed production is maximized when seedlings emerge before or shortly after crop emergence and weeds gain a competitive advantage over the crop, capturing resources required for growth (Fahad et al. 2015). Seed production of barnyardgrass was reduced by over 90% when seedlings emerged after the fourth leaf stage of corn (*Zea mays*) compared with plants emerging with corn (Bosnic and Swanton 1997). Similarly, barnyardgrass that emerged five weeks after cotton (*Gossypium hirsutum*) or rice (*Oryza sativa*) emergence had 90% and 60% fewer seeds, respectively, than those that emerged with the crop (Bagavathiannan et al. 2011).

Plant competition, whether inter- or intraspecific, can reduce seed production. As weed densities increase, seed production by an individual plant will decrease. However, seed production per unit area may increase with increasing weed density. For example, the number of barnyardgrass seeds produced per plant was 80 000 at 5 plants m^{-2} and 10 000 seeds per plant at a density of >50 plants m^{-2} (Norris et al. 2001), but on an area basis, this corresponds to 400 000 and 500 000 seeds m^2, respectively. Similarly, velvetleaf (*Abutilon theophrasti*) in a noncompetitive environment produced over 44 000 seeds per plant, but only 2710 seeds per plant when the weed density was 32 plant m^{-2} (Steinmaus and Norris 2002). Crop competition can also reduce weed seed production. Barnyardgrass growing in competition with tomato (*Lycopersicon esculentum*) reduced seed production by over 60% when weed densities were low but only 20%–30% at higher weed densities (Norris et al. 2001). Velvetleaf seed production was reduced by 90% when grown in the competition with kidney beans (*Phaseolus vulgaris*) and 99% when grown with corn (Steinmaus and Norris 2002). Weed seed production due to crop competition can also differ among varieties or hybrids of the same crop (Lindquist and Mortensen 1998).

Environmental conditions and agronomic practices also influence the number of seeds produced. Seed production of weed species differ in their response to moisture stress. For example, Palmer amaranth seed production was reduced by 55% under moderate moisture stress (75% field capacity) (Chahal et al. 2018), whereas junglerice (*Echinochloa colona*) seed production was not affected by moisture stress even at 50% field capacity (Chauhan and Johnson 2010). Soil fertility can also affect weed seed production. For example, seed production of giant foxtail (*Setaria faberi*) and velvetleaf increased upon addition of nitrogen; the responses differed with the amount and timing of nitrogen application (Harbur and Owen 2004). The fecundity of waterhemp (*Amaranthus tuberculatus*), however, was not affected by nitrogen application.

The environment under which seeds are produced can also influence seed characteristics. Seeds produced by Powell amaranth (*Amaranthus powellii*) plants growing in competition with broccoli (*Brassica oleracea* var. *italica*) had lower germination than the plants growing without competition (Brainard et al. 2005). Velvetleaf grown in competition with corn had fewer and lighter seeds with a lower number of dormant seeds compared with plants grown in a non-competitive environment (Nurse and DiTommaso 2005). Reduced tillage in some systems may result in reduction in growth of weeds, which in turn could reduce weed fecundity (Mashingaidze et al. 2017).

Agriculturally important weeds have traits that improve their probability of survival. For instance, under stress, many weed species will prioritize growth and reproduction instead of accumulating carbohydrate reserves. Under plant–plant competition, those plants that are more efficient in obtaining and using limited resources to allocate to reproduction will be able to produce relatively large numbers of seeds.

Preventing the production of viable seeds has become a focus of weed management practices, particularly when managing herbicide-resistant biotypes (Norsworthy et al. 2012). Stopping viable seed production requires killing the plant before or shortly after flowering. The number of viable seeds per plant increases as the plant reaches full maturity. Preventing viable seed production generally requires controlling the plant within two to three weeks after flowering (Bell and Tranel 2010).

Eliminating seeds returning to the seedbank can have dramatic effects on seed densities in the soil. Several studies have shown exponential loss of weed seed in the soil when no new seeds were added. The most dramatic loss occurs in the first and second year after seed return is eliminated (Schweizer and Zimdahl 1984; Egley and Williams 1990; Wilson and Lawson 1992; Buhler and Hartzler 2001).

2.3 Seed Dissemination

Seeds are disseminated within and between fields and over long distances. Dissemination can occur either with no external assistance (autochory) or with the aid of external vectors (allochory) (Benvenuti 2007a,b). The ability of a species to disperse seed to new environments is a trait that increases evolutionary fitness through increased likelihood of colonizing new environments (Table 2.1).

Many weed seeds are well adapted for dissemination, a phenomenon important for maintaining continuity where a weed population is already established, as well as increasing the size of the population and expanding the distribution. Dissemination includes two separate processes, dispersal (leaving the parent plant, often referred to as seed rain) and post-dispersal movement. Dispersal can occur immediately upon seed ripening or slowly over time. Many seeds have no special dispersal mechanism and simply fall from the parent plant when dislodged by a physical disturbance. The distance of seed dissemination from the parent plant can range from a few centimeters to meters depending on the plant height, seed density, and wind velocity.

Some plants in the Brassicaceae, Fabaceae, and Euphorbiaceae families eject seeds from their fruit by dehiscence. This phenomenon relies on differential drying of fruit parts that generate force to eject seeds. Seeds released by dehiscence can travel up to a few meters, depending on release height and angle (Howe and Westley 1986). However, the majority of seeds are not forcibly ejected, and simply fall from the parent plant (Hayashi et al. 2010).

Allochory, dissemination aided by external means, can be classified by the disseminating agent involved, e.g. water, wind, animals, and humans. Allochory does not require special adaptations, but some species do have structures or adaptations that facilitate seed movement.

2.3.1 Movement by Water

Flowing water can be a very efficient means of seed dissemination. Seeds can be disseminated with water flowing in streams, rivers, or irrigation channels or with surface runoff of water across a field. Many seeds can float on the water surface if their density is lower than water or when they have a flattened shape. However, once the surface tension of the water is broken, some seeds will sink. A few agriculturally important weed species have seed structures that facilitate movement by water. Curly dock (*Rumex crispus*), for example, has a papery ridge around the fruit, which increases fruit buoyancy. These types of structures are more common in plants that evolved near bodies of water, such as knotweeds (*Reynoutria* species), which have achene morphology designed to stay afloat for several days (Lamberti-Raverot et al. 2017)

Irrigation can disseminate weed seeds. For example, Palmer amaranth seeds have been shown to spread up to 114 m over the course of one year in furrow-irrigated cotton (Norsworthy et al. 2014). One of the primary methods of spread appeared to be along the furrows between the beds of cotton with water movement. Wilson (1980) sampled numerous irrigation canals in Nebraska and identified seeds of 77 weed species. There were 26 times more seeds in samples collected at the end of the irrigation canal than at the beginning. Most of these seeds floated on the water surface, with redroot pigweed (*Amaranthus retroflexus*) being the most common, accounting for 40% of the total seeds. Wilson (1980) estimated that 1.2 million seeds per hectare were introduced in canal-irrigated fields on an annual basis. Thus, managing weeds in close proximity of water reservoirs as well as irrigation canals is important for avoiding the introduction of weed seeds into irrigated fields.

2.3.2 Movement by Wind

Wind can disperse seeds by dislodging them from the parent plant and moving them before they reach the ground. The dispersal distance depends on release height, seed density, and wind speed; usually the distance is only a few meters. Many species in the Asteraceae family have structures or seed adaptations that enhance movement by wind (e.g. hairy pappus on yerba porosa [*Porophyllum ruderale*] achenes and pubescence on tansy ragwort [*Senecio jacobaea*] achenes) (DiTomaso and Healy 2007). Another example is horseweed, which has small, lightweight seeds with short pappus that allow very efficient ascent to hundreds of meters above the earth surface and low settling velocity. Shields et al. (2006) estimated that horseweed seed could travel over 500 km in a single dispersal event. It should be noted, however, that 99% of horseweed seeds were captured within 100 m of the seed source (Dauer et al. 2007).

Another form of seed movement is the tumbling of plants across the landscape by the wind. Kochia (*Kochia scoparia*) and Russian thistle (*Salsola tragus*) are two tumbleweed species common in the Western United States. At maturity, the stems break off at the base, along an abscission layer. These plants have a pyramidal or conical shape that facilitates tumbling across the landscape, disseminating seeds along its route. While it is difficult to document the distance to which seeds are spread as the plants roll across fields, approximately 90% or more seeds were dispersed over a distance of 1 km. As the tumbling speed of the plant increases, seeds are deposited more rapidly (Beckie et al. 2016). It should be noted that a large number of seeds could be dispersed prior to stem abscission.

2.3.3 Movement by Animals

Animals can also facilitate seed dispersion. When the propagule is transported by external attachment to the body of an animal, it is called epizoochory. Some weed species have appendages that allow their seeds to be entangled in the fur of animals and then be spread over great distances. These appendages include awns, hooked spines, or burs. Species with these adaptations include sandburs (*Cenchrus* spp.), bristly starbur (*Acanthospermum hispidum*), hairy beggarticks (*Bidens pilosa*), and common cocklebur (*Xanthium strumarium*). The distance these species are spread by animals is difficult to discern because the movement by animals that carry seeds are not well quantified. However, deer (*Odocoileus* spp.) are one of the most common large wildlife to visit crop fields in the Eastern United States, and based on their behavior, their territory seldom extends more than 10 km, with the exception of seasonal migration (Haus et al. 2019).

Dissemination of a weed propagule, which is ingested and transported inside the body of the animal and later expelled with the feces, is referred to as endozoochory. Ingested seeds must be able to survive the digestive system of the animal to be successfully disseminated. Bird species differ in their ability to devitalize seeds, with obligate seed-eating birds passing fewer viable seeds than seed-eating omnivores (Twigg et al. 2009). Some studies have reported that the digestive system of poultry inactivates most weed seeds (Harmon and Keim 1934; Cooper et al. 1960).

Birds can carry weed seeds over great distances. Migratory ducks (*Anas* spp.) were sampled during hunting seasons in Missouri. During one hunting season, 237 ducks were collected, and their digestive systems were examined for weed seeds. Over 14 000 plants emerged from the recovered seeds, representing over 47 species (Farmer et al. 2017). It was estimated that this migratory waterfowl could disperse weed seeds up to a distance of 1400 km.

2.3.4 Movement by Farm Equipment and Humans

Equipment is a major contributor to seed dispersal within a field. Tillage can move seeds 2–3 m in the direction of the tillage (Barroso et al. 2006). If weeds are mature and seeds are retained on the parent plant at harvest, combines can distribute seeds up to 30 m depending on seed size, density, and wind. However, 90% of the dispersed johnsongrass seeds fell within 5 m of the combine (Barroso et al. 2012).

Farm inputs, primarily soil amendments, are also a source of seeds for starting new weed infestations. Manures, compost, topsoil, and spoiled hay are examples of soil amendments that can serve as a source of weed seeds. A survey of bovine manure from New York farms found 48 different species, with one sample containing 400 000 seeds per ton of manure (Mt. Pleasant and Schlather 1994). Most commercial crop seed is thoroughly cleaned and contains little to no weed seeds. However, weed infestations have been traced back to use of weed seed-contaminated cover crop seed or to farmers' use of their own harvested grain to plant the following season's crop.

Human-mediated transport of livestock is another mechanism of weed seed transport. Moving cattle between pastures or transporting animals to new regions can lead to new weed infestations.

Humans are an extremely important vector of weed seed movement between different regions and countries. Many weed species were introduced as the result of transporting plants for ornamental or agronomic purposes. For example, Italian ryegrass (*Lolium perenne* L. spp. *multiflorum*) is native to the Mediterranean region and was introduced to many parts of the world as a forage crop (DiTomaso and Healy 2007). There are many examples (somewhat speculative) of weed introductions with the soil used as ballast for ships or dried plants used for bedding from the early colonization period of the western hemisphere. Once the ships sailed across the Atlantic Ocean, the soil or dried plants were offloaded as more cargo was added. The sand/soil or dried plant material used was often contaminated with weed seeds, and these species became established near the ports. Another example of human-mediated weed seed dispersal is in the genus *Echinochloa* spp., the most troublesome weed species complex in rice. Because of the similar seed shape of *Echinochloa* spp. and cultivated rice, contaminated rice seed lots presumably facilitated *Echinochloa* spp. seed movement over great distances (Huelma et al. 1996).

Many weed infestations can be traced back to poorly cleaned farm machinery. Cleaning harvest equipment to prevent the movement of seeds from infested fields to other fields is important to limit introduction of new weed species into previously uninfested fields.

In most cases, the number of weed seeds disseminated by wind, machinery, or contaminated equipment is relatively low, but the ability of weeds to produce a high number of seeds allows the species to infest new fields and become abundant in only two to three growing seasons if weed control strategies are not adequate (Norsworthy et al. 2014).

2.4 Weed Seedbank and Seedbank Dynamics

The term weed seedbank has been used in the international literature to describe viable seeds present in soil, on the soil surface, or in the plant residue within a field (Harper 1977; Thompson and Grime 1979; Roberts 1981; Carmona 1992). These seedbanks are the primary source of new infestations of annual weeds in crop production systems (Cavers 1983). They serve as a "memory" of a population (Cavers 1995). In the quest for trying to understand the persistence of weeds, it is important to understand weed seedbanks and their dynamics. In recent years, there is renewed interest in understanding seedbank ecology (e.g. Schwartz et al. 2015; Burton et al. 2016; Norsworthy et al. 2018), genetics (e.g. Shoemaker and Lennon 2018), and management (e.g. Walsh et al. 2013).

Weed seeds in soil seedbanks have four fates: germination, death, emigration, and persistence (Forcella 2003). The first three fates result in removal of weed seeds from the seedbank, whereas the last fate defines why weed management continues to be a continuous battle. Although seed dormancy, germination, longevity, and dynamics of weed seeds in the soil seedbank are covered in detail in other chapters of this book (Chapters 3, 5, and 6), we provide a brief overview of the four fates described by Forcella (2003).

2.4.1 Seed Germination

Weed seed germination is an important factor for seed loss from soil seedbanks. Seed germination and emergence periodicity are species dependent. Seedlings of summer annual species that germinate at lower soil temperatures emerge earlier in the spring compared with those with high base-temperature requirement for germination. In addition, some species exhibit a bi-model emergence pattern with a large "spike" in seedling emergence followed by a smaller "spike" (Figure 2.1). Similarly, winter annual weeds often have a fall emerging cohort followed by a smaller spring cohort.

Seedbanks can be either transient or persistent (Thompson and Grime 1979). In the transient seedbanks, the seeds generally do not live long and germinate by the second germination season, whereas in the persistent seedbank, the seeds live until or beyond the second germination season. The latter type of seedbank results in a high level of persistence, and the strategies weed seeds adopt to persist in this way are a major focus of this book.

Weed seed germination is known to be influenced by biotic (e.g. soil biota and allelopathic influences) and abiotic (e.g. temperature, light, moisture, soil salinity, and soil pH) factors and their interactions. Germination occurs when minimum thresholds for abiotic factors have been met (Bewley and Black 1994; Baskin and Baskin 2001). There is renewed interest in studying these biotic and abiotic factors and their interactions in recent years to understand seed germination behavior in soil seedbanks and the persistence of weeds (Kremer 1993; Benech-Arnold et al. 2000; Chee-Sanford et al. 2006; Singh et al. 2012; Chauhan et al. 2019; Frost et al. 2019).

Another fate that results in loss of weed seeds from soil banks is "fatal germination," which may involve germination of seeds at depths from which the seedlings cannot emerge (Benvenuti 2003; Davis and Renner 2007; Martinková and Honěk 2013).

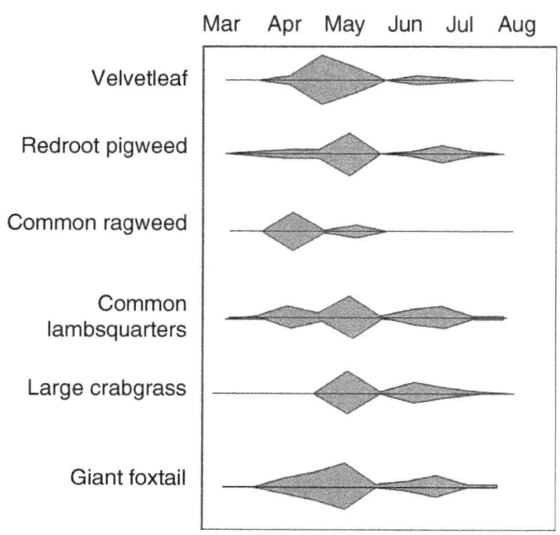

Figure 2.1 Representative emergence periodicity of summer annual weeds from the Mid-Atlantic region of the United States. *Source:* Adapted from Myers et al. (2005).

2.4.2 Seed Death

The major causes of seed death include physiological death, failed or fatal germination, predation, pathogen and microbial attacks, mechanical and physical damage by implements and animals, deep burial (Forcella 2003), and allelopathic influences (Cheng and Cheng 2015). However, seeds buried too deep can remain viable in a state of dormancy and could possibly germinate when brought to depths suitable for germination and seedling emergence.

Lutman et al. (2002) studied seed persistence of 16 common weed species under annually tilled soils and found that over six years seed decline ranged from 9% to 58% per year depending on the species. Davis et al. (2016) studied the seed half-life of 11 annual weed species and reported that it was species dependent; seed mortality over a 5 year period ranged from 13% in velvetleaf to 89% in kochia. They suggested factors such as seed coat rupture force, seed coat thickness, seed mass, and the presence of phenolic compounds contributed to seed loss. Long et al. (2015) have discussed the factors influencing seed death in their review of ecophysiology of seed persistence in the soil seedbank.

2.4.3 Seed Emigration

Weed seeds can emigrate from the soil seedbanks causing a decline in seed numbers. Means of emigration include transport by animals, irrigation water, and soil erosion (Forcella 2003), as well as agricultural machinery, flooding, and crop harvest. Seeds attached to or associated with plant material and seeds present on or near the soil surface are more likely to emigrate.

2.4.4 Seed Persistence

One primary reason for weed persistence is the ability of their seeds to remain dormant but viable (longevity) for extended periods (Benech-Arnold et al. 2000). Weed species differ in their ability to persist in the soil seedbank. Seeds generally are more persistent if they possess physical (e.g. hard seed coat) and physiological attributes that resist factors such as predation, chemical and microbial breakdown, and/or physical damage by agricultural implements (see Hulme 1998; Honda 2008). Persistence is also influenced by the burial depth of seeds in the soil and several other abiotic and biotic factors that are discussed in more detail in other chapters of this book.

2.4.5 Changes in Size and Composition of Weed Seedbanks

Changes in weed seedbank density and species diversity as a result of crop management has been reported from long-term trials (Cardina et al. 1991; Sosnoskie et al. 2006; Andreasen et al. 2018) as well as from surveys as short as three seasons (Smith and Gross 2006; Harbuck et al. 2009). Surveys of weed seedbanks have shown a wide range of seed densities. Carmona (1995) estimated the seedbank in four distinct agroecosystems: annual crop production (soybean, fallow, and bean), marshland, citrus

Table 2.2 Estimates of weed seedbank size in cultivated soils.

				Seeds m^{-2}		
Culture	**Local**	**Samples**	**Depth (cm)**	**Range**	**Mean**	**Sources**
Cereals	England	32	0–15	1 800–67 000	5500	Roberts (1981)
Vegetables	England	89	0–15	250–24 330	4 120	Roberts and Neilson (1982)
Barley, corn, and beet	USA	—	0–25	2 080–137 700	—	Schweizer and Zimdahl (1984)
Various	France	—	0–30	400–56 500	5 100	Barralis and Chadoeuf (1987)
Wheat	England	68	0–10	1 800–171 200	21 200	Carmona (1995)
Broad bean	England	36	0–10	21 800–132 200	49 800	Carmona (1995)
Broad bean	England	24	0–10	5 400–40 200	20 600	Carmona (1995)

Source: Adapted from Carmona (1995).

orchard, and permanent pasture. The average number of seeds per square meter was 22 313 in the marshland, 6768 in the annual crop rotation, 3595 in the citrus orchard, and 529 in the pasture. In cereals production in England, seed density ranged from 1800 to 67 000 seeds m^{-2} (Table 2.2).

Generally, seedbanks of agricultural fields are composed of many species, with only a few species accounting for 70–90% of the total seeds in the soil (Schweizer and Zimdahl 1984; Wilson 1988; Gulden et al. 2011). These few species are considered the most troublesome because they resist control measures and are well adapted to the soil and climatic conditions, as well as to the cultural practices (Buhler et al. 1997). The second group of species, comprising 10–20% of the seedbank, includes species that are adapted to the geographic region but are effectively controlled with the current cropping practices. The final group of species in the seedbank are recalcitrant from past management practices or are new introductions to the seedbank.

Seed longevity and seed decline are variable over time and among species. Seeds such as wild poinsettia (*Euphorbia heterophylla*), kochia, and hairy beggarticks generally have a high germination and emergence rate and a high percentage of seeds are depleted from the soil in about three to four years in the absence of reinfestation (Voll et al. 2001). On the other hand, velvetleaf or common mullein (*Verbascum thapsus*) seeds have been reported to maintain high viability after 17 years of burial in the soil (Burnside et al. 1996).

The dynamics of soil seedbanks is influenced by a number of factors. These are discussed in detail in various chapter of this book. Here we briefly discuss the influence of some weed management practices on seedbanks in relation to seed production, seed dissemination, and seedbank dynamics.

2.5 Weed Management and Seedbanks

2.5.1 Tillage

The effects of tillage practices on the seedbanks are complex and involve several factors. Tillage affects seed placement, kills seedlings, disrupts plant residue, and alters conditions that affect germination such as seed coat abrasion, gas exchange, soil temperature, and soil moisture level (Swanton et al. 1999). The depth of seed burial following tillage is an important factor influencing weed seedling emergence (Buhler et al. 1997 and Chapter 10 of this book).

Studies examining seedling emergence from seeds buried at various depths often report low emergence with seeds on the soil surface, increased emergence with shallow burial, and an exponential decrease from seeds buried at greater depths (Mohler and Galford 1997). Benvenuti et al. (2001) observed large variability in germination when evaluating the relationship between burial depth and weed seed size. In general, as seed size increased so did the potential to emerge from greater depths. Species with higher seed weights (Johnsongrass, velvetleaf, catchweed bedstraw [*Galium aparine*], and cutleaf geranium [*Geranium dissectum*]) were able to emerge from a 10-cm depth. In some species, as seed burial depth increased, seedling emergence was delayed. The study showed that reduced emergence of seeds placed deeper in the soil was due to induction of dormancy, rather than fatal germination. In contrast, seeds of horseweed need to be on the soil surface for successful germination and emergence. Emergence was reduced by 90% when seeds were buried 1 cm deep (Weaver 2001).

The number of emerged weeds is often higher in tilled fields or disturbed soils compared with no-till fields (Buhler 1997; Mohler and Galford 1997), with the percentage of seeds that germinate dependent on many factors, including soil moisture. Rainfall or irrigation shortly after tillage increases the likelihood of seedling emergence, and if no rainfall or irrigation occur and the soil is allowed to dry after disturbance, the impact of disturbance is reduced (Schutte et al. 2016).

The effect of tillage systems on seedling emergence may vary over time. Within a growing season, seedlings emergence occurred later and over a longer period in no-till compared with conventional tillage systems, especially for early emerging species like common ragweed (*Ambrosia artemisiifolia*) (Myers et al. 2005). Modeling weed seedling emergence over multiple seasons has shown that more emergence occurs in conventional tillage than in reduced tillage systems (Anderson 2005). There was a twofold and eightfold higher emergence in conventional tillage compared with no-tillage in the second and third years after seed rain, respectively.

The effect of tillage on seed density in the seedbank has been reported in a number of studies, but the results have not been consistent due to differences in management and environmental conditions. Some studies have reported higher weed seed density associated with no-tillage (Yenish et al. 1992), while others found higher densities with moldboard plowing (Cardina et al. 1991; Clements et al. 1996), and still others found no difference among tillage systems (Burnside et al. 1986; Egley and Williams 1990; Ruisi et al. 2015).

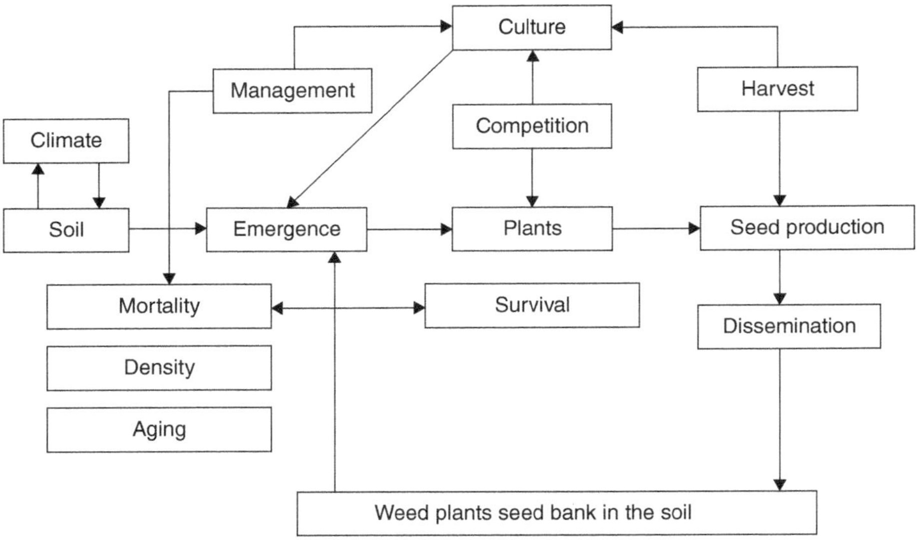

Figure 2.2 Components that affect the weed seedbank dynamics. *Source:* Gomes Jr. and Christoffoleti (2008).

Tillage may influence seedbank diversity. It is generally assumed that a highly disturbed site (for instance conventional tillage) would have a lower diversity of species given the stresses that tillage imposes on weed populations, but the research has not always borne this out. Still higher weed species diversity has often been recorded in soil seedbanks under no-tillage management compared with tilled soils (Derksen et al. 1993; Menalled et al. 2001; Cardina et al. 2002; Sosnoskie et al. 2006). Studies reporting no increase in species diversity in reduced tillage systems have often attributed this unexpected result to factors such as crop rotation or weather. This illustrates the complexity of identifying the factors that drive weed seedbank dynamics in agroecosystems (Figure 2.2).

Smith (2006) reported that spring tillage resulted in weed communities dominated by early-spring emerging broadleaf weeds and C_4 grasses, whereas fall tillage produced late-emerging broadleaves and C_3 grasses (Smith 2006). In a study comparing the effects of timing of tillage on weed communities in the Northeastern United States (Cordeau et al. 2017), it was suggested that tillage timing for summer crops could influence weed density, with early tillage, resulting in higher weed densities.

2.5.2 Effects of Weed Control Measures

Weed control measures impact weed seed dynamics significantly. Seedbank composition is influenced by herbicide use and mechanical and cultural weed control practices (Roberts 1981; Gulden et al. 2011). Ineffective weed control strategies allow weeds to complete their life cycle and contribute to the seedbank.

Cover crops and crop residues on the soil surface can limit weed emergence and weed seedling growth. They limit solar radiation and sunlight from reaching the soil surface, which, in turn, slows soil warming and reduces peak temperatures, soil water evaporation, and weed seed germination. Plant residues on the soil surface can also physically impede weed seedling emergence. In spite of the concentration of seeds in the upper layers of the soil, Teasdale and Mohler (1993) reported that the germination of several species was suppressed by the presence of straw on the soil surface. Weeds that grow through the cover crop layer can later benefit from improved moisture conservation by the cover crop.

Weed seeds near the soil surface are most vulnerable to weed management. Soil-applied herbicides are less effective on seeds buried deeper in the soil. Mechanical, interrow cultivation can also reduce weed seed density in soil banks. In addition to the seedling removal by interrow cultivation, this tillage operation can also stimulate seedling emergence, albeit to a lesser extent than the primary tillage (Ball and Miller 1990). Mechanical implements such as rotary hoes and harrows disturb only the upper layer of soil, and seeds below the tilled zone are not impacted (Clements et al. 1996).

2.6 Use of Chemicals to Deplete Soil Seedbanks: Potential and Limitations

Several different chemicals have been reported to alleviate physiological seed dormancy in several weed species (Roberts and Smith 1977). However, these chemicals may not affect seeds which are dormant because of the impermeability of seed coat to water or oxygen. Some substances that stimulate weed seed germination are briefly discussed in the following text.

2.6.1 Nitrogen-Containing Compounds

Nitrate, a common inorganic ion in the soil solution, can stimulate seed germination in some weed species (Espeby 1989). Under field conditions, nitrate concentration is usually higher near the soil surface due to the decomposition of organic matter and microbial activity. Nitrite is generally more effective in stimulating seed germination than nitrate, and it has been suggested that germination stimulating activity of nitrate could be due to its conversion to nitrite within the seed (Adkins et al. 1984).

Urea derivatives (e.g. thiourea) have been shown to stimulate seed germination of many weed species including barrelclover (*Medicago truncatula*) (Globerson 1978), common cocklebur (Esashi et al. 1979), rock bulrush (*Schoenoplectiella juncoides*) (Ishikura and Soga 1979), *Stylosanthes macrocephala* (Carmona et al. 1986), and common lambsquarters (Carmona 1995).

Some respiration inhibitors such as azide, cyanide, and hydroxylamine have been reported to stimulate weed seed germination (Roberts and Smith 1977; Fay et al. 1980). Hurtt and Taylorson (1986) concluded that sodium azide was a more promising agent for promoting grass and broadleaf species emergence compared with ammonium nitrate and butylate in the field.

2.6.2 Hormones

Phytohormone effects on weed seed germination are quite variable. The most consistent germination promoters appear to be gibberellins, cytokinins, and ethylene. The use of gibberellins and cytokinins under field conditions, however, is limited because of their high cost (Peters and West 1991).

Ethylene, used commercially for the control of flowering, fruit ripening, and induction of rooting in crops, also influences weed seed germination. In the United States, ethylene has been used to reduce the seedbanks of the parasitic weed *Striga* spp. (Egley and Dale 1970; Witt and Weber 1975; Babiker and Hamdoun 1982; Egley 1986).

2.6.3 Other Synthetic Compounds

Soil conditioners have been tested for stimulating weed seed germination in the field. Roberts (1981) observed as much as a 50% increase in weed seedling emergence after application of cellulose xanthate to the soil. It was suggested that this substance increases soil aeration by preserving soil surface micro-topography and preventing the formation of hardened crust after rainfall.

Fawcett and Slife (1975) demonstrated stimulation of germination of velvetleaf and common lambsquarters by certain carbamate herbicides. This observation was later confirmed for other broadleaf weeds by Hurtt and Taylorson (1986). Various peroxide compounds have also been reported to stimulate weed seed germination under controlled conditions (Hsiao and Quick 1985; Carmona 1995).

The synthetic growth regulator 1-(3-chlorophthalimido) cyclohexanecarboxamide with gibberellin-like properties has been shown to promote germination of 17 out of 24 weed species (Bond and Burch 1990). Peters and West (1991) were able to stimulate the germination of poverty brome (*Bromus sterilis*) seeds located on the soil surface by spraying this compound in the field. Its effect on buried seeds was significantly lower. Some synthetic analogs of strigol were effective in stimulating seed germination of purple witchweed (*Striga hermonthica*) (Babiker and Hamdoun 1982) and wild oats (Bradow et al. 1990). Thidiazuron, a synthetic regulator with cytokinin-like properties, induced germination and development in Asiatic witchweed (*Striga asiatica*) and cowpea witchweed (*Striga gesnerioides*) (Babiker and Hamdoun 1982). Ethanol has also been documented to promote seed germination (Corbineau et al. 1991; Taylorson and Hendricks 1972).

Despite numerous reports on stimulation of germination by chemicals in the laboratory, reports on their successful use under field conditions are limited. This is due in large part to the instability of the compounds in the soil (e.g. ethanol and phytohormones), species specificity (e.g. gibberellins), cost (e.g. phytohormones), toxicity (e.g. azide), and environmental effects on their action. Soil moisture, pH, and temperature can affect their efficacy in stimulating germination. There is also some evidence suggesting that some of these compounds are more effective when used in combination than alone.

2.7 Weed Seed Destruction or Devitalization of Seeds

Destroying seeds in the soil seedbank is difficult and expensive. Soil fumigants, such as methyl bromide, have been used in horticultural and high value crops for controlling a range of pests, including weeds (Duniway 2002). The use of methyl bromide has been drastically restricted due to its contribution to stratospheric ozone depletion. However, where still available, its use is limited due to its cost, requirements for specialized application equipment, and buffer zone requirement around treated areas. Alternatives, such as 1,3-dichloropropene and metam sodium, are not always effective in destroying weed seeds.

Soil solarization has also been used for high value crops to kill weed seeds. This practice involves covering the soil with a clear plastic, which increases surface soil temperature (>40 °C) by allowing the light to pass through but restricting heat dissipation (Chase et al. 1999). The plastic cover needs to remain in place for a long period to kill moist seeds on or near the soil surface. Effectiveness of seed destruction by soil solarization is species dependent, with some species being able to tolerate higher temperatures (Clements and Atwood 2012).

Combines customized with specialized mills can destroy mature weed seeds before they enter the soil seedbank. Currently, this technology is used primarily in Australia and is being adopted by other regions (Walsh et al. 2012; Schwartz-Lazaro et al. 2017; Shergill et al. 2020). These machines mechanically kill weed seeds using mill technology and can be highly effective on species that do not readily release their seeds at maturity. Since harvest weed seed control occurs at grain harvest, it does not eliminate weed competition with the crop and the resulting potential crop yield loss.

Stale seedbed procedure promotes weed seeds to germinate, and the seedlings produced are destroyed before planting the crop (Gallandt 2006). Typically, the field is prepared with a tillage operation, but planting is delayed three to five weeks to allow weed seed germination and emergence. Emerged seedlings are destroyed before planting with a light tillage operation, a postemergence herbicide, or some other means that minimally disturbs the soil. This approach is not as effective on species with long germination periodicity.

2.8 Soil Seedbank Research Methodology

There are two methods for sampling soil seedbanks; one relies on direct plant propagule counts and another on counts of emerged seedlings. The former method involves sieving the soil or seed floatation methodology to extract seeds followed by their identification. The latter uses exhaustive germination tests and identification of the emerged seedlings. Methodologies of sampling the weed seedbank are tedious and time consuming, and, in addition, the results can be variable and difficult to interpret (Forcella 1992; Cardina and Sparrow 1996; Reinhardt and Leon 2018). Furthermore, there is often a poor correlation between seedbank assessment and aboveground plant composition, and this can lead to different conclusions (Cardina and Sparrow 1996; Harbuck et al. 2009). It is important to consider the limitations of the methodology used to sample the seedbank and how this could influence data interpretation.

Studying seed movement in the soil can provide some useful information but is difficult. Plastic or ceramic beads have been used to track seed movement in the soil (Staricka et al. 1990; Colbach et al. 2000; Reid and Van Acker 2005; Mohler et al. 2006). They are relatively easy to distinguish from weed seeds and to separate from soil particles. Another approach involves using nondormant seeds that are positioned in a known area and then observing the location of emerged plants after tillage (Davis and Luschei 2009).

Interpretation of field trials is somewhat density dependent; with differences easier to observe with higher compared with lower seedbank densities (Buhler 1999). Seedbank composition at any point in time is highly influenced by the last cropping season (Smith and Gross 2006). Adding weed seeds to the soil for studying germination or emergence is a possibility, but can be challenging, since it may add unintended variables. Age of seed, preconditioning of the seed, soil disturbance, weed seed planting depth, seed density, and presence/absence of plant competition all add variables that need to be considered when interpreting the data (Myers et al. 2005).

2.9 Conclusions

Weed seeds and soil seedbanks are critical for weeds to survive and persist in an agroecosystem. Crop rotation, crop production, and management have very large impacts on weed seed production. Agroecosystems are managed to optimize crop production, and weed species that survive and thrive in those systems are expected to be highly represented in the soil seedbank.

Annual fluctuations in species composition and density are dependent on many factors. Preventing seed production by the weeds can reduce weed density in future seasons but requires sustained management over years. Weeds are prolific seed producers and can disseminate their seeds very effectively. Given the high seed fecundity of some species, one season of poor weed control can result in rapid increase of the soil seedbank. Understanding seed longevity, germination patterns, and seedling dynamics can improve weed management and thereby reduce weed persistence in agroecosystems.

Integrated weed management has renewed the emphasis on attempting to severely reduce or eliminate weed seed production. While this is a laudable goal, it is challenging for farmers to achieve. In addition, weeds have evolved mechanisms for efficient dispersal, and humans unwittingly play an active role in moving weed seeds within fields and over great distances. So while farmers can potentially achieve weed seedbank reductions in a relatively short period of time, eliminating weed seeds from the soil over the longterm is not achievable. Weed scientists, ecologists, and farmers need to increase their understanding of weed seedbanks to develop effective integrated weed management strategies and to reduce weed persistence.

References

Adkins, S.W., Simpson, G.M., and Naylor, J.M. (1984). The physiological basis of seed dormancy in *Avena fatua* III. Action of nitrogenous compounds. *Physiologia Plantarum* 60: 227–233.

Anderson, R.L. (2005). A multi-tactic approach to manage weed population dynamics in crop rotations. *Agronomy Journal* 97: 1579–1583.

Andreasen, C., Jensen, H.A., and Jensen, S.M. (2018). Decreasing diversity in the soil seedbank after 50 years in Danish arable fields. *Agriculture, Ecosystems and Environment* 259: 61–71.

Babiker, A.G.T. and Hamdoun, A.M. (1982). Factors affecting the activity of GR7 in stimulating germination of *Striga hermonthica* (Del.) Benth. *Weed Research* 22: 111–116.

Bagavathiannan, M.V., Norsworthy, J.K., Smith, K.L., and Neve, P. (2011). Seed production of barnyardgrass (*Echinochloa crus-galli*) in response to time and emergence in cotton and rice. *Journal of Agricultural Science* 150: 717–724.

Ball, D.A. and Miller, S.D. (1990). Weed seed population response to tillage and herbicide use in three irrigated cropping sequences. *Weed Science* 38: 511–517.

Barralis, G. and Chadoeuf, R. (1987). Weed seed banks of arable fields. *Weed Research* 27: 417–424.

Barroso, J., Navarrete, L., Sánchez del Arco, M. et al. (2006). Dispersal of *Avena fatua* and *Avena sterilis* patches by natural dissemination, soil tillage and combine harvesters. *Weed Research* 46: 118–128.

Barroso, J., Andujar, D., San Martin, C. et al. (2012). Johnsongrass (*Sorghum halepense*) seed dispersal in corn crops under Mediterranean conditions. *Weed Science* 60: 34–41.

Baskin, C.C. and Baskin, J.M. (2001). *Seeds: Ecology, Biogeography, and Evolution of Dormancy and Germination*. San Diego, CA: Academic Press.

Beckie, H.J., Francis, A., and Hall, L.M. (2012). The biology of Canadian weeds. 27. *Avena fatua* L. (Updated). *Canadian Journal of Plant Sciences* 92: 1329–1357.

Beckie, H.J., Blackshaw, R.E., Hall, L.M., and Johnson, E.N. (2016). Pollen- and seed-mediated gene flow in kochia (*Kochia scoparia*). *Weed Science* 64: 624–633.

Bell, M.S. and Tranel, P.J. (2010). Time requirements from pollination to seed maturity in waterhemp (*Amaranthus tuberculatus*). *Weed Science* 58: 167–173.

Benech-Arnold, R.L., Sánchez, R.A., Forcella, F. et al. (2000). Environmental control of dormancy in weed seed banks in soil. *Field Crops Research* 67: 105–122.

Benvenuti, S. (2003). Soil texture involvement in germination and emergence of buried weed seeds. *Agronomy Journal* 95: 191–198.

Benvenuti, S. (2007a). Weed seed movement and dispersal strategies in the agricultural environment. *Weed Biology and Management* 7: 141–157.

Benvenuti, S. (2007b). Natural weed seed burial: effect of soil texture, rain and seed characteristics. *Seed Science Research* 17: 211–219.

Benvenuti, S., MacChia, M., and Miele, S. (2001). Quantitative analysis of emergence of seedlings from buried weed seeds with increasing soil depth. *Weed Science* 49: 528–535.

Bewley, J.D. and Black, M. (1994). *Seeds: Physiology of Development and Germination*. New York, NY: Plenum Press.

Bond, W. and Burch, P.J. (1990). Responses of dry-stored seed of *Solanum nigrum* L. to the germination stimulant AC 94377. *Annals of Applied Biology* 116: 119–130.

Bosnic, A.C. and Swanton, C.J. (1997). Influence of barnyardgrass (*Echinochloa crus-galli*) time of emergence and density on corn (*Zea mays*). *Weed Science* 45: 276–282.

Bradow, J.M., Connick, W.L. Jr., Pepperman, A.B., and Wartelle, L.H. (1990). Germination stimulation in wild oats (*Avena fatua* L.) by synthetic strigol analogs and gibberellic acid. *Journal of Plant Growth Regulation* 9: 35–41.

Brainard, D.C., Bellinder, R.R., and DiTommaso, A. (2005). Effects of canopy shade on the morphology, phenology, and seed characteristics of Powell amaranth (*Amaranthus powellii*). *Weed Science* 53: 175–186.

Buhler, D.D. (1997). Effects of tillage and light environment on emergence of 13 annual weeds. *Weed Technology* 11: 496–501.

Buhler, D.D. (1999). Weed population responses to weed control practices. I. Seed bank, weed populations, and crop yields. *Weed Science* 47: 416–422.

Buhler, D.D. and Hartzler, R.G. (2001). Emergence and persistence of seed of velvetleaf, common waterhemp, woolly cupgrass, and giant foxtail. *Weed Science* 49: 230–235.

Buhler, D.D., Hartzler, R.G., and Forcella, F. (1997). Implications of weed seedbank dynamics to weed management. *Weed Science* 45: 329–336.

Burnside, O.C., Moomaw, R.S., Roeth, F.W. et al. (1986). Weed seed demise in soil in weed-free corn (Zea mays) production across Nebraska. *Weed Science* 34: 248–251.

Burnside, O.C., Wilson, R.G., Weisberg, S., and Hubbard, K.G. (1996). Seed longevity of 41 weed species buried 17 years in eastern and western Nebraska. *Weed Science* 44: 74–86.

Burton, N.R., Beckie, H.J., Willenborg, C.J. et al. (2016). Evaluating seed shatter of economically important weed species. *Weed Science* 64: 673–682.

Cardina, J. and Sparrow, D.H. (1996). A comparison of methods to predict weed seedling populations from soil seedbank. *Weed Science* 44: 46–51.

Cardina, J., Regnier, E., and Harrison, K. (1991). Long-term tillage effects on seed banks in three Ohio soils. *Weed Science* 39: 186–194.

Cardina, J., Herms, C.P., and Doohan, D.J. (2002). Crop rotation and tillage system effects on weed seedbanks. *Weed Science* 50: 448–460.

Carmona, R. (1992). Problemática e manejo de bancos de sementes de plantas daninhas em solos agrícolas (problematic and management of weed seed banks in agricultural soils). *Planta Daninha* 10: 5–16.

Carmona, R. (1995). Banco de sementes e estabelecimento de plantas daninhas em agroecossistemas (seed bank in the soil and the establishment of weeds in agro-ecosystems). *Planta Daninha* 13: 3–9.

Carmona, R., Ferguson, J.E., and Maia, M.S. (1986). Germinação de sementes em *Stylosanthes macrocephala M.B.,* Ferr. et Sousa Costa e *S. capitata* Vog. in Linnaea. *Revista Brasileira de Sementes* 8: 19–28.

Cavers, P. (1983). Seed demography. *Canadian Journal of Botany* 61: 3578–3590.

Cavers, P. (1995). Seed banks: memory in soil. *Canadian Journal of Plant Science* 75: 11–13.

Chahal, P.S., Irmak, S., Jugulam, M., and Jhala, A.J. (2018). Evaluating effect of degree of water stress on growth and fecundity of Palmer amaranth (*Amaranthus palmeri*) using soil moisture sensors. *Weed Science* 66: 738–745.

Chase, C.A., Sinclair, T.R., Chellemi, D.O. et al. (1999). Heat-retentive films for increasing soil temperature during solarization in a humid, cloudy environment. *HortScience* 34: 1085–1089.

Chauhan, B.S. and Johnson, D.E. (2010). Growth and reproduction of junglerice (*Echinochloa colona*) in response to water stress. *Weed Science* 58: 132–135.

Chauhan, B.S., Ali, H.H., and Florentine, S. (2019). Seed germination ecology of *Bidens pilosa* and its implications for weed management. *Scientific Reports* 9: 16004. (2019). https://doi.org/10.1038/s41598-019-52620-9.

Chee-Sanford, J.C., Williams, M.W., Davis, A.S., and Sims, G.K. (2006). Do microorganisms influence seed-bank dynamics? *Weed Science* 54: 575–587.

Cheng, F. and Cheng, Z. (2015). Research progress on the use of plant allelopathy in agriculture and the physiological and ecological mechanisms of allelopathy. *Frontiers in Plant Science* 6: 1020. https://doi.org/10.3389/fpls.2015.01020.

Clements, D.R. and Atwood, L.B. (2012). Protecting ecosystems from underground invasions – seed bank dynamics in a semi-arid shrub-steppe. In: *Diversity of Ecosystems* (ed. M. Ali), 393–416. Rijeka, Croatia: InTech.

Clements, D.R., Benoit, D.L., Murphy, S.D., and Swanton, C.J. (1996). Tillage effects on weed seed return and seedbank composition. *Weed Science* 44: 314–322.

Colbach, N., Roger-Estrade, J., Chauvel, B., and Caneill, C. (2000). Modelling vertical and lateral seed bank movements during mouldboard ploughing. *European Journal of Agronomy* 13: 111–124.

Concenço, G., Andres, A., Silva, A.F. et al. (2014). Ciência das plantas daninhas: Histórico, biologia, ecologia e fisiologia. In: *Aspectos da biologia e manejo das plantas daninhas* (ed. P.A. Monquero), 1–31. São Carlos: RiMa.

Cooper, J.B., Maxwell, T.L., and Owens, A.D. (1960). A study of the passage of weed seeds through the digestive tract of chickens. *Poultry Science* 39: 161–163.

Corbineau, F., Gouble, B., Lecat, S., and Come, D. (1991). Stimulation of germination of dormant oat (*Avena sativa* L.) seeds by ethanol and other alcohols. *Seed Science Research* 1: 21–28.

Cordeau, S., Smith, R.G., Gallandt, E.R. et al. (2017). Timing of tillage as a driver of weed communities. *Weed Science* 65: 504–514.

Dauer, J.T., Mortensen, D.A., and VanGessel, M.J. (2007). Temporal and spatial dynamics of long-distance *Conyza canadensis* seed dispersal. *Journal of Applied Ecology* 44: 105–114.

Davis, A.S. and Luschei, E.C. (2009). Living boundaries: tracking weed seed movement with nondormant seed. *Weed Science* 57: 163–168.

Davis, A.S. and Renner, K.A. (2007). Influence of seed depth and pathogens on fatal germination of velvetleaf (*Abutilon theophrasti*) and giant foxtail (*Setaria faberi*). *Weed Science* 55: 30–35.

Davis, A.S., Fu, X., Schutte, B.J. et al. (2016). Interspecific variation in persistence of buried weed seeds follows trade-offs among physiological, chemical, and physical seed defenses. *Ecology and Evolution* 6: 6836–6845.

Derksen, D.A., LaFond, G.P., and Thomas, A.G. (1993). Impact of agronomic practices on weed communities-tillage systems. *Weed Science* 41: 409–417.

DiTomaso, J.M. and Healy, E.A. (2007). *Weeds of California and Other Western States*, 1760. Oakland: University of Cahlifornia Division of Agriculture and Natural Resources.

Duniway, J.M. (2002). Status of chemical alternatives to methyl bromide for pre-plant fumigation of soil. *Phytopathology* 92: 1337–1343.

Egley, G.H. (1986). Stimulation of weed seed germination in soil. *Reviews of Weed Science* 2: 67–89.

Egley, G.H. and Dale, J.E. (1970). Ethylene, 2-chlorethylphosphonic acid, and witchweed germination. *Weed Science* 18: 586–589.

Egley, G.H. and Williams, R.D. (1990). Decline of weed seeds and seedling emergence over five years as effected by soil disturbances. *Weed Science* 38: 504–510.

Esashi, Y., Ohhara, Y., OKazaki, M., and Hishinuma, K. (1979). Control of cocklebur seed germination by nitrogenous compounds: nitrite, nitrate, hydroxylamine, thiourea, azide and cyanide. *Plant and Cell Physiology* 20: 349–361.

Espeby, L. (1989). Germination of weed seeds and competition in stands of weeds and barley – influences of mineral nutrients. *Crop Production Science* 6: 1–172.

Fahad, S., Hussain, S., Chauhan, B.S. et al. (2015). Weed growth and crop yield loss in wheat as influenced by row spacing and weed emergence times. *Crop Protection* 71: 101–108.

Farmer, J.A., Webb, E.B., Pierce, R.A., and Bradley, K.W. (2017). Evaluating the potential for weed seed dispersal based on waterfowl consumption and seed viability. *Pest Management Science* 73: 2592–2603.

Fawcett, R.S. and Slife, F.W. (1975). Germination stimulation properties of carbamate herbicides. *Weed Science* 23: 419–424.

Fay, P.K., Gorecki, R.S., and Fuerst, P.M. (1980). Coating sodium azide granules to enhance seed germination. *Weed Science* 28: 674–677.

Forcella, F. (1992). Prediction of weed seedling densities from buried seed reserves. *Weed Research* 32: 29–38.

Forcella, F. (2003). Debiting the seedbank: priorities and predictions. *Aspects of Applied Biology* 69: 151–162.

Frost, M.D., Haramoto, E.R., Renner, K.A., and Brainard, D.C. (2019). Tillage and cover crop effects on weed seed persistence: do light exposure and fungal pathogens play a role? *Weed Science* 67: 103–113.

Gallandt, E.R. (2006). How can we target the weed seedbank? *Weed Science* 54: 588–596.

Globerson, D. (1978). Germination and dormancy breaking by ethephon in mature and immature seeds of *Medicago truncatula* (Medic.) and *Trifolium subterraneum* (Clover.). *Australian Journal of Agricultural Research* 29: 43–49.

Gomes, F.G. Jr. and Christoffoleti, P.J. (2008). Biologia e manejo de plantas daninhas em áreas de plantio direto. *Planta Daninha* 26: 789–798.

Gulden, R.H., Lewis, D.W., Froese, J.C. et al. (2011). The effect of rotation and in-crop weed management on the germinable weed seedbank after 10 years. *Weed Science* 59: 553–561.

Harbuck, S.B., Menalled, F.D., and Pollnac, F.W. (2009). Impact of cropping systems on the weed seed bank in the northern Great Plains, USA. *Weed Biology and Management* 9: 160–168.

Harbur, M.M. and Owen, M.D.K. (2004). Response of three annual weeds to corn population density and nitrogen fertilization timing. *Weed Science* 52: 845–853.

Harmon, G.W. and Keim, F.D. (1934). The percentage and viability of weed seeds recovered in the feces of farm animals and their longevity when buried in manure. *Journal of American Society of Agronomy* 26: 762–767.

Harper, J.L. (1977). *Population Biology of Plants*. London: Academic Press, 892 pp.

Haus, J.M., Webb, S.L., Strickland, B.K. et al. (2019). Land use and dispersal influence mortality in white-tailed deer. *The Journal of Wildlife Management* 83: 1185–1196.

Hayashi, M., Gerry, S.P., and Ellerby, D.J. (2010). The seed dispersal of *Cardamine parviflora* (Brassicaceae) is efficient but unreliable. *American Journal of Botany* 97: 1595–1601.

Holm, L.G., Plucknett, D.L., Pancho, J.V., and Herberger, J.P. (1977). *Avena fatua* L. and other members of the 'wild oat' group. In: *World's Worst Weeds: Distribution and Biology*, 105–113. Honolulu: University of Hawaii.

Honda, Y. (2008). Ecological correlations between the persistence of the soil seed bank and several plant traits, including seed dormancy. *Plant Ecology* 196: 301–309.

Howe, H.F. and Westley, L.C. (1986). Ecology of pollination and seed dispersal. In: *Plant Ecology* (ed. M.G. Crawley), 262–283. Oxford: Blackwell.

Hsiao, A.I.S. and Quick, W.A. (1985). Actions of sodium hypochlorite and hydrogen peroxide on seed dormancy and germination of wild oats *(Avena fatua)*. *Weed Research* 25: 281–288.

Huelma, C.C., Moody, K., and Mew, T.W. (1996). Weed seeds in rice seed shipments: a case study. *International Journal of Pest Management* 42: 147–150.

Hulme, P.E. (1998). Post-dispersal seed predation and seed bank persistence. *Seed Science Research* 8: 513–519.

Hurtt, W. and Taylorson, R.B. (1986). Chemical manipulation of weed emergence. *Weed Research* 26: 259–267.

Ishikura, N. and Soga, Y. (1979). Studies on the ecology and control of perennial weeds, Scirpus in paddy fields: 2. Effects of temperature and light during the stratified period on the dormancy-awakening of seeds of *Scirpus juncoides* Roxb. Var. *ohwianus* T. Koyama. *Weed Research* 24: 28–32. (in Japanese with English summary).

Kaul, V., Sharma, N., and Koul, A.K. (2002). Reproductive effort and sex allocation strategy in *Commelina benghalensis* L., a common monsoon weed. *Botanical Journal of the Linnean Society* 140: 403–413.

Keeley, P.E., Carter, C.H., and Thullen, R.J. (1987). Influence of planting date on growth of Palmer amaranth (*Amaranthus palmeri*). *Weed Science* 35: 199–204.

Kremer, R.J. (1993). Management of weed seed banks with microorganism. *Ecological Applications* 3: 42–52.

Lamberti-Raverot, B., Piola, F., Thiébaut, M. et al. (2017). Water dispersal of the invasive complex *Fallopia*: the role of achene morphology. *Flora* 234: 150–157.

Lindquist, J.L. and Mortensen, D.A. (1998). Tolerance and velvetleaf (*Abutilon theophrasti*) suppressive ability of two old and two modern corn (*Zea mays*) hybrids. *Weed Science* 46: 569–574.

Long, R.L., Gorecki, M.J., Renton, M. et al. (2015). The ecophysiology of seed persistence: a mechanistic view of the journey to germination or demise. *Biological Reviews of the Cambridge Philosophical Society* 90: 31–59.

Lutman, P.J.W., Cussans, G.W., Wright, K.J. et al. (2002). The persistence of seeds of 16 weed species over six years in two arable fields. *Weed Research* 42: 231–241.

Martinková, Z. and Honěk, A. (2013). Fatal germination in barnyardgrass (*Echinochloa crus-galli*). *Plant Protection Science* 49: 193–197.

Mashingaidze, N., Twomlow, S., Madakadze, I.C. et al. (2017). Weed growth and crop yield responses to tillage and mulching under different crop rotation sequences in semi-arid conditions. *Soil Use and Management* 33: 311–327.

Menalled, F.D., Gross, K.L., and Hammond, M. (2001). Weed aboveground and seedbank community responses to agricultural management systems. *Ecological Applications* 11: 1586–1601.

Mohler, C.L. and Galford, A.E. (1997). Weed seedling emergence and seed survival: separating the effects of seed position and soil modification by tillage. *Weed Research* 37: 147–155.

Mohler, C.L., Frisch, J.C., and McCulloch, C.E. (2006). Vertical movement of weed seed surrogates by tillage implements and natural processes. *Soil and Tillage Research* 86: 110–122.

Mt. Pleasant, J.M. and Schlather, K.J. (1994). Incidence of weed seed in cow (*Bos* sp.) manure and its importance as a weed source for cropland. *Weed Technology* 8: 304–310.

Myers, M.W., Curran, W.S., VanGessel, M.J. et al. (2005). Effect of soil disturbance on annual weed emergence in the northeastern United States. *Weed Technology* 19: 274–282.

Norris, R.F. (2007). Weed fecundity: current status and future needs. *Crop Protection* 26: 182–188.

Norris, R.F., Elmore, C.L., Rajmánek, M., and Akey, W.C. (2001). Spatial arrangement, density, and competition between barnyardgrass and tomato: II. Barnyardgrass growth and seed production. *Weed Science* 49: 69–76.

Norsworthy, J.K., Ward, S.M., Shaw, D.R. et al. (2012). Reducing the risks of herbicide resistance: best management practices and recommendations. *Weed Science Special Issue* 1 (60): 31–62.

Norsworthy, J.K., Griffith, G., Griffin, T. et al. (2014). In-field movement of glyphosate-resistant Palmer amaranth (*Amaranthus palmeri*) and its impact on cotton lint yield: evidence supporting a zero-threshold strategy. *Weed Science* 62: 237–249.

Norsworthy, J.K., Korres, N.E., and Bagavathiannan, M.V. (2018). Weed seedbank management: revisiting how herbicides are evaluated. *Weed Science* 66: 415–417.

Nurse, R.E. and DiTommaso, A. (2005). Corn competition alters the germinability of velvetleaf (*Abutilon theophrasti*) seeds. *Weed Science* 53: 479–488.

Peters, N.C.B. and West, T.M. (1991). Stimulation of *Bromus sterilis* seed germination by 1-(3-chlorophthalimido) cyclohexanecarboxamide (AC 94377) or gibberellic acid (GA3). Brighton Crop Protection Conference, British Crop Protection Council. *Weeds*: 171–176.

Reid, D.J. and Van Acker, R. (2005). Seed burial by tillage promotes field recruitment of false cleavers (*Galium spurium*) and catchweed bedstraw (*Galium aparine*). *Weed Science* 53: 578–585.

Reinhardt, T. and Leon, R.G. (2018). Extractable and germinable seed bank methods provide different quantifications of weed communities. *Weed Science* 66: 715–720.

Roberts, H.A. (1981). Seed banks in the soil. In: *Advances in Applied Biology* (ed. H.A. Roberts), 1–55. Cambridge: Academic Press.

Roberts, H.A. and Neilson, J.E. (1982). Seed bank of soil under vegetable cropping in England. *Weed Research* 22: 13–16.

Roberts, E.H. and Smith, R.D. (1977). Dormancy and the pentose phosphate pathway. In: *The Physiology and Biochemistry of Seed Dormancy and Germination* (ed. A.A. Khan), 385–411. Amsterdam: Elsevier/North Holland Biomedical Press.

Ruisi, P., Frangipane, B., Amato, G. et al. (2015). Weed seedbank size and composition in a long-term tillage and crop sequence experiment. *Weed Research* 55: 320–328.

Schutte, B.J., Klypin, N., and Shukla, M.K. (2016). Influence of irrigation timing on disturbance-induced reductions in seedbank density. *Weed Science* 64: 613–623.

Schwartz, L.M., Gibson, D.J., Gage, K.L. et al. (2015). Seedbank and field emergence of weeds in glyphosate-resistant cropping systems in the United States. *Weed Science* 63: 425–439.

Schwartz-Lazaro, L.M., Norsworthy, J.K., Walsh, M.J., and Bagavathiannan, M.V. (2017). Efficacy of the integrated Harrington seed destructor on weeds of soybean and rice production systems in the southern United States. *Crop Science* 57: 2812–2818.

Schweizer, E.E. and Zimdahl, R.L. (1984). Weed seed decline in irrigated soil after six years of continuous corn (*Zea mays*) and herbicides. *Weed Science* 32: 76–83.

Shergill, L.S., Bejleri, K., Davis, A., and Mirsky, S.B. (2020). Fate of weed seeds after impact mill processing in Midwestern and mid-Atlantic United States. *Weed Science* 68: 92–97.

Shields, E.J., Dauer, J.T., VanGessel, M.J., and Neumann, G. (2006). Horseweed (*Conyza canadensis*) seed collected in the planetary boundary layer. *Weed Science* 54: 1063–1067.

Shoemaker, W.R. and Lennon, J.T. (2018). Evolution with a seed bank: the population genetic consequences of microbial dormancy. *Evolutionary Applications* 11: 60–75.

Singh, M., Ramirez, A.H.M., Sharma, S.D., and Jhala, A.J. (2012). Factors affecting the germination of tall morningglory (*Ipomoea purpurea*). *Weed Science* 60: 64–68.

Smith, R.G. (2006). Timing of tillage is an important filter on the assembly of weed communities. *Weed Science* 54: 705–712.

Smith, R.G. and Gross, K.L. (2006). Rapid change in the germinable fraction of the weed seed bank in crop rotations. *Weed Science* 54: 1091–1100.

Sosnoskie, L.M., Herms, N.P., and Cardina, J. (2006). Weed seedbank community composition in a 35-yr-old tillage and rotation experiment. *Weed Science* 54: 263–273.

Staricka, J.A., Burford, P.M., Allmaras, R.R., and Nelson, W.W. (1990). Tracing the vertical distribution of simulated shattered seeds as related to tillage. *Agronomy Journal* 82: 1131–1134.

Steinmaus, S.J. and Norris, R.F. (2002). Growth and analysis and canopy architecture of velvetleaf grown under light conditions representative of irrigated Mediterranean-type agroecosystems. *Weed Science* 50: 42–53.

Swanton, C.J., Shrestha, A., Knezevic, S.Z. et al. (1999). Effect of tillage systems, N, and cover crop on the composition of weed flora. *Weed Science* 47: 454–461.

Taylorson, R.B. and Hendricks, S.B. (1972). Phytochrome control of germination of *Rumex crispus* L. seeds induced by temperature shifts. *Plant Physiology* 50: 645–648.

Teasdale, J.R. and Mohler, C.L. (1993). Light transmittance, soil temperature, and soil moisture under residue of hairy vetch and rye. *Agronomy Journal* 85: 673–680.

Thompson, K. and Grime, J.P. (1979). Seasonal variation in the seed banks of herbaceous species in ten contrasting habitats. *Journal of Ecology* 67: 893–921.

Twigg, L.E., Lowe, T.J., Taylor, C.M. et al. (2009). The potential of seed-eating birds to spread viable seeds of weeds and other undesirable plants. *Austral Ecology* 34: 805–820.

Voll, E., Torres, E., Brichenti, A.M., and Gazziero, D.L.P. (2001). Dinâmica de um banco de sementes de plantas daninhas sob diferentes manejos do solo. *Planta Daninha* 19: 171–178.

Walsh, M.J., Harrington, R.B., and Powles, S.B. (2012). Harrington seed destructor: a new nonchemical weed control tool for global grain crops. *Crop Science* 52: 1343–1347.

Walsh, M., Newman, P., and Powles, S. (2013). Targeting weed seeds in-crop: a new weed control paradigm for global agriculture. *Weed Technology* 27: 431–436.

Weaver, S.E. (2001). The biology of Canadian weeds. 115. *Conyza canadensis*. *Canadian Journal of Plant Science* 81: 867–875.

Wilson, R.D. (1980). Dissemination of weed seed by surface irrigation in Western Nebraska. *Weed Science* 28: 87–92.

Wilson, R.G. (1988). Biology of weed seed in the soil. In: *Weed Management in Agroecosystem: Ecological Approaches* (eds. M.L. Altieri and M. Liebman), 25–39. Boca Raton, Florida: CRC Press.

Wilson, B.J. and Lawson, H.M. (1992). Seedbank persistence and seedling emergence of seven weed species in autumn-sown crops following a single year's seeding. *Annals of Applied Biology* 120: 105–116.

Witt, W.W. and Weber, B. (1975). Ethylene adsorption and movement in soils and adsorption by soil constituents. *Weed Science* 23: 302–307.

Yenish, J.P., Doll, J.D., and Buhler, D.D. (1992). Effects of tillage on vertical distribution and viability of weed seed in soil. *Weed Science* 40: 429–433.

Zimdahl, R.L. (2018). *Fundamentals of Weed Science*, 5e, 735. Fort Collins, EUA: Academic Press.

3

Weed Seed Dormancy and Persistence of Weeds

Mahesh K. Upadhyaya[1], *Steve W. Adkins*[2], *and Li Ma*[3]

[1] *Faculty of Land and Food Systems, University of British Columbia, Vancouver, BC, Canada*
[2] *School of Agriculture and Food Sciences, The University of Queensland, Brisbane, QLD, Australia*
[3] *Institute for Sustainable Horticulture, Kwantlen Polytechnic University, Surrey, BC, Canada*

3.1 Introduction: Seed Dormancy and Persistence of Weeds

The seed is an important stage in the life cycle of many weed species because it is a unit of reproduction and a unit of dissemination, and in a dormant state it can persist in the soil for a very long time. Seed dormancy, a form of developmental arrest (Foley 2001), plays an important role in persistence of weeds. It ensures seed germination when conditions are optimal for plant growth and development. Dormant seeds can survive under harsher environmental conditions compared with germinating seeds or seedlings. Because of genetic variation in the duration of seed dormancy and its modification by environmental factors (Benech-Arnold et al. 2000; Batlla and Benech-Arnold 2014; Long et al. 2015; Oreja et al. 2017; Chapter 4, this book), weeds are able to distribute their seed germination over time, which leads to the development of seedbanks in the soil.

The variability in germination behaviour of seeds in the soil seedbanks makes weed control very difficult. Some seeds from these banks germinate every year and the resulting plants can, under appropriate conditions, deposit a new crop of seeds to the soil seedbank. Seeds are removed from these banks mainly by germination, death, and predation. It is important to note that for many weeds, mature seeds are shed to the ground prior to crop harvest, which ensures their deposition into the soil seedbank. In addition to soil seedbanks, some weeds also have aerial seedbanks (seeds held by live or dead shoots, e.g. diffuse knapweed [*Centaurea diffusa*]), which also contribute to persistence of weeds. Unfortunately, while most researchers have focused on soil seedbanks, very little attention has been paid to the aerial seedbanks. In order to understand persistence strategies of specific weeds, a sound knowledge of the various aspects of their seed dormancy and germination regulation is important. Weed scientists and plant ecologists need to look at seed dormancy at the population level (species diversity, seedbanks, spatial distribution, etc.); at the individual plant level (dormancy polymorphism, effects of microsite environment and of the position of the seeds on the plant or in the inflorescence, maternal influences on seed dormancy, etc.); and

Persistence Strategies of Weeds, First Edition. Edited by Mahesh K. Upadhyaya, David R. Clements, and Anil Shrestha.

at the individual seed level (mechanism of dormancy, regulation for dormancy induction and release, response to biotic and abiotic environment, etc.) (Penfield 2017).

In this chapter we discuss dormancy and germination of weed seeds, types of seed dormancy, genotypic variability in duration of seed dormancy and its role in the persistence of weeds, effect of maternal environment on germination behaviour of seeds, and the dormancy cycle in soil seedbanks. Over the past several decades, a significant number of researches have focused on dormancy and germination of non-weed species. In this chapter, however, we will focus mostly on the work done on weed species.

3.2 Seed Dormancy and Germination

While seed dormancy has been extensively studied in weed and crop science, no unambiguous definition for this phenomenon is available. Seed dormancy has been defined in the literature differently by different authors. Simpson (1990) defined seed dormancy as a 'temporary failure of a viable seed to germinate, after a specified period of time, in a particular set of conditions that later evoke germination when the restrictive state has been terminated by either natural or artificial means'. Baskin and Baskin (2004a) suggested that 'a dormant seed (or other germination unit) is one that does not have the capacity to germinate in a specified period of time under any combination of normal environmental factors (temperature, light/dark, etc.) that otherwise is favourable for its germination, i.e., after the seed becomes non-dormant'. Bewley (1997) provided a simple definition of dormancy as 'the failure of an intact viable seed to complete germination under favourable conditions'. Dormancy has been suggested to vary on a continuous scale with its degree determining the range of conditions under which a seed will germinate (Vleeshouwers et al. 1995). Unfortunately, some studies in weed science classify a seed that fails to germinate as dormant, which may not be true. It is important to confirm that the seed that does not germinate is in fact viable and not dead.

Bewley et al. (1994) defined seed germination as 'the physiological process culminating in the emergence of the embryo from its enclosing coverings, which can include the endosperm, perisperm, testa, or pericarp'. While protrusion of the radicle is considered as an evidence for release of dormancy as well as germination in many studies involving weed seeds, it should be emphasized that the release of seed dormancy and seed germination are often considered two separate physiological processes. A seed can lose its dormancy and may not germinate if conditions are not favourable for germination. The term quiescent seed refers to a seed in which no germination process is occurring (Bewley and Black 1985). Quiescent seeds require only imbibition of water and conditions (e.g. for temperature and oxygen [O_2]) suitable for seed germination and should not be confused with dormant seeds, which are unable to germinate even under suitable conditions for germination.

3.3 Types of Seed Dormancy and Some Terminologies

Different systems and terms have been used in classification of seed dormancy, based on the time of occurrence, the process of induction, or the mechanism of dormancy. Harper (1977) stated 'some seeds are born dormant, some acquire dormancy and some have dormancy thrust upon them' and referred to these three dormancies as innate, induced, and enforced dormancies, respectively. He called induced and enforced dormancies as 'opportunistic'. The term primary dormancy is used to refer to the dormancy present when a seed separates from its parent plant. This dormancy is set in the seed while it is developing and maturing on the mother plant. Primary dormancy serves three functions: (i) prevents precocious seed germination while still attached to the parent plant, (ii) prevents immediate and synchronous germination so that seeds can disseminate away from the parent plant, and (iii) allows seeds to persist in soil seedbanks for varying periods. Certain biotic and abiotic factors can induce germination of primary dormant seeds. Primary dormant seeds gradually lose their dormancy with time and become nondormant. This poorly understood phase in a seed's life is called after-ripening. During after-ripening, seed's light requirement for germination may change, seed may lose nitrate requirement for germination, the temperature range for germination may widen, abscisic acid (ABA) level may decrease and/or sensitivity to this hormone and to gibberellic acid may change, and the germination velocity may increase (Long et al. 2015). A seed that has partially or fully lost its dormancy level can become dormant again when it encounters conditions unfavourable for germination (e.g. anaerobiosis, high temperature) and seedling growth. While air in the upper soil profile may deviate little from the ambient air (Murdoch and Roberts 1996), anaerobic microsites (e.g. in case of water logging) within a largely aerobic matrix may be enough to impose dormancy. This dormancy, induced by unfavourable environmental conditions, is called secondary or induced dormancy.

Several schemes of classification of weed seed dormancy have been proposed. In 1967 M.G. Nikolaeva proposed a classification system for seed dormancy considering seed morphological and physiological properties (Baskin and Baskin 2004a,b; Finch-Savage and Leubner-Metzger 2006). Baskin and Baskin (2004a,b) suggested a seed dormancy classification scheme, a modification of the scheme suggested by Nikolaeva, which classifies dormancy on the basis of its underlying mechanisms. Table 3.1 summarizes the major classes of this classification scheme with a brief description of each class.

Table 3.1 A classification system for seed dormancy.

Class	Description
Physiological dormancy	Can be deep, intermediate, or non-deep
Morphological dormancy	The embryo is underdeveloped but differentiated
Morphophysiological dormancy	Underdeveloped embryo plus some physiological dormancy
Physical dormancy	Water impermeability of the seed
Combinational dormancy	Water impermeability plus physiologically dormant embryo

Note: This classification system does not include seeds with undifferentiated embryos as well as mechanical and chemical dormancies.
Source: Modified Baskin and Baskin 2004a,b.

3.4 Dormancy Polymorphism

Many weed species exhibit variation in dormancy behaviour of seeds produced on the same plant or on different plants of the same species. For example, in many grass weeds (e.g. wild oat [*Avena fatua*], Harper 1977), seeds on different positions on inflorescence show different levels of dormancy. Cocklebur (*Xanthium strumarium*), a broadleaf weed, produces two seeds in a capsule, which show different levels of dormancy (Harper 1977). The upper, smaller seed is more dormant than the larger lower seed, with a difference of at least 12 months normally separating germination of the two seed types. In tansy ragwort (*Senecio jacobaea*), seeds from ray florets and from disc florets differ in their germination percentage and germination time (McEvoy 1984). Compared with the achenes produced by the disc florets at the centre, achenes from the ray florets are heavier and smoother, have greater dormancy, remain on the mother plant longer, and are dispersed close to the plant on which they are produced. In common lambsquarters (*Chenopodium album*), a small percentage of seeds are large, brown, thin-coated, and nondormant, in comparison with most other seeds which are smaller, black, thick-coated, and dormant (Williams and Harper 1965). This adaptive variation helps spread germination of seeds from the same plant or from a population of plants aiding to persistence of weeds. This variation in dormancy behaviour is due to both genotypic differences as well as environmental plasticity.

3.4.1 Genetic Variation in Seed Dormancy

Genetic variability in seed dormancy and its modification by environmental factors has been extensively demonstrated in a variety of weed species. The resultant variability in seed dormancy behaviour is an important factor in persistence of weeds. Natural populations of many weedy species are made up of diverse biotypes, which exhibit very different germination behaviour. This distributes seed germination over time and results in the maintenance of persistent seedbanks.

3.4.1.1 Wild Oat (*Avena fatua*) Case Study

Wild oat has been used extensively as a model system to study variability in seed dormancy and the influence of environmental factors during seed development on seed dormancy. Genetic variability in seed dormancy in field populations plays an important role in persistence of this species by helping it adapt to crop production and weed control practices (Naylor 1983). J.M. Naylor and his co-workers isolated genetically pure lines of wild oat from field populations in Saskatchewan, Canada, and Montana, USA. These lines differed significantly in their germination behaviour, and it was suggested that these differences may result in part from adaptation to agronomic practices (Naylor and Jana 1976). Using a parent–progeny regression method, Jana and Naylor (1980) estimated the heritability of dormancy to be approximately 50% within the primary seeds in four field populations of wild oat. From breeding studies employing genetically pure lines with different levels of primary seed dormancy, Jana et al. (1979, 1988) suggested that at least three interacting loci were involved in controlling after-ripening rates. It was suggested that at least two of the three genes influenced after-ripening at different times after seed maturation.

Wild oat is a highly inbreeding species and the 'genetically pure' lines of this weed isolated by Naylor and co-workers as described earlier produce seeds with characteristic dormancy behaviour without significant segregation. These lines have been extensively used in research on seed dormancy and its modification by environment and germination-stimulating chemicals. In addition to differences in primary dormancy (Adkins et al. 1986; Jana et al. 1988), seeds of these lines grown under controlled conditions have been shown to differ in their response to temperature (Naylor and Fedec 1978), duration of after-ripening for release of dormancy, and induction of secondary dormancy (Tilsner and Upadhyaya 1985). These lines have also been used in studies showing germination stimulation response to sodium azide (Upadhyaya et al. 1982a; Jana et al. 1988), nitrates (Adkins et al. 1984a), ethanol, ethephon (Adkins et al. 1984b; Adkins and Adkins 1994), and some substituted phthalimides (Upadhyaya et al. 1986). Using synthetic populations of these lines differing in their seed dormancy, Armstrong and Adkins (1998) reported that similar pure lines of Australian wild oats were quantitative long-day plants, taking a longer time to mature when the photoperiod was reduced.

3.4.1.2 Influence of Environment During Seed Development on Seed Dormancy

Many biochemical (e.g. hormonal, storage material, etc.) and structural changes occur in seeds during their development and maturation (Bewley and Black 1985). Environmental factors (e.g. temperature, soil moisture level, photoperiod, and light intensity and quality) and their interactions during seed development can influence dormancy and the subsequent germination behaviour of shed seeds (Gutterman 1985), which could influence persistence of weeds.

Sawhney and Naylor (1979, 1980) reported that exposure of plants of genetically pure lines of wild oat to various temperatures could influence the duration of primary dormancy in mature seeds and that the lines employed differed in their response to temperature. Adkins et al. (1987) also reported an influence of temperature during seed development on dormancy behaviour of mature seeds in pure lines of wild oat; duration of dormancy was longer in seeds developed at 15 °C compared with those developed at 25 °C. Sawhney and Naylor (1982) exposed plants to water stress during seed development and found that compared with well-watered plants, plants exposed to drought during seed development produced seeds with a shorter duration of primary dormancy. These studies show that seed dormancy in wild oat is a very plastic character. In addition to genotypic variation in duration of seed dormancy, the environment experienced during seed development as well as post-dispersal can also modify this trait. This allows wild oat to distribute its germination over time, an important strategy to maintain soil seedbanks and persist as a weed.

The position of seeds on the mother plant, in the inflorescence and in the fruit, and interactions with the stage of seed development and genotype are important in determining the influence of environmental factors during seed development on subsequent germination behaviour of mature seeds (Gutterman 1985, 1992).

Genetic variability in seed dormancy and its modification by the environment has also been reported in other weedy species (He et al. 2014, Kerdaffrec and Nordborg 2017). Kerdaffrec and Nordborg (2017) reported interaction between maternal environment and genotype in arabidopsis (*Arabidopsis thaliana*) and identified several possible candidate genes for this plasticity by genome-wide association study. The functional roles of these genes however remain to be determined (Kerdaffrec and Nordboug 2017). In a study involving 12 arabidopsis genotypes, He et al. (2014) also found influence of environmental factors on seed performance. They reported that temperature played the most important role and observed that there were genotype × environment interactions.

Exposure of imbibed seeds to light is known to influence seed germination in several weed species (Batlla and Benech-Arnold 2014). Light with a high red/far-red ratio stimulates and with a low red/far-red ratio inhibits germination of these photoblastic seeds (Nolan and Upadhyaya 1988). In a study on the effect of cold temperature (stratification) on light sensitivity of prostrate knotweed (*Polygonum aviculare*), Batlla and Benech-Arnold (2005) reported that light sensitivity of seeds increased as loss of seed dormancy progressed and the rate of increase in sensitivity was inversely correlated with soil temperature. This phytochrome-mediated control (Lymperopoulos et al. 2018) of germination is ecologically as well as agronomically important. The light requirement prevents germination of buried seeds or of seeds under a thick leaf cover, where the canopy leaf chlorophyll absorbs red light, thereby decreasing the red/far-red ratio of the canopy-transmitted light reaching the underlying seeds. Buried seeds germinate when they are brought to the soil surface (e.g. by cultivation) and are exposed to light. Seeds under thick leaf cover germinate when the cover is removed (e.g. by mowing or grazing) and the seeds are exposed to light with a high red/far-red ratio.

McCullough and Shropshire (1970) reported greater dark germination in arabidopsis seeds produced under light with a high red/far-red ratio compared with seeds from plants grown under a low red/far-red ratio. In a study involving several species, Cresswell and Grime (1981) studied the relationship between chlorophyll concentration, a measure of red light-filtering capacity, of investing structures during seed maturation, and the light requirement for germination of the mature seeds. Retention of chlorophyll in seed investing structures was found to strongly correlate with a light requirement for germination. The results suggest that when chlorophyll is present in structures investing a seed during maturation, the seed will receive light with lower red/far-red ratios and, therefore, will have most of its phytochrome in the Pr (inactive) form. These seeds will require light for germination when mature. On the other hand, a seed maturing in investing structures with little or no chlorophyll will receive light with higher red/far-red ratio and will germinate both in the presence of light or in darkness. Thus, differences in the rates of chlorophyll loss in seed investing structures and/or timing of drying of seeds within could cause differences in the light requirement for germination of mature seeds.

A light requirement for weed seed germination has important implications for management of weeds with photoblastic seeds (Batlla and Benech-Arnold 2014). Tillage can be timed when light sensitivity of buried seeds is high to stimulate greater germination, allowing more efficient weed control and greater seedbank depletion upon subsequent weed control (Batlla and Benech-Arnold 2014). Modification of the light environment below a canopy by the management of crop cover and by modifying plant architecture, planting density, and/or row spacing could provide options for weed management (Radosevich et al. 1997; Batlla and Benech-Arnold 2014). In a study on the role of light in stimulation of seed germination during tillage, Scopel et al. (1994) found that daytime tillage significantly increased germination of buried seeds compared with nighttime cultivation. The coverage of tillage equipment during daytime cultivation reduced seed germination, and the illumination of soil surface during nighttime cultivation increased germination significantly. Therefore, timing of tillage and manipulating the availability of light during tillage could offer a potential tool for management of weeds with photoblastic seeds.

While a significant amount of literature has accumulated on changes in weed seed dormancy with time and on the influence of environmental factors on dormancy and germination, most of the research has been done under uniform, controlled, laboratory, or greenhouse conditions using single species, employing on one or two factors. These studies do not allow us to understand the complex interactions among various factors that regulate the dynamics of the very diverse soil seedbanks of agroecosystems.

3.5 Mechanisms of Seed Dormancy

Despite a significant amount of work by plant ecologists, physiologists, geneticists, and breeders, our understanding of induction and regulation of seed dormancy is limited. This is particularly true for weedy species. The current understanding of weed seed dormancy regulation at the molecular level is not enough to allow us to develop effective means for depletion of soil seedbanks.

Seed coverings play important roles in persistence of weed seeds in soil seedbanks. They offer protection from soil-borne pathogens and pests and control seed germination by restricting water and O_2 uptake, mechanically resisting the embryo growth, preventing release of endogenous germination inhibitors, and/or blocking light penetration to embryo (Egley and Duke 1985). Mechanisms responsible for the inability of weed seeds to germinate under conditions otherwise favourable for germination are briefly described in the following text.

3.5.1 Impermeability to Water

Seeds of some weed species (e.g. annual morning glory [*Ipomoea purpurea*], clovers [*Trifolium* sp.], or vetch [*Vicia* sp.]) do not germinate even when an adequate amount of soil moisture is available because of the seed coat being impermeable to water. Impermeability of these 'hard-coated' seeds can be due to the presence of waxy cuticles, suberin depositions, osteosclerid layers, or thick-walled palisade cells (Bewley and Black 1985). Lignification and formation of water impervious polymers during the later stages of seed development have been suggested to be possible causes for this impermeability in some weed species (Egley and Duke 1985). Examples for this mechanism of dormancy regulation are found especially in weed species belonging to Chenopodiaceae, Convolvulaceae, Fabaceae, and Malvaceae families. These seeds will germinate when their seed coat or specific areas of the seed coat becomes permeable to water (LaCroix and Staniforth 1964; Egley and Paul 1981).

3.5.2 Impermeability to Gases

Dormant seeds of wild oats (and some other weed species) show stimulation of germination when the seed coat (pericarp) is damaged (by pricking or scarification) or when seeds are incubated under high O_2 partial pressure (Atwood 1914). Damage to the seed coat increases their O_2 uptake significantly. It was suggested that the intact seeds of wild oats did not germinate because of a restriction in supply of oxygen to the embryo. These seeds germinate when the seed coat either gets damaged or becomes permeable to O_2 by a yet unknown process. Insufficient O_2 uptake by seeds may also affect O_2-requiring processes other than respiration, such as the breakdown of germination inhibitors. Seed coat structure, the presence of water in the imbibed seed coat, or the existence of a mucilaginous coat may all contribute to the restriction of O_2 uptake by weed seeds (Egley and Duke 1985). Whether an impermeable seed coat also interferes with the release of carbon dioxide (CO_2) from the seed, which could influence seed respiration, is not known.

In a study on seed dormancy in some pioneer tree species, Zalamea et al. (2015) concluded that soil temperature fluctuations in the absence of microbial infection and soil abrasion were enough to break physical dormancy due to seed coat impermeability. To what extent this is true for agricultural weeds remains to be determined.

3.5.3 Mechanically Resistant Seed Coat

A balance of the mechanical force generated by the growth potential of the embryonic axis and the resistance offered by covering layers (e.g. endosperm, seed coat, pericarp) determines whether parts of the embryo will protrude and allow the seed to germinate (Chen and Thimann 1966; Steinbrecher and Leubner-Metzger 2017). The seed could germinate when the layer(s) covering the embryo become weaker or when the embryo is capable of generating greater force to break through. A mechanically hard seed coat, for example, could physically prevent seeds of some weeds (e.g. pigweed [*Amaranthus* sp.], shepherd's-purse [*Capsella bursa-pastoris*], and wild mustard [*Sinapis arvensis*]) from germinating by restricting embryo growth. These seeds would germinate when the coat is damaged physically, particularly near the radicle (Egley and Duke 1985), or is weakened by natural means, e.g. softening of a part of the hard covering by hydrolytic enzymes, alternate changes in soil temperature, or damage by pests and microbes in the soil seedbanks. Fungi have been reported to break seed dormancy in Westwood pear (*Opuntia streptacantha*) by weakening the seed coat and reducing mechanical resistance to germination (Delgado-Sánchez et al. 2011). Cocklebur fruit contains two seeds: a dormant smaller seed and a larger non-dormant seed. Esashi and Leopold (1968) studied forces generated by germinating seeds and the restraining force offered by testa of these two types of seeds. They found that thrust generated by the nondormant seed was enough to rupture the testa but by the dormant seed it was not, suggesting that dormancy in the latter was due to inadequacy of the force generated to rupture the testa.

Qi et al. (1993) and Stabell et al. (1996, 1998) have reported that seed coat plays an important role in control of germination of hound's-tongue (*Cynoglossum officinale*). The seed coat did not interfere with water uptake. However, its mechanical or acid scarification or its complete removal increased seed germination as well as O_2 uptake significantly. The water extracts of hound's-tongue embryo inhibited the germination of some grasses but not of its own seed (Rashid et al. 2005; Furness et al. 2008). In a time-course study of germination and role of the seed coat during seed development, which proceeded acropetally, it was found that the seeds were able to germinate at 44 days after anthesis but the seed coat removal did not improve germination or O_2 uptake by the embryo up to this stage of seed development. However, at 72 days after anthesis, mature seeds at the base of the cyme did not germinate unless the seed coat was removed. Coat removal at this age also stimulated O_2 uptake by the embryo. A study on the relationship between seed phenolics and seed coat-imposed dormancy indicated that as seeds of hound's-tongue developed, methanol-soluble phenolics in the embryo stabilized but in the seed coat they declined (Stabell et al. 1996). This decline was associated with an increase in thioglycolic acid-soluble phenolics, presumably lignins in the seed coat. These observations suggest that polymerization of methanol-soluble phenolics into lignins during the later stages of seed development makes the seed coat impermeable to O_2 and mechanically tough, preventing seed germination.

While in some weed species only one of the roles of the seed coat described earlier (impermeability to water or O_2 or a mechanically resistant coat) is responsible for seed dormancy, a combination of these roles can be involved in the regulation of dormancy in some species. A better understanding of how impermeable seed coats become permeable and how mechanically resistant coats becomes weak in nature may offer options in the future for managing weeds by promoting their germination in the soil seedbank.

3.5.4 Presence of Germination Inhibitors

The presence of inhibitory substances in a seed or its surrounding coverings can prevent it from germinating. These substances can be present in the embryo, testa, or pericarp. For example, hulls (palea and lemma) and caryopses of wild oat have been reported to contain germinating inhibitory substances (Black 1959). Seeds from weeds that fail to germinate because of the presence of inhibitors will germinate when the inhibitor is removed from the seed by washing (e.g. water-soluble substances upon a rainfall or irrigation event) or through metabolic breakdown, or upon an increase of the presence of a germination promoter within the seed. The release of water-soluble germination inhibitors from seeds by some weedy species (wild oat, cocklebur, hound's-tongue) has been demonstrated by comparing the germination of seeds on wet filter paper discs with that in a humid atmosphere, which prevents removal of inhibitors by water. Substances such as ABA, catechin, tannins, coumarin, phenolic substances, and short-chain fatty acids have all been suggested to act as inhibitors of seed germination (Bewley and Black 1985). The presence of an inhibitory chemical in a seed does not necessarily indicate its role in the inhibition of germination of that seed. It may instead have a role in defence against pests or pathogens (Hendry et al. 1994) or may have an inhibitory effect on the germination or growth of a neighbouring plant species (Rashid et al. 2005; Furness et al. 2008; Bainard et al. 2009), helping to reduce competition. Conclusions drawn from the demonstration of an inhibitory effect of an endogenous substance in a Petri dish bioassay can sometimes be incorrect. The inhibitor concentration applied in these bioassays, as compared to that found endogenously in the seed, must be considered. A whole seed extraction could underestimate the concentration of an inhibitor at its site of action if it is mainly compartmentalized in a different tissue. The values of inhibitor concentration based on whole seed extraction procedures therefore should be interpreted with caution.

3.5.5 Underdeveloped Embryo

Seeds of some weed species (e.g. bulrush [*Scirpus* sp.], smartweed [*Polygonum* sp.], or wild chervil [*Anthriscus sylvestris*]) have immature embryos when they are shed (Baskin et al. 2000; Zimdahl 2007). These seeds will not germinate (dormant) until the embryo has grown sufficiently to enable them to germinate.

3.5.6 Role of Hormones

ABA and gibberellins (GAs) and their interaction play important roles in the regulation of seed dormancy (Penfield and King 2009; Shu et al. 2016). It is generally agreed that these two plant growth regulators act antagonistically and the balance between ABA and GA is important in the regulation of seed dormancy in many species. Faster germination in mutants deficient in ABA and a deeper dormancy in transgenic plants constitutively expressing the gene for ABA biosynthesis suggest that ABA is important in the induction and maintenance of seed dormancy (Frey et al. 2012; Nonogaki et al. 2014). Both ABA biosynthesis and ABA signalling are important in this regard. GA also plays a major role in the regulation of seed dormancy and in the mobilization of food reserves during germination (Bewley and Black 1985; Graeber et al. 2012). Deeply dormant GA-deficient mutants germinate only after GA treatment (Shu et al. 2016) and a lower dormancy is exhibited in mutants with defective GA-deactivating enzymes (Yamauchi et al. 2007). Many genes involved in the regulation of seed dormancy and germination by influencing signalling or biosynthesis of GA and ABA have been identified (Shu et al. 2016).

In addition to ABA and GA, several other plant growth regulators (e.g. auxin, brassionsteroids, cytokinins, ethylene, jasmonic acid, salicylic acid, and strigolactones) are also involved in the regulation of seed dormancy (Shu et al. 2016). These plant growth regulators and their interactions also play an important role in the influence of various environmental factors (e.g. light, temperature, etc.) on seed germination behaviour (Shu et al. 2016). How a number of endogenous plant growth regulators, environmental factors, and their interactions influence seedbank dynamics of different weeds, differing in their dormancy mechanisms, in nature is an interesting and important area for future research.

3.5.7 Molecular Mechanisms of Seed Dormancy

Seed dormancy is a quantitative genetic trait. Because tissues of both maternal as well as zygotic origin play part in regulation of seed dormancy and germination, genetics of both of these tissues must be considered while analyzing the molecular basis of seed germination behaviour (Bentsink and Koornneef 2008). Many molecular mechanisms have been suggested for the regulation of seed dormancy, but how these mechanisms are employed in a complex dormancy cycling in a very fluctuating environment of soil seedbanks is far from clear (Finch-Savage and Footitt 2017). Several genes are involved in the regulation of dormancy and germination, their modification by environmental factors during seed development, and in dormancy cycling in soil seedbanks. While a significant amount of literature has accumulated on genetics of seed dormancy in crop species and in arabidopsis, little information is available for common agricultural weeds. Since genetics of seed dormancy is beyond the scope of this chapter, readers are directed to comprehensive articles on this topic (Finch-Savage et al. 2007; Bentsink and Koornneef 2008; Finkelstein et al. 2008; Holdsworth et al. 2008; Graeber et al. 2012; Dekkers and Bentsink 2015; Gu et al. 2015; Dekkers et al. 2016; Finch-Savage and Footitt 2017; Sohindji et al. 2020) and Chapter 4 of this book for more information.

3.6 Coadaptation of Seed Dormancy and Hormonal Regulation of Seed Reserve Mobilization

GA, which stimulates germination of dormant seeds of many grassy weeds (e.g. wild oat), also induces *de novo* synthesis of α-amylase in the aleurone layer, which reduces endosperm starch reserves into sugars to support the early seedling growth. Although release of seed dormancy and germination are two different events with germination following release of seed dormancy, it is expected that genetically more dormant seeds of a grassy weed will have stricter GA requirement for germination compared with genetically less dormant, or nondormant, seeds (Naylor 1983). In a comparative study of autonomous (independent of the germinating embryo) α-amylase production in endosperms of some genetically pure lines of wild oat differing in the duration of seed dormancy, Upadhyaya et al. (1982b) reported evidence for coadaptation of seed dormancy and a rigorous dependence of endosperm hydrolysis on GAs produced by the germinating embryo in genetically dormant lines. Endosperm segments (embryo removed) from six randomly selected dormant lines showed no α-amylase production unless GA was supplied. In contrast, endosperm segments from four out of six nondormant lines produced and released substantial amounts of α-amylase and reduced sugars. This study showed that long-term endosperm stability confers little advantage in nondormant wild oat seeds predisposed to brief periods of residence in soil seedbank, and therefore endosperm tissues of some of these lines could produce α-amylase in the absence of GAs produced by the embryo.

3.7 Duration of Seed Dormancy and Depletion of Seedbanks During Summer Fallow

The practice of 'summer fallowing' is used in crop rotations in some dry parts of the world (e.g. the Canadian Prairies) to conserve soil moisture for the following summer's crop. Although moisture conservation is the main purpose for adopting a summer fallow, weed control applied during this fallow season also depletes weed seedbanks. The effectiveness of weed control during summer fallow is dependent upon the length of dormancy of the weed species present. The practice is more effective on weeds with short-term dormancy because most of their seeds from the previous year's seed rain will germinate during the fallow season and the seedlings can be eliminated by cultivation or by herbicides. Since seeds of weeds with long-term dormancy extend their germination over several years, summer fallowing is less effective in depleting their seedbanks. In a study comparing effects of continuous cropping and a two-year rotation consisting of one year of summer fallow on two synthetic, heterogenous populations of different sets of genetically pure dormant and nondormant line mixtures of wild oat, Jana and Thai (1987) reported that the relative frequency of dormant lines increased in the propagation system involving summer fallowing.

3.8 Dormancy Cycling

Adaptive traits that have evolved in response to selective forces of agricultural practices have helped weedy species persist in those agroecosystems (Batlla et al. 2020). A sound understanding of the relationship between agricultural practices and germination behaviour of weedy species can help in the development of more effective, economical, and environmentally friendly weed management practices.

In soil seedbanks, seeds can change from the dormant to the nondormant state and vice versa (dormancy cycling). This dormancy cycling is determined not only by the genetic, physiological, and biochemical make-up of seeds, but also by the influence of several biotic (e.g. seed predation, influence of soil microflora and allelochemicals, live vegetation and trash cover, and seed deterioration) and abiotic (e.g. soil moisture, temperature, seed burial depth, light conditions, composition of soil air, and soil properties) factors and their interaction (Long et al. 2015).

When a new crop of primary dormant weed seeds is added to a soil seedbank, the seeds cycle through a series of physiological states with regard to their dormancy behaviour. The nature and the rate of these changes depend on the weed species, the environment experienced during seed development on the parent plant, the soil surface ecology (e.g. presence of trash and live canopy cover), seed burial depth, and other soil environmental features. Environmental factors such as temperature, soil moisture status, soil properties (physical, chemical, and biological), their temporal fluctuations, and their interactions are all significant in this regard. Primary dormant seeds can persist in the soil for varying periods depending on the species and their seed dormancy status. These seeds undergo the process of after-ripening, which gradually releases their dormancy (Figure 3.1). This can take varying periods of time depending on the species, the within-species variability in seed dormancy, and the environmental conditions surrounding the seed. Seeds that have lost their dormancy or were nondormant can become secondary dormant if environmental

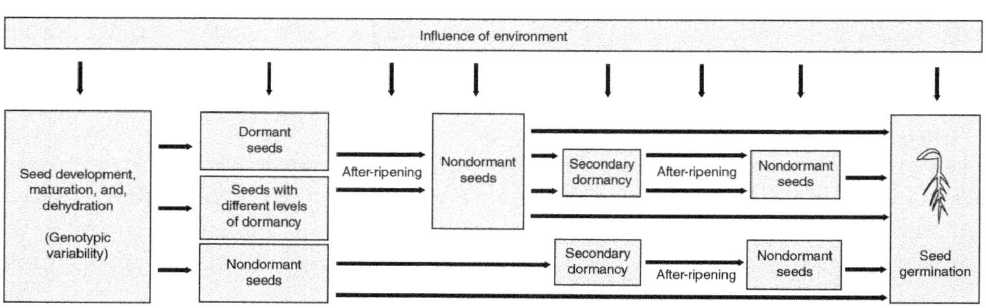

Figure 3.1 Changes in dormancy status of seeds during dormancy cycling in soil seedbanks. Downward pointing arrows show possibilities for environmental influences.

conditions are not favourable for germination and the subsequent seedling growth. Nondormant seeds will germinate if the environment is favourable for germination (Figure 3.1). If conditions are not favourable, they persist in seedbanks as 'quiescent' seeds or could become secondary dormant. Quiescence is an important state that enables seeds to remain within the seedbank, helping in the persistence of weeds. Secondary dormant seeds also undergo gradual after-ripening and slowly lose their dormancy (Figure 3.1). At any stage, secondary dormancy can be reinduced if the environment becomes unfavourable again. Several factors can stimulate the germination of primary or secondary dormant seeds. In order to understand the complex role of secondary dormancy in dormancy cycling in seedbanks, it is important to appreciate the role of between and within species variability and the influence of environmental factors in the induction and release of secondary dormancy.

Low O_2 (anoxia) (Tilsner and Upadhyaya 1985; Symons et al. 1986) and high temperature (thermodormancy) are important factors in the induction of secondary dormancy. Tilsner and Upadhyaya (1985) studied the induction of secondary dormancy in seeds of genetically pure lines of wild oats. When after-ripened dormant and nondormant seeds were immersed in deoxygenated water (anaerobiosis), 85% secondary dormancy could be induced in some lines of this species. Higher temperatures during anaerobiosis were more effective in this induction of dormancy. The authors also found that the induction of secondary dormancy was affected by seed age and the induced dormancy was slowly released when seeds were allowed to after-ripen in dry storage.

Phytochrome is a photoreceptor involved in the regulation of seed germination in many weedy species. In arabidopsis, there are five genes (PHYA-PHYE) in the phytochrome family. In an investigation of genetic pathways involved in induction of secondary dormancy by high temperatures in arabidopsis, Martel et al. (2018) reported a role of PHYD in modulating the induction of secondary dormancy by high (32 °C) temperature. They found that a functional PHYD gene was needed for the germination of seeds exposed to high temperature.

Temperature has been shown to influence the induction and release of secondary dormancy in many other weed species (Totterdell and Roberts 1979; Roberts and Totterdell 1981; Kebreab and Murdoch 1999). Higher temperatures influence the speed of induction of dormancy in imbibed seeds and the rate of dormancy loss in air dry seeds. Interestingly, in broomrape (*Orobanche* spp.) the rate of secondary dormancy induction was fastest at 10 °C and declined with increasing temperature up to 20 °C (Kebreab and Murdoch 1999). In addition to the factors described previously, darkness, prolonged far-red light, CO_2, and water stress have also been suggested as factors inducing secondary dormancy (Bewley and Black 1985).

As for primary dormancy, genetically pure lines of wild oat differ in their secondary dormancy behaviour (Tilsner and Upadhyaya 1985; Symons et al. 1986). The length of dormancy induced was greater in the after-ripened seeds of dormant compared with nondormant lines. Chemicals that are known to stimulate germination of primary dormant seeds (viz. kinetin, isopentenyl adenine, sodium azide, potassium nitrate, ethanol, and substituted phthalimides) also stimulated germination of secondary dormant seeds (Tilsner and Upadhyaya 1985; Symons et al. 1986; Upadhyaya et al. 1986). These results suggest some similarity in the regulation of these two types of dormancy. However, to what extent these dormancies are similar (or different) in their induction and regulation is not clear (Chahtane et al. 2017) and needs further investigation.

These cycles of induction and release of dormancy, by after-ripening or other germination promoting factors, can be repeated several times until a time when the right combination of seed germinability status and environmental conditions that can support germination are available. The seed germination event that follows produces plants that can produce more seeds that can add to the soil seedbank. This dormancy cycling is very important to ensure that seeds germinate when their seedlings have favourable conditions to grow and develop to complete their life cycle.

In some parts of the world, particularly in some developing countries, crop stubble is burned after harvest. This controversial practice and natural forest fires can significantly affect dormancy behaviour of seeds in soil seedbanks (Kępczyński 2018; Qasem 2019). High soil temperatures can decrease dormancy or induce dormancy depending on the species, the stage of seed dormancy, and the depth at which a seed is present in soil, which may influence the temperature exposure. Ooi et al. (2014) studied temperature thresholds of several physically dormant seeds and the plant's functional response to fire and found that six species had temperature cues (80 °C and above) that only occur naturally during a fire. In addition to direct effects on seed dormancy, fire can also affect seed dormancy and germination indirectly by removing the trash and/or live canopy cover, which can influence the light environment on the soil surface, allelopathic influence(s) and soil biota, and by influencing the soil chemistry and the soil moisture status. This could significantly influence the dormancy cycling in soil seedbanks. Plant-derived smoke, water-soluble extracts (smoke water), and a chemical present in smoke (karrikin) have been shown to stimulate germination in many weed species (Kępczyński 2018). Karrikin stimulated germination of both primary and secondary dormant seeds. Adkins and Peters (2001) have reported stimulation of seed germination of several monocotyledonous and dicotyledonous weed species by a commercially available smoke–water solution. The high germination stimulatory activity of karrikin at very low concentration suggests its potential use for weed control by promoting seed germination (Kępczyński 2018). Stevens et al. (2007) have reported stimulation of germination by a synthesized form of a chemical found in smoke under field conditions.

Seeds tune their dormancy cycle in response to environmental factors and determine the optimum time for seed germination when a seedling has the best chance for survival. A combination of dormancy cycling and seed longevity and their modification by endogenous and environmental signals is very important in the persistence strategy of many weedy species (Batlla and Benech-Arnold 2014; Long et al. 2015; Martel et al. 2018). Dormancy cycling ensures maintenance of the seedbank and leads to weed infestations over long periods of time. Footitt et al. (2011) found that depth of dormancy in arabidopsis was related to soil temperature with some influence of soil moisture level. Investigation of gene expression over the dormancy cycle in arabidopsis seeds in the field showed a correlation between gene expression patterns and seasonal changes in temperature.

While dormancy cycling is not considered as a way to prolong seed persistence in the seedbank, the seeds that undergo dormancy cycling often persist longer than they would normally do if dormancy loss was a unidirectional process that made all seeds nondormant and therefore able to germinate at any time throughout the year. Physically dormant seeds are incapable of undergoing dormancy cycling because once they are afforded a nondormant status, they cannot re-enter into a physical dormancy because the change in permeability of seed coverings cannot be overturned (Baskin and Baskin 2014).

The ability to predict the timing and extent of weed emergence is very important for land managers. Decisions to delay crop seeding to provide a window of time to control weeds ('delayed-seeding', Darwent and Smith 1985; Blackshaw et al. 2007), to plant a crop slightly earlier to give it a competitive advantage against the late-emerging weeds, and efficacy of weed control by the false seedbed technique (Travlos et al. 2020) are some examples where knowledge of the timing of weed seedling emergence can help in making some important decisions about weed management. This however is a very complex issue because seedbanks are dynamic, both spatially and temporally, and the agroecosystem environment is constantly fluctuating (Jha et al. 2017). Furthermore, seedbanks are very diverse, consisting of many different species, many with several biotypes having different dormancy characteristics. Because seed dormancy is an important factor in the persistence of weeds, a sound understanding of the factors affecting dormancy cycling is essential. Information on quantitative functional relationship between these factors and changes in the germination behaviour of weed seeds can be very helpful in making weed management decisions (Batlla and Benech-Arnold 2010). Many agronomic practices influence weed seed germination behaviour by affecting the seed's biotic and abiotic environment. Sound understanding of these influences may allow manipulation of weed seed dormancy and germination by agronomic practices (Dyer 1995).

3.9 Conclusions

Seed dormancy is an adaptive and very plastic trait that plays an important role in persistence of weeds. Genetic variability in seed dormancy and its modification by the environment during seed development, maturation and post-dispersal distribute weed seed germination over time, which helps in the establishment and maintenance of soil seedbanks. The presence of soil seedbanks makes weed control very difficult and is an important factor in the persistence strategy of weeds. Recruitment of new seedlings from the soil seedbank and the deposition of a new batch of seeds to these banks in every growing season make weeds persistent and difficult to control.

Several mechanisms are involved in the regulation of seed dormancy and some weed species use more than one of these mechanisms in regulation of their germination behaviour. Weeds tune their seed dormancy cycle in response to environmental factors so that they germinate when seedlings have an optimum chance to establish. A sound understanding of these mechanisms is important to understand how seed dormancy is released and induced during dormancy cycling in soil seedbanks and how agricultural practices and environmental factors influence these processes. This knowledge is important in predicting emergence of weed seedlings, which is necessary for developing effective, economical, and environmentally friendly weed management strategies.

Most weed control methods in use at present aim to control or kill weed seedlings. Practical options to deplete soil seedbanks directly are largely unavailable. Many chemicals that stimulate seed germination in laboratory tests are not effective, or their use is not practical under field conditions. Some recent developments, e.g. stimulation of germination by a synthesized form of a chemical found in smoke has shown some promise in this regard. More research is needed in this area.

Research on the molecular mechanisms of seed dormancy has identified several genes and molecular processes involved in dormancy regulation and its modification by environmental factors during seed development and dormancy cycling. Because tissues of both maternal and zygotic origin can be involved in the regulation of seed dormancy, genetics of both these tissues must be considered in analyzing seed dormancy. Further research in this area could increase our understanding of the dormancy cycle in agroecosystems enough to develop improved weed management strategies and could potentially identify targets to effectively deplete soil seedbanks.

References

Adkins, S.W. and Adkins, A.L. (1994). Effect of potassium nitrate and ethephon on fate of wild oat (*Avena fatua*) seed in soil. *Weed Science* 42: 353–357.

Adkins, S.W. and Peters, N.C.B. (2001). Smoke derived from burnt vegetation stimulates germination of arable weeds. *Seed Science Research* 11: 213–222.

Adkins, S.W., Simpson, G.M., and Naylor, J.M. (1984a). The physiological basis of seed dormancy in *Avena fatua* IV. Alternative respiration and nitrogenous compounds. *Physiologia Plantarum* 60: 234–238.

Adkins, S.W., Simpson, G.M., and Naylor, J.M. (1984b). The physiological basis of seed dormancy in *Avena fatua* VI. Respiration and the stimulation of germination by ethanol. *Physiologia Plantarum* 62: 148–152.

Adkins, S.W., Loewen, M., and Symons, S.J. (1986). Variation within pure lines of wild oats (*Avena fatua*) in relation to degree of primary dormancy. *Weed Science* 34: 859–864.

Adkins, S.W., Loewen, M., and Symons, S.J. (1987). Variation within pure lines of wild oats (*Avena fatua*) in relation to temperature of development. *Weed Science* 35: 169–172.

Armstrong, L.J. and Adkins, S.W. (1998). Variation in photoperiod response of different isogenic lines of wild oats (*Avena fatua*). *Weed Science* 46: 39–47.

Atwood, W.M.A. (1914). A physiological study of the germination of *Avena fatua*. *Botanical Gazette* 57: 386–414.

Bainard, L.D., Brown, P.D., and Upadhyaya, M.K. (2009). Inhibitory effect of tall hedge mustard (*Sisymbrium loeselii*) allelochemicals on rangeland plants and arbuscular mycorrhizal fungi. *Weed Science* 57: 386–393.

Baskin, J.M. and Baskin, C.C. (2004a). A classification system for seed dormancy. *Seed Science Research* 14: 1–16.

Baskin, J.M. and Baskin, C.C. (2004b). On criteria to use in studies of seed evolution. Research Opinion.*Seed Science Research* 14: 315–320.

Baskin, C.C. and Baskin, J.M. (2014). *Seeds: Ecology, Biogeography, and Evolution of Dormancy and Germination*. San Diego, CA: Academic Press.

Baskin, C.C., Milberg, P., Andersson, L., and Baskin, J.M. (2000). Deep complex morphological dormancy in seeds of *Anthriscus sylvestris* (Apiaceae). *Flora* 195: 245–251.

Batlla, D. and Benech-Arnold, R.L. (2005). Changes in light sensitivity of buried *Polygonum aviculare* seeds in relation to cold-induced dormancy loss: development of a predictive model. *New Phytologist* 165: 445–452.

Batlla, D. and Benech-Arnold, R.L. (2010). Predicting changes in dormancy level in nature seed soil banks. *Plant Molecular Biology* 73: 3–13.

Batlla, D. and Benech-Arnold, R.L. (2014). Weed seed germination and the light environment: implications for weed management. *Weed Biology and Management* 14: 77–87.

Batlla, D., Ghersa, C.M., and Benech-Arnold, R.L. (2020). Dormancy, a critical trait for weed success in crop production systems. *Pest Management Science* 76: 1189–1194.

Benech-Arnold, R.L., Sanchez, R.A., Forcella, F. et al. (2000). Environmental control of dormancy in weed seed banks in soil. *Field Crops Research* 67: 105–122.

Bentsink, L. and Koornneef, M. (2008). Seed dormancy and germination. In: *The Arabidopsis Book*, The American Society of Plant Biologists, vol. 2008, Issue 6, 18. https://doi.org/10.1199/tab.0119.

Bewley, J.D. (1997). Seed germination and dormancy. *The Plant Cell* 9: 1055–1066.

Bewley, J.D. and Black, M. (1985). *Seeds: Physiology of Development and Germination*, 367. New York, NY, USA: Plenum Press.

Bewley, J.D., Bradford, K., Hilhorst, H., and Nonogaki, H. (1994). *Seeds: Physiology of Development, Germination and Dormancy*, 3e. New York, NY USA: Springer.

Black, M. (1959). Dormancy studies in *Avena fatua*. 1. The possible role of germination inhibitors. *Canadian Journal of Botany* 37: 393–402.

Blackshaw, R.E., Anderson, R.L., and Lemerle, D. (2007). Cultural weed management. In: *Non-chemical Weed Management: Principles, Concepts and Technology* (eds. M.K. Upadhyaya and R.E. Blackshaw). Oxfordshire, UK: CAB International.

Chahtane, H., Kim, W., and Lopez-Molina, L. (2017). Primary seed dormancy: a temporally multilayered riddle waiting to be unlocked. *Journal of Experimental Botany* 68: 857–869.

Chen, S.S.C. and Thimann, K.V. (1966). Nature of seed dormancy in *Phacelia tanacetifolia*. *Science* 153: 1537–1539.

Cresswell, E.G. and Grime, J.P. (1981). Induction of a light requirement during seed development and its ecological consequences. *Nature* 291: 583–585.

Darwent, A.L. and Smith, J.H. (1985). Delayed seeding for wild oat control in rapeseed in northwest Alberta. *Canadian Journal of Plant Science* 65: 1101–1106.

Dekkers, B.J.W. and Bentsink, L. (2015). Regulation of seed dormancy by abscisic acid and DELAY OF GERMINATION 1. *Seed Science Research* 25: 82–98.

Dekkers, B.J.W., Pearce, S.P., van Bolderen-Veldkamp, R.P.M. et al. (2016). Dormant and after-ripened *Arabidopsis thaliana* seeds are distinguished by early transcriptional differences in the imbibed state. *Frontiers of Plant Science* 7, article 1323. doi: https://doi.org/10.3389/fpls.2016.01323.

Delgado-Sánchez, P., Ortega-Amaro, M.A., Jiménez-Bremont, J.F., and Flores, J. (2011). Are fungi important for breaking seed dormancy in desert species? Experimental evidence in *Opuntia streptacantha* (Cactaceae). *Plant Biology* 13: 154–159.

Dyer, W.E. (1995). Exploiting weed seed dormancy and germination requirements through agronomic practices. *Weed Science* 43: 498–503.

Egley, G.H. and Duke, S.O. (1985). Physiology of seed dormancy and germination. In: *Seed Physiology*, Reproduction and Ecophysiology, vol. 1 (ed. S. Duke), 28–64. Boca Raton, FL, USA: CRC Press.

Egley, G.H. and Paul, R.N. Jr. (1981). Morphological observations on the early imbibition of water by *Sida spinosa* (Malvaceae) seed. *American Journal of Botany* 68: 1056–1065.

Esashi, Y. and Leopold, A.C. (1968). Physical force in dormancy and germination of Xanthium seeds. *Plant Physiology* 43: 871–876.

Finch-Savage, W.E. and Footitt, S. (2017). Seed dormancy cycling and the regulation of dormancy mechanisms to time germination in variable field environments. *Journal of Experimental Botany* 68: 843–856.

Finch-Savage, W.E. and Leubner-Metzger, G. (2006). Seed dormancy and control of germination. *New Phytologist* 171: 501–523.

Finch-Savage, W.E., Cadman, C.S., Toorop, P.E. et al. (2007). Seed dormancy release in Arabidopsis Cvi by dry after-ripening, low temperature, nitrate and light shows common quantitative patterns of gene expression directed by environmentally specific sensing. *The Plant Journal* 51: 60–78.

Finkelstein, R., Reeves, W., Ariizumi, T., and Steber, C. (2008). Molecular aspects of seed dormancy. *Annual Review of Plant Biology* 59: 387–415.

Foley, M.E. (2001). Seed dormancy: an update on terminology, physiological genetics, and quantitative trait loci regulating germinability. *Weed Science* 49: 305–317.

Footitt, S., Douterelo-Soler, I., Clay, H., and Finch-Savage, W.E. (2011). Dormancy cycling in Arabidopsis seeds is controlled by seasonally distinct hormone-signaling pathways. *Proceedings of the National Academy of Sciences of the United States of America* 108: 20236–20241.

Frey, A., Effroy, D., Lefebvre, V. et al. (2012). Epoxycarotenoid cleavage by NCED5 fine-tunes ABA accumulation and affects seed dormancy and drought tolerance with other NCED family members. *Plant Journal* 70: 501–512.

Furness, N.H., Adomas, B., Dai, Q. et al. (2008). Allelopathic influence of houndstongue (*Cynoglossum officinale*) and its modification by UV-B radiation. *Weed Technology* 22: 101–107.

Graeber, K., Nakabayashi, K., Miatton, E. et al. (2012). Molecular mechanisms of seed dormancy. *Plant Cell Environment* 35: 1769–1786.

Gu, X.Y., Zhang, J., Ye, H. et al. (2015). Genotyping of endosperms to determine seed dormancy genes regulating germination through embryonic, endospermic, or maternal tissues in rice. *Genes, Genomes, Genetics* 5: 183–193.

Gutterman, Y. (1985). Flowering, seed development and the influences during seed maturation on seed germination of annual weeds. In: *Seed Physiology, Vol 1, Reproduction and Ecophysiology* (ed. S. Duke), 2–23. Boca Raton, FL, USA: CRC Press.

Gutterman, Y. (1992). Environmental conditions during seed maturation affecting seed germination. *Acta Horticulturae* 314: 179–187.

Harper, J.L. (1977). *Population Biology of Plants*, 892 pp. New York, USA: Academic Press.

He, H., de Souza Vidigal, D., Snoek, L.B. et al. (2014). Interaction between parental environment and genotype affects plant and seed performance in *Arabidopsis*. *Journal of Experimental Botany* 65: 6603–6615.

Hendry, G.A.F., Thompson, K., Moss, C.J. et al. (1994). Seed persistence: a correlation between seed longevity in the soil and *ortho*-dihydroxyphenol concentration. *Functional Ecology* 8: 658–664.

Holdsworth, M.J., Bentsink, L., and Soppe, W.J. (2008). Molecular networks regulating Arabidopsis seed maturation, after-ripening, dormancy and germination. *New Phytologist* 179: 33–54.

Jana, S. and Naylor, J.M. (1980). Dormancy studies in seed of *Avena fatua*. 11. Heritability for seed dormancy. *Canadian Journal of Botany* 58: 91–93.

Jana, S. and Thai, K.M. (1987). Patterns of changes of dormant genotypes in *Avena fatua* populations under different agricultural conditions. *Canadian Journal of Botany* 65: 1741–1745.

Jana, S., Acharya, S.N., and Naylor, J.M. (1979). Dormancy studies in seed of *Avena fatua*. 10. On the inheritance of germination behaviour. *Canadian Journal of Botany* 57: 1663–1667.

Jana, S., Upadhyaya, M.K., and Acharya, S.N. (1988). Genetic basis of dormancy and differential response to sodium azide in *Avena fatua* seeds. *Canadian Journal of Botany* 66: 635–641.

Jha, P., Godara, R.K., and Jhala, A.J. (2017). Targeting weed seedbanks: implications for weed management. In: *Biology, Physiology and Molecular Biology of Weeds* (ed. M. Jugulam), 12–26. Boca Raton, FL, USA: CRC Press.

Kebreab, E. and Murdoch, A.J. (1999). A quantitative model for loss of primary dormancy and induction of secondary dormancy in imbibed seeds of *Orobanche* spp. *Journal of Experimental Botany* 50: 211–219.

Kępczyński, J. (2018). Induction of agricultural weed seed germination by smoke and smoke-derived karrikin (KAR1), with a particular reference to *Avena fatua* L. *Acta Physiologiae Plantarum* 40, article 87. https://doi.org/10.1007/s11738-018-2663-2.

Kerdaffrec, E. and Nordborg, M. (2017). The maternal environment interacts with genetic variation in regulating seed dormancy in *Arabidopsis thaliana*. *PLoS One* https://doi.org/10.1371/journal.pone.0190242.

LaCroix, L.J. and Staniforth, D.W. (1964). Seed dormancy in velvetleaf. *Weeds* 12: 171–174.

Long, R.L., Gorecki, M.J., Renton, M. et al. (2015). The ecophysiology of seed persistence: a mechanistic view of the journey to germination or demise. *Biological Reviews* 90: 31–59.

Lymperopoulos, P., Msanne, J., and Rabara, R. (2018). Phytochrome and phytohormones: working in tandem for plant growth and development. *Frontiers in Plant Science* 9: 1037.

Martel, C., Blair, L.K., and Donohue, K. (2018). PHYD prevents secondary dormancy establishment of seeds exposed to high temperature and is associated with lower PIL5 accumulation. *Journal of Experimental Botany* 69: 3157–3169.

McCullough, J.M. and Shropshire, W. Jr. (1970). Physiological predetermination of germination responses in *Arabidopsis thaliana* (L.) Heynh. *Plant and Cell Physiology* 11: 139–148.

McEvoy, P.B. (1984). Dormancy and dispersal in dimorphic achenes of tansy ragwort (*Senecio jacobaea*) (Compositae). *Oecologia* 61: 160–168.

Murdoch, A.J. and Roberts, E.H. (1996). Dormancy cycle of *Avena fatua* seeds in soil. In: *Proceedings of the Second International Weed Control Congress Copenhagen, Denmark*, vol. 1, 147–152. Flakkebjerg, Denmark: Department of Weed Control and Pesticide Ecology.

Naylor, J.M. (1983). Studies on genetic control of some physiological processes in seeds. *Canadian Journal of Botany* 61: 3561–3567.

Naylor, J.M. and Fedec, P. (1978). Dormancy studies in seed of *Avena fatua*. 8. Genetic diversity affecting response to temperature. *Canadian Journal of Botany* 56: 2224–2229.

Naylor, J.M. and Jana, S. (1976). Genetic adaptation for seed dormancy in *Avena fatua*. *Canadian Journal of Botany* 54: 306–312.

Nolan, D.G. and Upadhyaya, M.K. (1988). Primary seed dormancy in diffuse and spotted knapweed. *Canadian Journal of Plant Science* 68: 775–783.

Nonogaki, M., Sall, K., Nambara, E., and Nonogaki, H. (2014). Amplification of ABA biosynthesis and signalling through a positive feedback mechanism in seeds. *Plant Journal* 78: 527–539.

Ooi, M.K.J., Denham, A.J., Santana, V.M., and Auld, T.D. (2014). Temperature thresholds of physically dormant seeds and plant functional response to fire: variation among species and relative impact of climate change. *Ecology and Evolution* 4: 656–671.

Oreja, F.H., de la Fuente, E.B., and Batlla, D. (2017). Role of seed environment and covering structures on large crabgrass germination. *South African Journal of Botany* 111: 170–175.

Penfield, S. (2017). Seed dormancy and germination. *Current Biology* 27: 874–878.

Penfield, S. and King, J. (2009). Towards a systems biology approach to understanding seed dormancy and germination. *Proceedings of Royal Society B: Biological Sciences* 276: 3561–3569.

Qasem, J.R. (2019). Weed seed dormancy: the ecophysiology and survival strategies. In: *Seed Dormancy and Germination* (ed. J. Jimenez-Lopez). London, UK: IntechOpen https://doi.org/10.5772/intechopen.88015.

Qi, M.Q., Upadhyaya, M.K., Furness, N.H., and Ellis, B.E. (1993). Mechanism of seed dormancy in *Cynoglossum officinale*. *Journal of Plant Physiology* 142: 325–330.

Radosevich, S., Holt, J., and Ghersa, C. (1997). *Weed Ecology, Implications for Management.* New York, NY: Wiley.

Rashid, A., Furness, N.H., Ellis, B.E., and Upadhyaya, M.K. (2005). Inhibition of seed germination and seedling growth by hound's-tongue (*Cynoglossum officinale* L.) seed leachate. *Weed Biology and Management* 5: 143–149.

Roberts, E.H. and Totterdell, S. (1981). Seed dormancy in *Rumex* species in response to environmental factors. *Plant Cell and Environment* 4: 97–106.

Sawhney, R. and Naylor, J.M. (1979). Dormancy studies in seed of *Avena fatua*. 9. Demonstration of genetic variability affecting the response to temperature during seed development. *Canadian Journal of Botany* 57: 59–63.

Sawhney, R. and Naylor, J.M. (1980). Dormancy studies in seed of *Avena fatua*. 12. Influence of temperature on germination behavior of nondormant families. *Canadian Journal of Botany* 58: 578–581.

Sawhney, R. and Naylor, J.M. (1982). Dormancy studies in seed of *Avena fatua*. 13. Influence of drought stress during seed development on duration of seed dormancy. *Canadian Journal of Botany* 60: 1016–1020.

Scopel, A.L., Ballare, C.L., and Radosevich, S.R. (1994). Photostimulation of seed germination during soil tillage. *New Phytologist* 126: 145–152.

Shu, K., Liu, X.-D., Xie, Q., and He, Z.-H. (2016). Two faces of one seed: hormonal regulation of dormancy and germination. *Molecular Plant* 9: 34–45.

Simpson, G.M. (1990). *Seed Dormancy in Grasses*, 297 pp. New York, NY: Cambridge University Press.

Sohindji, F.S., Sogbohossou, D.E.O., Zohoungbogbo, H.P.F. et al. (2020). Understanding molecular mechanisms of seed dormancy for improved germination in traditional leafy vegetables: an overview. *Agronomy* 10: 57. https://doi.org/10.3390/agronomy10010057.

Stabell, E., Upadhyaya, M.K., and Ellis, B.E. (1996). Development of seed coat-imposed dormancy during *Cynoglossum officinale* L. seed development. *Physiologia Plantarum* 97: 28–34.

Stabell, E., Upadhyaya, M.K., and Ellis, B.E. (1998). Role of seed coat in regulation of seed germination in *Cynoglossum officinale* L. *Weed Science* 46: 344–350.

Steinbrecher, T. and Leubner-Metzger, G. (2017). The biomechanics of seed germination. *Journal of Experimental Botany* 68: 765–783.

Stevens, J.C., Merritt, D.J., Flematti, G.R. et al. (2007). Seed germination of agricultural weeds is promoted by the butenolide 3-methyl-2i/-furo[2,3-c]pyran-2-one under laboratory and field conditions. *Plant and Soil* 298: 113–124.

Symons, S.J., Naylor, J.M., Simpson, G.M., and Adkins, S.W. (1986). Secondary dormancy in *Avena fatua*: induction and characteristics in genetically pure dormant lines. *Physiologia Plantarum* 68: 27–33.

Tilsner, H.R. and Upadhyaya, M.K. (1985). Induction and release of secondary seed dormancy in genetically pure lines of *Avena fatua*. *Physiologia Plantarum* 64: 377–382.

Totterdell, S. and Roberts, E.H. (1979). Effects of low temperature on the loss of innate dormancy and the development of induced dormancy in seeds of *Rumex obtusifolius* L. and *Rumex crispus* L. *Plant, Cell & Environment* 2: 131–138.

Travlos, I., Gazoulis, I., Kanatas, P. et al. (2020). Key factors affecting weed seeds' germination, weed emergence, and their possible role for the efficacy of false seedbed technique as weed management practice. *Frontiers in Agronomy* 2, article 1: 9. https://doi.org/10.3389/fagro.2020.00001.

Upadhyaya, M.K., Naylor, J.M., and Simpson, G.M. (1982a). The physiological basis of seed dormancy in *Avena fatua* L. I. Action of the respiratory inhibitors sodium azide and salicylhydroxamic acid. *Physiologia Plantarum* 4: 419–424.

Upadhyaya, M.K., Naylor, J.M., and Simpson, G.M. (1982b). Co-adaptation of seed dormancy and hormonal requirement of α-amylase production in endosperm segments of some genetically pure lines of *Avena fatua* L. *Canadian Journal of Botany* 60: 1142–1147.

Upadhyaya, M.K., Hsiao, A.I., and Bonsor, M.E. (1986). Differential response of pure lines of *Avena fatua* L. to substituted phthalimides: germination and endosperm-mobilization studies. *Annals of Botany* 58: 455–463.

Vleeshouwers, L.M., Bouwmeester, H.J., and Karssen, C.M. (1995). Redefining seed dormancy: an attempt to integrate physiology and ecology. *Journal of Ecology* 83: 1031–1037.

Williams, J.T. and Harper, J.L. (1965). Seed polymorphism and germination. 1. The influence of nitrates and low temperatures on the germination of *Chenopodium album*. *Weed Research* 5: 141–150.

Yamauchi, Y., Takeda-Kamiya, N., Hanada, A. et al. (2007). Contribution of gibberellin deactivation by AtGA2ox2 to the suppression of germination of dark-imbibed *Arabidopsis thaliana* seeds. *Plant & Cell Physiology* 48: 555–561.

Zalamea, P.-C., Sarmiento, C., Arnold, A.E. et al. (2015). Do soil microbes and abrasion by soil particles influence persistence and loss of physical dormancy in seeds of tropical pioneers? *Frontiers in Plant Science* 5, article 799: 1–14.

Zimdahl, R.L. (2007). *Fundamentals of Weed Science*, 3e, 666. Oxford, UK: Academic Press.

4

Seed Dormancy Genes and Their Associated Adaptive Traits Underlie Weed Persistence: A Case Study of Weedy Rice

Xing-You Gu

Agronomy, Horticulture, and Plant Science Department, Seed Molecular Biology Laboratory, South Dakota State University, Brookings, SD, USA

4.1 Introduction

Weed persistence relies on genetic diversity for an array of adaptive traits that evolved in agroecosystems (Dekker 1997). The weed adaptive traits can be wild-like or crop-mimic, based on phenotypic similarities to wild or crop species. The wild-like traits are characteristics that are present in wild species and make plants weedy. Most wild-like traits are seed related, such as easy shattering, extended dormancy and longevity, high fecundity, dark/red pigments, and extra appendages for dispersal. The crop-mimic traits are characteristics resembling crop cultivars at specific stages of the life cycle of a weed species, such as seedling morphologies, plant growth habit, and flowering time (FT) (Barrett 1983). These two groups of traits are complementary in regulation of weed persistence. For example, the seed-related characteristics contribute to the size and survivability of weed soil seedbanks and the spatial and temporal distributions of germination, whereas the crop-mimic characteristics contribute to weed competition with crops and avoidance of eradication by agronomic practices. Knowledge of genetic, evolutionary, and molecular mechanisms regulating the trait variations is of fundamental importance in weed biology and weed management.

Natural variation for each of the adaptive traits is presumably regulated by allelic variants of genes accumulated in a species during evolution. Genes for many of the traits have been identified in model and crop plants using forward genetic and genomic approaches (Alonso-Blanco et al. 2009). However, the genetic and genomic information has been collected from detached research on single traits for dissimilar purposes. For instance, *DELAY OF GERMINATION 1* was cloned from *Arabidopsis thaliana* to identify molecular mechanisms of seed dormancy (SD) (Bentsink et al. 2006), and allelic variants of *sh4* were cloned

Persistence Strategies of Weeds, First Edition. Edited by Mahesh K. Upadhyaya, David R. Clements, and Anil Shrestha.

from cultivated (*Oryza sativa*) and wild (*Oryza nivara*) rice to understand the crop domestication history for the seed shattering (SH) trait (Li et al. 2006a). It is unknown how genes for different traits interact with each other in a system to influence weed adaptation. Thus, it is necessary to synthesize the detached information to infer synergistic effects of the adaptive traits on weed persistence.

Dormancy is critical for the persistence of weed soil seedbanks. SD can be primary or secondary. Primary dormancy is genetically determined and developed on the plant before seed maturation or dispersal by shattering to environments. The primary dormancy is released during a physiological process known as after-ripening (Simpson 1990). Secondary dormancy is induced to after-ripened seeds by adverse environmental conditions. A soil seedbank may alternate between a dormant and an after-ripened status in correspondence with seasonal changes in temperature or moisture (Hilhorst 2007). Therefore, seeds with primary dormancy have a greater chance than nondormant or weakly dormant seeds to enter a soil seedbank and survive across the next season(s) before germination to complete the life cycle.

SD has been associated with other adaptive traits in the past century. Early research observed associations of SD with grain colors in common wheat (*Triticum aestivum*; Nilsson-Ehle 1914) or spikelet morphologies in oats (*Avena fatua* and *Avena sativa*; Johnson 1935). Recent research discovered associations of SD with additional wild-like and crop-mimic traits in weedy rice (Gu et al. 2003; Mispan et al. 2013). The phenotypic associations observed in different populations imply the presence of coadaptive relationship between SD and some other traits in a weed species to enhance plant adaptations to agro-ecosystems frequently disturbed by agronomic practices.

Genotypic variation for SD in a species is controlled by multiple genes, each with a relatively small effect on germinability. The naturally occurring genes for SD have been mapped as quantitative trait loci (QTLs or QTL for singular) in several species since the 1990s. A QTL is a marker-delimited genomic region, which may contain genes controlling different adaptive traits. For such a species as rice with high quality genome sequence annotation, it is possible to detect gene(s) responsible for the trait associations through pleiotropy or linkage disequilibrium (LD). LD for a dormancy QTL-containing region in weedy rice (*O. sativa*) could be strong enough to maintain a haplotype for two or more genes controlling different adaptive traits across many generations (Gu et al. 2005a). Currently, weedy rice is the only agricultural weed, to my knowledge, which has been mapped for most QTLs controlling SD and its associated traits (Table 4.1).

This chapter, as a case study of weedy rice, aims to review the genetic and genomic data in the context of weed persistence. Topics discussed in this chapter include a brief review of the model system of weedy rice used for the study (Section 4.2); genetics of SD, including its natural variation, QTLs, QTL epistatic and genotype-by-environment (G-by-E) interactions, and QTL underlying genes (Section 4.3); genes/QTLs and haplotypes responsible for the associations with the other wild-like (Section 4.4) or crop-mimic (Section 4.5) traits; and conclusions and implications of the case study (Section 4.6). It is noted that the model system was developed from the conspecific weedy and cultivated rice (*O. sativa*), and both share the same wild ancestors but are divergent in many domestication-related traits. Thus, some information from wild and cultivated relatives is also discussed within these topics.

Table 4.1 List of seed dormancy QTLs and their associated adaptive traits in weedy rice.

QTL	Associated traits
qSD1-1	Flowering time (Gu et al. 2018)
qSD1-2	Plant height, tiller number, plant growth rate, and seedbank longevity (Ye et al. 2015; Mispan et al. 2013; Pipatpongpinyo et al. 2020)
qSD4	Seed shattering, black hull, and awn length (Gu et al. 2005b; Mispan et al. 2013; Zhang et al. 2017)
qSD6	Flowering time and awn length (Mispan et al. 2013)
qSD7-1	Red pericarp color, seedbank longevity, and flowering time (Gu et al. 2011, 2018; Pipatpongpinyo et al. 2020)
qSD7-2	Plant height, seedbank longevity, and flowering time (Ye et al. 2013; Gu et al. 2018; Pipatpongpinyo et al. 2020)
qSD8	Seed shattering and awn length (Gu et al. 2005b; Mispan et al. 2013; Zhang et al. 2017)
qSD10	Flowering time (Gu et al. 2018)
qSD12	Seedbank longevity (Pipatpongpinyo et al. 2020)

4.2 Weedy Rice

Weedy rice refers to various forms of plants that belong to the *Oryza* genus and infest rice fields (Delouche et al. 2007). The genus consists of two cultivated species (Asian and African) and about 20 wild species (Khush and Brar 2002). Asian rice (*O. sativa*) is grown worldwide and accounts for >90% of rice production, while African rice (*Oryza glaberrima*) is limited to part of West Africa and comprises a small proportion of the crop production (Global Rice Science Partnership [GRiSP] 2013). Weedy rice is not an independent species in the genus; instead, it is basically conspecific with rice cultivars grown in the same area (Oka 1988). Genotypes of weedy rice discussed in the following sections belong to the same species as Asian rice.

Origins of weedy rice (*O. sativa*) are associated with domestication and differentiation of Asian cultivated rice. Rice was domesticated from the wild ancestors *Oryza rufipogon* or *O. nivara* in South to Southeast Asia, and cultivars were differentiated into the *indica* and *japonica* subspecies in tropical to subtropical and temperate areas, respectively (Khush and Brar 2002). Thus, weedy rice is classified into *indica*- and *japonica*-like groups, each consisting of wild-like and crop-mimic subgroups based on diagnostic characters for the wild ancestors or cultivated subspecies (Suh et al. 1997; Tang and Morishima 1997). The *indica*-like group of populations in the areas with the wild ancestors could have originated from natural hybridization between cultivated and the wild species or from natural variants of wild rice (Oka 1988; Hussain et al. 2010; Vigueira et al. 2018). The *japonica*-like and some *indica*-like populations in the areas historically absent of wild rice (e.g. Europe and the United States) have complex origins. Hypotheses include origins from feral plants of old cultivars (or de-domestication), recombinants between ecotypes of cultivars, and contaminants of weedy or wild rice in imported seeds, based on genome-wide polymorphic patterns of DNA sequences in collections of weed accessions (Kanapeckas et al. 2016; Li et al. 2017; Qiu et al. 2017, 2020; Chai et al. 2018; Sun et al. 2019; De Leon et al. 2019).

Populations of weedy rice, regardless of their origins, are distinguished from cultivars in wild-like traits at the physiological and morphological levels. The physiological characteristics include early shattering, moderate to strong SD, and extended seedbank longevity (Delouche et al. 2007). The morphological traits include those of qualitative nature in genetics, such as pericarp color (PC) (red or white), hull color (HC) (black or straw), and awn (present or absent). The morphological traits are often combined to describe biotypes, such as black-/straw-hulled awned/awnless types of red rice. Red rice is characterized by red/brown pigments on the pericarp tissue and is the most common type of weedy rice in all rice-growing areas.

Weedy rice has become a serious problem in rice production for both cultural and biological reasons. The cultural reason is the transition from transplanting to direct seeding in more rice-growing areas to reduce labor costs in recent years (Nadir et al. 2017). The transition apparently increases the time window for weed seeds to germinate together with cultivars and consequently the population size of weed plants in direct-seeding systems. The biological reason is the conspecific nature between the weed and crop relatives, which makes weedy rice difficult to control using regular techniques developed based on morphological or physiological properties of the rice plant. In addition, cross-pollination, although the rate is relatively low, could mediate gene flow from conventional or genetically engineered (GE) cultivars to weedy rice, which could accelerate the evolution of weedy rice in the cropping system (Chen et al. 2004; Shivrain et al. 2008). For example, Clearfield® rice (CL) is a non-GE technique to control weedy rice, which uses an induced acetolactate synthase mutant to provide CL cultivars with resistance to the imidazolinone herbicides (Croughan 2003). However, several years after adoption of the CL technique, 21–79% plants sampled from weedy rice populations in 26 fields had gained the same type of herbicide resistance (Burgos et al. 2014). The herbicide-resistant plants were mainly genetic recombinants between the CL and weedy rice, as they had the wild-like characteristics such as shattering, dormancy, awn, red/brown pericarp, and dark HC.

4.3 Genetics of Primary Seed Dormancy

4.3.1 Natural Variation in Wild, Weedy, and Cultivated Rice

SD can be defined as the temporal failure of a viable seed to complete germination under favorable conditions (Bewley et al. 2013). Thus, the ability of seeds from a genotype to germinate (i.e. germinability) is determined by the dormancy degree at a time point. The "favorable" conditions for genetic research on SD vary with species, such as ~30 °C for rice (Roberts 1962) and 15 °C for wild oats (*A. fatua*) (Fennimore et al. 1998). The "temporally" dormant seeds from a genotype (line or plant) require a certain period of storage under warm dry (after-ripening), or cool moist (stratification) conditions depending on species, to release the dormancy. Thus, germination testing for a population of genotypes can be conducted at a series of time points of after-ripening when the seed quantity is not a limiting factor. The germination data can be used to model the period required for a genotype to reach a certain level of germination (e.g. 50%) and to estimate genetic parameters or map QTLs for SD.

About half of the species of wild rice, including annuals, perennials, or ancestors of the two cultivated species, have been evaluated for SD under controlled conditions. Both interspecific and intraspecific variation was observed in two independent experiments (Oka and Morishima 1967; Veasey et al. 2004). SD was stronger in the annuals than in the perennials, and the dormancy degree was correlated negatively with the plant regeneration capacity from stem segments (vegetative propagule) and positively with the number of seeds per plant (or fecundity) and awn length (AL) (Oka and Morishima 1967). The interspecific variation was also related to local ecological conditions. For example, the species adapted to the distinct wet and dry seasons in the tropical regions usually have a longer period of SD than those in the Amazon basin, where there is a little change in temperature and moisture across a year (Veasey et al. 2004).

Domestication of cereal crops tended to select natural variants with reduced SD to promote rapid uniform germination, and this is true for rice. Compared with accessions of wild rice, 115 cultivars of rice had an extremely low (76.5%), low (21.7%), or moderate (1.7%) degree of SD (Oka and Morishima 1967). In a recent experiment where 350 cultivars from different ecotypes of rice were evaluated for germination percentages of freshly harvested seeds, the mean was ~75%, and about 75% of the cultivars had weak or no dormancy (60–100% germination) at maturation (Magwa et al. 2016). Results from these and other researches (Roberts 1962; Lu et al. 2018) indicate that a relatively low level of genotypic variation for SD exists in cultivated rice. In general, SD is stronger in the *indica* than in the *japonica* subspecies and stronger in traditional than in modern cultivars. It is noted that lack of primary dormancy in modern cultivars often causes a preharvesting sprouting (PHS) problem in cereal crop (including rice) production (Rodríguez et al. 2015).

Weedy rice varies in SD from strong to moderate or weak, depending on populations or genotypes. A collection of 17 accessions of weedy rice, which were collected from eight countries and grown in a controlled environment, displayed <5% (12) to 90% (1) germination for freshly harvested seeds (Gu et al. 2005b). The 12 strongly dormant accessions were all black-hulled, awned, red rice. Both inter- and intrapopulation variation in SD was observed in US weedy rice sampled from 17 fields in different counties of the Arkansas state (Tseng et al. 2013). Intrapopulation variation accounted for most of the variance in days of after-ripening (DAR) to achieve 50% germination, as compared with the other factors (field, biotype, and county) in two experiments of this study. Thus, a single weed population is capable of harboring great genetic diversity for the dormancy trait.

4.3.2 Quantitative Trait Loci

Both linkage and association mapping have been used to identify QTLs responsible for the natural variation of SD in rice. Linkage mapping involves developing a segregating population from a controlled cross and a linkage map to scan for SD QTLs along the map. Three types of crosses (cultivated/cultivated, cultivated/wild, and cultivated/weedy) have been used to map SD QTLs for different purposes. For example, the cultivar/cultivar crosses were mainly used to map SD genes that could be used to improve varieties for resistance to PHS (Lin et al. 1998; Guo et al. 2004; Wan et al. 2006; Li et al. 2011; Rathi et al. 2011); the crosses of cultivated/wild rice were used to map QTLs associated with domestication-related traits including SD, but the experiments were not designed specifically for the dormancy trait (Cai and Morishima 2000; Thomson et al. 2003; Lee et al. 2005; Li et al. 2006b; Sun et al. 2018), and the crosses of cultivated/weedy rice were used to identify the genetic architecture evolved for SD as a key adaptive trait (Gu et al. 2004; Jing et al. 2008; Subudhi et al. 2012; Mispan et al. 2013; Zhang et al. 2017). Collectively, more than 20 SD QTLs (excluding those mapped to similar positions) have been detected from the three types of crosses. About half of the QTLs were validated with different populations or isolated as single Mendelian factors (Takeuchi et al. 2003; Gu et al. 2006a; Subudhi et al. 2012; Nguyen et al. 2019, 2020). The validated QTLs could explain most genotypic variation for SD in the lineage of cultivated, weedy, and wild rice.

Populations of weedy rice retain dormancy-enhancing alleles at most of the validated QTLs. The lines "SS18-2" from Thailand (7.1 °N Lat.) and "Ludao" from China (34.3 °N Lat.) represent tropical and temperate ecotypes of weedy rice, respectively (Tang and Morishima 1997). A backcross approach was used to introduce genomes of SS18-2 and Ludao into the same genetic background of a weakly dormant cultivated line (EM93-1) to compare the difference or similarity in the number and position of SD genes between the ecotypes. A total of 10 and 12 SD QTLs were identified in the SS18-2 and Ludao systems, respectively; the weedy rice parents donated dormancy-enhancing alleles to 8 (80%) of the 10 or 9 (75%) of the 12 loci (Ye et al. 2010; Zhang et al. 2017). Nine of the QTLs mapped in the two systems are located on the same genomic intervals. The comparative research suggested that SD genes were basically conserved across geographically isolated weed populations.

Genome-wide association mapping for SD in rice has been reported for three panels of >800 varieties selected from different ecotypes. Nine SD QTLs were identified in one panel (Lu et al. 2018), and ~40 loci (including those clustered on short segments of the physical map) were detected in the other two panels (Magwa et al. 2016; Lee et al. 2017). Some of the loci were located near the validated SD QTL. The association and linkage mapping data revealed that domestication retained some SD genes in different ecotypes. These genes scattered in cultivars form a part of the genetic basis for feral plants to survive in croplands.

4.3.3 QTL Epistatic and Genotype-by-Environment (G-by-E) Interactions

SD in weed populations is characterized by the phenotypic plasticity in germination. QTL analysis for SD in weedy rice provided evidence that both epistatic and G-by-E interactions contribute to the phenotypic plasticity. All the six QTLs identified for SD in a population derived from SS18-2 were involved in two to four orders of epistasis (Gu et al. 2004). Subudhi et al. (2012) reported 27 or 17 pairs of digenic epistasis for SD in two populations, which were developed by crossing an accession of US weedy rice with different cultivars. Each of the epistatic interactions acted to delay or promote germination and accounted for 1–11% of the total phenotypic variance in the mapping populations.

A precise evaluation of epistatic interactions was conducted for four SD QTLs in an isogenic background. A set of 16 isogenic lines (ILs) for the *qSD1-2*, *7-1*, *7-2*, and *12* loci, each with two alleles from weedy and cultivated rice, respectively, was used to model component main (additive) and epistatic effects on the degree of SD in a controlled environment.

A total of 15 components, including 11 types of epistasis at two to four orders, were identified; the epistatic interactions contributed to dynamic changes of the 16 genotypes in germinability at 7–150 DAR in nonlinear patterns (Pipatpongpinyo et al. 2020). This research also revealed that epistatic interactions could cause the variation not only in the effect size but also in the direction of an SD gene's effect on germination. For example, the dormancy-enhancing allele of *qSD1-2* inhibited germination at 7 DAR and promoted germination at 150 DAR when the dormancy-enhancing alleles were absent and present at the other three loci, respectively.

In general, low temperatures during seed development tend to enhance primary dormancy (Simpson 1990). However, low seed development temperatures could enhance or reduce the dormancy, depending on the QTLs, in weedy rice. An F_2-like population was developed by introducing the *qSD1-1*, *7-1*, and *12* alleles from SS18-2 into the EM93-1 background. The population was duplicated by a split-tiller (cloning) technique at the seedling stage, and the two genetically identical populations were grown in a high (26.3 ± 1.8 °C) and a low (22.0 ± 2.8 °C) temperature environment from flowering to harvest, respectively, to model G-by-E interactions. The two populations displayed similar patterns of phenotypic variation for germinability at 10, 30, or 50 DAR. Of the three genes, *qSD12* contributed more (30–80%) to phenotypic variation in the high than in the low temperature environment, while both *qSD1-1* and *7-1* contributed more (23–150%) to variation in the low than in the high temperature environment (Gu et al. 2006b). The donor parent SS18-2 of the three dormancy genes originated from Southern Thailand where there are only wet (cool) and dry (hot) seasons. The divergent responses of three loci to temperatures suggest that the tropical ecotype of weedy rice evolved a homeostatic mechanism to maintain the dormancy level in different seasons.

Biological implications of the epistatic and G-by-E interactions include that they are both involved in modifications of gene expression in different genetic backgrounds or environments, and produce additional proportion of variation in germination that cannot be explained by main effects of SD genes and major environment factors. In natural populations or dormant soil seedbanks, the interaction effects would be major contributors to germination phenotypic plasticity and provide additional opportunity of selection for locally adapted genotypes. The presence of epistatic and G-by-E interactions is expected to reduce the predictability of germinability for weed soil seedbanks.

4.3.4 Genes Underlying Seed Dormancy QTLs

Several SD genes have been cloned from rice using a map-based approach. This approach involves fine mapping a genomic region containing an SD QTL to delimit its candidate gene(s) and characterization of the gene for molecular function, functional mutation(s), and allelic distributions in the lineage of wild, weedy, and cultivated rice.

The *Sdr4* QTL on the long arm of chromosome 7 was identified from a cross between an *indica* landrace and a *japonica* cultivar, with the dormancy-enhancing allele from the landrace (Lin et al. 1998). *Sdr4* was cloned as a protein-coding gene, although the protein is unknown for molecular function (Sugimoto et al. 2010). The dormancy gene was transcribed in the embryo tissue of developing seeds and may act downstream of *OsVP1*, an ortholog of the maize mutant *Viviparous-1* involved in the regulation of seed maturation programs. The dormancy alleles of *Sdr4* are present in wild rice (*O. rufipogon*) and *indica* cultivars, but are rare in *japonica* cultivars. Functional mutation(s) in the dormancy-reducing alleles of *Sdr4* remain undetermined (Sugimoto et al. 2010; Subudhi et al. 2012).

The *qSD1-2*, *7-1*, *7-2*, and *12* QTLs were identified from a cross between SS18-2 and EM93-1 in different generations (Gu et al. 2004; Ye et al. 2010). *qSD7-1* was mapped on the short arm of chromosome 7, with the dormancy-enhancing allele from SS18-2. This QTL was delimited to the *Rc* locus using intragenic recombinants (Gu et al. 2011). *Rc* encodes a basic helix-loop-helix (bHLH) family transcription factor for proanthocyanidin production (Sweeney et al. 2006; Furukawa et al. 2007). Expression of *Rc/SD7-1* in early developing caryopses promoted biosynthesis and accumulation of the abscisic acid (ABA) hormone for dormancy induction and also activated the flavonoid biosynthesis pathway for production of red pigments in the lower epidermal cell layer of the pericarp tissue (Gu et al. 2011). The dormancy-enhancing allele of *SD7-1* is present in all populations of weedy and wild "red" rice. In contrast, the loss-of-function mutation of *SD7-1* reduced the seed ABA content and primary dormancy and blocked the pigment biosynthesis. The dormancy-reducing allele of *SD7-1* is present in all cultivars of white pericarp.

The *qSD1-2* QTL was mapped on the long arm of chromosome 1, and its dormancy-reducing ($SD1\text{-}2^d$) and dormancy-enhancing ($SD1\text{-}2^D$) alleles were donated by SS18-2 and EM93-1, respectively. Fine mapping of *qSD1-2* delimited the QTL to a 15-Kb region containing the *semidwarf1* (*sd1*) locus (Ye et al. 2013, 2015). The *sd1* natural mutant is the "Green Revolution" gene prevailing in semi-dwarf high-yielding varieties of rice, and its functional alleles (*Sd1*) encode the gibberellin (GA) synthase OsGA20ox-2 and promote stem elongation in tall plant varieties (Sasaki et al. 2002). $SD1\text{-}2^d$ and $SD1\text{-}2^D$ are functional and mutant (loss-of-function) alleles, respectively, at *OsGA20ox-2* (Ye et al. 2015). Expression of *Sd1* ($SD1\text{-}2^d$) increased the GA content in early developing seeds, which promoted the tissue morphogenesis and maturation programs resulting in reduced primary dormancy. In contrast, mutation of the GA gene (*sd1* or $SD1\text{-}2^D$) dramatically reduced the hormone content, which in turn delayed seed maturation programs, including the ABA accumulation, dehydration, physiological maturity, and acquisition of germinability, or the onset of primary dormancy. Thus, the association of *qSD1-2* with SD is a pleiotropic effect of *OsGA20ox-2*. The GA gene mutation only delayed the peak time of ABA accumulation in developing seeds, but did not reduce the ABA concentration at the peak time (Ye et al. 2015). The developmental mechanism of primary dormancy regulated by *SD1-2* is different from that by *SD7-1*, in which the mutation lowered the ABA content, but did not change the peak time (Gu et al. 2011).

The *qSD7-2* QTL was mapped on the long arm of chromosome 7, with the dormancy-enhancing allele donated by SS18-2. By fine mapping, *qSD7-2* was located near (<2 megabases) but different from *Sdr4* (Figure 4.1d). *qSD7-2* was narrowed to a genomic region containing only one predicted gene encoding a protein kinase. This kinase gene was expressed in imbibed, but not developing, seeds resulting in germination inhibition, and the inhibiting effect could not be offset by GA application (Ye et al. 2013). Research on *qSD7-2* is being conducted to identify substrates of the kinase, their relations with the GA signaling core elements, and allelic distributions of the dormancy gene in wild, weedy, and cultivated rice.

The *qSD12* QTL was mapped on the long arm of chromosome 12, with the dormancy-enhancing allele also donated by SS18-2. *qSD12* has the largest effect on SD in rice and

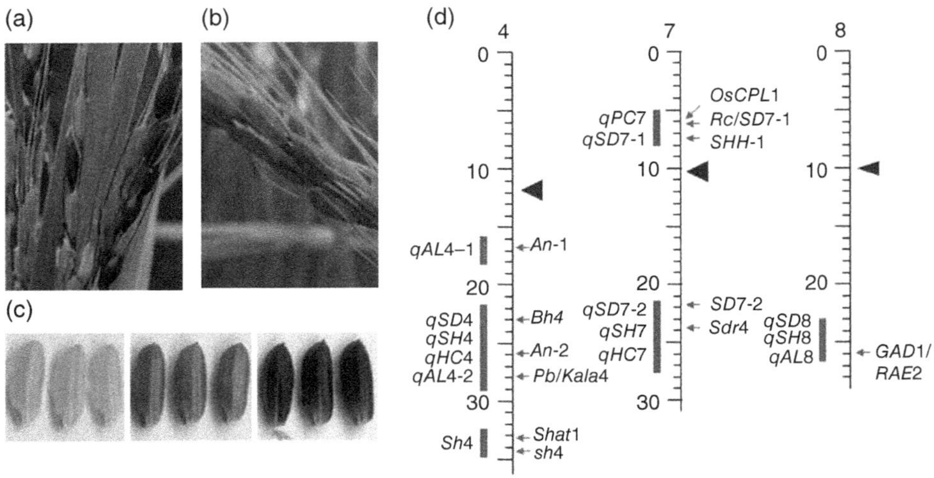

Figure 4.1 Seed-related adaptive traits and haplotypes in weedy rice. (a) Black hull color and awns. The image shows developed (dark and filled) and sterile (gray and empty) spikelet on a plant of SS18-2, a weedy rice line carrying the dominant gene *Bh4* for black hull. (b) Straw hull color and awns. The image shows developed (gray and filled) spikelet on a CRISPR/Cas9-induced mutant, a hybrid F_1 plant from a cross between SS18-2 and a transgenic line containing a *Bh4* guiding RNA::*Cas9* construct in the background of the straw-hulled cultivar Nipponbare. (c) Pericarp colors associated with the *Rc/rc* and *Pb/pb* loci. The images show caryopses from isogenic lines (IL) for the *RcRcpbpb* red (IL_{Rcpb}; middle), *rcrcPbPb* purple (IL_{rcPb}; right), or *rcrcpbpb* white (IL_{rcpb}; left) genotypes. (d) Physical maps for chromosomes 4, 7, and 8. Vertical bars left to the chromosomes indicate adaptive haplotypes encompassing one or more of QTLs for awn length (*qAL*), hull color (*qHC*), pericarp color (*qPC*), seed dormancy (*qSD*), and seed shattering (*qSH4* and *8*). Arrows right to the chromosomes indicate known genes for awn (*An-1*, *An-2*, and *RAE2/GLA*), hull color (*Bh4*), pericarp color (*Rc* and *Pb/Kala4*), seed dormancy (*Sdr4*, *SD7-1*, and *SD7-2*), or shattering (*sh4*, *Shat1*, *SHH-1*, and *OsCPL*). Triangles indicate approximate positions of centromeres.

was associated specifically with embryo dormancy (Gu et al. 2015a). Map-based cloning of *qSD12* identified three underlying genes (*SD12a*, *b*, and *c*) in a genomic region of <100 kilobases; each of the three had an additive effect, while two or three together had cumulative effects on germination inhibition, and the haplotype for all of the three dormancy genes is absent in a pool of ~100 cultivars selected from different ecotypes (Gu et al. 2010; Gu and Feng 2017). Research is being conducted to characterize *SD12a*, *b*, and *c* for molecular functions and to explore the evolutionary history of the three linked loci in the *Oryza* genus.

Molecular characterization of SD QTLs is providing new insights into developmental and evolutionary mechanisms of the adaptive trait. The SD genes cloned from rice are involved in the regulation of seed development/maturation programs through the embryo (*Sdr4* and *SD12s*), endosperm (*SD1-2*), or maternal (*SD7-1*) tissues (Gu et al. 2015a). Some of the genes are pleiotropic and influence the development of primary dormancy by direct or indirect regulations of the GA (*sd1*/*SD1-2*) or ABA (*Rc*/*SD7-1*) accumulation patterns. A major QTL for SD (e.g. *qSD12*) may consist of >1 gene. These discoveries also raise new questions about molecular genetics and evolution of the dormancy trait. These include what genes underlie the other validate SD QTLs, if orthologs of the rice genes were also selected to control SD in the other grass weeds, and if a multigenic SD QTL also exists in the other weed species.

4.4 Genes/QTLs Responsible for Associations of Wild-Like Traits with SD

4.4.1 Seed Shattering

SH is a quantitative trait, which varies from easy/early shattering in wild and weed species to non-shattering at maturation in cereal crop plants. Weed seeds that shatter before the crop matures can avoid being harvested. Association between SH and dormancy was discovered in four experimental populations developed from different lines of weedy rice (Gu et al. 2005b; Mispan et al. 2013). SH rate at 40 days after flowering was negatively correlated with germination rate ($r = -0.20$ to -0.57, $P < 0.0001$) in the four populations under greenhouse conditions. The negative correlations indicate that genotypes (plants) with more shattered seeds tend to have stronger primary dormancy than those with less or no shattered seeds. Thus, the trait association could increase the chance for shattered seeds to enter soil seedbanks when the field conditions are conducive for germination during or immediately after harvesting.

Variation in SH is regulated by genes involved in the formation and degradation of an abscission layer between the glume of a spikelet and the pedicel. Several QTLs have been associated with SH in weedy and cultivated rice (Gu et al. 2005c; Thurber et al. 2013; Mispan et al. 2013; Subudhi et al. 2014). Three of the SH QTLs were collocated with SD loci, i.e. *qSD4/qSH4*, *qSD7-2/qSH7*, and *qSD8/qSH8* (Figure 4.1d). For each of the three clusters, the haplotype from weedy lines enhance both shattering and dormancy, which accounted for the trait associations in the mapping populations (Gu et al. 2005c; Mispan et al. 2013). Interestingly, phenotypic selection for plants with strong SD to backcross with the non-shattering line EM93-1 in each of the five generations (starting from F_2) retained both *qSD4/qSH4* and *qSD8/qSH8* haplotypes from the weed line SS18-2 in the introgression lines (Gu et al. 2005a). These observations can be explained by strong linkage disequilibria between the *qSD4* and *qSH4* and the *qSD8* and *qSH8* loci.

Two (*sh4* and *qSH1*) SH QTLs (Li et al. 2006a; Konishi et al. 2006) and four (*Shat1*, *OsCPL1*, *SHH-1*, and *SH5*) induced SH mutants (Ji et al. 2010; Zhou et al. 2012; Yoon et al. 2014; Jiang et al. 2019) have been cloned from wild or cultivated rice. The *sh4* and *Shat1* loci locate as a cluster at the position about 5 Mb from the *qSH4/qSD4* haplotype on chromosome 4 (Figure 4.1d), and *qSH1* is 1.9 Mb away from *SD1-2* on the long arm of chromosome 1. Thus, the linkage between the *sh4* and *qSD4* or the *qSH1* and *SD1-2* QTLs has a potential to result in the SH–SD association in some populations. In addition, the *OsCPL1-Rc/SD7-1-SHH-1* genomic region of <2 Mb on the short arm of chromosome 7 could also contribute the trait association because of the tight linkage (Figure 4.1d).

4.4.2 Pericarp Colors

Pericarp, the coat of a caryopsis (single-seeded fruit) in grass species, is fused with the testa (seed coat) as a physical structure to protect the seed. In addition, the pericarp tissue replaces a function of the testa to synthesize and store pigment chemicals. PC is genetically a qualitative trait, which varies from colored (red, brown, purple, or black) to white, with the colored dominant over the white. Adaptive significance of PC must also include its association with SD. The PC–SD association was reported first for common wheat (Nilsson-Ehle 1914), then for weedy proso millet (*Panicum miliaceum*, Khan et al. 1996), and weedy rice (Gu et al. 2003), with red-colored pericarp genotypes being more dormant than white ones.

There are two PC systems in rice, red and purple (Figure 4.1c). The red pigments are oxidized proanthocyanidins and the purple pigments are anthocyanidins. Both proanthocyanidins and anthocyanidins are synthesized through a conserved flavonoid pathway in plants and involved in plant defense against microbes, herbivores, and ultraviolet radiation (Dixon and Pasinetti 2010).

The red PC system is controlled by the *Rc* and *Rd* loci in rice. Nonallelic recombination of the two loci produces three phenotypes: red, brown, and white (Kinoshita 1984). One major (*qPC7*, R^2 = 73–87%) and two minor (*qPC6* and *12*, R^2 = 3–4%) QTLs were associated with the red pigment system in weedy red rice (Gu et al. 2005b; Mispan et al. 2013). *qPC7* and *qSD7-1* collocate at *Rc*, and its pleiotropic effects underlie the PC–SD association (Gu et al. 2011). The *Rc*-containing genomic region received much stronger selection for the mutant (*rc*) in cultivated rice than for the wild-type (*Rc*) alleles in wild and weedy rice (Sweeney et al. 2007; Gross et al. 2010). Thus, the haplotypes containing a functional *Rc* could retain linked genes for plant adaptation from the wild ancestors. This could help explain the high frequency of weedy red rice in tropical to temperate areas.

In addition, a reverse mutation of the *rc* allele could produce a functional *Rc* and the red phenotype (Brooks et al. 2008; Lee et al. 2009). This indicates that weedy red rice could also originate from rare spontaneous mutation in cultivars.

The purple PC system is controlled by the *Pp* (*Rd* or *Kala1*) and *Pb* (*Kala4*) loci in rice. Nonallelic recombination of the two loci could produce four PC phenotypes: dark purple, medium purple, brown, and white (Rahman et al. 2013; Maeda et al. 2014). Like *Rc* in the red PC system, *Pb* plays a major role in the purple pigment system. *Pb* was also identified as a bHLH family transcription factor gene (*Ra* or *Kala4*) regulating anthocyanidin biosynthesis (Wang and Shu 2007; Oikawa et al. 2015). To determine if *Pb* has an effect on SD, a functional *Pb* allele, together with a *Rc* allele, was introduced into the same genetic background of a white (double mutant *rcpb*) cultivar as a set of ILs: IL_{Rcpb}, IL_{rcPb}, and IL_{rcpb} (Figure 4.1c). Mean germination percentages of non-after-ripened seeds from IL_{Rcpb}, IL_{rcPb}, and IL_{rcpb} were 45% (±11%), 74% (±7%), and 95% (±3%), respectively, evaluated under a controlled environment. These data indicate that *Pb* also influenced SD, although its effect on germination was smaller than that of *Rc*. The *Pb*, *qSD4*, and *qSH4* loci are located on the same genomic segment on chromosome 4 (Figure 4.1d). It is unknown if a functional *Pb* is present in weedy rice or if some "red" rice is actually "purple" rice.

4.4.3 Hull Color

A hull comprises a lemma and a palea, and both enclose a caryopsis to protect the seed in grass species. The hull tissue of a fertile seed may synthesize non-anthocyanin phenolic compounds and show black, brown, or golden colors in rice (Kinoshita 1984). A range of variation in HC was observed in hybrid populations from crosses between black-hulled weedy rice and straw-hulled cultivars (Gu et al. 2003, 2005a; Mispan et al. 2013; Zhang et al. 2017). The HCs, scored by visualization or quantified with a reflection spectrometer, were negatively correlated with germinability of non-after-ripened seeds ($r = -0.20$ to -0.62; $P<0.001$), with dark-colored genotypes having stronger primary dormancy than light-colored genotypes, in the hybrid populations. Thus, the adaptive significance of the HC trait must include its association with SD.

One major (*qHC4*, $R^2 = 54–69\%$) and three minor (*qHC1*, *qHC7*, and *qHC8*, $R^2 = 1–8\%$) QTLs were identified for the HC trait in three mapping populations (Gu et al. 2005b; Mispan et al. 2013; Zhang et al. 2017). The *qHC4* and *qHC7* loci were collocated with *qSD4* and *qHC7-2*, respectively, with the *qHC4/qSD4* and *qHC7/qSD7-2* haplotypes from weedy rice enhancing hull blackness and SD. In addition to SD, both the *qHC4* and *qHC7* loci also contribute to the associations with SH in weedy rice (Figure 4.1d).

The *qHC4/qSD4* region contains the *Bh4* gene (Figure 4.1d). *Bh4* was cloned from a line of black-hulled wild rice (*O. rufipogon*), which encodes an amino acid transporter family protein and was expressed in the hull tissue (Zhu et al. 2011). *qHC4* is a major QTL, and *Bh4* is a dominance gene for the black hull system. To determine the allelic relationship between *Bh4* and *qHC4*, a CRISPR/Cas9 system targeting *Bh4* (CRISPR/Cas9::gRNA*Bh4*) was used to transform the straw-hulled (*bh4bh4*) cultivar Nipponbare, and a transgenic line crossed with the black-hulled (*Bh4Bh4*) line SS18-2 (Figure 4.1a) to induce mutations in the F_1 hybrid (*Bh4bh4*) plants (Kena 2017). A CRISPR/Cas9-induced mutation silenced the dominance gene in the hybrid plants, resulting in the straw-hulled phenotype (Figure 4.1b). This result indicates that *qHC4* and *Bh4* are identical. Phylogenetic analysis of the *Bh4* locus detected greater genetic diversity for mutant alleles in straw-hulled cultivars than for wild-type alleles in wild and US weedy rice (Zhu et al. 2011; Vigueira et al. 2013). This suggests that natural selection for a black-hulled genotype could also retain gene(s) linked to *Bh4*, as an adaptive haplotype, from wild or weedy rice. Research is being conducted to determine if *Bh4* has an effect on SD and seed longevity.

4.4.4 Awns

An awn in grass species is a needlelike structure extending from the top of the lemma (Figures 4.1a,b) and functions to facilitate seed dispersal and movement into the soil. The awn trait is of both qualitative and quantitative natures in rice. As a qualitative trait, the awn can be absent in all seeds (spikelet) on a plant (largely for cultivars) or present in some or all the seeds (mainly for wild and weedy rice). However, the awned plants may vary in mean AL or in awned seed percentage (AP). It is noted that the AL and AP parameters are complementary in quantifying the trait variation because their relationship is nonlinear (Mispan et al. 2013). Both AL and AP were negatively correlated with germinability at maturation ($r = -0.22$ to -0.44; $P<0.001$), with long-awned genotypes having stronger dormancy than genotypes with no awn or short awns (Gu et al. 2003, 2005c; Mispan et al. 2013). Thus, the adaptive significance of the awn trait must also include its association with SD, i.e. long-awned dormant seeds could be distributed farther from the mother plant and survive in the soil before germination.

Variation in the awn trait is controlled by multiple genes, which were identified as Mendelian factors by genetic analysis, in rice (Kinoshita 1984). Several awn QTLs were mapped in different populations of weedy rice, including those (*qAL4-1*, *qAL4-2*, and *qAL8*) having a major effect on AL (Gu et al. 2005c; Mispan et al. 2013; Zhang et al. 2017). *qAL4-1* and *qAL4-2* are located about 10 Mb (~30 cM) apart on the long arm of chromosome 4; *qAL4-2* and *qAL8* collocate with *qSD4* and *qSD8*, respectively (Figure 4.1d). The *qAL4-2/qSD4* and *qAL8/qSD8* haplotypes from weedy rice both promote AL and increase SD, which accounted for the trait associations in the previously mentioned mapping populations.

Awn genes in the genomic regions encompassing *qAL4-1* (*An-1*), *qAL4-2* (*An-2/LABA1*), and *qAL8* (*GLA1/RAE2*) were cloned from wild rice (*O. rufipogon*; Luo et al. 2013; Gu et al. 2015b; Hua et al. 2015; Bessho-Uehara et al. 2016; Jin et al. 2016). The three genes were associated with seed number and size, but not with SD or germinability in the reported research. It is possible that the *qSD4* and *qSD8* underlying genes are closely linked to *An-2/LABA1* and *GLA1/RAE2*, respectively, and the haplotypes for the linked loci in weedy rice were derived from the wild ancestors.

4.4.5 Seedbank Longevity

A soil seedbank for a species is a natural repository of seeds consisting of genotypes accumulated in the past generation(s). The life span (longevity) of a weed soil seedbank varies with species and genotypes, in addition to environments. For example, seeds of weedy red rice could survive in the soil for a few to >10 years, while cultivated rice usually loses seed viability in fields within several months (Goss and Brown 1939; Noldin et al. 2006; Abeysekara et al. 2010; Baek and Chung 2012). Thus, domestication of rice reduced SD and might have also reduced soil seedbank longevity (SL). Genetic research has been conducted for seed longevity in storage (storability), in *A. thaliana* and some crops (barley [*Hordeum vulgare*], rice, and wheat [*T. aestivum*]), but not in field environments.

Primary dormancy is required for seeds to enter a soil seedbank, particularly for seeds that mature in humid environments. Association between primary dormancy and soil SL was evaluated in a genetic system of rice. The system consists of ILs for 16 recombinant genotypes at the *qSD1-2*, *qSD7-1* (*Rc*), *qSD7-2*, and *qSD12* loci, each with two alleles from weedy and cultivated rice, respectively. The ILs were grown in a controlled environment. A subset of seed samples was evaluated for SD by germination (%) at 7–150 DAR, and the other subset of samples was evaluated for seed decay rate and survivability in a burial trial in a rice field for seven months after harvesting (Pipatpongpinyo et al. 2020). Most of the lines that contain two or more of the SD genes had a lengthened dormancy duration so that seeds survived in the soil (20 cm deep) remained dormant at the time of excavation. Germination levels at 7–150 DAR were correlated positively with the seed decay rate ($r = 0.41$ to 0.53, $P < 0.0001$) and negatively with the survivability ($r = -0.45$ to -0.28, $P = 0.01$ to < 0.0001). Both main and epistatic effects of the four loci contributed to 62% and

55% of the phenotypic variances for the seed decay rate and survivability, respectively, in the field trial. Although *qSD7-1/Rc* played a major role in the genetic system, the genotype containing only the functional *Rc* gene completely lost seed viability in the burial trial. These results indicate that natural genes controlling SD are also involved in regulation of soil SL, and accumulation of dormancy genes in a weed population could result in the formation of a persistent seedbank.

4.5 Genes/QTLs Responsible for Associations of Crop-Mimic Traits with Seed Dormancy

4.5.1 Plant Height

Plant height (PH) influences the competitiveness of a genotype for photosynthetic resources (sunlight, H_2O, and CO_2) and subsequently the growth rate of vegetative tissues and fecundity. The plant growth rate during the vegetative phase and the early reproductive phase was found to be greater in the weedy rice line US1 than in the semi-dwarf line EM93-1, and this rapid growth character was heritable from the parental to hybrid F_2 generations (Mispan et al. 2013). In a population of approximately 500 F_2 plants, the PH at maturation was correlated negatively with the number of tillers per plant ($r = -0.42$), but positively with the plant growth rate ($r = 0.26$) and the number of seeds per plant ($r = 0.28$).

Association between PH and SD has been reported for rice and wheat. The association was identified when allelic variants of a gene/QTL for PH or SD were introduced into the same genetic background, such as *reduced height 3* (*Rht3*) and *QPhs.ocs-4B.1* in common wheat (Flintham and Gale 1982; Wang et al. 2019) and *qSD1-2* and *qSD7-2* in rice (Ye et al. 2013). For all these loci, the alleles that reduce PH also enhanced SD or resistance to PHS.

Some of the PH–SD associations are known to be pleiotropic effects of a GA biosynthesis or signaling gene. The GA hormones promote stem elongation and germination through the GA signaling pathway negatively regulated by the DELLA proteins (Nelson and Steber 2016). In rice, *qSD1-2* was mapped as an SD QTL collocated with the PH QTL *qPH1* in populations derived from different weedy lines (Ye et al. 2013; Mispan et al. 2013). *qSD1-2* is identical to *OsGA20-ox2* at the *sd1* locus, as discussed in Section 4.3.4. No additional gene for PH was identified in the *qSD1-2* region by fine mapping of the QTL; a functional analysis of *OsGA20-ox2* using an artificial mutant confirmed pleiotropic effects of the GA synthase gene on PH and SD (Ye et al. 2015). In common wheat, the wild-type *Rht3* and its homeologs encode the DELLA proteins for normal GA responses in traditional (tall) varieties, while the gain-of-function mutants *Rht-B1* and *Rht-D1* alter the protein structure, which blocks the signal transduction resulting in reduced PH (Peng et al. 1999). Thus, the effects of *Rht3* on reduced PH and enhanced resistance to PHS (Flintham and Gale 1982) can be explained by pleiotropy of the GA signaling gene.

Molecular mechanisms underlying the association of *qSD7-2* or *QPhs.ocs-4B.1* with PH are unknown. *qSD7-2* was collocated with *qPH7*, with the *qSD7-2/qPH7* haplotype from SS18-2 enhancing SD and reducing PH. Nonallelic recombination between the *qSD7-2/qPH7* and *qSD1-2/qPH1* loci (genomic regions) produced dwarf and tall extremes, which increased phenotypic variations for both traits and the degree of negative correlation between PH and germination percentage ($r = -0.75$, $R^2 = 0.57$; Ye et al. 2013). However, genes in these two regions differ in GA response. For example, the inhibitory effects on PH and germination were compensated by the exogenous GA application for genotypes of the *qSD1-2/qPH1*, but not the *qSD7-2/qPH7* haplotype (Ye et al. 2013). Thus, associations between PH and SD must involve different genetic (pleiotropy or tight linkage) and molecular (GA signaling or other physiological pathways) mechanisms.

The dwarf and tall extremes observed in experimental populations may have no selective advantage in weed populations, particularly in fields planted to semi-dwarf high-yield cultivars. In fact, natural mutants of PH genes scatter in different populations and contribute to plant adaptation. For example, the *sd1* mutation likely occurred in a population of wild rice (*O. rufipogon*) (Nagano et al. 2005), and the *qSD7-2/qPH7* haplotype that enhances SD and reduces PH was identified from the weedy accession SS18-2. The "Green Revolution" gene *sd1* has been found in some newly established populations of US weedy rice, likely due to gene flow from semi-dwarf cultivars (Reagon et al. 2011).

4.5.2 Flowering Time

FT is an adaptive trait regulating the timing of seed maturation to complete the life cycle in local ecosystems. In addition, FT also determines environmental conditions, such as temperatures and day lengths or photoperiods, for the development and release of SD (Simpson 1990). Association between FT and SD has been observed in oats (*Avena* spp.), rice, shepherd's purse (*Capsella bursa-pastoris*), and Arabidopsis (Chang and Tagumpay 1973; Simpson 1990; Toorop et al. 2012; Debieu et al. 2013; Vidigal et al. 2016). The association may generate a synergy for weed adaptation to environments where the plant growth duration is critical for seed survival in the soil to contribute to the next plant population (Gu et al. 2018).

There are negative and positive associations between FT and SD in weedy rice, depending on ecotypes or populations. The negative or positive association is defined based on directions of correlation between time (days) to flowering and germination percentage of non-after-ripened or partially after-ripened seeds. Various degrees of negative correlation ($r = -0.24$ to -0.45) were detected in seven (F_2 or BC_1F_1) populations developed from crosses between cultivated and weedy lines of rice (Gu et al. 2018). The negative correlations were accounted for by one or more sets of LD between SD and FT QTLs. The parental lines of weedy rice carry wild-type haplotypes, which contain alleles that enhance dormancy or delay flowering under long-day conditions at the linked loci. Rice is a short-day plant and sensitive to photoperiod for floral initiation. The wild-type haplotypes are found in tropical to temperate ecotypes of weedy rice and some cultivars with a moderate or high sensitivity to photoperiod. An adaptive advantage of the negative association would be that late flowering genotypes tend to have more dormant seeds than early flowering genotypes to enter soil seedbanks at the end of a growing season. Such a synergistic effect is important for weed populations to survive in the following winter or dry season.

Positive correlations between time to flowering and germination percentage were identified when the *qSD1-1* ($r = 0.47$) or *qSD10* ($r = 0.59$) allele from SS18-2 was introduced into the EM93-1 background (Gu et al. 2018). In this background, two FT QTLs, *qFT1* and *qFT10*, were identified in the *qSD1-1* and *qSD10* regions, respectively. The haplotype from SS18-2 for the *qSD1-1/qFT1* region of 300 Kb or the haplotype from EM93-1 for the *qSD10/qFT10* region of <700 Kb promoted flowering and reduced germination. In populations segregating for the *qSD1-1/qFT1* or *qSD10/qFT10* region, early flowering genotypes had stronger SD than late flowering genotypes. However, the positive association was hidden in the previously stated populations where the negative association was present. Rice can be grown all year round for two or three crops in some tropical or subtropical areas. The positive association may have an adaptive advantage in such a multiple cropping system where early flowering weed genotypes required dormant seeds to survive in a wet warm environment.

4.6 Conclusions and Implications

SD is a highly variable trait, based on the number of validated QTLs responsible for the trait variation in the lineage of wild, weedy, and cultivated rice. Theoretically, more than 10 SD loci functionally differentiated between weedy and cultivated rice could produce a large quantity of recombinant genotypes (e.g. 1024 homozygotes for 10 loci each with 2 alleles) for natural selection in cropping systems. Some of the QTLs are known to be involved in epistatic and G-by-E interactions, contributing to phenotypic plasticity of germinability. Several genes underlying the dormancy QTLs were cloned from weedy and cultivated rice, which are involved in the regulation of seed development/maturation programs, or in biosynthesis of the GA or ABA hormones, or flavonoid pigments in the pericarp tissues. In addition, the mapped SD loci are distributed nonrandomly on the rice genome and tend to cluster with genes or QTLs for other wild-like or crop-mimic traits to form adaptive haplotypes and trait associations. These lines of information from the case study suggest that the SD trait might have played a central role in the evolution of weeds in agroecosystems (Gu et al. 2005a,b,c).

The associations of SD with a series of wild-like or crop-mimic traits could produce synergistic effects on weed success in cropping systems. For example, the dormancy trait functions to ensure more seeds to supply soil seedbanks and distribute germination over time, while the associated traits expressed after germination regulate plant growth habit and rate, PH or flowering/maturation time to increase weed competitive ability, or function to enhance seed dispersal (shattering and long awn), seed protection (antioxidant flavonoids in pigmented hull or pericarp), or seed longevity in the soil. Importantly, there are linked or pleiotropic genes underlying the trait associations, supporting the hypothesis that annual plants evolved the SD gene-oriented haplotypes to enhance plant adaptations to ever-changing environments. It is likely that haplotypes for the pleiotropic and/or linked genes have been maintained in tropical to temperate ecotypes of weedy rice since the very beginning of domestication.

The importance of the associations between SD and the other adaptive traits in weed persistence has been underestimated mainly because genetic studies have been focused on individual traits. Because of the gene-controlled associations, dormant genotypes in a population could be products of direct selection for delayed germination or indirect selection for associated characters, such as easy shattering, long awn, black hull, or red pericarp in weedy rice. Without the associated characters, dormant seeds may not be produced at the right time to enter the soil seedbank or cannot be preserved well within the bank. The knowledge of the dormancy gene-oriented adaptive haplotypes could be used to help (i) estimate the persistence of soil seedbanks for different biotypes of weedy rice that are classified based on the presence/absence of awn, black hull, and red pericarp; (ii) devise strategies to reduce the size of dormant soil seedbanks by managing the wild-like characters; or (iii) conduct comparative genomic research on weediness by testing synteny of these haplotypes in the other grass weeds.

Genetic research on SD using natural variants has advanced our knowledge to the level of QTLs/genes in several species. Further research on the trait should include (i) modeling the genes/QTLs (or the dormancy gene-oriented haplotypes) for G-by-E interaction effects on the persistence of soil seedbanks in commonly used cropping systems, such as till, no-till, and different crop rotations and (ii) molecular characterization of the additional QTLs for underlying genes and their regulatory mechanisms for SD development/release and germination. These lines of knowledge are required to improve integrated weed management strategies or to develop biotechniques for weed control using RNAi, genome editing, or gene-drive systems to silence major genes for SD (Kena 2017).

Acknowledgements

This research was supported by grants from the United States Department of Agriculture – National Institute of Food and Agriculture (2013-33522-21097, 2018-33522-28795, and 2019-67014-29238) and National Research Initiative (2008-35301-19058), Natural Science Foundation (IOS 1021382 and 0641376), and the American Seed Research Foundation, and by South Dakota Agricultural Extension Station.

References

Abeysekara, A.S.K., Herath, H.M.S., Wickrame, U.B., Nugaliyadde, L., Johnson, D.E. (2010) Germinability, viability, and longevity of weedy and cultivated rice in Sri Lanka. *Proceedings of International Rice Research Conference on Pest, Disease, and Weed Management*, Hanoi, Vietnam.

Alonso-Blanco, C., Aarts, M.G., Bentsink, L. et al. (2009). What has natural variation taught us about plant development, physiology, and adaptation? *Plant Cell* 21: 1877–1896.

Baek, J.S. and Chung, N.J. (2012). Seed wintering and deterioration characteristics between weedy and cultivated rice. *Rice* 5: 21.

Barrett, S.H. (1983). Crop mimicry in weeds. *Economic Botany* 37: 255–282.

Bentsink, L., Jowett, J., Hanhart, C.J., and Koornneef, M. (2006). Cloning of *DOG1*, a quantitative trait locus controlling seed dormancy in Arabidopsis. *Proceedings of the National Academy of Sciences of the United States of America* 103: 17042–17047.

Bessho-Uehara, K., Wang, D.R., Furuta, T. et al. (2016). Loss of function at *RAE2*, a previously unidentified EPFL, is required for awnlessness in cultivated Asian rice. *Proceedings of the National Academy of Sciences of the United States of America* 113: 8969–8974.

Bewley, J.D., Bradford, K., Hilhorst, H., and Nonogaki, H. (2013). *Seeds: Physiology of Development, Germination and Dormancy*, 3e. New York, NY: Springer.

Brooks, S.A., Yan, W., Jackson, A.K., and Deren, C.W. (2008). A natural mutation in *rc* reverts white-rice-pericarp to red and results in a new, dominant, wild-type allele: *Rc*-g. *Theoretical and Applied Genetics 117*: 575–580.

Burgos, N.R., Singh, V., Tseng, T.M. et al. (2014). The impact of herbicide-resistant rice technology on phenotypic diversity and population structure of United States weedy rice. *Plant Physiology* 166: 1208–1220.

Cai, H.W. and Morishima, H. (2000). Genomic regions affecting seed shattering and seed dormancy in rice. *Theoretical and Applied Genetics* 100: 840–846.

Chai, C., Shankar, R., Jain, M., and Subudhi, P.K. (2018). Genome-wide discovery of DNA polymorphisms by whole genome sequencing differentiates weedy and cultivated rice. *Scientific Reports* 8: 14218.

Chang, T.T. and Tagumpay, O. (1973). Inheritance of grain dormancy in relation to growth duration in 10 rice crosses. *SABRAO Newsletter* 5: 87–94.

Chen, L.J., Lee, D.S., Song, Z.P. et al. (2004). Gene flow from cultivated rice (*Oryza sativa*) to its weedy and wild relatives. *Annals of Botany* 93: 67–73.

Croughan, T.P. (2003). Clearfield rice: it's not a GMO. *Louisiana Agriculture* 46: 24–26.

De Leon, T.B., Karn, E., Al-Khatib, K. et al. (2019). Genetic variation and possible origins of weedy rice found in California. *Ecology and Evolution* 9: 5835–5848.

Debieu, M., Tang, C., Stich, B. et al. (2013). Co-variation between seed dormancy, growth rate and flowering time changes with latitude in *Arabidopsis thaliana*. *PLoS One* 8: e61075.

Dekker, J. (1997). Weed diversity and weed management. *Weed Science* 45: 357–363.

Delouche, J.C., Burgos, N.R., Gealy, D.R. et al. (2007). *Weedy Rices: Origin, Biology, Ecology and Control*. Rome: Food and Agriculture Organization of the United Nations.

Dixon, R.A. and Pasinetti, G.M. (2010). Flavonoids and isoflavonoids: from plant biology to agriculture and neuroscience. *Plant Physiology* 154: 453–457.

Fennimore, S., Nyquist, W., Shaner, G. et al. (1998). Temperature response in wild oat (*Avena fatua* L.) generations segregating for seed dormancy. *Heredity* 81: 674–682.

Flintham, J.E. and Gale, M.D. (1982). The tom thumb dwarfing gene, *Rht3* in wheat. I. Reduced pre-harvest damage to breadmaking quality. *Theoretical and Applied Genetics* 62: 121–126.

Furukawa, T., Maekawa, M., Oki, T. et al. (2007). The *Rc* and *Rd* genes are involved in proanthocyanidin synthesis in rice pericarp. *The Plant Journal* 49: 91–102.

Global Rice Science Partnership, GRiSP (2013). *Rice Almanac*, 4e. Los Baños, Philippines: International Rice Research Institute.

Goss, W.L. and Brown, E. (1939). Buried red rice. *Agronomy Journal* 31: 633–637.

Gross, B.L., Reagon, M., Hsu, S.C. et al. (2010). Seeing red: the origin of grain pigmentation in US weedy rice. *Molecular Ecology* 19: 3380–3393.

Gu, X.-Y. and Feng, J. (2017). Evolution of embryo dormancy involved local gene duplication, recombination hot spot formation, and adaptation to changing seed development temperatures in *Oryza*. *The 12th International Society of Seed Science Conference*, Monterey, CA (10–14 September 2017).

Gu, X.-Y., Chen, Z.X., and Foley, M.E. (2003). Inheritance of seed dormancy in weedy rice. *Crop Science* 43: 835–843.

Gu, X.-Y., Kianian, S.F., and Foley, M.E. (2004). Multiple loci and epistases control genetic variation for seed dormancy in weedy rice (*Oryza sativa*). *Genetics* 166: 1503–1516.

Gu, X.-Y., Kianian, S.F., and Foley, M.E. (2005a). Phenotypic selection for dormancy introduced a set of adaptive haplotypes from weedy into cultivated rice. *Genetics* 171: 695–704.

Gu, X.-Y., Kianian, S.F., and Foley, M.E. (2005b). Seed dormancy imposed by covering tissues interrelates to shattering and seed morphological characteristics in weedy rice. *Crop Science* 45: 948–955.

Gu, X.-Y., Kianian, S.F., Hareland, G.A. et al. (2005c). Genetic analysis of adaptive syndromes interrelated with seed dormancy in weedy rice (*Oryza sativa*). *Theoretical and Applied Genetics* 110: 1108–1118.

Gu, X.-Y., Kianian, S.F., and Foley, M.E. (2006a). Isolation of three dormancy QTLs as Mendelian factors in rice. *Heredity* 96: 93–99.

Gu, X.-Y., Kianian, S.F., and Foley, M.E. (2006b). Dormancy genes from weedy rice respond divergently to seed development environments. *Genetics* 172: 1199–1211.

Gu, X.-Y., Liu, T., Feng, J. et al. (2010). The *qSD12* underlying gene promotes abscisic acid accumulation in early developing seeds to induce primary dormancy in rice. *Plant Molecular Biology* 73: 97–104.

Gu, X.-Y., Foley, M.E., Horvath, D.P. et al. (2011). Association between seed dormancy and pericarp color is controlled by a pleiotropic gene that regulates abscisic acid and flavonoid synthesis in weedy red rice. *Genetics* 189: 1515–1524.

Gu, X.-Y., Zhang, J., Ye, H. et al. (2015a). Genotyping of endosperms to determine seed dormancy genes regulating germination through embryonic, endospermic, or maternal tissues in rice. *G3 (Bethesda)* 5: 183–193.

Gu, B., Zhou, T., Luo, J. et al. (2015b). *An-2* encodes a cytokinin synthesis enzyme that regulates awn length and grain production in rice. *Molecular Plant* 8: 1635–1650.

Gu, X.-Y., Pipatpongpinyo, W., Zhang, L. et al. (2018). Two contrasting patterns and underlying genes for coadaptation of seed dormancy and flowering time in rice. *Scientific Reports* 8: 16813.

Guo, L., Zhu, L., Xu, Y. et al. (2004). QTL analysis of seed dormancy in rice (*Oryza sativa* L.). *Euphytica* 140: 155–162.

Hilhorst, H.W.M. (2007). Definitions and hypotheses of seed dormancy. In: *Seed Development, Dormancy and Germination* (eds. K.J. Bradford and H. Nonogaki), 50–71. Blackwell Publishing.

Hua, L., Wang, D.R., Tan, L. et al. (2015). *LABA1*, a domestication gene associated with long, barbed awns in wild rice. *Plant Cell* 27: 1875–1888.

Hussain, Z.P., Man, A., and Othman, A.S. (2010). Morphological study of the relationships between weedy rice accessions (*Oryza sativa* Complex) and commercial rice varieties in Pulau Pinang Rice Granary Area. *Tropical Life Sciences Research* 21: 27–40.

Ji, H., Kim, S.R., Kim, Y.H. et al. (2010). Inactivation of the CTD phosphatase-like gene *OsCPL1* enhances the development of the abscission layer and seed shattering in rice. *The Plant Journal* 61: 96–106.

Jiang, L., Ma, X., Zhao, S. et al. (2019). The APETALA2-like transcription factor SUPERNUMERARY BRACT controls rice seed shattering and seed size. *Plant Cell* 31: 17–36.

Jin, J., Hua, L., and Zhu, Z. (2016). *GAD1* encodes a secreted peptide that regulates grain number, grain length, and awn development in rice domestication. *Plant Cell* 28: 2453–2463.

Jing, W., Jiang, L., Zhang, W. et al. (2008). Mapping QTL for seed dormancy in weedy rice. *Acta Agronomica Sinica* 34: 737–742.

Johnson, L.P.V. (1935). The inheritance of delayed germination in hybrids of *Avena fatua* and *A. sativa*. *Canadian Journal of Research* 13: 367–387.

Kanapeckas, K.L., Vigueira, C.C., Ortiz, A. et al. (2016). Escape to ferality: the endoferal origin of weedy rice from crop rice through de-domestication. *PLoS One* 11: e0162676.

Kena, A. (2017). Silencing seed dormancy genes to mitigate risk of transgene flow to weedy rice. Dissertation. South Dakota State University, Brookings, SD, USA.

Khan, M., Cavers, P.B., Kane, M., and Thompson, K. (1996). Role of the pigmented seed coat of proso millet *(Panicum miliaceum* L.) in imbibition, germination and seed persistence. *Seed Science Research* 7: 21–25.

Khush, G.S. and Brar, D.S. (2002). Rice. In: *Evolution and Adaptation of Cereal Crops* (eds. V.L. Chopra and S. Prakash), 1–41. Enfield, USA/Plymouth, UK: Science Publishers, Inc.

Kinoshita, T. (1984). Genetics. In: *Biology of Rice* (eds. S. Tsunoda and N. Takahashi), 187–274. Tokyo: JSSP/Elsevier.

Konishi, S., Izawa, T., Lin, S.Y. et al. (2006). An SNP caused loss of seed shattering during rice domestication. *Science* 312: 1392–1396.

Lee, S.J., Oh, C.S., Suh, J.P. et al. (2005). Identification of QTLs for domestication-related and agronomic traits in an *Oryza sativa* × *O. rufipogon* BC1F7 population. *Plant Breeding* 124: 209–219.

Lee, D., Lupotto, E., and Powell, W. (2009). G-string slippage turns white rice red. *Genome* 52: 490–493.

Lee, G.-A., Jeon, Y.-A., Lee, H.-S. et al. (2017). New genetic loci associated with preharvest sprouting and its evaluation based on the model equation in rice. *Frontiers in Plant Science* 8: 1393.

Li, C., Zhou, A., and Sang, T. (2006a). Rice domestication by reducing shattering. *Science* 311: 1936–1939.

Li, C., Zhou, A., and Sang, T. (2006b). Genetic analysis of rice domestication syndrome with the wild annual species, *Oryza nivara*. *New Phytologist* 170: 185–193.

Li, W., Xu, L., Bai, X., and Xing, Y. (2011). Quantitative trait loci for seed dormancy in rice. *Euphytica* 178: 427–435.

Li, L.F., Li, Y.L., Jia, Y. et al. (2017). Signatures of adaptation in the weedy rice genome. *Nature Genetics* 49: 811–814.

Lin, S., Sasaki, T., and Yano, M. (1998). Mapping quantitative trait loci controlling seed dormancy and heading date in rice, *Oryza sativa* L., using backcross inbred lines. *Theoretical and Applied Genetics* 96: 997–1003.

Lu, Q., Niu, X., Zhang, M. et al. (2018). Genome-wide association study of seed dormancy and the genomic consequences of improvement footprints in rice (*Oryza sativa* L.). *Frontiers in Plant Science* 8: 2213.

Luo, J., Liu, H., Zhou, T. et al. (2013). *An-1* encodes a basic helix-loop-helix protein that regulates awn development, grain size, and grain number in rice. *Plant Cell* 25: 3360–3376.

Maeda, H., Yamaguchi, T., Omoteno, M. et al. (2014). Genetic dissection of black grain rice by the development of a near isogenic line. *Breeding Science* 64: 134–141.

Magwa, R.A., Zhao, H., and Xing, Y. (2016). Genome-wide association mapping revealed a diverse genetic basis of seed dormancy across subpopulations in rice (*Oryza sativa* L.). *BMC Genetics* 17: 28.

Mispan, M.S., Zhang, L., Feng, J., and Gu, X.-Y. (2013). Quantitative trait locus and haplotype analyses of wild and crop-mimic traits in U.S. weedy rice. *G3 (Bethesda)* 3: 1049–1059.

Nadir, S., Xiong, H., Zhu, Q. et al. (2017). Weedy rice in sustainable rice production. A review. *Agronomy for Sustainable Development* 37: 46.

Nagano, H., Onishi, K., Ogasawara, M. et al. (2005). Genealogy of the "Green Revolution" gene in rice. *Genes & Genetic Systems* 80: 351–356.

Nelson, S.K. and Steber, C.M. (2016). Gibberellin hormone signal perception: down-regulating DELLA repressors of plant growth and development. In: *Annual Plant Reviews, The Gibberellins*, vol. 49 (eds. P. Hedden and S.G. Thomas), 153–187. Chichester, UK: Wiley Blackwell.

Nguyen, T., Zhou, C., Zhang, T. et al. (2019). Identification of QTL for seed dormancy from weedy rice and its application to elite rice cultivar 'Ninggeng 4'. *Molecular Breeding* 39: 123.

Nguyen, T., Fu, K., Mou, C. et al. (2020). Fine mapping of *qSdr9*, a novel locus for seed dormancy (SD) in weedy rice, and development of NILs with a strong SD allele. *Molecular Breeding* 40: 81.

Nilsson-Ehle, H. (1914). Zur Kenntnis der mit der Keimungsphy siologie des Weizens in Zusammenhang stehenden inneren Faktoren. *Zeitschrift für Pflanzenzüchtung* 2: 153–187.

Noldin, J.A., Chandler, J.M., and McCauley, G.N. (2006). Seed longevity of red rice ecotypes buried in soil. *Planta Daninha* 24: 611–620.

Oikawa, T., Maeda, H., Oguchi, T. et al. (2015). The birth of a black rice gene and its local spread by introgression. *Plant Cell* 27: 2401–2414.

Oka, H.I. (1988). *Origin of Cultivated Rice*. Tokyo: Japan Science Society Press.

Oka, H.I. and Morishima, H. (1967). Variations in the breeding systems of a wild rice, *Oryza perennis*. *Evolution* 21: 249–258.

Peng, J., Richards, D.E., Hartley, N.M. et al. (1999). 'Green Revolution' genes encode mutant gibberellin response modulators. *Nature* 400: 256–261.

Pipatpongpinyo, W., Korkmaz, U., Wu, H. et al. (2020). Assembling seed dormancy genes into a system identified their effects on seedbank longevity in weedy rice. *Heredity* 124: 135–145.

Qiu, J., Zhou, Y., Mao, L. et al. (2017). Genomic variation associated with local adaptation of weedy rice during de-domestication. *Nature Communications* 8: 15323.

Qiu, J., Jia, L., Wu, D. et al. (2020). Diverse genetic mechanisms underlie worldwide convergent rice feralization. *Genome Biology* 21: 70.

Rahman, M.M., Lee, K.E., Lee, E.S. et al. (2013). The genetic constitutions of complementary genes *Pp* and *Pb* determine the purple color variation in pericarps with cyanidin-3-*O*-glucoside depositions in black rice. *Journal of Plant Biology* 56: 24–31.

Rathi, S., Baruah, A.R., Chowdhury, R.K., and Sarma, R.N. (2011). QTL analysis of seed dormancy in indigenous rice of Assam, India. *Cereal Research Communications* 39: 137–146.

Reagon, M., Thurber, C.S., Olsen, K.M. et al. (2011). The long and the short of it: *SD1* polymorphism and the evolution of growth trait divergence in U.S. weedy rice. *Molecular Ecology* 20: 3743–3756.

Roberts, E.H. (1962). Dormancy in rice seeds. III. The influence of temperature, moisture, and gaseous environment. *Journal of Experimental Botany* 13: 75–94.

Rodríguez, M., Barrero, J., Corbineau, F. et al. (2015). Dormancy in cereals (not too much, not so little): about the mechanisms behind this trait. *Seed Science Research* 25: 99–119.

Sasaki, A., Ashikari, M., Ueguchi-Tanaka, M. et al. (2002). A mutant gibberellin-synthesis gene in rice. *Nature* 416: 701–702.

Shivrain, V.K., Burgos, N.R., Gealy, D.R. et al. (2008). Maximum outcrossing rate and genetic compatibility between red rice (*Oryza sativa*) biotypes and Clearfield™ rice. *Weed Science* 56: 807–813.

Simpson, G.M. (1990). *Seed Dormancy in Grasses*. Cambridge: Cambridge University Press.

Subudhi, P.K., Parco, A., Singh, P.K. et al. (2012). Genetic architecture of seed dormancy in U.S. weedy rice in different genetics backgrounds. *Crop Science* 52: 2564–2575.

Subudhi, P.K., Singh, P.K., DeLeon, T. et al. (2014). Mapping of seed shattering loci provides insights into origin of weedy rice and rice domestication. *Journal of Heredity* 105: 276–287.

Sugimoto, K., Takeuchi, Y., Ebana, K. et al. (2010). Molecular cloning of *Sdr4*, a regulator involved in seed dormancy and domestication of rice. *Proceedings of the National Academy of Sciences of the United States of America* 107: 5792–5797.

Suh, H.S., Sato, Y.I., and Morishima, H. (1997). Genetic characterization of weedy rice (*Oryza sativa* L.) based on morphophysiology, isozymes and RAPD markers. *Theoretical and Applied Genetics* 94: 316–321.

Sun, A.L., Wu, H.M., Chen, G.M. et al. (2018). Mapping of QTLs for seed dormancy in *Oryza rufipogon* Griff. *Acta Agronomica Sinica* 44: 15–23.

Sun, J., Ma, D., Tang, L. et al. (2019). Population genomic analysis and de novo assembly reveal the origin of weedy rice as an evolutionary game. *Molecular Plant* 12: 632–647.

Sweeney, M.T., Thomson, M.J., Pfeil, B.E., and McCouch, S. (2006). Caught red-handed: *Rc* encodes a basic helix-loop-helix protein conditioning red pericarp in rice. *Plant Cell* 18: 283–294.

Sweeney, M.T., Thomson, M.J., Cho, Y.G. et al. (2007). Global dissemination of a single mutation conferring white pericarp in rice. *PLoS Genetics* 3: e133.

Takeuchi, Y., Lin, S.Y., Sasaki, T., and Yano, M. (2003). Fine linkage mapping enables dissection of closely linked quantitative trait loci for seed dormancy and heading in rice. *Theoretical and Applied Genetics* 107: 1174–1180.

Tang, L.H. and Morishima, H. (1997). Genetic characterization of weedy rices and the inference on their origins. *Breeding Science* 47: 153–160.

Thomson, M.J., Tai, T.H., McClung, A.M. et al. (2003). Mapping quantitative trait loci for yield, yield components and morphological traits in an advanced backcross population between *Oryza rufipogon* and the *Oryza sativa* cultivar Jefferson. *Theoretical and Applied Genetics* 107: 479–493.

Thurber, C.S., Jia, M.H., Jia, Y., and Caicedo, A.L. (2013). Similar traits, different genes? Examining convergent evolution in related weedy rice populations. *Molecular Ecology* 22: 685–698.

Toorop, P.E., Cuerva, R.C., Begg, G.S. et al. (2012). Co-adaptation of seed dormancy and flowering time in the arable weed *Capsella bursa-pastoris* (shepherd's purse). *Annals of Botany* 109: 481–489.

Tseng, T.M., Burgos, N.R., Shivrain, V.K. et al. (2013). Inter and intrapopulation variation in dormancy of *Oryza sativa* (weedy red rice) and allelic variation in dormancy-linked loci. *Weed Research* 53: 440–451.

Veasey, E.A., Karasawa, M.G., Santos, P.P. et al. (2004). Variation in the loss of seed dormancy during after-ripening of wild and cultivated rice species. *Annals of Botany* 94: 875–882.

Vidigal, D.S., Marques, A.C.S.S., Willems, L.A.J. et al. (2016). Altitudinal and climatic associations of seed dormancy and flowering traits evidence adaptation of annual life cycle timing in Arabidopsis thaliana. *Plant, Cell & Environment* 39: 1737–1748.

Vigueira, C.C., Li, W., and Olsen, K.M. (2013). The role of *Bh4* in parallel evolution of hull colour in domesticated and weedy rice. *Journal of Evolutionary Biology* 26: 1738–1749.

Vigueira, C.C., Qi, X., Song, B.K. et al. (2018). Call of the wild rice: *Oryza rufipogon* shapes weedy rice evolution in Southeast Asia. *Evolutionary Applications* 12: 93–104.

Wan, J.M., Jiang, L., Tang, J.Y. et al. (2006). Genetic dissection of the seed dormancy trait in cultivated rice (*Oryza sativa* L.). *Plant Science* 170: 786–792.

Wang, C. and Shu, Q. (2007). Fine mapping and candidate gene analysis of purple pericarp gene *Pb* in rice (*Oryza sativa* L.). *Chinese Science Bulletin* 52: 3097–3104.

Wang, X., Liu, H., Liu, G. et al. (2019). Phenotypic and genotypic characterization of near-isogenic lines targeting a major 4BL QTL responsible for pre-harvest sprouting in wheat. *BMC Plant Biology* 19: 348.

Ye, H., Foley, M.E., and Gu, X.-Y. (2010). New seed dormancy loci detected from weedy rice-derived advanced populations with major QTL alleles removed from the background. *Plant Science* 179: 612–619.

Ye, H., Beighley, D.H., Feng, J., and Gu, X.-Y. (2013). Genetic and physiological characterization of two clusters of quantitative trait loci associated with seed dormancy and plant height in rice. *G3 (Bethesda)* 3: 323–331.

Ye, H., Feng, J., Zhang, L. et al. (2015). Map-based cloning of *seed dormancy1-2* identified a gibberellin synthesis gene regulating the development of endosperm-imposed dormancy in rice. *Plant Physiology* 169: 2152–2165.

Yoon, J., Cho, L.H., Kim, S.L. et al. (2014). The BEL1 type homeobox gene *SH5* induces seed shattering by enhancing abscission zone development and inhibiting lignin biosynthesis. *The Plant Journal* 79: 717–728.

Zhang, L., Lou, J., Foley, M.E., and Gu, X.-Y. (2017). Comparative mapping of seed dormancy loci between tropical and temperate ecotypes of weedy rice (*Oryza sativa* L.). *G3 (Bethesda)* 7: 2605–2614.

Zhou, Y., Lu, D., Li, C. et al. (2012). Genetic control of seed shattering in rice by the APETALA2 transcription factor shattering abortion. *Plant Cell* 24: 1034–1048.

Zhu, B.F., Si, L., Wang, Z. et al. (2011). Genetic control of a transition from black to straw-white seed hull in rice domestication. *Plant Physiology* 155: 1301–1311.

5

Environmental Regulation of Weed Seedbanks and Seedling Emergence

Roberto L. Benech-Arnold and Diego Batlla

Facultad de Agronomía, Cátedra de Cultivos Industriales and Cátedra de Cerealicultura, IFEVA (UBA/CONICET), Buenos Aires, Argentina

5.1 Introduction

5.1.1 The Seedbank: Dispersal over Time

Weedy species that rely on sexual reproduction perpetuate themselves via seeds. Seeds dispersed to the soil from the mother plant can germinate immediately, or they can remain as non-germinated for an indefinite time period. Seeds in the soil, whether on the surface or buried at different depths, form what is called "the seedbank." Depending on the physiological attributes of the seeds that form them (i.e. dormancy, longevity, requirements for germination, etc.), these banks can function as highly efficient sources of seedling emergence and dispersion over time. Seedbanks have been classified based on different aspects, which determine their function. The most used classification has been that of Thompson and Grime (1979), which identifies four types of seedbanks; Type I and Type II are considered transient (i.e. no seed lasts more than a year in the soil) and normally correspond to species that germinate in early fall or early spring, respectively. The seeds of species that form Type I banks are present in the soil during the summer only, while those of species that form Type II banks remain only during winter. Seeds that form both types of bank normally exhibit dormancy, which is relieved by high temperatures in summer in the case of species that form Type I banks and by the cold winter temperatures for species that form Type II seedbanks. In Types III and IV banks, a lesser or greater fraction of the seed population becomes part of a persistent bank and survives for more than a year (Thompson and Grime 1979). Thompson et al. (1997) proposed a new classification with only three categories, which better takes into account the longevity of the seeds: (i) species whose seeds last less than one year in the soil form **transient banks**; (ii) species whose seeds survive one to five years in the soil form banks of **short persistence**; and (iii) species whose seeds last for at least five years form banks of **long persistence**.

The size of a seedbank (e.g. the number and species composition of seeds within a given volume of soil) can be quantified at any given time with any of the sampling methods

Persistence Strategies of Weeds, First Edition. Edited by Mahesh K. Upadhyaya, David R. Clements, and Anil Shrestha.

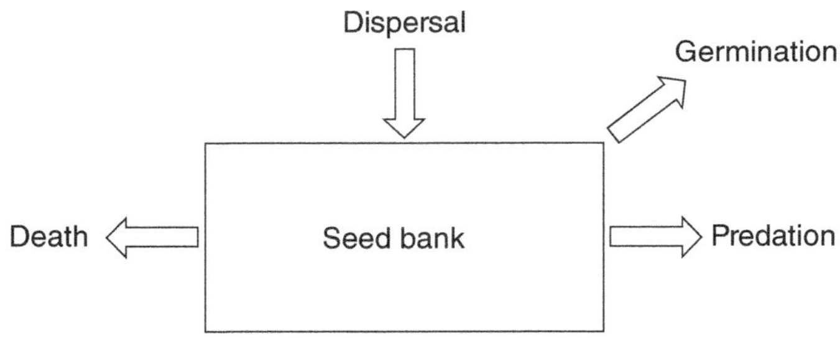

Figure 5.1 Schematic representation of seedbank dynamics. Inflows are through dispersal from mother plants. Outflows are through germination, predation, and death. *Source:* Adapted from Benech-Arnold et al. (2014).

proposed in the literature (see Fenner and Thompson (2005) for a review). The size of the seedbank is the result of the intensity with which the different processes that intervene in the addition and loss of seeds to and from the bank act. Indeed, the addition of seeds into the bank occurs through the process of dispersal from the mother plant. In some cases, seeds may be dispersed by animals via passage through their digestive tract. Losses, on the other hand, are due to germination, predation, and death or loss of viability as a result of physiological deterioration (Figure 5.1). The difference between additions and losses determines the size of the seedbank at a given time.

5.1.2 Scope of the Chapter

The possibility of forming persistent banks is related to the physiological and morphological attributes of the seeds and to characteristics of the habitat in which these species are found. In this chapter we outline characteristics that, in relation to each of the processes that determine losses from the seedbank, favor the formation of persistent banks.

5.2 Germination

Seed germination, whether it leads to successful seedling emergence or not, will lead to depletion of the seedbank. However, if the seeds exhibit dormancy (i.e. a physical or physiological constraint that prevents germination even under otherwise favorable water, temperature, and aeration conditions), germination will not occur immediately after the seeds have entered the soil bank. Dormancy and its control by the environment will be discussed in greater detail in later sections of this chapter, even though it is worth emphasizing at this instance that the presence of dormancy (imposed either prior to dispersal or after entrance into the soil) is a necessary condition (although not sufficient) for a seed population to form a persistent bank. Moreover, the possibility of entering secondary dormancy (i.e. the restoration of the constraint after it had been removed) increases the ability of a population to form a persistent seedbank greatly. Indeed, seeds that can be induced into secondary dormancy frequently exhibit a cyclic dormancy pattern that allows them to remain ungerminated for long periods throughout the year, only being "germinable" when environmental conditions favor the successful establishment of the resultant plants. However, if after dormancy alleviation, germination takes place without any further requirements being fulfilled (i.e. light), those seeds that had entered the bank with the last dispersal would germinate in large proportion, thus precluding the formation of a persistent bank. Light is often a "dormancy terminating" factor (i.e. removes the last impediment to germination once dormancy level has been sufficiently lowered) and can only be perceived by the seeds if they are located at the soil surface or few millimeters below (Bliss and Smith 1985) or somehow exposed to the sunlight (even if the exposure has been for milliseconds) by tillage (Benech-Arnold et al. 2000; Kruk et al. 2006). Therefore, if the seeds are buried, the light requirement will not be satisfied, and the germination process will not be triggered, thus disabling their ability to exit the seedbank. Hence, a cyclic pattern of dormancy, coupled with light requirement for dormancy termination once dormancy is at a minimum level, is a characteristic that favors the formation of persistent seedbanks. The value of these characteristics for persistent seedbank formation will be enhanced in seeds that have efficient burial mechanisms and in habitats where soil disturbance is not a recurrent feature.

5.3 Predation

As highlighted in Section 5.2, burial is a crucial step in the development of persistent seedbanks (Fenner and Thompson 2005). Indeed, seeds located at the soil surface have two primary fates: to germinate readily or to be predated. Scursoni et al. (1999), working with wild oat (*Avena fatua*) growing with barley (*Hordeum vulgare*), determined that if the weed seeds were left at the soil surface after the crop harvest (i.e. no soil disturbance after harvest), the weed seedbank on the surface was reduced to 50% of its initial size after six months. More than 80% of this reduction was attributed to predation. Once buried, the probability of being predated decreases dramatically since seed predators (rodents, birds, and insects) feed primarily on the soil surface. The size of seeds is an important factor influencing the potential for seeds to be predated. On the one hand, small seeds tend to be less preferred by and apparent to animals and on the other hand, smaller seeds have a greater probability of ending up buried in the soil (Fenner and Thompson 2005). In fact, these relationships between seed size, burial capacity, and predation lead to an almost universal relationship between small seed size and persistence in the soil (Leck 1989; Tsuyuzaki 1991; Thompson et al. 1993; Dalling and Hubbell 2002; Pakeman et al. 2002). Interestingly, the few examples of habitats where this relationship does not hold are those where the relationship between small seed size and burial is disrupted; agricultural habitats where the soil is disturbed in a recurrent way constitutes one of those examples since seed burial by tillage is independent of seed size.

In addition to size, there may be other seed characteristics that prevent predation. For example, in a survey carried out with 80 species, Hendry et al. (1994) found a significant relationship between persistence of the seeds in the seedbank and the concentration of *ortho*-dihydroxyphenol in the seed. Other authors have also found similar correlations between seed chemistry and physical defenses and the persistence of seeds in the soil (Davis et al. 2008; Dalling et al. 2011). Consistent with these results, Hulme (1998) determined that the seeds of species that tend to form persistent banks were less preferred by predators than those of species that form transient banks. Together, these results suggest that the seeds of species that form persistent seed banks could have, in addition to the other characteristics already mentioned, a chemical composition that repels predators.

5.4 Loss of Viability as a Result of Physiological Deterioration

Once a part of a seedbank is buried, seeds face two dangers that might preclude persistence: (i) attack by bacteria and fungi and (ii) the loss of viability as a result of physiological deterioration, a process that begins immediately after seed dispersal. Soil is rich in microorganisms that can be potentially pathogenic for seeds. Seed coats can act as an effective physical barrier to the entrance of those pathogens. However, there is evidence that, at least in some species, seeds can defend themselves from pathogens by means of antifungal chemicals (Pérez-García et al. 1992; Siemens et al. 1992; Davis et al. 2008; Dalling et al. 2011).

Seeds that form persistent banks in temperate agricultural habitats remain imbibed at least during some time after the last precipitation event. Despite this, dormancy and/or the lack of a light stimulus for dormancy termination prevents germination. The longevity of orthodox seeds (i.e. those that tolerate desiccation drying and consequently freezing) when stored can be increased by decreasing the storage temperature and/or the moisture content under which they are stored (Ellis and Roberts 1980). This low temperature and humidity condition during storage slows down the advancement of physiological deterioration and, consequently, the accumulation of membrane and genetic damage that leads to loss of vigor and, eventually, viability. This contrast between the conditions under which the seeds are "stored" in a persistent seedbank (i.e. high seed water content and high soil temperatures) and those known to maximize longevity in seeds stored dry (i.e. low water content and low temperatures) makes explanation of extended seed viability in persistent banks difficult. However, it must be considered that beyond a certain threshold, in terms of seed water content (threshold that is exceeded in imbibed seeds forming part of a persistent bank), longevity begins to increase (Ellis and Roberts 1980). With high seed water content, physiological deterioration (a process that in dry seeds does not depend on metabolism but is a-enzymatic) also happens, but it is repaired quickly by the metabolically active seed (repair does depend on metabolism and is enzymatic) (Villiers 1974) even if the imbibition is intermittent (Villiers and Edgecumbe 1975), as in natural seedbanks. In other words, in humid climates the seeds are continuously "repairing" themselves, and their longevity depends on this process occurring actively.

The adaptive advantage of persistent seedbanks is that they allow the risks associated with successful seedling emergence and survival, and therefore survival of the population, to be spread across longer time periods (i.e. years). Essentially, it is a mechanism for seed dispersal across time rather than space. From a practical weed control perspective, for example, persistent seedbanks ensure that weeds remain a challenge year after year even when weed management practices have been successful in preventing the establishment of new individuals and, consequently, the incorporation of new seeds to the bank. Transient banks, on the other hand, will exhaust the reservoir of propagules during the first year after dispersal; therefore, perpetuation of the population will depend on new seed inputs. That said, the seeds that form transient banks possess mechanisms to germinate and emerge over shorter time periods (i.e. intra-annual). Hence, even in species that form transient banks, emergence is not necessarily concentrated in a short time period but can occur at different times, something that in itself constitutes a characteristic with high adaptive value in diverse or variable habitats, including agricultural habitats. Moreover, these mechanisms allow these species to time their emergence in ways that maximize the probability of their survival.

Section 5.5 and its subsections will discuss in detail the mechanisms behind seed persistence and ways in which they are controlled by the environment.

5.5 Dormancy in Seedbanks and Its Control by the Environment

As mentioned previously, dormancy can be defined as an attribute of the seed that prevents germination under moisture, thermal, and gaseous conditions, which otherwise would be suitable for germination (Egley 1986; Benech-Arnold et al. 2000). This "impedance" to germination can be caused by morphological and/or physiological properties of the seed (Nikolaeva 1967). In addition, seed (or fruit) dormancy may be imposed on the embryo or by the seed/fruit coats. Dormancy imposed on the embryo is generally determined by the presence of endogenous inhibitors (particularly abscisic acid), while the seed or fruit coat can impose dormancy by different means: by restricting the release of inhibitors, the entry of water, or gaseous exchange, by mechanically restricting the expansion of the embryo, or through the presence of chemical inhibitors in the coat.

Dormancy can also be classified into primary and secondary dormancy (Karssen 1982). Primary dormancy refers to the dormant state of the seeds at dispersal, while secondary dormancy results from the re-induction of dormancy in the case of already dispersed seeds that had been previously released from primary dormancy or had attained a low dormancy level (Figure 5.2). Many weed seeds possess a high level of primary dormancy at dispersal

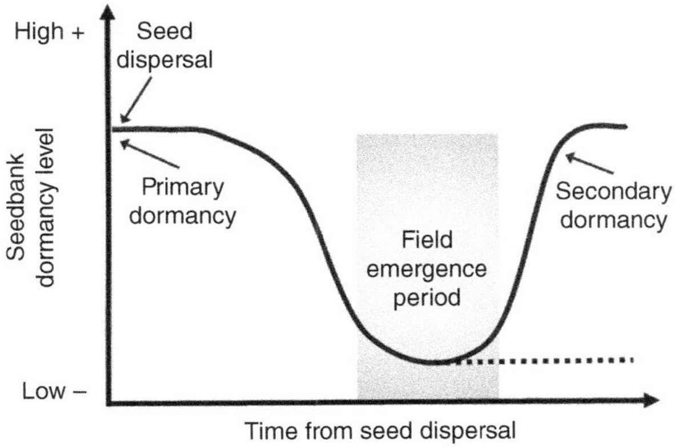

Figure 5.2 Schematic representation of changes in seedbank dormancy level after seed dispersal. The dotted line represents seeds that do not enter secondary dormancy, such as those exhibiting physical dormancy. *Source:* Adapted from Kruk (2002).

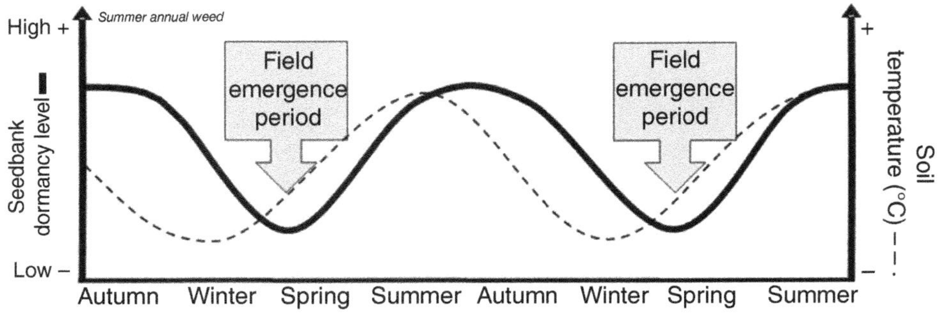

Figure 5.3 Schematic representation of cyclic seasonal changes in seedbank dormancy level for summer annual species. Solid line indicates seedbank dormancy level, and dotted line indicates soil temperature. *Source:* Adapted from Batlla and Benech-Arnold (2007).

from the mother plant. This high primary dormancy level decreases with time after dispersal, determining when environmental conditions are favorable for germination (Figure 5.2). If environmental conditions are not favorable for germination, the seeds may enter a state of secondary dormancy driven by the participation of some environmental feature that induce dormancy (i.e. nonfulfillment of light requirement in seeds from a summer annual, followed by exposure to high soil temperatures) (Bewley and Black 1982). In species possessing physical dormancy (for example, where the seed coats are impermeable to water), secondary dormancy is not possible, since once the impedance has been removed, it cannot be reimposed (Figure 5.2).

Exit from primary dormancy followed by subsequent re-inductions into secondary dormancy may result in cyclic changes in the level of dormancy in the seedbank. These cycles have been documented in seeds of many weed species under field conditions (Baskin and Baskin 1988). For example, in the case of many summer annual species, the level of dormancy of the seedbank usually decreases during winter, determining a minimum level of dormancy at the beginning of spring and then increasing again at the end of spring–beginning of summer (Figure 5.3). This mechanism ensures that the seasonal window of emergence takes place during spring, allowing the resultant plants to avoid the effect of winter frosts and thus ensuring the possibility of reproductive success. On the other hand, winter annual species generally show the reverse dormancy pattern: the dormancy level decreases during summer and increases during winter. These observations highlight the adaptive value of dormancy, determining the occurrence of germination under environmental conditions that will ensure the growth and perpetuation of the species under a given environment (Karssen 1982).

5.5.1 Environmental Factors that Modify the Dormancy Level of Seedbanks

The environmental factors that affect dormancy of buried seeds can be divided into two classes: those that regulate the level of dormancy and those that terminate dormancy (Benech-Arnold et al. 2000). The factors that regulate the level of dormancy of the seedbank are those responsible for the seasonal pattern of changes in the seed dormancy level, ensuring emergence takes place at the "right" season. These factors alter the dormancy level of seedbanks, producing seasonal changes in their germinative behavior through the alteration of the sensitivity of the seeds to environmental signals (Finch-Savage and Leubner-Metzger 2006).

Temperature has been referred to as the main factor regulating changes in dormancy level in seeds of many species (Koller et al. 1974; Karssen 1980–1981; Baskin and Baskin 1985, Batlla and Benech-Arnold 2010). However, the way temperature affects seed dormancy level can differ between species. In summer annuals, low winter temperatures decrease dormancy level, setting minimum dormancy at the onset of spring, while high summer temperatures increase dormancy level, leading to secondary dormancy (Figure 5.3). This type of behavior has been observed in many spring-emerging species, such as common lambsquarters (*Chenopodium album*), hedge mustard (*Sisymbrium officinale*), ladysthumb (*Polygonum persicaria*) (Bouwmeester 1990), prostrate knotweed (*Polygonum aviculare*) (Kruk and Benech-Arnold 1998) and common ragweed (*Ambrosia artemisiifolia*) (Baskin and Baskin 1980). In contrast, high summer temperatures usually trigger dormancy release in winter annual species, while low winter temperatures induce seed into secondary dormancy (Karssen 1982). This type of behavior was observed in many winter annual species, including shepherd's purse (*Capsella bursa-pastoris*) (Baskin and Baskin 1989), wild oat (Baskin and Baskin 1998), rigid ryegrass (*Lolium rigidum*) (Steadman et al. 2003), downy brome (*Bromus tectorum*) (Christensen et al. 1996), and corn gromwell (*Lithospermum arvense*) (Chantre et al. 2009). This effect of temperature is modulated by the seed moisture content, which in turn depends on the soil moisture content. While the process of dormancy release through low temperatures in summer annual species generally requires that the seeds are partially imbibed (commonly called stratification), the process of dormancy release through exposure of seeds to high temperatures, typical of winter annual species, usually requires that the seeds have a low level of moisture (generally called dry post-ripening). For example, in the case of downy brome, release from dormancy takes place within a specific range of seed moisture values (it does not occur in imbibed seeds or extremely dry seeds), and within that range, the level of hydration of the seeds modulates the effect of the temperature on the rate of dormancy release (Bair et al. 2006). Beyond these general response patterns, there are species in which, depending on their seed moisture content, low and high temperatures can both provoke dormancy release, although at different rates (i.e. *Arabidopsis thaliana*; Finch-Savage et al. 2007).

In many weed species, the termination of dormancy requires the action of certain specific environmental cues. From an ecological point of view, these factors indicate if "safe site" conditions are present for germination and establishment (Finch-Savage and Leubner-Metzger 2006). Among these dormancy termination factors, the most common in the field are light and alternating temperatures; nevertheless, there are many others that can elicit termination of dormancy under specific conditions, such as the concentrations of nitrates, ethylene, carbon dioxide, etc. in the soil (Benech-Arnold et al. 2000).

In relation to the effect of light, there is plenty of evidence showing that light is one of the main factors that terminates dormancy in some species, thus triggering germination of seeds from the seedbank (Bewley and Black 1982; Cone and Kendrick 1986; Scopel et al. 1991; Pons 1992; Batlla and Benech-Arnold 2014). Seeds perceive the light stimulus through photoreceptors, mainly those belonging to the phytochrome family. Phytochromes have two interconvertible forms: Pfr (considered the active form for germination) that presents its maximum absorption around 730 nm (far-red light) and Pr, which presents its maximum absorption around 660 nm (red light) (Borthwick et al. 1954). Phytochromes are synthesized in the seed in the Pr form, and the proportion that is converted to the active form (Pfr) depends on the light environment (spectral composition and irradiance) to which the seeds are exposed. For example, exposures to a light environment characterized by a high red/far-red ratio result in a high proportion of the molecule being in its active form (Pfr), which will trigger a cascade of events ending in dormancy termination.

Soil temperature fluctuations is another factor that terminates dormancy in seedbanks in the field (Bewley and Black 1982). In fact, many seeds require temperature fluctuations as a stimulus for dormancy termination and thus enabling germination. Several attributes of fluctuating temperature cycles that could be responsible for its stimulating effect have been identified. Among the most relevant is the thermal amplitude of the cycle. For example, common lambsquarters seeds display a significant increase in the proportion of seeds released from dormancy with increases in the amplitude of the cycle between 2.4 and 15 °C (Murdoch and Roberts 1997). On the other hand, it has been observed that the stimulating effect of the cycles is cumulative since an increase in the number of cycles determines the termination of dormancy in a greater proportion of the seed population. For example, 10 cycles of alternating temperatures of a defined composition (i.e. 15/35 °C) caused termination of dormancy in a proportion of a johnsongrass (*Sorghum halepense*) seed population that was twice the proportion released from dormancy when exposed to only five cycles of fluctuating temperatures (Benech-Arnold et al. 1990a,b). Similar results were reported by Batlla et al. (2003) for seeds of prostrate knotweed. From an ecological point of view, the requirement of thermal fluctuations and light to terminate dormancy and elicit germination has been related to the possibility of detecting depth of burial, soil disturbance, and vegetation gaps (Ghersa et al. 1992; Benech-Arnold et al. 2000; Thompson et al. 1977).

5.5.2 Dormancy Level Relates to the Width of the Environmental Range Permissive for Germination

Dormancy is not a qualitative characteristic of the seed; rather there is a continuous scale of dormancy levels (Vegis 1964; Bouwmeester and Karssen 1992; Vleeshouwers et al. 1995). Changes in dormancy level are expressed through changes in the range of environmental conditions that allow seed germination (Benech-Arnold et al. 2000; Batlla and Benech-Arnold 2010). Seeds with a high dormancy level would germinate under a narrow range of environmental conditions while seeds with a low dormancy level would germinate under a wide range of environmental conditions. Vegis (1964) observed that as the dormancy level of the seeds decreased, the temperature range within which germination could proceed broadened, whereas as seed dormancy level increased, the range of temperatures within which they could germinate narrowed. In most summer annual species, it has been reported that changes in the range of temperatures permissive for seed germination are mainly a result of changes in the lower temperature of this range (T_l) (minimum temperature at which dormancy is not expressed). In this case, dormancy release can be measured as an increase in the permissive thermal range for seed germination as a consequence of a decrease in T_l. On the contrary, during dormancy induction, there is a narrowing of this thermal range due to an increase in T_l. In the case of winter annual species, broadening of

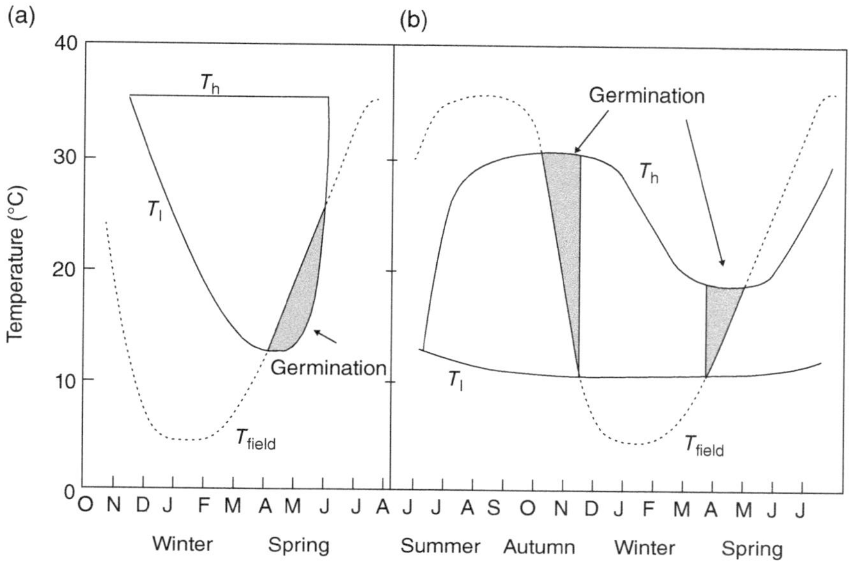

Figure 5.4 Seasonal changes in the thermal range permissive for seed germination and its relation with seasonal dynamics of soil temperature. Solid lines indicate lower (T_l) and higher (T_h) limit temperatures allowing germination. Broken lines show mean daily maximum temperature in the field. The shaded area represents the period when germination takes place due to overlapping of required and actual temperature. (a) Strict summer annual and (b) facultative winter species. *Source:* originally from Karssen (1982), adapted from Benech-Arnold et al. (2000).

the thermal range during dormancy release and narrowing during dormancy induction mainly results from an increase or decrease in the temperature of the upper limit (T_h) (maximum temperature at which dormancy is not expressed) of the thermal range permissive for germination, respectively (Kruk and Benech-Arnold 1998, 2000; Benech-Arnold et al. 2000; Batlla and Benech-Arnold 2003). Thus, in species that do not require the effect of dormancy terminating factors (i.e. light and/or alternating temperatures), the broadening of the range of permissive temperatures for germination as a result of dormancy release determines the germination of the seed population in the field when the soil temperature enters that range (Figure 5.4). If the latter does not occur, or if moisture conditions are insufficient, germination will be inhibited, and the population might be induced into secondary dormancy, depending on the prevailing thermal environment, thus restarting the cycle (Probert 1992).

It has been shown that changes in dormancy levels are not only related to changes in the permissive temperatures for germination, but also to changes in the sensitivity of the seeds to many other factors, for example, seed responses to water availability, light, and alternating temperatures. Working with the summer annual, prostrate knotweed, Batlla and Agostinelli (2017) showed that dormancy release and induction were associated with an increase and a decrease in the width of the range of water potential (Ψ) permissive for seed germination, respectively. These changes were related to changes in the base water potential (Ψ_b) for germination of the seed population. The concept of Ψ_b was originally proposed by Gummerson (1986) to describe the germination behavior of seeds to water availability, and can be defined as the water potential (Ψ_b (g)) below which a certain fraction (g) of the seed population would not germinate. Christensen et al. (1996) and Bauer et al. (1998) also observed a progressive decrease in Ψ_b during dormancy release in downy brome seeds. Therefore, analogously to what was mentioned previously about the thermal range permissive for germination, the dormancy level of a seed population could also be evaluated through changes in the range of water potentials permissive for germination (Bradford 1995; Batlla and Benech-Arnold 2004; Batlla et al. 2004; Allen et al. 2007).

Finally, in those species where dormancy requires the effect of light and/or alternating temperatures for its termination, changes in the level of dormancy can also be assessed through changes in the sensitivity to these dormancy terminating factors.

In relation to light, several authors have reported increases in the sensitivity of seeds to light during the release from dormancy and a decrease in this sensitivity as a result of induction of secondary dormancy (Derkx and Karssen 1993, 1994; Batlla and Benech-Arnold 2005, 2014). These results show that changes in dormancy level are related to parallel changes in the sensitivity to light. In certain species, this increase in the sensitivity to light is such that seeds that have attained a low dormancy level display an extremely high sensitivity to light and respond in a very-low-fluence response (VLFR) mode (Scopel et al. 1991). The acquisition of this type of response has been documented for soil seedbanks of a number of species after some months of burial and may lead to the induction of germination after very brief exposures to sunlight, such as those resulting from soil disturbance through tillage operations (Scopel et al. 1991; Botto 1998).

Studies conducted by Benech-Arnold et al. (1990a,b) in Johnsongrass and Batlla et al. (2003) in prostrate knotweed showed that, as in the case of light, dormancy changes in seeds can also be related to changes in sensitivity to the stimulatory effect of fluctuating temperatures. These can be expressed as changes in the proportion of the seed population capable of responding to the stimulus of temperature fluctuations and in the response to the different attributes that characterize the cycles (for example, thermal amplitude of the cycle and the number of cycles required to complete dormancy release).

5.5.3 Conceptual Model

On the basis of the concepts discussed previously, it can be concluded that the dormancy level of a seed population can be measured through the range of temperatures and water potentials permissive for seed germination and through the seed's sensitivity to the effect of the dormancy terminating factors (Figure 5.5). Therefore, seeds with a high level of dormancy will only germinate under a narrow range of temperature and/or water potential and will present low sensitivity to light and/or to alternating temperatures, or they may not germinate directly under any environmental conditions (i.e. full dormancy). On the other hand, seeds that have a low dormancy level will be able to germinate under a wide range of temperature and water potential and will present high sensitivity to light and/or to alternating temperature. As mentioned previously, changes in dormancy level (from high to low and vice versa) are regulated by soil temperature and modulated by the hydration level of the seeds, which in turn is determined by the soil moisture content. The germination and consequent emergence in the field will take place when the prevailing environmental conditions coincide with those required for germination. The latter will be determined by the level of dormancy of the seeds in the seedbank. It is important to note that in this

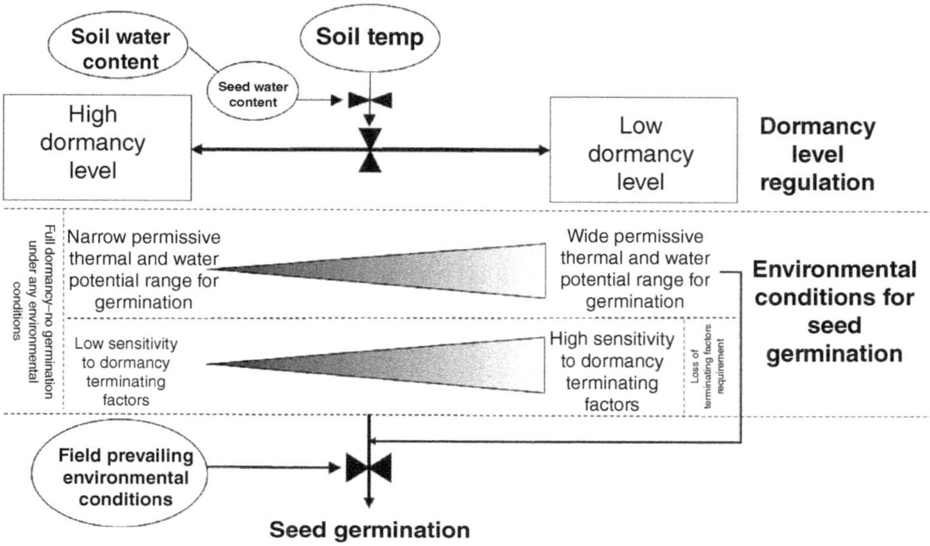

Figure 5.5 Flowchart representing most relevant environmental factors regulating dormancy level and changes in the range of environmental conditions for seed germination in soil seedbanks. *Source:* Karssen, C.M. (1982).

conceptual model the effects of temperature and its modulation by seed moisture content will be different for summer and winter annual species. In summer annual species the passage from a high to a low level of dormancy is mainly determined by exposure to low temperatures in partially imbibed seeds, and the reverse (passage from a low to a high level of dormancy) is determined by exposure to high temperatures. In winter annuals the passage from a high to a low level of dormancy is determined by exposure to high temperatures of partially dry seeds, while the passage from a low to a high level of dormancy is determined by exposure to low temperatures. This difference in relation to the effect of soil temperature in the regulation of dormancy levels is responsible for the different seasonal patterns of emergence observed in both types of weed species in the field. Finally, it should be taken into account that in species that do not require light or alternating temperatures to terminate dormancy, seed germination only depends on soil temperature and water potential to be within the range permissive for germination. In light-requiring species, or in those that need to perceive alternating temperatures to terminate dormancy, a low dormancy level could determine not only an increase in the sensitivity of the seeds to the effect of those factors, but even the loss of this requirement in a fraction of the population (i.e. germination in the dark or under constant temperatures).

5.6 Germination as Affected by Temperature and Water Availability

In seeds of weed species that do not possess dormancy, or in those that have been "released" from dormancy (where dormancy is not expressed because the prevailing environmental conditions are within the range of environmental conditions permissive for germination), temperature and water availability are the main environmental factors that control germination under conditions where oxygen is not limiting.

In the thermal range within which germination is possible, temperature modulates seed germination velocity or germination rate (Bewley and Black 1982; Washitani 1985; Batlla and Benech-Arnold, 2015). The germination rate, defined as the inverse of the time required to reach a certain percentage of germination, increases linearly with temperature from a base temperature (T_b; defined as that temperature below which the germination rate becomes zero) until it reaches a maximum at a temperature that is consequently defined as optimum temperature (T_o). From T_o onward, germination rate decreases linearly with temperature until it becomes zero at the maximum temperature (T_m; defined as the temperature above which the germination rate becomes zero) (Figure 5.6). The range between T_b and T_o is considered the suboptimal thermal range, while the range between T_o and T_m is considered the supraoptimal thermal range. The inverse of the slope of the relationship between the germination rate and temperature in both ranges (sub- and supraoptimal) is regarded as the required thermal time for germination of fraction g ($\theta_T(g)$) with units being degree days (°C day) or degree hours (°C hour). Figure 5.6 shows the relationship between temperature and the germination rate for 50% of the seed population. The inverse of the slope would be the θ_T required for 50% of the population to germinate ($\theta_T(50)$). Different fractions of the population (10%, 20%, 30%, etc.) will need to accumulate different values of θ_T to germinate as the relationship between temperature and germination rate for each fraction has a different slope (not shown in this graph; see Bradford (2002) for a more detailed scheme).

Parameters T_b, T_o, T_m, and θ_T, defined previously, can be used to describe or eventually predict the dynamics of germination of a seed population as a function of time and temperature within the thermal range in which germination is possible using the thermal time

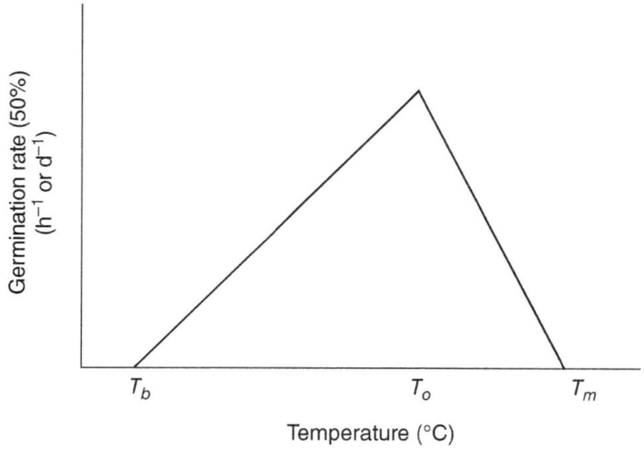

Figure 5.6 Schematic representation of the relationship between germination rates (GRg = 1/tg) and temperature at the suboptimal and the supraoptimal thermal range for the 50% seed fraction. Cardinal temperatures for seed germination are indicated in *x*-axis as T_b (base temperature), T_o (optimum temperature), and T_m (maximum temperature).

model (Washitani 1987; Benech-Arnold and Sánchez 1995). For example, in the case of the suboptimal thermal range for seed germination (between T_b and T_o), if we know the values of T_b and $\theta_T(50)$ for germination of a particular weed species, we can predict the time required to achieve 50% of germination using the following equation:

$$\theta_T(50) = (T_s - T_b)t_{(50)} \tag{5.1}$$

where T_s is the soil temperature and $t_{(50)}$ is the time for germination of 50% of the seed population. The time for germination of other seed fractions can also be calculated knowing their corresponding θ_T. This thermal time approach was successfully used to analyze and model the germinative response and emergence of different species as a function of time and temperature (Garcia-Huidobro et al. 1982; Covell et al. 1986; Ferrari and Leguizamón 2006; Leguizamón et al. 2009). A similar model can be used to predict germination as a function of time and temperature in the supraoptimal thermal range (between T_o and T_m); for details see Bradford (2002).

As mentioned previously, another factor that has a marked influence on germination is water availability. Variations in soil water content can affect the percentage of germinating seeds and germination velocity. The response of the seeds to water availability can be analyzed in a similar way to that as mentioned before for the analysis of temperature effects on seed germination. So analogously to the thermal time model, seed germination response to water availability can be described or predicted using the hydrotime model equation:

$$\theta_H = (\Psi_s - \Psi_b(50))t_{(50)}, \tag{5.2}$$

where (θ_H) is the hydrotime (MPa-hours or MPa-days) that is required for the seed population to germinate, Ψ_s is the soil water potential in MPa, $\Psi_b(50)$ is the average water potential of the seed population in MPa (the water potential of the incubation medium below which 50% of the seed population would not germinate), and $t_{(50)}$ is the time required for 50% of the seed population to germinate.

Analogous to the thermal time model, in the hydrotime model seeds require a certain amount of hydrotime to germinate, and that amount of hydrotime is accumulated above a threshold value called base water potential (Ψ_b). Unlike the temperature model, in which the seeds accumulate thermal time above a common base temperature for the entire seed population and the thermal time required for germination is different for each fraction of the population, in the hydrotime model the Ψ_b above which the seeds accumulate hydrotime is distributed in the population (usually following a normal distribution), while the amount of hydrotime required to germinate is equal for all the seeds of the population.

A third model, called hydrothermal model, was developed based on the two models explained previously (the thermal time [Eq. (5.1)] and the hydrotime [Eq. (5.2)]). The hydrothermal time model allows description of the joint effects of temperature and water potential on seed germination (Gummerson 1986; Bradford 1995).

5.7 The Functional Ecology of Weed Seedbanks: Concluding Remarks

From the previous sections, it can be concluded that the dynamics of weed seedbanks result from a complex interaction among seed attributes (i.e. morphology, physiology, and chemical composition), the surrounding biota (i.e. predators and microflora), and the environment. Either persistency or transiency will result from the outcome of this complex, fine-tuned interaction. It is clear that without dormancy, seed persistence is not possible. For that reason, a thorough understanding of dormancy in seedbanks and the way in which changes in dormancy are driven by the environment is essential for predicting the behavior of weed populations and their dynamics. However, dormancy in seeds without effective mechanisms for sensing their environment in the short term (i.e. light requirement for dormancy termination) would be useless as a means for seedbank persistence, even in the case of seeds with efficient devices for burial. Similarly, dormancy in seeds that lack defenses against predators or microflora attack, or with deficient repair mechanisms to ensure seed longevity, would preclude persistence.

But on the other hand, persistence is by no means a necessary condition for successful plant population establishment or invasion. Indeed, examples of very noxious and problematic weeds with transient seedbanks are abundant in the literature. Prolific reproductive output coupled with very efficient dispersal and germination mechanisms to optimize emergence and dispersion in time and space may be as effective for invasion as any persistent seedbank. Such an efficiency in germination can be assessed from the value of those parameters that can be derived from the thermal time and the hydrotime analysis as discussed in previous sections.

Overall, it is very difficult to allocate a seedbank in an air-tight category regarding its persistence or transience. However, the assessment of the different aspects that have been discussed in this chapter can give us clues about seedbank functioning in species of interest. This knowledge can be used, not only to manage weed seedbanks toward a weed population size compatible with sustainable agricultural production, but also to forecast weed persistence amidst changes in production systems. An emblematic current example of this kind of change is the switch from conventional tillage to no-tillage. As a result of this switch, some species exhibited reduced importance as weeds while others became very noxious. This scenario could have been predicted from seedbank dynamics considering known attributes such as seed dormancy (and its environmental control), seed longevity, and seed defense as discussed throughout this chapter.

References

Allen, P.S., Benech-Arnold, R.L., Batlla, D., and Bradford, K.J. (2007). Modeling of seed dormancy. In: *Seed Development, Dormancy and Germination*, vol. 27 (eds. K. Bradford and H. Nonogaki), 72–112. Oxford (RUGB): Blackwell Publishing.

Bair, N.B., Meyer, S.E., and Allen, P.S. (2006). A hydrothermal after–ripening time model for seed dormancy loss in *Bromus tectorum* L. *Seed Science Research* 16: 17–28.

Baskin, C.C. and Baskin, J.M. (1980). Ecophysiology of secondary dormancy in seeds of *Ambrosia artemisiifolia*. *Ecology* 61: 475–480.

Baskin, C.C. and Baskin, J.M. (1985). The annual dormancy cycle in buried weed seeds: a continuum. *BioScience* 35: 492–498.

Baskin, C.C. and Baskin, J.M. (1988). Germination ecophysiology of herbaceous plant species in temperate region. *American Journal of Botany* 75: 286–305.

Baskin, J.M. and Baskin, C.C. (1989). Germination responses of buried seeds of *Capsella bursa-pastoris* exposed to seasonal temperature changes. *Weed Research* 29: 205–212.

Baskin, C.C. and Baskin, J.M. (1998). *Seed Dormancy and Germination. Ecology, Biogeography and Evolution*, 666 pp. San Diego, CA (EEUU): Academic Press.

Batlla, D. and Agostinelli, A. (2017). Thermal regulation of secondary dormancy induction in *Polygonum aviculare* seeds: a quantitative analysis using the hydrotime model. *Seed Science Research* 27: 231–242.

Batlla, D. and Benech-Arnold, R.L. (2003). A quantitative analysis of dormancy loss dynamics in *Polygonum aviculare* L. seeds. Development of a thermal time model based on changes in seed population thermal parameters. *Seed Science Research* 13: 55–68.

Batlla, D. and Benech-Arnold, R.L. (2004). Seed dormancy loss assessed by changes in *Polygonum aviculare* L. population hydrotime parameters. Development of a predictive model. *Seed Science Research* 14: 277–286.

Batlla, D. and Benech-Arnold, R.L. (2005). Changes in the light sensitivity of buried *Polygonum aviculare* seeds in relation to cold-induced dormancy loss: development of a predictive model. *New Phytologist* 165: 445–452.

Batlla, D. and Benech-Arnold, R.L. (2007). Predicting changes in dormancy level in weed seed soil banks: implications for weed managament. *Crop Protection* 26: 189–197.

Batlla, D. and Benech-Arnold, R.L. (2010). Predicting changes in dormancy level in natural seed soil banks. *Plant Molecular Biology* 73: 3–13.

Batlla, D. and Benech-Arnold, R.L. (2014). Weed seed germination and the light environment. Implications for weed management. *Weed Biology and Management* 14: 77–87.

Batlla, D. and Benech-Arnold, R.L. (2015). A framework for the interpretation of temperature effects on dormancy and germination in seed populations showing dormancy. *Seed Science Research* 25: 147–148.

Batlla, D., Verges, V., and Benech-Arnold, R.L. (2003). A quantitative analysis of seed responses to cycle-doses of fluctuating temperatures in relation to dormancy level. Development of a thermal-time model for *Polygonum aviculare* L. seeds. *Seed Science Research* 13: 197–207.

Batlla, D., Kruk, B.C., and Benech-Arnold, R.L. (2004). Modelling changes in dormancy in weed soil seed banks: implications for the prediction of weed emergence. In: *Handbook of Seed Physiology: Applications to Agriculture* (eds. R.L. Benech-Arnold and R.A. Sánchez), 245–264. New York, USA: Haworth Press.

Bauer, M.C., Meyer, S.E., and Allen, P.S. (1998). A simulation model to predict seed dormancy loss in the field for *Bromus tectorum* L. *Journal of Experimental Botany* 49: 1235–1244.

Benech-Arnold, R.L. and Sánchez, R.A. (1995). Modelling weed seed germination. In: *Seed Development and Germination* (eds. J. Kigel and G. Gallili), 545–566. New York, USA: Marcel Dekker Inc.

Benech-Arnold, R.L., Ghersa, C.M., Sanchez, R.A., and Insausti, P. (1990a). A mathematical model to predict *Sorghum halepense* (L.) Pers. seedling emergence in relation to soil temperature. *Weed Research* 30: 91–99.

Benech-Arnold, R.L., Ghersa, C.M., Sanchez, R.A., and Insausti, P. (1990b). Temperature effects on dormancy release and germination rate in *Sorghum halepense* (L.) Pers. seeds: a quantitative analysis. *Weed Research* 30: 81–89.

Benech-Arnold, R.L., Sánchez, R.A., Forcella, F. et al. (2000). Environmental control of dormancy in weed seed soil banks. *Field Crops Research* 67: 105–122.

Benech-Arnold, R.L., Batlla, D., Guglielmini, A.C., and Kruk, B.C. (2014). Ecología de Malezas II: la reanudación del crecimiento y el aumento del área. In: *Malezas e Invasoras de la Argentina: su identificación, biología y manejo*, Tomo I: Ecología y Manejo (eds. O. Fernández, E. Leguizamón and H. Aciaressi), 139–169. Editorial Universidad Nacional del Sur.

Bewley, J.D. and Black, M. (1982). *Physiology and Biochemistry of Seeds in Relation to Germination*, 375 pp, vol. 2. Berlin, Germany: Springer Verlag.

Bliss, D. and Smith, H. (1985). Penetration of light into soil and its role in the control of seed germination. *Plant, Cell and Environment* 8: 475–483.

Borthwick, H.A., Hendricks, S.B., Toole, E.H., and Toole, V.K. (1954). Action of light on lettuce seed germination. *Botanical Gazette* 115: 205–225.

Botto, J.F. (1998). La germinación de las semillas por la luz y su relación con la emergencia de plántulas de malezas, 110 pp. PhD thesis. Universidad de Buenos Aires, Facultad de Ciencias Exactas y Naturales.

Bouwmeester, H.J. (1990). The effect of environmental conditions on the seasonal dormancy pattern and germination of weed seeds. 157 pp. PhD thesis. Agricultural University, Wageningen, The Netherlands.

Bouwmeester, H.J. and Karssen, C.M. (1992). The dual role of temperature in the regulation of the seasonal changes in dormancy and germination of seeds of *Polygonum persicaria* L. *Oecologia* 90: 88–94.

Bradford, K.J. (1995). Water relations in seed germination. In: *Seed Development and Germination* (eds. J. Kigel and A. Galili), 351–396. New York, USA: Marcel Dekker Inc.

Bradford, K.J. (2002). Applications of hydrothermal time to quantifying and modeling seed germination and dormancy. *Weed Science* 50: 248–260.

Chantre, G., Batlla, D., Sabbatini, M., and Orioli, G. (2009). Germination parameterization and development of an after-ripening thermal-time model for primary dormancy release of *Lithospermum arvense* seeds. *Annals of Botany* 103: 1291–1301.

Christensen, M., Meyer, S., and Allen, P.S. (1996). A hydrothermal time model of seed after-ripening in *Bromus tectorum* L. *Seed Science Research* 6: 147–153.

Cone, J.W. and Kendrick, R.E. (1986). Photocontrol of seed germination. In: *Photomorphogenesis in Plants* (eds. R.E. Kendrick and G.H.M. Kronenberg), 443–465. Dordrecht, The Netherlands: Martinus Nijhoff.

Covell, S., Ellis, R.H., Roberts, E.H., and Summerfield, R.J. (1986). The influence of temperature on seed germination rate in grain legumes.1. A comparison of chickpea, lentil, soybean and cowpea at constant temperatures. *Journal of Experimental Botany* 37: 705–715.

Dalling, J.W. and Hubbell, S.P. (2002). Seed size, growth rate and gap microsite conditions as determinants of recruitment success for pioneer species. *Journal of Ecology* 90: 557–568.

Dalling, J.W., Davis, A.S., Schutte, B.J., and Arnold, A.E. (2011). Seed survival in soil: interacting effects of predation, dormancy and the soil microbial community. *Journal of Ecology* 99: 89–95.

Davis, A.S., Schutte, B.J., Iannuzzi, J., and Renner, K.A. (2008). Chemical and physical defenses of weed seeds in relation to soil seedbank persistence. *Weed Science* 56: 676–684.

Derkx, M.P.M. and Karssen, C.M. (1993). Changing sensitivity to light and nitrate but not to gibberellins regulates seasonal dormancy patterns in *Sisymbrium officinale* seeds. *Plant, Cell and Environment* 16: 469–479.

Derkx, M.P.M. and Karssen, C.M. (1994). Are seasonal dormancy patterns in *Arabidopsis thaliana* regulated by changes in seed sensitivity to light, nitrate and gibberellin? *Annals of Botany* 73: 129–136.

Egley, G. (1986). Stimulation of weed seed germination in soil. *Reviews of Weed Science* 2: 67–89.

Ellis, R.H. and Roberts, E.H. (1980). Improved equations for the prediction of seed longevity. *Annals of Botany* 45: 13–30.

Fenner, M. and Thompson, K. (2005). *The Ecology of Seeds*, 250 pp. Cambridge, Reino Unido: Cambridge University Press.

Ferrari, G. and Leguizamón, E.S. (2006). Requerimientos germinativos y modelización de la emergencia de plántulas de *Portulaca oleracea* L. ("Verdolaga"). U.N.R. Editora. *Revista de Investigaciones de la Facultad de Ciencias Agrarias* 9: 23–33.

Finch-Savage, W.E. and Leubner-Metzger, G. (2006). Seed dormancy and the control of germination. *New Phytologist* 171: 505–523.

Finch-Savage, W.E., Cadman, C.S.C., Toorop, P.E. et al. (2007). Seed dormancy release in Arabidopsis Cvi by dry after-ripening, low temperature, nitrate and light shows common quantitative patterns of gene expression directed by environmentally specific sensing. *The Plant Journal* 51: 60–78.

Garcia-Huidobro, J., Monteith, J.L., and Squire, G.R. (1982). Time, temperature and germination of pearl millet (*Pennisetum typhoides* S & H). I. Constant temperature. *Journal of Experimental Botany* 33: 288–296.

Ghersa, C.M., Benech-Arnold, R.L., and Martinez Ghersa, M.A. (1992). The role of fluctuating temperatures in germination and establishment of *Sorghum halepense* (L.) Pers. II. Regulation of germination at increasing depths. *Functional Ecology* 6: 460–468.

Gummerson, R.J. (1986). The effect of constant temperatures and osmotic potentials on the germination of sugar beet. *Journal of Experimental Botany* 37: 729–741.

Hendry, G.A.F., Thompson, K., Moss, C.J. et al. (1994). Seed persistence: a correlation between seed longevity in the soil and *ortho*-dihydroxyphenol concentration. *Functional Ecology* 8: 658–664.

Hulme, P.E. (1998). Post-dispersal seed predation and seed bank persistence. *Seed Science Research* 8: 513–519.

Karssen, C.M. (1980–1981). Patterns of change in dormancy during burial of seeds in soil. *Israel Journal of Botany* 29: 65–73.

Karssen, C.M. (1982). Seasonal patterns of dormancy in weed seeds. In: *The Physiology and Biochemistry of Seed Development, Dormancy and Germination* (ed. A.A. Khan), 243–270. Amsterdam, The Netherlands: Elsevier.

Koller, D., Kigel, J., Ofir, M., et al. (1974). Environmental control of weed physiology. USDA. PL 480. Project Number: A10–CR–80, Grand Number: FGIs–277. Final Report.

Kruk, B. (2002). Análisis funcional y cuantitativo de la emergencia de especies maleza en el cultivo de trigo, 190 pp. PhD tesis. Escuela para graduados Alberto Soriano, Facultad de Agronomía, Universidad de Buenos Aires, Argentina.

Kruk, B.C. and Benech-Arnold, R.L. (1998). Functional and quantitative analysis of seed thermal responses in prostate knotweed (*Polygonum aviculare*) and common purslane (*Portulaca oleracea*). *Weed Science* 46: 83–90.

Kruk, B.C. and Benech-Arnold, R.L. (2000). Evaluation of dormancy and germination responses to temperature in *Carduus acanthoides* and *Anagallis arvensis* using a screening system, and relationship with field-observed emergence patterns. *Seed Science Research* 10: 77–88.

Kruk, B., Insausti, P., Razul, A., and Benech-Arnold, R.L. (2006). Light and thermal environments as modified by a wheat crop: effects on weed seed germination. *Journal of Applied Ecology* 43: 227–236.

Leck, M.A. (1989). Wetland seed banks. In: *Ecology of Soil Seed Banks* (eds. M.A. Leck, V.T. Parker and R.L. Simpson), 283–305. San Diego (EEUU): Academic Press.

Leguizamón, E.S., Rodríguez, N., Rainero, H. et al. (2009). Modelling the emergence pattern of six summer annual weed grasses under no tillage systems in Argentina. *Weed Research* 49: 98–106.

Murdoch, A.J. and Roberts, E.H. (1997). Temperature and the rate of germination of dormant seeds of *Chenopodium album*. In: *Basic and Applied Seed Biology* (eds. R.H. Ellis, M. Black, A.J. Murdoch and T.D. Hong), 547–553. Dordrecht, The Netherlands: Kluwer Academic Publishers.

Nikolaeva, M.G. (1967). *Physiology of Deep Dormancy in Seeds*. Washington, USA: National Science Foundation.

Pakeman, R.J., Digneffe, G., and Small, J.L. (2002). Ecological correlates of endozoochory by herbivores. *Functional Ecology* 16: 296–304.

Pérez-Garcia, F., Ceresuela, J.L., Gonzalez, A.E., and Aquinagalde, I. (1992). Flavonoids in seed coats of *Medicago arbórea* and *M. strasseri* (Leguminosae): ecophysiological aspects. *Journal of Basic Microbiology* 32: 241–248.

Pons, T.L. (1992). Seed responses to light. In: *Seeds. The ecology of regeneration in plant niches. C.A.B* (ed. M. Fenner), 259–284. Wallingford, UK: International.

Probert, R.J. (1992). The role of temperature in germination ecophysiology. In: *Seeds. The Ecology of Regeneration in Plant Communities* (ed. M. Fenner), 285–325. Wallingford, UK: C.A.B. International.

Scopel, A.L., Ballaré, C.L., and Sánchez, R.A. (1991). Induction of extreme light sensitivity in buried weed seeds and its role in the perception of soil cultivation. *Plant, Cell and Environment* 14: 501–508.

Scursoni, J.A., Benech-Arnold, R.L., and Hirchoren, H. (1999). Demography of wild oat (*Avena fatua* L.) in barley crops: effect of crop, sowing rate and herbicide treatment. *Agronomy Journal* 91: 478–485.

Siemens, D.H., Johnson, C.D., and Ribardo, K.J. (1992). Alternative seed defense mechanisms in congeneric plants. *Ecology* 73: 2152–2166.

Steadman, K.J., Crawford, A.D., and Gallagher, R.S. (2003). Dormancy release in *Lolium rigidum* seeds is a function of thermal after-ripening time and seed water content. *Functional Plant Biology* 30: 345–352.

Thompson, K. and Grime, J.P. (1979). Seasonal variation in the seed banks of herbaceous species in ten contrasting habitats. *Journal of Ecology* 67: 893–921.

Thompson, K., Grime, J.P., and Mason, G. (1977). Seed germination in response to diurnal fluctuations of temperature. *Nature* 267: 147–149.

Thompson, K., Band, S.R., and Hodgson, J.G. (1993). Seed size and shape predict persistence in soil. *Functional Ecology* 7: 236–241.

Thompson, K., Bakker, J.P., and Bekker, R.M. (1997). *The Soil Seed Banks of North West Europe: Methodology, Density and Longevity*, 276 pp. Cambridge, UK: Cambridge University Press.

Tsuyuzaki, S. (1991). Survival characteristics of buried seeds 10 years after the eruption of the Usu volcano in northern Japan. *Canadian Journal of Botany* 69: 2251–2256.

Vegis, A. (1964). Dormancy in higher plants. *Annual Review of Plant Physiology* 15: 185–224.

Villiers, T.A. (1974). Seed aging: chromosome stability and extended viability of seeds stored fully imbibed. *Plant Physiology* 53: 875–878.

Villiers, T.A. and Edgecumbe, D.J. (1975). On the cause of seed deterioration in dry storage. *Seed Science and Technology* 3: 761–774.

Vleeshouwers, L.M., Bouwmeester, H.J., and Karssen, C.M. (1995). Redefining seed dormancy: an attempt to integrate physiology and ecology. *Journal of Ecology* 83: 1031–1037.

Washitani, I. (1985). Germination-rate dependency on temperature of *Geranium carolinianum* seeds. *Journal of Experimental Botany* 36: 330–337.

Washitani, I. (1987). A convenient screening test system and a model for thermal germination responses of wild plant seeds: behaviour of model and real seed in the system. *Plant, Cell & Environment* 10: 587–598.

6

Longevity of Weed Seeds in Seedbanks

Ali A. Bajwa[1], Fernanda C. Beveridge[2], Mahesh K. Upadhyaya[3], and Steve W. Adkins[2]

[1] *Weed Research Unit, New South Wales Department of Primary Industries, Wagga Wagga, NSW, Australia*
[2] *School of Agriculture and Food Sciences, The University of Queensland, Brisbane, QLD, Australia*
[3] *Faculty of Land and Food Systems, University of British Columbia, Vancouver, BC, Canada*

6.1 Introduction

The unique and resilient seed characteristics are among the most important features of weed species in terms of their establishment, competitive ability, and spread in a range of environments. Seed persistence and seed longevity are key features of weed seedbanks. Seed persistence is the ability of a seed to survive from the time it is formed until the time it is lost either through germination, predation, decay, or its age-related death. Seed persistence is influenced by several traits including dormancy and longevity and the interplay of these traits with the environment. In contrast, seed longevity refers to the time span during which a seed remains alive and is determined by a multifaceted expression of physiological characteristics involving those that are related to cell metabolism, protection, and repair. Many studies have shown that these two attributes of seeds, persistence and longevity, are closely associated. Although it is difficult to discuss these two traits independently, in this chapter we primarily emphasize the role of seed longevity in weed seedbank development. The topic of seed dormancy is covered in Chapter 3 of this book.

From an ecological point of view, seed longevity in combination with seed dormancy permits weeds to disperse their seeds through time, while evading germination under certain conditions, which is an important trait particularly of weed species. Thus, a knowledge of the mechanisms that control longevity and the ability to predict the accurate duration of weed seed longevity is important in formulating decisions for prioritizing which weed species to manage and the time duration that the management needs to be applied to substantially deplete the weed seedbank.

This chapter will throw some light on the plant seed and the environmental factors (including climatic and soil factors) that contribute to seed longevity in the soil seedbank. This discussion of longevity will also briefly consider the roles of seed vigour and

Persistence Strategies of Weeds, First Edition. Edited by Mahesh K. Upadhyaya, David R. Clements, and Anil Shrestha.

dormancy, and will evaluate the role of physiological, biochemical, and ecological factors. The focus is primarily on seeds existing in soil seedbanks; however, seed longevity will also be discussed when addressing *ex situ* seed conservation.

Our objective is to provide a comprehensive account of biological phenomena regulating the longevity of weed seeds in seedbanks in relation to a range of internal and external factors. This knowledge will improve our understanding of seed longevity in association with weed seedbank dynamics. The discussion supported by various examples from a range of ecological and management scenarios will provide the reader with a thorough understanding of the key biological relationships relevant to seedbank longevity and their management implications. Based on information in the literature and our own analysis, we also provide some suggestions on potential future research needs in the area of weed seed longevity and seedbank management.

6.2 Seeds and Seedbanks as Survival Mechanisms

The success of many weed species depends on their ability to rapidly infiltrate and establish successfully in new environments, and this is facilitated by their mode of reproduction. For many weed species, sexual reproduction by seed aids spread and survival in the invaded environment. Weed species may vary in the amount of seed produced, the size and quality of their seed, and the anatomical features that aid seed dispersal (Naylor 2003). Seedbank size varies greatly across weed species and geographic and/or climatic conditions (Table 6.1).

Table 6.1 The size of the soil weed seedbanks in arable and rangelands selected from several countries.

Location	Depth sampled	Seeds m^{-2}	Sampling technique
Australia Rice field	10	646 000	Soil sample germination
Denmark Agricultural field	20	496 000	Seed extraction
United States of America Irrigated cereal crops	25	137 700	Seed extraction
England Agricultural fields, cereals, and vegetables	10	65 580	Seed extraction
Argentina Agricultural fields, oats, sunflower, and maize	7	57 750	Seed extraction
Australia Weedy pasture	3.5	44 639	Soil sample germination
Australia Buffel grass pasture	3.5	5 094	Soil sample germination

Soil samples were collected after the weed seed dispersal and seedbanks were determined by either seed extraction or seedling grow-out methods
Source: Modified from Baskin and Baskin (1998), and other sources.

Therefore, it would be very useful to know the actual composition of the soil seedbank in a landscape requiring weed management. Apart from increased dispersal, reproduction via seeds confers other survival advantages. Sexual reproduction results in greater genetic diversity conferring a higher predisposition to local adaptation and aid in survival. Dormancy coupled with longevity allows seeds to persist a long time, even after they have been shed from the parent plant. This allows germination to occur in different cohorts over an extended period of time (Baskin and Baskin 2006; Long et al. 2015). When seeds are disseminated from the parent plant, it is common for a proportion of the seeds to be buried in the soil and become part of the seedbank that can sporadically produce seedlings over the coming years. If these seeds remain in the seedbank, they maintain their dormant state (Long et al. 2015). Over time, seeds from different generations accumulate in the soil, and although capable, they do not necessarily germinate (i.e. they persist). The result of this process is a long-term soil seedbank, which is a dynamic entity receiving regular deposits (further seed rain) and withdrawals (germination or death) over time (see Chapters 2 and 5 of this book for further details).

6.3 Role of Seed Longevity in Seedbank Regulation

Seed longevity is regulated by several physiological traits related to cell metabolism, protection, and repair. It is a distinct trait and contributes directly to seed persistence, which, as defined before, is the life expectancy of seeds in their natural environment. Due to the highly variable nature, the longevity of seeds of different species, and the mechanisms that underpin it, is relatively poorly understood. However, seed longevity is known to be influenced by conditions during seed maturation and during seed fall (Hay et al. 2006), and the progression of biochemical 'ageing' reactions within the seeds due to moisture and temperature impacts (Walters et al. 2005). While some examples of extreme longevity do exist, in typical seedbanks longevity is much shorter than seen in these examples.

An important character of any seedbank is the period over which its seeds can remain viable. Evidence of remarkable seed longevity is shown by studies undertaking germination tests on seeds of known age collected from sites, which include archaeological sites, from dated herbarium specimens, or from buried seed experiments. There have been some interesting claims for the extended longevity of buried seeds. For instance, an extreme case is of Arctic lupin (*Lupinus arcticus*) viable seeds discovered in the Yukon Territory, Canada, in an ancient lemming burrow dating back to the Pleistocene (>12 000 years ago). The seeds were able to germinate and produce plants, causing them to be proclaimed as the oldest viable seeds ever discovered (Porsild et al. 1967). Uncertainty surrounds the claims for the longevity of seeds of the East Indian lotus (*Nelumbo nucifera*). The hard-coated fruits of this species were recovered from an ancient lakebed peat deposit at Pulantien in southern Manchuria. The initial radiocarbon study dated some of the seeds to >1000 years old. However, more recent radiocarbon dating of a single viable seed suggests a probable age of about 466 years (Priestley and Posthumus 1982). More recently Russian scientists have grown flowering plants from seed apparently stored by Arctic ground squirrels more than 30 000 years ago and safeguarded by the Siberian permafrost. Embryos isolated from the frozen seeds of the narrow-leaved campion (*Silene stenophylla*) were successfully grown into reproductively active plants producing their own viable seed (Yashina et al. 2012). In each of these three examples, environmental conditions were involved, which enabled the seeds to remain in a dormant state that unnaturally prolonged their viability, whereby the establishment of seedlings would be very unlikely. Studies on seeds that have been exposed to the temperature, moisture, and gas composition typically found in soils under field conditions will clearly be of more relevance to understanding longevity for practical purposes.

6.4 Classical Ecological Experiments on Weed Seed Longevity

Seed longevity for a species under field conditions can be determined by putting into the ground a known number of seeds into the soil and monitoring their germinability over time. It has been shown that the numbers of seeds remaining viable tends to fall-off exponentially with time. The longest study on the longevity of seeds under field conditions is that of William Beal. In 1879, Beal chose seeds of 23 widely found species in East Lansing, Michigan. He placed 50 seeds of each species into moist sand contained within open narrow-necked glass bottles and buried them 46 cm deep with the mouth slanting downward in a sandy knoll. At regular intervals, bottles were recovered and the viability of the seeds was determined by germination. The year 2000 marked the 120-year anniversary of Beal's seed viability study (Telewski and Zeevaart 2002) (Table 6.2). Under the germination conditions employed, only seeds of moth mullein (*Verbascum blattaria*) and common mullein (*Verbascum thapsus*), a common weed throughout the United States remained viable after 120 years of burial. The next most long-lived species was common mallow (*Malva sylvestris*). Only one of those seeds germinated in 1980 after 100 years of burial, although many species did survive for 25–40 years, and a limited number of seeds of common evening primrose (*Oenothera biennis*) and curly dock (*Rumex crispus*) remained viable for 80 years. Thus, longevity for buried seeds maintained under semi-natural conditions is much shorter than for those with viability unnaturally prolonged. Longevity assessment using methodology involving burying of seeds in bags or containers could overestimate seed longevity because seeds may be protected against natural processes or it may be underestimated if seed viability is assessed by measuring germination and dormant seeds are classified as nonviable (Masin et al. 2006). Higher than natural density of seeds in bags or containers could also provide incorrect estimation of seed longevity by affecting seed-to-seed contamination by pathogens (Von Mourik et al. 2005).

One of the largest experiments undertaken on the longevity of seeds under field conditions was that of Joseph Duvel. In 1902 in Arlington, Virginia, Duvel started an experiment in which he buried batches of seeds mixed into sterilized soil and placed them into covered flower pots buried at three depths (20, 56, and 107 cm) in soil. Of the 107 species buried, 36 survived for 39 years, with over 80% survival of black nightshade (*Solanum nigrum*), jimsonweed (*Datura stramonium*), and pokeweed (*Phytolacca americana*). There was negligible survival of 22 crop species even for one year; whereas seeds of the small-seeded species, viz. tobacco (*Nicotiana tabacum*), celery (*Apium graveolens*), red clover (*Trifolium pratense*), and Kentucky bluegrass (*Poa pratensis*), persisted for 39 years to the end of the experiment (Toole and Brown 1946). There was a general tendency for seeds buried at the greater

Table 6.2 Results of Beal's buried seed study.

Time elapsed (years)	Weed species		Number of seeds retrieved	Germination % of seeds first sown
	Common name	Scientific name		
120	Moth mullein	*Verbascum blattaria*	23	46
	Common mullein	*Verbascum thapsus*	2	4
100	Moth mullein	*Verbascum blattaria*	21	42
	Common mullein	*Verbascum thapsus*	1	2
	Common mallow	*Malva sylvestris*	1	2
90	Moth mullein	*Verbascum blattaria*	10	20
80	Evening primrose	*Oenothera biennis*	5	10
	Curly dock	*Rumex crispus*	1	2
	Moth mullein	*Verbascum blattaria*	35	70
70	Evening primrose	*Oenothera biennis*	7	14
	Curly dock	*Rumex crispus*	4	8
	Moth mullein	*Verbascum blattaria*	37	74
40	Common purslane	*Portulaca oleracea*	1	2
	Redroot pigweed	*Amaranthus retroflexus*	33	66
	Black mustard	*Brassica nigra*	19	38
	Curly dock	*Rumex crispus*	9	18
	Common ragweed	*Ambrosia artemisiifolia*	2	4
	Evening primrose	*Oenothera biennis*	19	38
	Virginia pepper weed	*Lepidium virginicum*	1	2
	Broad-leaf plantain	*Plantago major*	5	10

Source: Modified from Telewski and Zeevaart (2002).

depth (107 cm) to produce more seedlings than seeds buried in the shallower depth (20 cm). Both Beal's and Duvel's experiments indicate that seeds of many agricultural weeds can remain in the soil seedbank for several decades.

6.5 Factors Affecting Weed Seed Longevity

6.5.1 Genetics

The genetics of weed seed longevity is poorly understood; however, some studies have shown that genes related to seed structural composition and hormonal regulation are involved in regulating seed longevity (Long et al. 2015). The actual expression of seed longevity is complex and is governed by genetics in combination with maternal environmental conditions prevailing during seed production (Mondoni et al. 2014). Seed longevity can be affected by phylogenetic relationships as similar taxonomic groups often show similar seed longevity and associated characteristics such as seed dormancy (Long et al. 2015). However, it is a complex phenomenon, and broad generalizations cannot be made, especially from the management point of view. One important feature linked with seed longevity in seedbanks is a high level of genetic diversity within seed traits in most arable weed species (Kochanek et al. 2011). For example, two closely related weed subspecies, boneseed (*Chrysanthemoides monilifera* ssp. *monilifera*) and bitou bush (*C. monilifera* ssp. *rotundata*), significantly differed in terms of their seed longevity in seedbank (Schoeman et al. 2010). Greater genetic diversity within a weed species is generally responsible for greater seed longevity or persistence for that matter.

The wild relatives (often acting as problematic associated weeds) of domesticated plant species have been found to have higher genetic diversity related to their high seed longevity. Presotto et al. (2020) reported that populations of wild sunflower (*Helianthus annuus*) and those of a hybrid of wild and domesticated sunflower exhibited high levels of genetic diversity, which was correlated with the longevity of their seeds in seedbanks in their invasive range. The developmental relationship between seed dormancy and seed longevity also has a genetic basis. Pipatpongpinyo et al. (2020) assembled seed dormancy genes of domesticated rice (*Oryza sativa*) and weedy rice (*O. sativa* f. *spontanea*) into a tetragenic system and evaluated the germination response. Their results suggested that seed dormancy genes in weedy rice were strongly responsible for regulating the seed longevity in soil seedbanks. Hence, genetics plays an important role in inherent seed longevity, which noticeably affects the overall longevity and persistence of weed seeds in the soil seedbank.

6.5.2 Seed Characteristics

The formation of long-lived seedbanks with persistent seeds is regulated by some complex processes. However, the initial characteristics that drive this are the seed load entering the seedbank and the seed's viability and vigour (Figure 6.1). In addition, the duration of primary dormancy of the seed, the responsiveness of the seed to the induction of secondary dormancy, and the ability of the seed to resist ageing, decay, and predation all play an important role in determining seed longevity and persistence in the soil. The production of seeds can differ widely among species and among biotypes of the same species, from year to year. The number of seeds produced is determined by the interaction between the weed's genetic potential and the environmental conditions experienced at the time of seed production. Local climatic conditions and land management practices will influence the number of plants of weedy species reaching maturation and the probability of seed development on those plants. The environmental conditions experienced during seed development can influence the number of seeds produced and their quality, including their viability, vigour, and dormancy status. Persistent annual weeds generally have a large seed output that is sustained even under relatively adverse conditions.

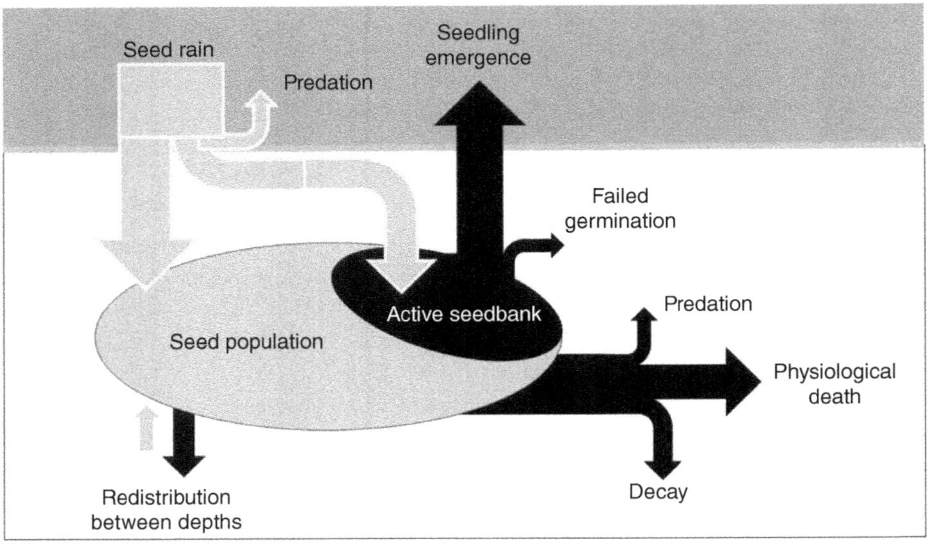

Figure 6.1 A model of the soil seedbank budget showing the non-dormant active seedbank and the dormant or quiescent seed population. The arrows indicate entry, exit, and fate of seeds.

6.5.2.1 Seed Vigour and Viability

The term vigour when used in relation to seeds refers to the performance of a seed lot during the process of germination and seedling emergence (Smith et al. 2003). In contrast, viability is the ability of a seed to continue to exist as a living entity and able to elicit some sort of germination response in the absence of dormancy under suitable conditions (Smith et al. 2003). Evaluating the quality of seed that leaves the parent plant is a fundamental step in determining its persistence in the seedbank. High viability and vigour at the time of seed shedding are common traits of many weed species. Maintenance of these traits for most weed seeds while in the seedbank is primarily dependent upon their low soil moisture content. While seed viability cannot be reinstated once lost, seed vigour may be re-established by the wetting and drying cycles that occur in soil as the higher moisture phase enables repair reactions to occur inside the seed.

6.5.2.2 Seed Morphology and Anatomy

Seed longevity can be influenced by the morphology and anatomy of the seed. Seeds often undergo morphological changes during their time in the soil seedbank, and understanding the biology of these changes can give insight into their longevity. Soil characteristics such as its chemistry and the microbiota present can alter the external morphology of the seed (Martinková et al. 2006). Saska et al. (2020) observed that during an artificial seedbank experiment using 26 herbaceous species (most of them weeds), gradual loss of seed viability over time was related to the seed volume and mass loss and, in some cases, also with seed shape and density. The authors linked their results to seed shrinking and the deterioration of the embryo and seed coat over time.

Various studies have investigated the relationship between seed mass and longevity (Probert et al. 2009; Merritt et al. 2014; Long et al. 2015), with results differing depending on the species studied and the environmental factors present. Studies on Australian species have related increased longevity with greater seed mass (Merritt et al. 2014; Satyanti et al. 2018), and larger seed size has been associated with increased viability (Moles et al. 2007). In contrast, longevity has also been negatively related with seed mass (Guzzon et al. 2018), while other studies have found no correlation (Probert et al. 2009). Smaller seeds may have higher longevity but they must persist longer in the soil before germination. Moreover, studies have suggested that larger seeds or seeds with a large surface-area-to-volume ratio are not integrated into the seedbank as easily as their smaller and smoother counterparts (Bekker et al. 1998a). Small round seeds might be more fire-resistant, having higher persistence in fire-prone environments (Gómez-González et al. 2011). Contradictory results around the world suggest that environmental influences (such as soil disturbance, altitude, nature of the biome, presence of granivores, etc.) may be overriding factors in determining whether seed morphology influences seed persistence.

Seed coat structure also plays an important role in seed longevity, as it is the layer that protects the embryo from the biotic and abiotic stresses. Gardarin et al. (2010) found that decreasing seed coat thickness increased seed mortality in 13 weed species studied and related coat thickness with slowing down decay, preventing seed attack in the soil and protecting the embryo from temperature and moisture fluctuations. Phenolic components present in the seed coat could also prevent seed attack by deterring insect and microorganism damage (Hendry et al. 1994). Surface texture of the seed coat can also influence microorganism attack, as rough surfaces could potentially provide more favourable microsites for microorganism colonization (Chee-Sanford et al. 2006). Moreover, hard-coated seeds are protected by a densely packed layer of water-repellent cells, which shields the seed from environmental moisture fluctuations that can reduce their longevity (Baskin and Baskin 2006).

Diaspore structures such as awns can also influence seed persistence, by aiding seed burial in the soil in the right orientation for germination (Humphreys et al. 2011). Hygroscopic awns present in many Poaceae weeds, for example, can guide the floret into the soil profile. Seeds with awns can move across the soil surface and then fall into cracks placing them deeper in the soil profile. Passage of seeds into the soil profile can protect them from predators and provide a stable environment for germination. Deep burial can also stop seeds from germinating, preventing the loss of seeds in the soil seedbank due to germination (Bekker et al. 1998a). Seed shape and coat texture have also been suggested to influence burial, with flat and/or rough seed coats having less burial pre-disposition (Benvenuti 2007). In addition, seeds that are enclosed by strong, spiny tissues are protected from predation and often from disease. Those with fleshy fruit tissues can be consumed by birds and mammals and when passed through the animals' digestive systems can be deposited in a protected environment, thus promoting their germination.

The absence of endosperm tissue in seeds has also been correlated with longevity. Probert et al. (2009) studied the longevity of 195 species from different environments, and Merritt et al. (2014) investigated the longevity of 172 Australian species. Both studies identified a positive correlation between non-endospermic seeds and longer-lived seeds. Additionally, high embryo-to-seed ratio and embryos with a folded, large spatulate shape were found to be longer-lived in the Australian flora (Merritt et al. 2014). Short-lived endospermic seeds have been related to early angiosperms that evolved in moist environments. Larger embryos could also be related to faster germination in arid and hot environments (Parsons 2012).

6.5.2.3 Seed Dormancy

Dormancy is perhaps one of the most distinctive features of weed species, credited as a key to their success. It has a complex relationship with seed longevity, especially in soil seedbanks. Dormancy is a physiological process that prevents seed germination during unfavourable environmental conditions, when the likelihood of seedling survival is dubious. Seed dormancy is caused by an impediment to the process of germination and is defined as the failure of a viable seed to germinate when given appropriate moisture, oxygen, and temperature conditions that would normally support radicle emergence and early seedling growth (Bewley 1997). Seed dormancy contributes to a species persistence within the environment, especially in those that are variable in climate, and is an important trait to be considered when determining management strategies. Some studies have claimed that physiological seed dormancy is not at all necessary for seed longevity in the seedbank (e.g. Thompson and Ooi 2010). This view centres upon the notion that dormancy breakage is a separate process from germination, so that seeds can become non-dormant prior to a germination stimulation event. However, this model has not been adopted by recent literature, which supports a concept of a continuum between the dormant state and germination (Finch-Savage and Footitt 2012).

The longevity of weed seeds in the soil seedbank is determined by a multitude of different factors, including the number and quality of the seeds entering the seedbank, the depth and duration of primary dormancy, the ability of the seeds to resist ageing, and the environmental conditions to which the seeds are exposed. Seed dormancy is often classified by the origin of the dormancy or by the putative mechanism (Baskin and Baskin 1998). When considering the longevity of seeds, a distinction needs to be made between primary and secondary dormancy. Primary (or innate) dormancy is acquired during seed maturation on the mother plant, while secondary dormancy is developed after shedding (Bewley and Black 1985). Physiologically dormant seeds of several weed species in the seedbank are known to go through an annual cycle of both non-dormant and dormant phases, with the dormant phase being reintroduced by environmental conditions such as high temperature and moisture variation. By nature, dormancy cycling is not thought to be a way to prolong seed life in the seedbank, but a way to ensure that germination occurs in conditions that are favourable for germination and seedling establishment (Long et al. 2015). However, seeds that undergo dormancy cycling often live longer than they would normally do if dormancy loss was a unidirectional process that made all seeds non-dormant and therefore able to germinate at any time throughout the year. Physically dormant seeds are incapable of undergoing dormancy cycling, as once they reach a non-dormant status, they cannot re-enter into the physical dormancy condition because the change in seed covering permeability cannot be overturned (Baskin and Baskin 2006). It is often the interface of weed seed dormancy, inherent seed longevity, and ambient conditions that define the overall dynamics of weed seedbanks. Therefore, understanding of such ecological interactions is of utmost importance for management programs.

6.5.2.4 Seed Ageing and Its Biochemistry

The degree of hydration and ambient temperature influence the seed's ageing process (Walters et al. 2005). Various biochemical adjustments occur during the ageing processes that result in the reduction of seed vigour and eventually reduce viability. These adjustments include downregulation of genetic expression, protein synthesis, and enzyme activity, and the modification of membrane functionality. Other molecules that are also thought to be affected by seed ageing include various carbohydrates, nucleic acids, and lipids (Smith and Berjak 1995). As seeds age in the soil seedbank, their ability to balance normal metabolism along with the reactions required for cell maintenance and repair becomes poor.

A fundamental causal agent for the reduction of seed vigour are the reactive oxygen species (ROS), which are a form of unstable, oxygenated molecules that are highly reactive with other cellular components. They are generated during mitochondrial oxidative metabolism, which is enhanced under conditions of stress (Long et al. 2015). The ROS may be generated in both dry and imbibed seeds (Vertucci and Farrant 1995); however the damage caused by ROS is generally higher in the imbibed state when water-requiring enzymatic antioxidant defence systems are active. As seeds age, their ability to generate antioxidant molecules (such as glutathione) to protect themselves against ROS activity gradually decreases with time. Some of the major consequences of ROS formation, in the absence of antioxidant protection, are the direct damage to DNA and genetic integrity, the deactivation of certain enzymes, the promotion of lipid peroxidation and protein degradation, and the disruption of cellular membrane function (Smith and Berjak 1995; Long et al. 2015). Thus, the integrity and function of cellular membranes, including mitochondrial membranes, are negatively affected, resulting in measurable ion leakage and changes in the intracellular redox environment (Schafer and Buettner 2001). This in turn is believed to cause the onset of reduced seed vigour and viability loss (Kranner et al. 2006).

The damage that accumulates within seeds during a period of dehydration-induced quiescence in the soil can be repaired, or partly repaired, upon rehydration so that reinvigourated seeds can live on in the soil for longer periods of time (Long et al. 2011). Many proteins involved with cell resistance to oxidative stress and repair are re-established during the imbibition process and become important to the reinstatement of seed vigour. In addition, seeds in the soil undergo cycles of dehydration and rehydration that sequentially repair and re-establish seed longevity by reinstating antioxidant competence including antioxidant enzyme activity (Long et al. 2015). This ability to regain antioxidant competence rapidly upon imbibition will promote longevity. Seeds can also repair DNA, which becomes degraded during the dry state in the soil seedbank. DNA repair is activated very soon after imbibition of the seed. These repair mechanisms are thought to be involved in keeping the seed in a healthy state during the cycles of dehydration–rehydration that occur in a typical seedbank (Long et al. 2015).

6.5.3 External Factors

As discussed earlier, environmental conditions can have a significant impact on weed seed production, seed characteristics, and seed dispersal. Similarly, climatic conditions, edaphic conditions, land management practices, and biological activities can all affect weed seed longevity. Since the specific impacts of the factors discussed in the following text on various aspects of weed seedbanks are discussed in different chapters of this book, we have limited ourselves to discussion of relevant to weed seed longevity herein.

6.5.3.1 Moisture and Temperature

The longevity of weed seeds in seedbanks is determined not only by their genetic, physical, and biochemical make-up but also by the environment and the interaction between the seed and the environment. Seeds in the seedbank come into equilibrium with features of the environment, notably those of soil moisture and temperature. These two factors influence the rate of ageing and therefore persistence. In the seedbank, seeds experience diurnal and seasonal fluctuations in temperature, particularly in the upper soil layers. Seeds are also affected by hydration phases, when rainfall moistens soils and periods without rain allow soils to dry. Hydration phases in soils may prepare seeds for germination or, if dormant, allow biochemical repair to take place. As seeds often persist longer in dry, cool soils than in soils subject to higher rainfall and warmer temperatures (Burnside et al. 1996), climate must be considered as a major influence on persistence. The hydrothermal time often decides the viability, vigour, and longevity of weed seeds within the seedbanks, while light and physical environment hold the key for germination and emergence (Long et al. 2015). Much research has focused on weed seed germination and emergence as affected by hydrothermal conditions, but the actual impact on seed longevity within seedbanks is still unclear.

6.5.3.2 Soil Characteristics

Soil is a complex and dynamic medium encompassing a range of physical, biological, and chemical processes. These processes not only affect the physical quality of weed seeds but also govern seed longevity by regulating the impact of other environmental conditions, such as temperature, moisture, light, and aeration. Soil temperature and moisture content and light transmissibility are all known to affect weed seed longevity, while they can also modify rates of germination, dormancy loss, and ageing. Cooler, dryer soils boost longevity (Pakeman et al. 2012) by supplying conditions that reduce the rates of seed dormancy loss and ageing and reduce microbial activity. Light-textured soils allow for greater light penetration and aeration, which promote the germination of light-stimulated seeds (Long et al. 2015). The gaseous environment deeper within soils is different than the air above and is modified by soil structure, water content, and the level of invertebrate and microbial activity within that soil. The major gases found in the soil environment that can modify dormancy status and germination include oxygen, carbon dioxide, and ethylene. Seed responses to soil gases differ significantly between species and depend on the nature of their dormancy mechanism. Oxygen is required for germination, so in its absence seeds remain quiescent and secondary dormancy can be imposed in some species (Corbineau 2012). Ethylene can alleviate dormancy, while other volatile organic chemicals from smoke created by fire may promote or inhibit germination (Long et al. 2015).

The chemical nature of soils is also known to affect weed seed longevity. One study considered the possibility that enhanced nutrient availability would promote soil organism activity and decomposition processes, which would in turn reduce seed longevity (Bekker et al. 1998b). However, nitrogen, phosphorus, and potassium fertilizer application failed to cause any significant effects on the persistence of seeds from 17 species during a two-year study. In another study about seed persistence, it was shown that soils with a lower carbon-to-nitrogen ratio could promote seed persistence (Pakeman et al. 2012). In a study where three different soil types were compared at a single site, seed persistence of three weed species (*Avena sterilis* ssp. *ludoviciana*, *Gomphocarpus physocarpus*, and *Ligustrum lucidum*) was lowest in the soil with the highest organic matter (Long et al. 2009). The salinity status of a soil can also modify germination and elevate viability loss by lowering the water potential of the soil solution. Lastly, soil pH may further modify nutrient availability for plants and microorganisms, and acidic conditions can reduce seed persistence (Pakeman et al. 2012). Thus, the chemical nature of the soil can vary considerably between sites and contribute to the variation seen in seed persistence within a species, across different sites.

6.5.3.3 Land Use and Management Operations

Land use and management operations have a significant impact on weed seed distribution and seed longevity belowground. This impact is more substantial and prominent in crop production systems, where continuous disturbance in the form of tillage or other crop management operations keeps altering the physical and biological conditions in the seedbanks. Weed seeds can be found at various depths in the soil due to self-burial, or environmental disturbance including the impact of rain and overland water movement and animal and agricultural land disturbance. The location in which seeds are buried can influence their longevity in the seedbanks (Benvenuti 2007). While seeds normally live longer when buried deep in the soil, seed longevity also depends upon soil type, seed size, and texture (Benvenuti 2007). At greater depths, light is better excluded, and temperature and moisture conditions are more stable, generally slowing the processes of seed ageing and dormancy release. Seed longevity is further assisted at depth because seeds are better protected from decay and predation (Baraibar et al. 2017). Seeds of many species may attempt to germinate from greater depths in the soil but fail to establish seedlings due to insufficient seed food reserves, thus ending their persistence in the soil seedbank (Saatkamp et al. 2011). Presotto et al. (2020) reported that soil disturbance did not have a major impact on seed longevity of wild sunflower in the seedbank *per se*, instead it showed the potential of facilitating the further spread of this weed.

There is a strong positive relationship between land disturbance and weed seed longevity (Albrecht and Auerswald 2009). Increased longevity acts as a key adaptive strategy in highly managed ecosystems such as commercial crop production systems. Numerous studies have reported the potential impacts of tillage intensity, type, frequency, and timing on weed seed longevity and seedbank dynamics (e.g. Clements et al. 1996; Chauhan et al. 2006). The transition from conventional to conservation tillage systems in many developed countries has led to a paradigm shift in weed dynamics. The major parameter of this shift is the change in weed seed distribution in the seedbanks. Less frequent, shallower, and site-specific tillage in conservation systems often avoids deep burial of weed seeds and therefore reduces their longevity in seedbanks (Bajwa et al. 2015). However, it causes greater in-crop weed infestations as an adaptive strategy. For example, most growers in Australia have adopted conservation tillage, which concentrates weed seeds, especially those of annual ryegrass (*Lolium rigidum*), in the top 0–1 cm of soil and, further, leaves seeds (10–95%) on the soil surface (no tillage or zero tillage, Chauhan et al. 2006). This has led to an increase in crop infestations by surface germinating weed species. Crop rotations, sowing techniques, residue management, and other crop management operations also affect weed seed longevity (Cardina et al. 2002; Weisberger et al. 2019). Some of these aspects are discussed in Chapter 10 of this book. Therefore, land use and management operations should be carefully selected when weed seedbank management is a major priority.

6.5.3.4 Seed Predation and Deterioration

The seed's ability to protect itself from predation and microorganism decay will influence its longevity in the soil. Both insects (Martinková et al. 2006) and microorganisms (Blaney and Kotanen 2002) have been implicated in influencing seed longevity, primarily through seed predation, decay, or relocation. Organisms within the soil can cause premature seed death, relocate seeds within the soil profile, and accelerate the process of seed ageing (Forcella 2003). Predation of seeds by granivores will be determined by different seed traits. Studies have shown that granivores can discriminate between seeds according to their physical attributes such as seed size, density, and structural strength (Lundgren and Rosentrater 2007). The seed's nutritional value may also influence seed predation (Long et al. 2015), with granivores usually favouring seeds with higher oil and energy content (Dechaine et al. 2010). Therefore, different seed traits can influence the rate of consumption by granivores (Martinková et al. 2006). To resist predation, seeds have developed sophisticated mechanisms, such as hard coats, indehiscent fruits, mucilage (Dalling et al. 2011), woody endocarps (Schoeman et al. 2010), and repellent chemicals in the testa (Xiao et al. 2009). The topic of seed predation is discussed in detail in Chapter 8 of this book.

Microorganisms in the soil can also influence seed longevity of plant species recruiting from soil seedbanks (Dalling et al. 2011). High microbial activity is recognized to cause increased seed mortality in soils; however soil microorganisms are very diverse and so are their interactions with seeds. For example, the soil-borne fungus *Fusarium oxysporum* can overwhelm and kill healthy seeds (Thomas et al. 1999). Seed vulnerability to microbial decay and ageing can be critically influenced by the seed coat's chemical composition (Long et al. 2015). In contrast, the collembolan *Protaphorura fimata* predates upon seed-rotting microbes and acts as a seed protectant (Mitschunas et al. 2006). Certain seed surface-associated bacteria, too, have disease-suppressive properties that prevent infection and decay by pathogenic species (Nelson 2004). Soil microbes can be an integral part of seed germination, breaking down the seed coat so that germination can occur (Delgado-Sánchez et al. 2011) and degrading allelochemicals in the soil that would otherwise inhibit germination (Zhu et al. 2011). Certain imbibed and germinating seeds have been found to exude chemical compounds used by host-specific pathogens to detect the presence of seeds in the soil (Long et al. 2015). Effects of soil microbes on weed seeds are discussed in detail in Chapter 13 of this book.

In addition to causing deterioration and decay, seeds can be relocated through the soil profile by animals. Even though many relocation events do not result in seed death, these events can contribute to soil seedbank loss. Lengyel et al. (2010) observed that seed underground dispersal by ants was an adaptation that has developed in 4.5% of elaiosome-bearing angiosperms. Seeds can also be dispersed throughout the soil profile by the movement of earthworms or buried by livestock hooves in muddy pastures. Some seeds are covered by a fleshy tissue that can be ingested by animals such as birds and mammals

(Vivian-Smith and Gosper 2010). During animal ingestion, seeds are usually scarified and/or inhibiting chemicals from their fleshy tissues are removed. By the time these seeds are dispersed, they have overcome physical dormancy and are ready to germinate. Seed predation after dispersal is also a deterrent on seed longevity, as seeds can be left on the soil surface, exposed to predation by insects, birds, mammals, and other generalist animals. Seed predation may occur rapidly post-dispersal and can kill many seeds (Salazar et al. 2012).

6.6 Implications of Seedbank Longevity for Weed Management

The knowledge of seed longevity of different weed species in soil seedbanks in conjunction with seedbank composition (magnitude and diversity), emergence patterns, and aboveground weed population dynamics is imperative for sustainable weed management. Unfortunately, effort is often invested in scouting and managing the aboveground weed infestations in crop production systems with little or no attention paid to soil weed seedbank longevity. This leads to ineffective and short-term weed control, and long-term weed seedbank development. Studies have shown that knowing the seed longevity in a seedbank under actual field conditions provides a real-time estimate of time and effort required for seedbank depletion to a threshold acceptable from the economic point of view. For instance, in northern Australia, as it was known that seed longevity was relatively short for the invasive weed species, *Calotropis procera,* rigorous weed control was able to deplete the seedbank to a reasonable level within just two years (Bebawi et al. 2015). On the other hand, another highly invasive weed species, parthenium weed (*Parthenium hysterophorus*), was found to develop rich seedbanks with long life in pastures and grazing lands of Queensland, Australia, within a short period after its introduction. It required decades of continuous management effort to even exhaust half of the seedbank (Nguyen et al. 2017).

Management tactics focused to expose weed seeds to biotic and abiotic elements of the environment also help reduce the weed seedbank size. For example, Baraibar et al. (2017) reported that delaying the post-harvest tillage in arable fields increased seed predation and therefore reduced seed return to the soil seedbank. Therefore, knowledge of seed longevity of any weed species in the soil seedbank can be very useful for its effective and long-term management.

6.7 Conclusions and Future Research Directions

Weed seed longevity in seedbanks is clearly a very complex process and is a function of several seed and species attributes that interplay with the environment. These attributes influence a seed's ability to resist germination and to defy death, while exposure to the local climate, soil features, and other characteristics of a seed's pre- and post-dispersal environment will ultimately control when and how the soil-borne seeds will germinate or lose viability. Seed dormancy and the seed's biological defence mechanisms all contribute to the ability to resist loss from the soil seedbank (prolonging seed longevity). Genetics, biotic, and abiotic components of the pre-dispersal parental environment and the post-dispersal environment can alter seed longevity through compositional changes and therefore impact the seedbank make-up. Similarly, land use changes and management operations result in altered micro-environments that impact weed seed longevity. Although seed longevity indices are important to assess the strength of seedbanks, population dynamics, intra- and inter-specific variations in seed longevity, and climate change should be considered to develop reliable predictions and weed management.

Future studies on the seed and species characteristics that determine seed longevity, and the mechanisms that control longevity, are required. This will help predict the duration of weed seed persistence accurately, which in turn will be important in making decisions regarding the prioritization of weed species for management and the duration of the management process required to exhaust the weed seedbank.

References

Albrecht, H. and Auerswald, K. (2009). Seed traits in arable weed seed banks and their relationship to land-use changes. *Basic and Applied Ecology* 10: 516–524.

Bajwa, A.A., Mahajan, G., and Chauhan, B.S. (2015). Nonconventional weed management strategies for modern agriculture. *Weed Science* 63: 723–747.

Baraibar, B., Canadell, C., Torra, J. et al. (2017). Weed seed fate during summer fallow: the importance of seed predation and seed burial. *Weed Science* 65: 515–524.

Baskin, C.C. and Baskin, J.M. (1998). *Seeds: Ecology, Biogeography, and Evolution of Dormancy and Germination*. San Diego: Academic Press.

Baskin, C.C. and Baskin, J.M. (2006). The natural history of soil seed banks of arable land. *Weed Science* 54: 549–557.

Bebawi, F.F., Campbell, S.D., and Mayer, R.J. (2015). Seed bank longevity and age to reproductive maturity of *Calotropis procera* (Aiton) WT Aiton in the dry tropics of northern Queensland. *The Rangeland Journal* 37: 239–247.

Bekker, R., Bakker, J., Grandin, U. et al. (1998a). Seed size, shape and vertical distribution in the soil: indicators of seed longevity. *Functional Ecology* 12: 834–842.

Bekker, R.M., Knevel, I., Tallowin, J. et al. (1998b). Soil nutrient input effects on seed longevity: a burial experiment with fen meadow species. *Functional Ecology* 12: 673–682.

Benvenuti, S. (2007). Natural weed seed burial: effect of soil texture, rain and seed characteristics. *Seed Science Research* 17: 211–219.

Bewley, J.D. (1997). Seed germination and dormancy. *The Plant Cell* 9: 1055–1066.

Bewley, J.D. and Black, M. (1985). *Seeds: Physiology of Development and Germination*, 367. New York, NY, USA: Plenum Press.

Blaney, S. and Kotanen, P.M. (2002). Persistence in the seed bank: the effects of fungi and invertebrates on seeds of native and exotic plants. *Ecoscience* 9: 509–517.

Burnside, O.C., Wilson, R.G., Weisbert, S., and Hubbard, K.G. (1996). Seed longevity of 41 weed species buried 17 years in eastern and western Nebraska. *Weed Science* 44: 74–86.

Cardina, J., Herms, C.P., and Doohan, D.J. (2002). Crop rotation and tillage system effects on weed seedbanks. *Weed Science* 50: 448–460.

Chauhan, B.S., Gill, G., and Preston, C. (2006). Influence of tillage systems on vertical distribution, seeding recruitment and persistence of rigid ryegrass (*Lolium rigidum*) seed bank. *Weed Science* 54: 669–676.

Chee-Sanford, J.C., Williams, M.M., Davis, A.S., and Sims, G.K. (2006). Do microorganisms influence seed-bank dynamics? *Weed Science* 54: 575–587.

Clements, D.R., Benott, D.L., Murphy, S.D., and Swanton, C.J. (1996). Tillage effects on weed seed return and seedbank composition. *Weed Science* 44: 314–322.

Corbineau, F. (2012). Markers of seed quality: from present to future. *Seed Science Research* 22: 61–68.

Dalling, J.W., Davis, A.S., Schutte, B.J., and Elizabeth Arnold, A. (2011). Seed survival in soil: interacting effects of predation, dormancy and the soil microbial community. *Journal of Ecology* 99: 89–95.

Dechaine, J.M., Burger, J.C., and Burke, J.M. (2010). Ecological patterns and genetic analysis of post-dispersal seed predation in sunflower (*Helianthus annuus*) crop-wild hybrids. *Molecular Ecology* 19: 3477–3488.

Delgado-Sánchez, P., Ortega-Amaro, M., Jiménez-Bremont, J., and Flores, J. (2011). Are fungi important for breaking seed dormancy in desert species? Experimental evidence in *Opuntia streptacantha* (Cactaceae). *Plant Biology* 13: 154–159.

Finch-Savage, W.E. and Footitt, S. (2012). To germinate or not to germinate: a question of dormancy relief not germination stimulation. *Seed Science Research* 22: 243–248.

Forcella, F. (2003). Debiting the seedbank: priorities and predictions. *Aspects of Applied Biology* 69: 151–162.

Gardarin, A., Dürr, C., Mannino, M.R. et al. (2010). Seed mortality in the soil is related to seed coat thickness. *Seed Science Research* 20: 243–256.

Gómez-González, S., Torres-Díaz, C., Bustos-Schindler, C., and Gianoli, E. (2011). Anthropogenic fire drives the evolution of seed traits. *Proceedings of the National Academy of Sciences of the United States of America* 108: 18743–18747.

Guzzon, F., Orsenigo, S., Gianella, M. et al. (2018). Seed heteromorphy influences seed longevity in *Aegilops*. *Seed Science Research* 28: 277–285.

Hay, F.R., Klin, J., and Probert, R.J. (2006). Can a post-harvest ripening treatment extend the longevity of *Rhododendron* L. seeds? *Scientia Horticulturae* 111: 80–83.

Hendry, G., Thompson, K., Moss, C. et al. (1994). Seed persistence: a correlation between seed longevity in the soil and *ortho*-dihydroxyphenol concentration. *Functional Ecology* 8: 658–664.

Humphreys, A.M., Antonelli, A., Pirie, M.D., and Linder, H.P. (2011). Ecology and evolution of the diaspore "burial syndrome". *Evolution: International Journal of Organic Evolution* 65: 1163–1180.

Kochanek, J., Steadman, K.J., Probert, R.J., and Adkins, S.W. (2011). Parental effects modulate seed longevity: exploring parental and offspring phenotypes to elucidate pre-zygotic environmental influences. *New Phytologist* 191: 223–233.

Kranner, I., Birtic, S., Anderson, K.M., and Pritchard, H.W. (2006). Glutathione half-cell reduction potential: a universal stress marker and modulator of programmed cell death? *Free Radical Biology and Medicine* 40: 2155–2165.

Lengyel, S., Gove, A.D., Latimer, A.M. et al. (2010). Convergent evolution of seed dispersal by ants, and phylogeny and biogeography in flowering plants: a global survey. *Perspectives in Plant Ecology, Evolution and Systematics* 12: 43–55.

Long, R.L., Steadman, K.J., Panetta, F.D., and Adkins, S.W. (2009). Soil type does not affect seed ageing when soil water potential and temperature are controlled. *Plant and Soil* 320: 131–140.

Long, R.L., Kranner, I., Panetta, F.D. et al. (2011). Wet-dry cycling extends seed persistence by re-instating antioxidant capacity. *Plant and Soil* 388: 511–519.

Long, R.L., Gorecki, M.J., Renton, M. et al. (2015). The ecophysiology of seed persistence: a mechanistic view of the journey to germination or demise. *Biological Reviews* 90: 31–59.

Lundgren, J.G. and Rosentrater, K.A. (2007). The strength of seeds and their destruction by granivorous insects. *Arthropod-Plant Interactions* 1: 93–99.

Martinková, Z., Saska, P., and Honek, A. (2006). Consumption of fresh and buried seed by ground beetles (Coleoptera: Carabidae). *European Journal of Entomology* 103: 361–364.

Masin, R., Zuin, M.C., Otto, S., and Zanin, G. (2006). Seed longevity and dormancy of four summer annual grass weeds in turf. *Weed Research* 46: 362–370.

Merritt, D.J., Martyn, A.J., Ainsley, P. et al. (2014). A continental-scale study of seed lifespan in experimental storage examining seed, plant, and environmental traits associated with longevity. *Biodiversity and Conservation* 23: 1081–1104.

Mitschunas, N., Wagner, M., and Filser, J. (2006). Evidence for a positive influence of fungivorous soil invertebrates on the seed bank persistence of grassland species. *Journal of Ecology* 94: 791–800.

Moles, A.T., Ackerly, D.D., Tweddle, J.C. et al. (2007). Global patterns in seed size. *Global Ecology and Biogeography* 16: 109–116.

Mondoni, A., Orsenigo, S., Dona, M. et al. (2014). Environmentally induced transgenerational changes in seed longevity: maternal and genetic influence. *Annals of Botany* 113: 1257–1263.

Naylor, R.E.L. (2003). Weed seed biology. In: *Encyclopedia of Applied Plant Sciences* (eds. D.J. Murphy, B. Thomas and B.G. Murray), 1500–1508. USA: Elsevier/Academic Press.

Nelson, E.B. (2004). Microbial dynamics and interactions in the spermosphere. *Annual Review of Phytopathology* 42: 271–309.

Nguyen, T.L.T., Bajwa, A.A., Navie, S.C. et al. (2017). The soil seedbank of pasture communities in central Queensland invaded by *Parthenium hysterophorus* L. *Rangeland Ecology & Management* 70: 244–254.

Pakeman, R.J., Small, J.L., and Torvell, L. (2012). Edaphic factors influence the longevity of seeds in the soil. *Plant Ecology* 213: 57–65.

Parsons, R. (2012). Incidence and ecology of very fast germination. *Seed Science Research* 22: 161–167.

Pipatpongpinyo, W., Korkmaz, U., Wu, H. et al. (2020). Assembling seed dormancy genes into a system identified their effects on seedbank longevity in weedy rice. *Heredity* 124: 135–145.

Porsild, A.E., Harington, C.R., and Mulligan, G.A. (1967). *Lupinus arcticus* Wats. grown from seeds of Pleistocene age. *Science* 158: 113–114.

Presotto, A., Hernández, F., Casquero, M. et al. (2020). Seed bank dynamics of an invasive alien species. *Helianthus annuus* L. *Journal of Plant Ecology* 13: 313–322.

Priestley, D.A. and Posthumus, M.A. (1982). Extreme longevity of lotus seeds from Pulantien. *Nature* 299: 148–149.

Probert, R.J., Daws, M.I., and Hay, F.R. (2009). Ecological correlates of ex situ seed longevity: a comparative study on 195 species. *Annals of Botany* 104: 57–69.

Saatkamp, A., Affre, L., Baumberger, T. et al. (2011). Soil depth detection by seeds and diurnally fluctuating temperatures: different dynamics in 10 annual plants. *Plant and Soil* 349: 331–340.

Salazar, A., Goldstein, G., Franco, A.C., and Miralles-Wilhelm, F. (2012). Seed limitation of woody plants in Neotropical savannas. *Plant Ecology* 213: 273–287.

Saska, P., Foffová, H., Martinková, Z., and Honěk, A. (2020). Persistence and changes in morphological traits of herbaceous seeds due to burial in soil. *Agronomy* 10: 448.

Satyanti, A., Nicotra, A.B., Merkling, T., and Guja, L.K. (2018). Seed mass and elevation explain variation in seed longevity of Australian alpine species. *Seed Science Research* 28: 319–331.

Schafer, F.Q. and Buettner, G.R. (2001). Redox environment of the cell as viewed through the redox state of the glutathione disulfide/glutathione couple. *Free Radical Biology and Medicine* 30: 1191–1212.

Schoeman, J., Buckley, Y., Cherry, H. et al. (2010). Inter-population variation in seed longevity for two invasive weeds: *Chrysanthemoides monilifera* ssp. *monilifera* (boneseed) and ssp. *rotundata* (bitou bush). *Weed Research* 50: 67–75.

Smith, M.T. and Berjak, P. (1995). Deteriorative changes associated with the loss of viability of stored desiccation-tolerant and desiccation-sensitive seeds. In: *Seed Development and Germination* (eds. J. Kigel and G. Galili), 701–746. New York, NY: Marcel Dekker, Inc.

Smith, R., Dickie, J.B., Linington, S.H. et al. (2003). *Seed Conservation: Turning Science into Practice*. Kew, England: Royal Botanic Gardens.

Telewski, F.W. and Zeevaart, J.A. (2002). The 120-year period for Dr. Beal's seed viability experiment. *American Journal of Botany* 89: 1285–1288.

Thomas, H., Heller, A., Sauerborn, J., and Muller-Stover, D. (1999). *Fusarium oxysporum* f. sp. orthoceras, a potential mycoherbicide, parasitizes seeds of *Orobanche cumana* (Sunflower Broomrape): a Cytological Study. *Annals of Botany* 83: 453–458.

Thompson, K. and Ooi, M. (2010). To germinate or not to germinate: More than just a question of dormancy. *Seed Science Research* 20: 209–211.

Toole, E.H. and Brown, E. (1946). Final results of the Duvel buried seed experiment. *Journal of Agricultural Research* 72: 201–210.

Vertucci, C. and Farrant, J.M. (1995). Acquisition of desiccation tolerance. In: *Seed Development and Germination* (eds. J. Kigel and G. Galili), 237–271. New York, NY: Marcel Dekker Press Inc.

Vivian-Smith, G.E. and Gosper, C.R. (2010). Comparative seed and dispersal ecology of three exotic subtropical Asparagus species. *Invasive Plant Science and Management* 3: 93–103.

Von Mourik, T.A., Stomph, T.J., and Murdoch, A.J. (2005). Why high seed densities within buried mesh bags may overestimate depletion rates of soil seed banks. *Journal of Applied Ecology* 42: 299–305.

Walters, C., Wheeler, L.M., and Grotenhuis, J.M. (2005). Longevity of seeds stored in a genebank: species characteristics. *Seed Science Research* 15: 1–20.

Weisberger, D., Nichols, V., and Liebman, M. (2019). Does diversifying crop rotations suppress weeds? A meta-analysis. *PLoS One* 14: e0219847. https://doi.org/10.1371/journal.pone.0219847.

Xiao, Z., Gao, X., Jiang, M., and Zhang, Z. (2009). Behavioral adaptation of Pallas's squirrels to germination schedule and tannins in acorns. *Behavioral Ecology* 20: 1050–1055.

Yashina, S., Gubin, S., Maksimovich, S. et al. (2012). Regeneration of whole fertile plants from 30,000-year-old fruit tissue buried in Siberian permafrost. *Proceedings of the National Academy of Sciences of the United States of America* 109: 4008–4013.

Zhu, X., Zhang, J., and Ma, K. (2011). Soil biota reduce allelopathic effects of the invasive *Eupatorium adenophorum*. *PLoS One* 6 (9): e25393. https://doi.org/10.1371/journal.pone.0025393.

7

Evolution and Persistence of Herbicide-Resistant Weeds

Hugh J. Beckie[1], *Linda M. Hall*[2], *Roberto Busi*[1], *and Michael B. Ashworth*[1]

[1] *Australian Herbicide Resistance Initiative, School of Agriculture and Environment, The University of Western Australia, Perth, WA, Australia*
[2] *Department of Agricultural, Food and Nutritional Science, University of Alberta, Edmonton, Alberta, Canada*

7.1 Introduction

The evolution of herbicide resistance in arable weed populations markedly impacts their persistence, both spatially and temporally. Persistence can be defined as the length of time a population remains extant. Herein, we consider the effects of herbicide resistance and consequent altered management practices on population demographics of herbicide-resistant (HR) weeds.

Enhanced fitness of HR weeds under recurrent herbicide selection increases the probability of survival of successive generations. Herbicide efficacy, soil-residual activity, and application frequency influence the level of population exposure to herbicide selection and thus the duration of fitness-enhancing conditions (Beckie 2006). In the absence of herbicide selection, pleiotropic effects associated with HR traits may confer a fitness cost that could be exacerbated when mechanical or cultural weed management practices are employed. These practices that are aimed at increasing the heterogeneity of human-mediated selection, reducing migration of weed propagules or decreasing environmental uniformity could reduce the persistence of HR weeds.

Population demography can be a useful approach in quantifying persistence of weed populations with HR traits. By quantifying fitness over the entire life cycle through estimation of population growth rate (lambda, λ), population demography, or matrix modeling enables prediction of the probability of populations to be persistent or invasive ($\lambda \geq 1$). Additionally, the analytical tools of prospective and retrospective analyses can inform the underlying factors affecting population persistence (Caswell 2001).

Persistence of weed populations is affected by spatial movement of herbicide resistance alleles via seed or pollen. Occurrence of HR weeds is often attributed to independent selection, with little attention paid to the role of HR allele movement. Rates of gene flow are generally greater than rates of mutation, which reduces the time before herbicide

Persistence Strategies of Weeds, First Edition. Edited by Mahesh K. Upadhyaya, David R. Clements, and Anil Shrestha.

resistance is predominant in weed populations (Jasieniuk et al. 1996). Seed dispersal through natural or anthropogenic vectors can rapidly spread HR alleles over long distances across agroecoregions.

Temporal persistence is achieved through vegetative reproduction (e.g. rhizomes) or seedbank survival. Seedbank longevity and persistence are facilitated by primary and secondary seed dormancy and can vary widely among species. Enhanced seed dormancy can result in a slower rate of herbicide resistance evolution due to the smaller number of weed cohorts exposed to recurrent herbicide selection and thus impact weed population dynamics and persistence. Moreover, pleiotropic effects associated with HR traits may influence seedbank dynamics and alter fitness (Vila-Aiub et al. 2009). Seedbank longevity is also influenced by management. For example, the seedbank in low soil disturbance no-tillage systems tend to be less dormant with a greater turnover rate than intensive-tillage regimes, which can hasten the evolution of herbicide resistance (Beckie et al. 2008).

Weeds often mature in synchrony with crop plants (crop mimicry), with their seeds admixed with crop seed at harvest. The use of harvested crop seed in subsequent sowing allows for potential weed reintroduction and can contribute to persistence of HR weed populations. Increasing adoption of management practices to compensate for fewer effective postemergence herbicides may alter selection pressures on HR weed populations, resulting in physiological, morphological, or phenological adaptation to escape or mitigate control. The classic example is recurrent hand weeding in rice (*Oryza sativa*), which resulted in barnyard grass (*Echinochloa crus-galli* var. *oryzicola*) mimicking the crop in phenology and appearance (Barrett 1983).

Case studies of five economically important weeds are examined to exemplify how the evolution of herbicide resistance has influenced population persistence: (i) transgenic canola (*Brassica napus*) volunteers or feral populations in North America and Australia, (ii) kochia (*Kochia scoparia*, *Bassia scoparia*) in the Great Plains of North America, (iii) wild oat (*Avena fatua*) in the Northern Great Plains of North America, (iv) wild radish (*Raphanus raphanistrum*) in Western Australia, and (v) annual ryegrass (*Lolium rigidum*) in Australia. As detailed in Section 7.3, population persistence is the culmination of the interactions among the extent of weedy traits of the species, evolved fitness-enhancing HR traits under recurrent herbicide selection, degree of HR gene dispersal via seed and pollen, management practices, and the environment.

7.2 How Evolution of Herbicide Resistance Influences Persistence of Weed Populations

7.2.1 Enhanced Fitness Under Recurrent Herbicide Selection Pressure

An understanding of the fitness consequences of herbicide resistance alleles in the presence and absence of a herbicide is important for predicting the evolutionary dynamics of herbicide resistance (Neve et al. 2003). Fitness of an HR biotype at various life stages is assessed in the absence of herbicide exposure under non-competitive and competitive (intra- and interspecific) conditions. A fitness cost may result from mutations in herbicide target enzymes that compromise or interfere with normal plant function or metabolism, defense mechanisms diverting resources away from growth and reproduction, or altered ecological interactions (Vila-Aiub et al. 2009). Initial studies in the 1970s consistently documented a fitness cost associated with target-site resistance to triazine herbicides that inhibit photosystem II. However, this cost of resistance has often been minimal or not detectable in biotypes resistant to herbicides of other sites of action. The best evidence for lack of measurably reduced fitness in HR biotypes is the absence of significant decline in the proportion of resistant to herbicide-susceptible (HS) individuals in a field population over time after use of the selecting herbicide has been discontinued (e.g. Andrews and Morrison 1997).

Under recurrent herbicide selection, however, enhanced fitness is clearly bestowed on HR individuals at the expense of HS individuals – the proportion of HR individuals in the population increases. Recurrent selection pressure on weed populations can arise from frequent application of non-residual herbicides such as glyphosate or from exposure of cohorts to persistent soil-residual herbicides such as imazethapyr. Therefore, potentially exploiting pleiotropic effects that result in plant fitness costs of an HR weed biotype through altered management will not be effective if the population is exposed to near continual herbicide selection.

7.2.2 Using Population Demography to Quantify Persistence

Population demographic or matrix modeling, although not widely used in agriculture, is an established method of assessing population fitness in ecological studies by estimating population growth rate (lambda, λ) (Caswell 2001). Population growth rate is a function of both survival and fecundity, referred to as vital rates or "lower level parameters" (LLP). A λ of <1 (log $\lambda<0$) indicates a decreasing population, while $\lambda \geq 1$ (log $\lambda \geq 0$) indicates a stable or increasing population. Focusing on the components of fitness, rather than fitness over the entire life cycle, could lead to erroneous conclusions regarding population demographics (Caswell 2001).

Therefore, population demography can be employed to estimate potential population persistence (the length of time a population remains extant) or invasiveness (ability of a population to increase and spread). In the absence of selection pressure, populations will not likely be persistent or invasive if λ is <1. For example, the potential persistence or invasiveness of feral transgenic canola (oilseed rape, *B. napus*) in non-crop areas is considered transient or of low probability, as analyzed by population demography (Crawley et al. 1993; Knispel and McLachlan 2010; Alexander 2016).

Comparisons of growth rates of populations can ascertain differences and their magnitude, but do not address why the differences occur or which LLP might influence population growth in the future. Demographic modeling provides the analytical tools of prospective (sensitivity or elasticity) and retrospective analyses (life table response experiments [LTRE]), which are important in identifying LLP with the potential to have a large influence on λ and are also responsible for the observed variability in λ (Caswell 2001). Thus, elasticity and LTRE analyses can be used to investigate LLP that can cause (elasticity) or did cause (LTRE) differences in population growth rates. If the population growth rate is highly elastic to a LLP, then even small changes made to the LLP will cause proportionally large changes to the population growth rate. The utility of prospective and retrospective analyses using a population matrix model parameterized with wild oat demographic data from the Northern Great Plains is described in Section 7.3.3.

7.2.3 Spatial Movement of Herbicide Resistance Alleles via Seed and Pollen

Incidence of HR weeds in a region is often attributed to independent evolution through herbicide selection. The role of HR allele movement via seeds and pollen is often underappreciated and underestimated (Beckie et al. 2019b). In genotypic studies of HR weed populations, the contribution of gene flow to incidence of resistance is often similar to or greater than that of independent evolution (Délye et al. 2010; Molin et al. 2018). Simulation models have consistently predicted that frequent applications of highly effective herbicides (e.g. acetolactate synthase (ALS) inhibitors, glyphosate) provide connective "high-fitness" habitats across the landscape, which facilitates a rapid increase in the frequency and movement of an HR trait within and among populations (Brûlé-Babel et al. 2006; Dauer et al. 2009). Gene flow through pollen or seed movement from HR weed populations provide a source of HR alleles in previously mentioned HS populations.

It is difficult to control the spread of herbicide resistance via pollen flow, especially when resistance is often endowed by a single nuclear dominant or semi-dominant gene. Once HR alleles are established in a weed population, pollen-mediated stacking of HR traits will further limit and confound herbicide control options. Seed movement has the potential to influence HR gene spread on a much larger scale than pollen flow. Whereas pollen movement is important in relatively short distance HR gene dispersal, seed movement is probably responsible for the majority of gene flow in weed populations (Jasieniuk et al. 1996; Diggle and Neve 2001). Wind dispersal of weed species with lightweight seeds, such as prickly lettuce (*Lactuca serriola*), *Conyza*, and *Sonchus* spp., can spread herbicide resistance rapidly over long distances (Thill and Mallory-Smith 1997; Shields et al. 2006). Epidemiological studies have shown that regional interpopulation movement of HR alleles of weed species with efficient pollen and seed dispersal mechanisms can be extremely rapid. However, movement of HR seed via equipment (especially grain harvesters), non-composted manure, silage, or contaminated commercial seed stocks or feed can be even greater than natural seed dispersal, crossing international borders (Carvalho et al. 2015). Weed sanitation practices that limit the spread of HR seed can slow the occurrence of herbicide resistance (Légère et al. 2000).

7.2.4 Temporal Persistence: Seedbank Dynamics

Seedbanks allow species to survive unfavorable environmental conditions or control measures through staggered germination within or across growing seasons. Seedbank longevity can vary greatly among weed species. Average seed viability in the soil profile can range from transient [<1–2 years; e.g. dandelion (*Taraxacum officinale*)] to persistent [>5 years; e.g. common lambsquarters (*Chenopodium album*)] (Van Acker 2009). Seedbank longevity may also be influenced by management. In no-tillage systems, most seeds occur at depths shallower than 10 cm. This microclimate is conducive to more rapid turnover of the seedbank compared with tillage-intensive systems, impacting the rate of resistance evolution (Moss 2002; Beckie et al. 2008).

Seedbank persistence is facilitated by seed dormancy. Many weed seeds at the time of maturity exhibit primary (innate) dormancy. Additionally, secondary or induced dormancy may occur in some species where seeds cycle in and out of a dormant state due to specific environmental conditions. Dormancy may aid weed fitness by increasing the probability of achieving genetic representation in future generations. Pleiotropic effects that result in altered fitness of HR weeds may influence seedbank dynamics. For example, triazine-resistant black nightshade (*Solanum nigrum*) emerges earlier than HS plants because of germination at lower soil temperatures (Kremer and Lotz 1998).

There has been increasing attention to mitigating weed seedbank replenishment because of the increasing prevalence of multiple-HR weed populations concomitant with declining availability of effective herbicides. Various harvest weed seed control (HWSC) techniques, now becoming routine across Australia, are being evaluated in other cropping regions globally. The efficacy of these techniques depend upon a high level of seed retention on the plant at harvest time, with seeds positioned above cutting height for collection by combine harvesters.

7.2.5 Phenological Adaptation of Herbicide-Resistant Weeds to Management

Alteration in phenology of HR weeds at any stage in their life cycle may be due to a pleiotropic effect influencing plant fitness or an evolutionary response to any recurrent selective force. With increasing reliance upon non-herbicidal practices to prolong the longevity of remaining effective herbicides, further phenological adaptation of HR weeds can be anticipated. Weed populations can adapt to any constant or recurrent selection pressure, such as similar date of crop seeding, frequent cutting, or weed seed capture at harvest (Jordan and Jannink 1997). Repetitive use of chemical, physical, or cultural techniques can select physiological, morphological, and phenological traits in weeds to evade control (Mortimer 1997).

Seed dispersal and return to the soil seedbank are key factors in the persistence of many weed populations. In response to repetitive HWSC, it would be advantageous for weeds to evolve more prostrate forms or earlier-flowering or seed-shedding ecotypes. Alternatively, weed populations now resistant to most postemergence herbicides may possess enhanced seed dormancy and delayed germination to escape soil-residual activity of preemergence herbicides (Owen et al. 2011).

7.3 Case Studies

7.3.1 Transgenic Canola Volunteers or Feral Populations in North America and Australia

Canola (oilseed rape) is a member of the Brassicaceae family. In Western Canada where canola is widely grown (9.3 million ha in 2017), volunteers are abundant. In preharvest weed surveys conducted in the Prairie provinces between 2010 and 2016, the relative abundance ranking of volunteer canola averaged fifth place, significantly higher than previous surveys (e.g. Leeson et al. 2017). This increased ranking reflects the yearly trends in the canola area and its frequency in annual crop rotations, as well as prevailing herbicide use and management practices.

Weediness of canola volunteers is aided by high fecundity, seed loss before or during harvest, and secondary seed dormancy. Although pod-shattering tolerance is being bred into commercial cultivars, potential seed loss before and during harvest is still substantial. Canola seedbank additions through harvest losses range from 2500 to >8000 seeds m^{-2} (4–10% of seed yield) in Western Canada (Gulden et al. 2003; Haile et al. 2014; Cavalieri et al. 2016). While canola has almost no primary seed dormancy, it possesses secondary dormancy that is influenced by environmental conditions, burial and genotype (Gulden et al. 2004). Most volunteers emerge in the first year following the canola crop, although seeds may persist and seedlings emerge at low densities throughout a three- or four-year rotation (Harker et al. 2006). In a study in Western Canada, seedbank persistence was 44%, 14%, and 0.2% of the original seedbank over the first three years, respectively, following canola (Gulden et al. 2003). Seedbanks are depleted by predation, fatal germination, expiration, or pathogens.

The HR trait in canola, endowing resistance to either glyphosate, glufosinate, or imidazolinone herbicides, does not impact fitness or persistence of volunteers in the absence of herbicide selection. However, glyphosate-HR crop volunteers can persist where glyphosate is used alone for burndown (e.g. no-tillage regime) or in-crop weed control (reviewed in Beckie and Owen 2007). Even if the cultivar grown in a field is not glyphosate-HR, pollen flow from adjacent glyphosate-HR canola fields or admixtures in planted seedlots can result in trait stacking in volunteers (Hall et al. 2000; Beckie et al. 2003; Friesen et al. 2003; Figure 7.1), which the grower may not anticipate. Uncontrolled volunteers contribute to seedbank persistence. In a study in Western Canada, transgenic *B. napus* volunteers persisted for 7 years in an arable field around wet depressional areas that did not permit crop cultivation or herbicide application (Beckie and Warwick 2010).

In ruderal areas, seed dispersal from adjacent fields or spillage via grain transportation can result in abundant roadside or railside feral populations (Yoshimura et al. 2006;

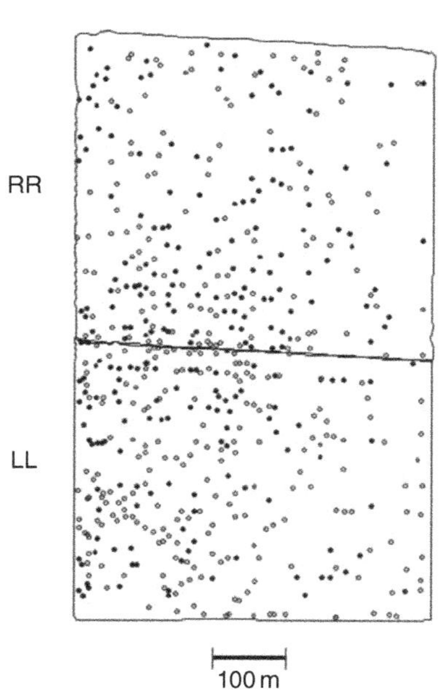

Figure 7.1 The occurrence of double herbicide-resistant (glyphosate (RR) + glufosinate (LL)) canola volunteers in adjacent fields in Saskatchewan, Canada in 2000 as a result of bidirectional pollen flow in the previous year. *Source:* Adapted from Beckie et al. (2003). Circles with fills indicate a single plant, and circles without fills indicate more than one plant.

Knispel and McLachlan 2010; Busi and Powles 2016). These studies have shown that populations can persist in ruderal areas with periodic seed replenishment or application of glyphosate for nonselective weed control but are otherwise ephemeral (population growth rate, $\lambda < 1$) if seed immigration is prevented (Alexander 2016). Nevertheless, future canola cultivars with abiotic stress tolerance (e.g. drought, salinity) have potential for greater weediness, persistence, and invasiveness in ruderal or natural areas (Beckie et al. 2010).

7.3.2 Kochia in the Great Plains of North America

Kochia is an annual broadleaf weed species native to Eurasia and introduced as an ornamental to the Americas in the mid- to late 1800s (reviewed in Friesen et al. 2009). This naturalized species is a common and economically important weed in crop production systems and ruderal areas in semiarid to arid regions of Western Canada and the United States and has expanded northward in the Canadian Prairies during the past 40 years. Kochia competitiveness and persistence in cropping systems and ruderal areas are aided by its ability to germinate at low soil temperatures and emerge early; grow rapidly; tolerate heat, drought, and salinity; and exert allelopathic effects on neighboring species.

Herbicidal control of kochia has been compromised by the widespread evolution of herbicide resistance in this species, which directly impacted its persistence. Virtually all populations are resistant to ALS inhibitors, as confirmed by surveys and random testing. Occurrence of resistance to glyphosate and synthetic auxin herbicides across the western Great Plains has increased rapidly over the past decade. A 2017 random survey of 305 kochia populations in southern Alberta, Canada found that 50% were resistant to glyphosate and 18% were resistant to dicamba, an auxinic herbicide (Beckie et al. 2019c). In contrast with those results, 5% of populations were resistant to glyphosate and non-resistant to dicamba five years previously (Hall et al. 2014). Similarly, the incidence of glyphosate resistance in about 300 kochia populations in Manitoba, Canada increased from 1% in 2013 to 59% in 2018 (Geddes et al. 2019). In the 2017 Alberta survey, 10% of populations were resistant to three sites of action: ALS inhibitors, enolpyruvylshikimate-3-phosphate synthase (EPSPS) inhibitors (glyphosate), and synthetic auxins (dicamba). There are no in-crop herbicide options to control triple-HR kochia in mustard (*Brassica juncea* or *Sinapis alba*), field pea (*Pisum sativum*), or lentil (*Lens culinaris*) (Beckie et al. 2019c). In Colorado, United States in 2014, 52%, 28%, and 20% of kochia populations surveyed were resistant to glyphosate, dicamba, and both, respectively (Westra et al. 2019).

Although primarily self-pollinated, substantial pollen-mediated gene flow and efficient seed dispersal (tumbleweed) aids both short- and long-distance spread of kochia (Figure 7.2). One tumbleweed can drop nearly 100 000 seeds over 1 km in about five minutes. Once herbicide resistance evolves in a field population, the regional spread of multiple-HR alleles in both seed and pollen is extremely rapid. In a regional genotypic study across Western Canada, glyphosate-resistant kochia populations showed little genetic differentiation (no correlation between genetic and geographic distance, i.e. high connectivity) and maintained similar genetic diversity as a result of high rates of gene flow (Martin et al. 2018; Beckie et al. 2019b). Results from the study suggested that any HR alleles will be able to spread quickly throughout all kochia populations and that the spread of these

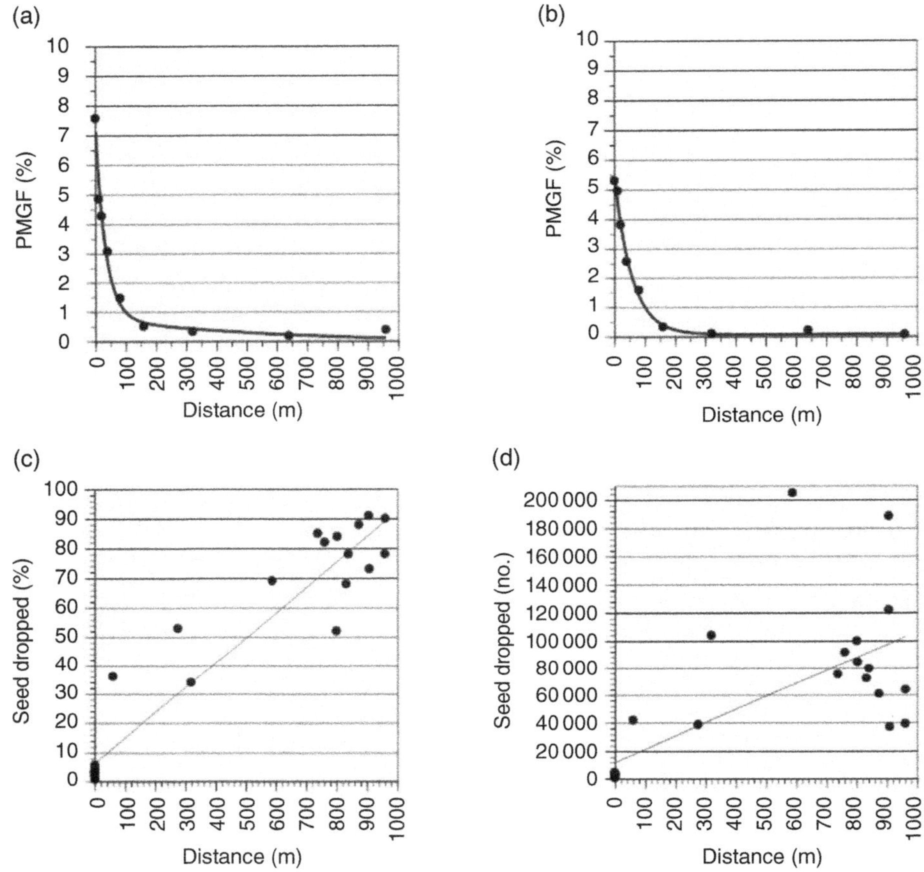

Figure 7.2 Pollen-mediated gene flow (PMGF) (% outcrossing) from glyphosate-resistant (GR) kochia pollen donor to varying distances of non-GR receptor plants at a field site at Saskatoon, Saskatchewan, Canada in 2014 (a) and 2015 (b). Kochia tumbleweed seed dropped at a field site at Lethbridge, Alberta, Canada, expressed as a percentage (c) or number (d) with increasing distance. *Source:* Adapted from Beckie et al. (2016).

alleles appears unlikely to reduce genetic variation. The tremendous magnitude of recurrent, unrelenting gene flow of HR alleles will result in frequent, widespread multiple-HR trait stacking in kochia individuals and populations, which compromises herbicidal control and facilitates persistence of the species across semiarid regions of the Great Plains of North America.

One potential weakness in the life cycle of kochia is its short seedbank longevity (<1 year), which can be exploited to reduce population abundance (Beckie et al. 2018). However, this exploitation is conditional upon mitigating HR seed movement into fields and seedbank replenishment. Glyphosate-resistant kochia has a detectable fitness penalty, which manifests as delayed germination, emergence, and flowering, as well as reduced competitiveness and fecundity (Martin et al. 2017; Beckie et al. 2018). Therefore, crop seeding and preemergence herbicide application dates may be altered to preferentially control this biotype. However, the continued lack of a coordinated community or regional response to mitigate herbicide resistance evolution and abundance of field and ruderal populations will hinder management efforts by individual growers.

7.3.3 Wild Oat in the Northern Great Plains of North America

As a widespread species originating in Eurasia, wild oat is one of the 10 worst annual weeds of temperate agricultural regions of the world (reviewed in Beckie et al. 2012). The characteristics of an "ideal" weed were described by Baker (1974) as follows: (i) has germination requirements fulfilled in many environments; (ii) has discontinuous germination and great longevity of seed; (iii) has rapid growth through vegetative phase to flowering; (iv) has continuous seed production for as long as growing conditions permit; (v) has self-compatibility but is not completely autogamous or apomictic; (vi) when cross-pollinated, has unspecialized visitors or wind-utilized; (vii) has very high seed output in favorable environmental circumstances; (viii) produces some seed under a wide range of environmental conditions, is tolerant, and is plastic; (ix) has adaptations for short- and long-distance dispersal; and (x) has the ability to compete interspecifically by special means (rosette, choking growth, allelochemicals). Key weediness traits of this highly selfing species include rapid growth, high fecundity, propensity for seed shatter, and a persistent seedbank with variable degrees of primary seed dormancy. First cohort emergence of wild oat coincides with planting and emergence of crops, although additional cohorts can emerge throughout the growing season. Wild oat is well adapted to both semiarid and subhumid climates.

Prospective (elasticity) and retrospective (LTRE) analyses were conducted using a population matrix model parameterized with wild oat demographic data from Western Canada (Tidemann et al. 2016). Population growth rates were consistently greater than one ($\lambda > 1$). Elasticity analyses indicated that λ was most highly elastic to the overwinter seedbank, followed by seedling survival, fecundity, and survival of newly shed seed. LTRE showed the stochasticity of wild oat population growth rates among years and the ability of populations to take advantage of lapses in control.

The many weedy traits and widespread herbicide resistance in wild oat have challenged effective management in some major crops and aided population persistence. The continual evolution of multiple-HR wild oat populations, seed spread via farm machinery, and limited herbicide sites of action for its control threaten sustained annual field crop production in many temperate agricultural areas. For example, prevalence of ALS plus acetyl-CoA carboxylase (ACCase) inhibitor resistance in wild oat populations in the Great Plains of North America (up to 40% of fields in some jurisdictions) renders most herbicides registered for its control ineffective. In wheat (*Triticum aestivum*) and barley (*Hordeum vulgare*), only preemergence-applied trifluralin and triallate remain for its control.

Therefore, further adoption and integration of multiple non-herbicidal weed management practices, such as enhanced crop seeding rates, competitive crops and cultivars, and precision (site-specific) management (Figure 7.3), is needed to mitigate wild oat interference with crop productivity. Recent research has focused on the potential of HWSC for reducing immigration into the weed seedbank (Tidemann et al. 2017). Depending on crop and environment, seed shatter of wild oat before harvest can vary from 13% to 76%, based on studies in Western Canada (Beckie et al. 2017). Therefore, the efficacy of HWSC to

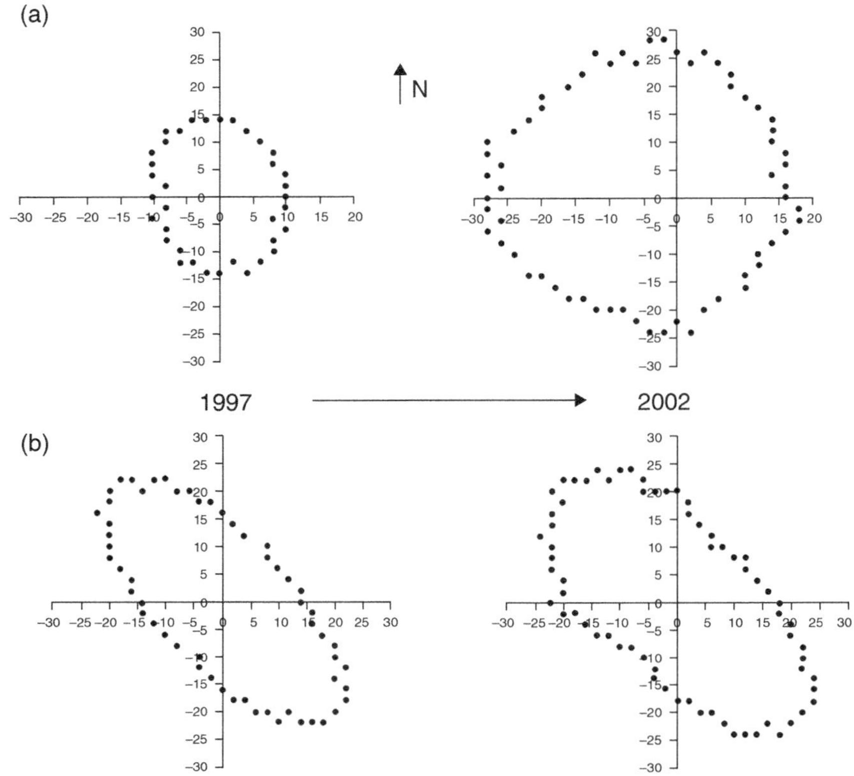

Figure 7.3 Multiple herbicide-resistant wild oat in a field in Saskatchewan, Canada: non-treated patch over a six-year period (a) and treated patch (b) through panicle clipping before seed shed (axes in meters). *Source:* Adapted from Beckie et al. (2005).

reduce population abundance and persistence will likely also be variable. Silaging annual crops such as barley before crop and weed maturity markedly reduces wild oat population abundance (Harker et al. 2016). Farming systems with significant crop diversity, which includes perennials such as alfalfa (*Medicago sativa*) or sweet clover (*Melilotus officinalis*), effectively control wild oat and mitigate herbicide resistance by reduced herbicide use and seedbank depletion (Beckie et al. 2014). However, the issues of profitability and simplicity remain challenges to their widespread adoption.

7.3.4 Wild Radish in Western Australia

Wild radish was introduced into Australia in the late 1800s as a contaminant in grain and hay (Woolls 1867). With the movement of grain, machinery, and livestock, it has colonized every state of Australia, thriving in acid sands and loams that dominate cropping regions (Gardner 1930). Wild radish is now considered the most problematic dicot weed species in the Western Australian grain belt (Alemseged et al. 2001; Borger et al. 2012).

Herbicide resistance (defined as ≥20% survival of plants tested with the recommended herbicide dose) in populations of this outcrossing species is widespread. In a random survey in the

Western Australian grain belt in 2010 (Owen et al. 2014a), 73% of wild radish populations were ALS inhibitor resistant and 38% were auxinic (2,4-D) resistant. Phytoene desaturase (PDS) (diflufenican) inhibitor resistance was found in 5% of populations. Of concern is the widespread occurrence of multiple-resistant wild radish populations (ALS inhibitor, synthetic auxin, PDS inhibitor, photosystem-II inhibitor). In a 2003 random survey, 58% of populations were resistant to two sites of action, 18% to three sites, and 7% to four sites (Walsh et al. 2007). The most prevalent multiple resistance pattern was ALS inhibitor plus synthetic auxin.

The evolutionary response of a species to selection is directly related to the genetic diversity within populations and the number of individuals under selection (Jasieniuk et al. 1996). Significant genetic variation is exhibited both within and among wild radish populations (Madhou et al. 2005). This genetic diversity is expressed by high morphological and phenological variation (Conner and Via 1993; Cheam and Code 1995).

Even low wild radish plant density can result in significant crop yield loss (Cheam and Code 1995; Code and Donaldson 1996). Moreover, survivors can produce a prolific amount of seed to rapidly replenish the soil seedbank. In a wheat crop, wild radish density of 50 plants m^{-2} produced 17 000 seeds m^{-2} (Reeves et al. 1981). In addition to having a protracted germination period, wild radish seed can remain dormant in the soil seedbank for up to seven years (Young 2001) with as much as 90% of freshly shed, pod-enclosed seed not germinating the following season (Cheam 1986). The protracted germination and multi-season dormancy of wild radish is due to the combined effects of a leachable chemical inhibitor on the seed coat and the presence of the pod around the seed that physically restricts imbibition of moisture and restricts the expansion of the embryonic root (Cheam 1986, Young 2001). The staggered germination and protracted dormancy of wild radish allows the survival of herbicide-unselected individuals, which help to maintain the genetic diversity of populations under herbicide selection (Orr and Coyne 1992). When combined with its prolific seed production, the germination and emergence pattern of the species necessitates multiple herbicide applications within or across growing seasons to reduce population abundance (Cheam 1986; Young 2001). Because of lack of seed shatter before harvest, HWSC practices are highly effective and should be employed for sustainable wild radish management (Walsh et al. 2013). However, the potential for selection of earlier-flowering ecotypes increases the risk of greater seed abscission before harvest, resulting in HWSC evasion (Ashworth et al. 2016).

7.3.5 Annual Ryegrass in Australia

Annual ryegrass is an adaptable annual grass weed found in the Mediterranean area of Southern Europe and North Africa (Bennett et al. 2000). It is a genetically variable, obligate outcrossing diploid species. Highly productive varieties of annual ryegrass were consciously selected and introduced into southern Australia in the late 1800s as a pasture species (Goggin et al. 2012). With a shift to continuous cropping in the 1980s, annual ryegrass became a major crop weed. In the last 40 years, it has been the top weed in Australia causing annual revenue losses of nearly AUD$ 100 million (Llewellyn et al. 2016).

Annual ryegrass is a prolific seed producer, and field infestations of the species can produce up to 45 000 seeds m^{-2} (Gill 1996; Stanton et al. 2002). Seed transport is easily facilitated by wind, water, animals, or farm machinery. Seeds have a short period (about two months) of primary dormancy (Steadman 2004). On average, 15–30% of the seedbank can persist into the following year (Chauhan et al. 2006), and studies estimate seedbank longevity to a maximum of four years (Narwal et al. 2008).

Its biological characteristics, a propensity to evolve herbicide resistance, widespread use of no-tillage, and an almost exclusive reliance on herbicidal control, have collectively resulted in rapid evolution of herbicide resistance in the species. Annual ryegrass in Australia is often present in crop fields, and herbicide resistance that was initially reported in 1982 (Heap and Knight 1982) has evolved to herbicides in many different groups and is widespread across the Australian grain-growing region (Boutsalis et al. 2012; Owen et al. 2014b). In Western Australia, overreliance on postemergence herbicides has led to the rapid selection of high levels of multiple resistance in populations (>95% frequency) as documented by periodic random surveys (Figure 7.4a) (Llewellyn and Powles 2001; Owen et al. 2007, 2014b).

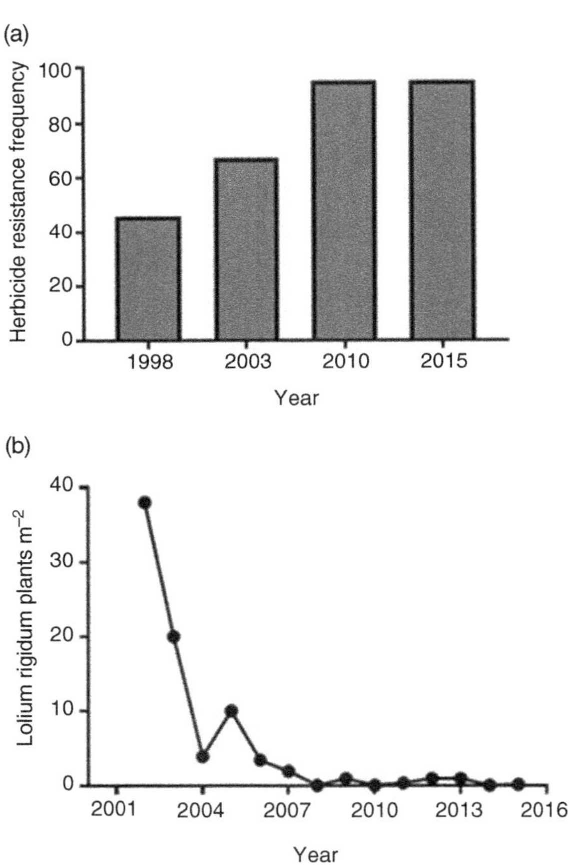

Figure 7.4 (a) Increase of the percentage frequency of herbicide (diclofop) resistance in annual ryegrass populations collected in four random surveys in Western Australia. *Source:* Adapted from Llewellyn and Powles (2001) and Owen et al. (2007, 2014b) and (b) effect of long-term adoption of integrated weed management (herbicide-based control plus harvest weed seed control) on field infestations of herbicide-resistant annual ryegrass in Western Australia. *Source:* Adapted from Walsh et al. (2013).

Because annual ryegrass populations are widespread and abundant across the southern cropping regions of Australia, the risk for pollen and seed dispersal is high. In a study conducted in Western Australia, interpopulation pollen-mediated flow of HR genes was documented at a distance of 3000 m (Busi et al. 2008). Therefore, the ability of this ubiquitous weed to spread HR alleles over long distances increases the likelihood of population persistence across agricultural landscapes frequently subjected to herbicide selection (Neve et al. 2014). Regional seed movement of annual ryegrass via contaminated crop seed or livestock feed can also be a recurring problem (Michael et al. 2010), and grain contaminated with HR annual ryegrass even impacts internationally traded wheat (Shimono et al. 2010).

Because of the rapid evolution of multiple herbicide resistance in annual ryegrass and the significant cost associated with herbicide resistance management, alternative chemical and nonchemical means to control the weed have been developed and deployed in HR population-infested fields. Growers have responded to weed resistance to postemergence herbicides by relying upon soil-applied preemergence herbicides, such as trifluralin, propyzamide, prosulfocarb, and pyroxasulfone (Busi 2014). The incidence of resistance to these preemergence herbicides is currently low relative to postemergence herbicides (Broster et al. 2011; Boutsalis et al. 2012; Owen et al. 2014b).

Similar to other troublesome weeds, there has been increasing focus on depletion of the seedbank as the best strategy to manage HR annual ryegrass. Effective control tactics have been developed based on HWSC practices (Walsh et al. 2018). Overall, HWSC can deplete annual ryegrass seedbank inputs, resulting in reduced recruitment by about 60% (Walsh et al. 2017). In a long-term study, the combination of conventional preemergence herbicide use with HWSC resulted in weed seedbank decline to very low levels (Figure 7.4b). Future monitoring is required for early detection of evolved plant traits, which may decrease the efficacy of HWSC techniques.

7.4 Conclusions

Persistence of arable weed populations is determined by the interplay among a number of factors: abundance and immigration of propagules (propagule pressure), survival and fecundity, seed shatter, seedbank persistence, efficiency of gene dispersal, effectiveness and timeliness of management practices, and environmental heterogeneity and limitations (Cousens and Mortimer 1995). These five case studies show that selection of HR traits in weed populations can increase their likelihood of persistence in agricultural fields and adjacent ruderal areas subject to recurrent herbicide application. The case studies include a volunteer HR crop (canola), two of the most common grass weeds of temperate or Mediterranean regions (wild oat and annual ryegrass), and two common broadleaf weeds found in temperate or Mediterranean regions: kochia from North America and wild radish from Australia. The proliferation of single- or multiple-HR populations of these annual weed species in different global agroecosystems reinforces the concept that herbicides are a powerful evolutionary force within cropping systems. Widespread, recurrent herbicide usage across the landscape facilitates contiguous gene flow of HR alleles and promotes population persistence.

Reducing the abundance, persistence, and spread of HR weed populations is urgently required for sustainable crop production. A mitigation strategy based on reduced weed population abundance in fields and adjacent ruderal areas by minimizing seedbank replenishment is recommended to help deter population persistence and spread. Deployment of multiple tactics or practices that encompass crop diversification, competitive crops and cultivars, judicious use of preemergence herbicides or herbicide mixtures/sequences, and weed sanitation including weed seed control are recommended by academia to sustainably manage HR weed populations and limit their persistence and spread. Periodic monitoring to assess the risk of weed population adaptation to recurrent non-herbicidal practices such as HWSC is prudent and advised. Prolonged use of any selecting agent has the potential for genotypic and phenotypic changes in weed populations to ensure survival of future generations.

Reducing the abundance, persistence, and spread of HR weed populations will become more difficult with increasingly widespread multiple herbicide site-of-action resistance in many crop weed populations and legislated pesticide-use restrictions or bans (e.g. glyphosate) in a growing number of countries worldwide. Both factors will force growers and land managers to implement alternative weed management practices, strategies, or systems (Blackshaw et al. 2008; Beckie et al. 2019a). This transition will require extensive research and development for identification of additive or synergistic integrated weed management (IWM) elements and then validation of their effectiveness and profitability in different agroecoregions. Precision or site-specific weed management technologies can facilitate the automation and adoption of IWM practices or systems. Transdisciplinary research leading to successful areawide management of mobile weeds is essential to adequately address and mitigate HR gene movement across the landscape (Bagavathiannan et al. 2019). With the involvement of numerous stakeholders including growers, land managers, agronomists, agri-businesses, weed practitioners/scientists, sociologists, economists, and local government administrators, some community-based programs in different countries have recently been initiated, which may prove to be transformative in the history of HR weed management.

References

Alemseged, Y., Jones, R.E., and Medd, R.W. (2001). A farmer survey of weed management and herbicide problems on winter crops in Australia. *Plant Protection Quarterly* 16: 21–25.

Alexander, B.C.S. (2016). Design and analysis of population demographic experiments for use in environmental risk assessments for genetically modified crops, 157 pp. MSc thesis, University of Alberta, Edmonton, AB, Canada.

Andrews, T.S. and Morrison, I.N. (1997). The persistence of trifluralin resistance in green foxtail (*Setaria viridis*) populations. *Weed Technology* 11: 369–372.

Ashworth, M.B., Walsh, M.J., Flower, K.C. et al. (2016). Directional selection for flowering time leads to adaptive evolution in *Raphanus raphanistrum* (wild radish). *Evolutionary Applications* 9: 619–629.

Bagavathiannan, M.V., Graham, S., Ma, Z. et al. (2019). Considering weed management as a social dilemma bridges individual and collective interests. *Nature Plants* 5: 343–351.

Baker, H.G. (1974). The evolution of weeds. *Annual Review of Ecology and Systematics* 5: 1–24.

Barrett, S.C.H. (1983). Crop mimicry in weeds. *Economic Botany* 37: 255–282.

Beckie, H.J. (2006). Herbicide-resistant weeds: management tactics and practices. *Weed Technology* 20: 793–814.

Beckie, H.J. and Owen, M.D.K. (2007). Herbicide-resistant crops as weeds in North America. *CAB Reviews: Perspectives in Agriculture, Veterinary Science, Nutrition and Natural Resources* 2: 1–22.

Beckie, H.J. and Warwick, S.I. (2010). Persistence of an oilseed rape transgene in the environment. *Crop Protection* 29: 509–512.

Beckie, H.J., Warwick, S.I., Nair, H., and Séguin-Swartz, G. (2003). Gene flow in commercial fields of herbicide-resistant canola (*Brassica napus*). *Ecological Applications* 13: 1276–1294.

Beckie, H.J., Hall, L.M., and Schuba, B. (2005). Patch management of herbicide-resistant wild oat (*Avena fatua*). *Weed Technology* 19: 697–705.

Beckie, H.J., Leeson, J.Y., Thomas, A.G. et al. (2008). Risk assessment of weed resistance in the Canadian prairies. *Weed Technology* 22: 741–746.

Beckie, H.J., Hall, L.M., Simard, M.-J. et al. (2010). A framework for postrelease environmental monitoring of second-generation crops with novel traits. *Crop Science* 50: 1587–1604.

Beckie, H.J., Francis, A., and Hall, L.M. (2012). The biology of Canadian weeds. 27. *Avena fatua* L. (updated). *Canadian Journal of Plant Science* 92: 1329–1357.

Beckie, H.J., Johnson, E.N., Leeson, J.Y. et al. (2014). Selection and evolution of acetyl-CoA carboxylase (ACC)-inhibitor resistance in wild oat (*Avena fatua* L.) in a long-term alternative cropping systems study. *Canadian Journal of Plant Science* 94: 727–731.

Beckie, H.J., Blackshaw, R.E., Hall, L.M., and Johnson, E.N. (2016). Pollen- and seed-mediated gene flow in kochia (*Kochia scoparia*). *Weed Science* 64: 624–633.

Beckie, H.J., Blackshaw, R.E., Harker, K.N., and Tidemann, B.D. (2017). Weed seed shatter in spring wheat in Alberta. *Canadian Journal of Plant Science* 98: 107–114.

Beckie, H.J., Blackshaw, R.E., Leeson, J.Y. et al. (2018). Seedbank persistence, germination and early growth of glyphosate-resistant *Kochia scoparia*. *Weed Research* 58: 177–187.

Beckie, H.J., Ashworth, M.B., and Flower, K.C. (2019a). Herbicide resistance management: recent developments and trends. *Plants* 8 (6): 161. https://doi.org/10.3390/plants8060161.

Beckie, H.J., Busi, R., Bagavathiannan, M.V., and Martin, S.L. (2019b). Herbicide resistance gene flow in weeds: under-estimated and under-appreciated. *Agriculture, Ecosystems & Environment* 283: e106566. https://doi.org/10.1016/j.agee.2019.06.005.

Beckie, H.J., Hall, L.M., Shirriff, S.W. et al. (2019c). Triple-resistant kochia [*Kochia scoparia* (L.) Schrad.] in Alberta. *Canadian Journal of Plant Science* 99: 281–285.

Bennett, S.J., Hayward, M.D., and Marshall, D.F. (2000). Morphological differentiation in four species of the genus *Lolium*. *Genetic Resources and Crop Evolution* 47: 247–255.

Blackshaw, R.E., Harker, K.N., O'Donovan, J.T. et al. (2008). Ongoing development of integrated weed management systems on the Canadian prairies. *Weed Science* 56: 146–150.

Borger, C.P.D., Michael, P.J., Mandel, R. et al. (2012). Linking field and farmer surveys to determine the most important changes to weed incidence. *Weed Research* 52: 564–574.

Boutsalis, P., Gill, G., and Preston, C. (2012). Incidence of herbicide resistance in rigid ryegrass (*Lolium rigidum*) across southeastern Australia. *Weed Technology* 26: 391–398.

Broster, J., Koetz, E., and Wu, H. (2011). Herbicide resistance levels in annual ryegrass (*Lolium rigidum* Gaud.) in southern New South Wales. *Plant Protection Quarterly* 26: 22–28.

Brûlé-Babel, A.L., Willenborg, C.J., Friesen, L.F., and Van Acker, R.C. (2006). Modeling the influence of gene flow and selection pressure on the frequency of a GE herbicide-tolerant trait in non-GE wheat and wheat volunteers. *Crop Science* 46: 1704–1710.

Busi, R. (2014). Resistance to herbicides inhibiting the biosynthesis of very-long-chain fatty acids. *Pest Management Science* 70: 1378–1384.

Busi, R. and Powles, S. (2016). Transgenic glyphosate-resistant canola (*Brassica napus*) can persist outside agricultural fields in Australia. *Agriculture, Ecosystems & Environment* 220: 28–34.

Busi, R., Yu, Q., Barrett-Lennard, R., and Powles, S. (2008). Long distance pollen-mediated flow of herbicide resistance genes in *Lolium rigidum*. *Theoretical and Applied Genetics* 117: 1281–1290.

Carvalho, S.J.P., Gonçalves Netto, A., Nicolai, M. et al. (2015). Detection of glyphosate-resistant Palmer amaranth (*Amaranthus palmeri*) in agricultural areas of Mato Grosso, Brazil. *Planta Daninha* 33: 579–586.

Caswell, H. (2001). *Matrix Population Models: Construction, Analysis, and Interpretation*, 792 pp, 2e. Sunderland, Maryland, USA: Sinauer Associates.

Cavalieri, A., Harker, K.N., Hall, L.M. et al. (2016). Evaluation of the causes of on-farm harvest losses in canola in the Northern Great Plains. *Crop Science* 56: 2005–2015.

Chauhan, B.S., Gill, G., and Preston, C. (2006). Influence of tillage systems on vertical distribution, seedling recruitment and persistence of rigid ryegrass (*Lolium rigidum*) seed bank. *Weed Science* 54: 669–676.

Cheam, A.H. (1986). Seed production and seed dormancy in wild radish (*Raphanus raphanistrum*) and some possibilities for improving control. *Weed Research* 26: 405–413.

Cheam, A.H. and Code, G.R. (1995). The biology of Australian weeds: 24 *Raphanus raphanistrum* L. *Plant Protection Quarterly* 10: 2–13.

Code, G.R. and Donaldson, T.W. (1996). Effect of cultivation, sowing methods and herbicides on wild radish populations in wheat crops. *Australian Journal of Experimental Agriculture* 36: 437–442.

Conner, J. and Via, S. (1993). Patterns of phenotypic and genetic correlations among morphological and life history traits in wild radish, *Raphanus raphanistrum*. *Evolution* 47: 704–711.

Cousens, R. and Mortimer, M. (1995). *Dynamics of Weed Populations*, 331 pp. UK: Cambridge University Press.

Crawley, M.J., Hails, R.S., Rees, M. et al. (1993). Ecology of transgenic oilseed rape in natural habitats. *Nature* 363: 620–623.

Dauer, J.T., Luschei, E.C., and Mortensen, D.A. (2009). Effects of landscape composition on spread of an herbicide-resistant weed. *Landscape Ecology* 24: 735–747.

Délye, C., Clément, J.A.J., Pernin, F. et al. (2010). High gene flow promotes the genetic homogeneity of arable weed populations at the landscape scale. *Basic and Applied Ecology* 11: 504–512.

Diggle, A.J. and Neve, P. (2001). The population dynamics and genetics of herbicide resistance – a modeling approach. In: *Herbicide Resistance and World Grains* (eds. S.B. Powles and D.L. Shaner), 61–99. New York, NY: CRC Press.

Friesen, L.F., Nelson, A.G., and Van Acker, R.C. (2003). Evidence of contamination of pedigreed canola (*Brassica napus*) seedlots in western Canada with genetically engineered herbicide resistance traits. *Agronomy Journal* 95: 1342–1347.

Friesen, L.F., Beckie, H.J., Warwick, S.I., and Van Acker, R.C. (2009). The biology of Canadian weeds. 138. *Kochia scoparia* (L.) Schrad. *Canadian Journal of Plant Science* 89: 141–167.

Gardner, C.A. (1930). *A Systematic Census of Plants Occurring in Western Australia*, 150 pp. Perth, WA: Government Printers.

Geddes, C.M. Ostendorf, T.E., Gulden, R.H., et al. (2019). Rapid spread of glyphosate-resistant kochia [*Bassia scoparia* (L.) A.J. Scott] in Manitoba. https://36f.606.myftpupload.com/wp-content/uploads/2020/02/CWSS-SCM-2019-Kelowna_Meeting-package-v6.pdf (accessed 21 June 2021).

Gill, G.S. (1996). Why annual ryegrass is a problem in Australian agriculture. *Plant Protection Quarterly* 11: 193–195.

Goggin, D.E., Powles, S.B., and Steadman, K.J. (2012). Understanding *Lolium rigidum* seeds: the key to managing a problem weed? *Agronomy* 2: 222–239.

Gulden, R.H., Shirtliffe, S.J., and Thomas, A.G. (2003). Harvest losses of canola (*Brassica napus*) cause large seedbank inputs. *Weed Science* 51: 83–86.

Gulden, R.H., Thomas, A.G., and Shirtliffe, S.J. (2004). Secondary dormancy, temperature, and burial depth regulate seedbank dynamics in canola. *Weed Science* 52: 382–388.

Haile, T.A., Gulden, R.H., and Shirtliffe, S.J. (2014). On-farm seed loss does not differ between windrowed and direct-harvested canola. *Canadian Journal of Plant Science* 94: 785–789.

Hall, L., Topinka, K., Huffman, J. et al. (2000). Pollen flow between herbicide-resistant *Brassica napus* is the cause of multiple-resistant *B. napus* volunteers. *Weed Science* 48: 688–694.

Hall, L.M., Beckie, H.J., Low, R. et al. (2014). Survey of glyphosate-resistant kochia (*Kochia scoparia* L. Schrad.) in Alberta. *Canadian Journal of Plant Science* 94: 127–130.

Harker, K.N., Clayton, G.W., Blackshaw, R.E. et al. (2006). Persistence of glyphosate-resistant canola in western Canadian cropping systems. *Agronomy Journal* 98: 107–119.

Harker, K.N., O'Donovan, J.T., Turkington, T.K. et al. (2016). Diverse rotations and optimal cultural practices control wild oat (*Avena fatua*). *Weed Science* 64: 170–181.

Heap, I. and Knight, R. (1982). A population of ryegrass tolerant to the herbicide diclofop-methyl. *The Journal of the Australian Institute of Agricultural Science* 48: 156–157.

Jasieniuk, M., Brûlé-Babel, A.L., and Morrison, I.N. (1996). The evolution and genetics of herbicide resistance in weeds. *Weed Science* 44: 176–193.

Jordan, N.R. and Jannink, J.L. (1997). Assessing the practical importance of weed evolution: a research agenda. *Weed Research* 37: 237–246.

Knispel, A.L. and McLachlan, S.M. (2010). Landscape-scale distribution and persistence of genetically modified oilseed rape (*Brassica napus*) in Manitoba, Canada. *Environmental Science and Pollution Research* 17: 13–25.

Kremer, E. and Lotz, L.A.P. (1998). Germination and emergence characteristics of triazine-susceptible and triazine-resistant biotypes of *Solanum nigrum*. *Journal of Applied Ecology* 35: 302–310.

Leeson, J.Y., Gaultier, J., and Grenkow, L. (2017). Manitoba weed survey of annual crops in 2016. Weed Survey Series Publ. 17-2, 213 pp. Agriculture and Agri-Food Canada, Saskatoon, SK, Canada.

Légère, A., Beckie, H.J., Stevenson, F.C., and Thomas, A.G. (2000). Survey of management practices affecting the occurrence of wild oat (*Avena fatua*) resistance to acetyl-CoA carboxylase inhibitors. *Weed Technology* 14: 366–376.

Llewellyn, R.S. and Powles, S.B. (2001). High levels of herbicide resistance in rigid ryegrass (*Lolium rigidum*) in the wheat belt of Western Australia. *Weed Technology* 15: 242–248.

Llewellyn, R.S. et al. (2016). Impact of weeds on Australian grain production: the cost of weeds to Australian grain growers and the adoption of weed management and tillage practices. 112 pp. Report for GRDC, CSIRO, Australia

Madhou, P., Wells, A., Pang, E., and Stevenson, T. (2005). Genetic variation in populations of Western Australian wild radish. *Australian Journal of Agricultural Research* 56: 1079–1087.

Martin, S.L., Benedict, L., Sauder, C.A. et al. (2017). Glyphosate resistance reduces kochia fitness: comparison of segregating resistant and susceptible F2 populations. *Plant Science* 261: 69–79.

Martin, S.L., Benedict, L., Sauder, C.A. et al. (2018). Kochia population genetics: panmictic prairie populations. Proceedings of 2018 Canadian Weed Science Society Annual Meeting, Pinawa, MB, Canada (19–22 November 2018). pp. 34. http://www.weedscience.ca (accessed 30 November 2018).

Michael, P.J., Owen, M.J., and Powles, S.B. (2010). Herbicide-resistant weed seeds contaminate grain sown in the Western Australian grainbelt. *Weed Science* 58: 466–472.

Molin, W.T., Wright, A.A., VanGessel, M.J. et al. (2018). Survey of the genomic landscape surrounding the 5-enolpyruvylshikimate-3-phosphate synthase (EPSPS) gene in glyphosate-resistant *Amaranthus palmeri* from geographically distant populations in the USA. *Pest Management Science* 74: 1109–1117.

Mortimer, A.M. (1997). Phenological adaptation in weeds – an evolutionary response to the use of herbicides? *Pesticide Science* 51: 299–304.

Moss, S.R. (2002). Herbicide-resistant weeds. In: *Weed Management Handbook* (ed. R.E.L. Naylor), 225–252. Oxford, UK: British Crop Protection Council.

Narwal, S., Sindel, B.M., and Jessop, R.S. (2008). Dormancy and longevity of annual ryegrass (*Lolium rigidum*) as affected by soil type, depth, rainfall, and duration of burial. *Plant and Soil* 310: 225–234.

Neve, P., Diggle, A.J., Smith, F.P., and Powles, S.B. (2003). Simulating evolution of glyphosate resistance in *Lolium rigidum* I: population biology of a rare resistance trait. *Weed Research* 43: 404–417.

Neve, P., Busi, R., Renton, M., and Vila-Aiub, M.M. (2014). Expanding the eco-evolutionary context of herbicide resistance research. *Pest Management Science* 70: 1385–1393.

Orr, H.A. and Coyne, J.A. (1992). The genetics of adaptation: a reassessment. *The American Naturalist* 140: 725–742.

Owen, M.J., Walsh, M.J., Llewellyn, R.S., and Powles, S.B. (2007). Widespread occurrence of multiple herbicide resistance in Western Australian annual ryegrass (*Lolium rigidum*) populations. *Australian Journal of Agricultural Research* 58: 711–718.

Owen, M.J., Michael, P.J., Renton, M. et al. (2011). Towards large-scale prediction of *Lolium rigidum* emergence: II. Correlation between dormancy and herbicide resistance levels suggests an impact of cropping systems. *Weed Research* 51: 133–141.

Owen, M.J., Martinez, N.J., and Powles, S.B. (2014a). Multiple herbicide-resistant wild radish (*Raphanus raphanistrum*) populations dominate Western Australian cropping fields. *Crop & Pasture Science* 66: 1079–1085.

Owen, M.J., Martinez, N.J., and Powles, S.B. (2014b). Multiple herbicide-resistant *Lolium rigidum* (annual ryegrass) now dominates across the Western Australian grain belt. *Weed Research* 54: 314–324.

Reeves, T.G., Code, G.R., and Piggin, C.M. (1981). Seed production and longevity, seasonal emergence and phenology of wild radish (*Rahanus raphanistrum*). *Australian Journal of Experimental Agriculture* 21: 524–530.

Shields, E.J., Dauer, J.T., VanGessel, M.J., and Neumann, G. (2006). Horseweed (*Conyza canadensis*) seed collected in the planetary boundary layer. *Weed Science* 54: 1063–1067.

Shimono, Y., Takiguchi, Y., and Konuma, A. (2010). Contamination of internationally traded wheat by herbicide-resistant *Lolium rigidum*. *Weed Biology and Management* 10: 219–228.

Stanton, R., Piltz, J., Pratley, J. et al. (2002). Annual ryegrass (*Lolium rigidum*) seed survival and digestibility in cattle and sheep. *Australian Journal of Experimental Agriculture* 42: 111–115.

Steadman, K.J. (2004). Dormancy release during hydrated storage in *Lolium rigidum* seeds is dependent on temperature, light quality, and hydration status. *Journal of Experimental Botany* 55: 929–937.

Thill, D.C. and Mallory-Smith, C.A. (1997). The nature and consequence of weed spread in cropping systems. *Weed Science* 45: 337–342.

Tidemann, B.D., Hall, L.M., Harker, K.N., and Alexander, B.C.S. (2016). Identifying critical control points in the wild oat (*Avena fatua*) life cycle and the potential effects of harvest weed-seed control. *Weed Science* 64: 463–473.

Tidemann, B.D., Hall, L.M., Harker, K.N., and Beckie, H.J. (2017). Factors affecting weed seed devitalization with the Harrington Seed Destructor. *Weed Science* 65: 650–658.

Van Acker, R.C. (2009). Weed biology serves practical weed management. *Weed Research* 49: 1–5.

Vila-Aiub, M.M., Neve, P., and Powles, S.B. (2009). Fitness costs associated with evolved herbicide resistance alleles in plants. *New Phytologist* 184: 751–767.

Walsh, M.J., Owen, M.J., and Powles, S.B. (2007). Frequency and distribution of herbicide resistance in *Raphanus raphanistrum* populations randomly collected across the Western Australian wheatbelt. *Weed Research* 47: 542–550.

Walsh, M., Newman, P., and Powles, S. (2013). Targeting weed seeds in-crop: a new weed control paradigm for global agriculture. *Weed Technology* 27: 431–436.

Walsh, M.J., Aves, C., and Powles, S.B. (2017). Harvest weed seed control systems are similarly effective on rigid ryegrass. *Weed Technology* 31: 178–183.

Walsh, M.J., Broster, J.C., Schwartz-Lazaro, L.M. et al. (2018). Opportunities and challenges for harvest weed seed control in global cropping systems. *Pest Management Science* 74: 2235–2245.

Westra, E.P., Nissen, S.J., Getts, T.J. et al. (2019). Survey reveals frequency of multiple resistance to glyphosate and dicamba in kochia (*Bassia scoparia*). *Weed Technology* 33: 664–672.

Woolls, F.L.S. (1867). Notes on introduced plants occurring in the neighbourhood of Sydney. *Botanical Journal* 10: 35–42.

Yoshimura, Y., Beckie, H.J., and Matsuo, K. (2006). Transgenic oilseed rape along transportation routes and port of Vancouver in western Canada. *Environmental Biosafety Research* 5: 67–75.

Young, K.R. (2001). Germination and emergence of wild radish (*Raphanus raphanistrum* L.). PhD dissertation. University of Melbourne, Australia.

8

Seed Predation and Weed Seedbanks

Pavel Saska and Alois Honěk

Functional Biodiversity Group, Crop Research Institute, Praha, Czech Republic

8.1 Introduction

Although mortality can happen at any phase during a weed's life cycle, in many species the greatest mortality occurs during the seed and seedling establishment phases (Turnbull et al. 2000; Myers and Sarfraz 2017). The causes of seed death include imperfect or failed germination; seed mortality due to microorganisms, pathogens, and predators; or physiological death due to senescence in soil conditions unsuitable for seed germination (Dalling et al. 2011). Seed predation, i.e. seed consumption by vertebrates and invertebrates, is considered to be one of the drivers of plant population dynamics (Larios et al. 2017) and selection pressure for seed defence (Dalling et al. 2011). Other terms for seed predation used in the literature are 'seed feeding', 'granivory', and 'spermophagy'. However, we prefer using the term 'seed predation' because it unambiguously describes the functional aspect of the interaction of the seed as the prey item that is killed and therefore removed from the population by the predator.

Seed predation is a multi-faceted ecological phenomenon that involves evolution of protective physical and chemical adaptations of plant/seed on one side and behavioural, morphological, and physiological adaptations of predators related to seed location, handling, and digestion on the other side. Seed predation represents an important aspect of the natural weed regulation, which may become an important part of integrated weed regulating systems (Petit et al. 2018). Seed predation research has become interdisciplinary, integrating not only botany with zoology/entomology but also incorporating agronomy, agroecology, landscape ecology, behavioural ecology and evolutionary ecology, plant physiology and phytochemistry, and microbiology.

In this chapter we examine the ecology of seed predation, with regard to weed persistence. This chapter is not meant to be an exhaustive review, but rather represents an overview of topics that need to be understood, before an input of weed seeds to the soil bank in arable settings can be managed through seed predation.

Persistence Strategies of Weeds, First Edition. Edited by Mahesh K. Upadhyaya, David R. Clements, and Anil Shrestha.

8.2 Predators and Seed Predation Windows in the Life Cycle of a Weed

There are several phases in the life cycle of a weed (or any plant), when seeds are subjected to predation. In this chapter we call these phases 'seed predation windows'. We identify three seed predation windows: (i) pre-dispersal seed predation on plant, (ii) post-dispersal seed predation on soil surface, and (iii) predation on seeds that are in the seedbank.

8.2.1 Pre-dispersal Seed Predation

Seeds can be attacked on the plant before they are fully mature or after maturation but before dispersal from the mother plant. The guild of pre-dispersal seed predators is diverse and includes both vertebrates (mainly granivorous birds and rodents) and invertebrates (Kolb et al. 2007). The impact of pre-dispersal seed predation on seed mortality has been more frequently studied in natural than in agricultural settings (Kolb et al. 2007). Much of the research on weed seed predation has focused on invertebrates. Most invertebrate pre-dispersal seed predators are specialists in the orders Coleoptera, Diptera, Lepidoptera, and Hymenoptera (Kolb et al. 2007). As an example, in one study at the time of seed maturation, the guild of invertebrate seed predators of *Taraxacum* spp. comprised 10 species of Coleoptera, Diptera, Lepidoptera, and Thysanoptera (Honek and Martinkova 2005). Although most of pre-dispersal predation can be attributed to larvae, adult insects also contribute to seed mortality on plants (Koprdova et al. 2015). The impact of the predators on the seed crop differs according to their feeding manner. Insects with chewing mouthparts, such as beetles (Figure 8.1a) and larvae of Lepidoptera and Hymenoptera, ultimately

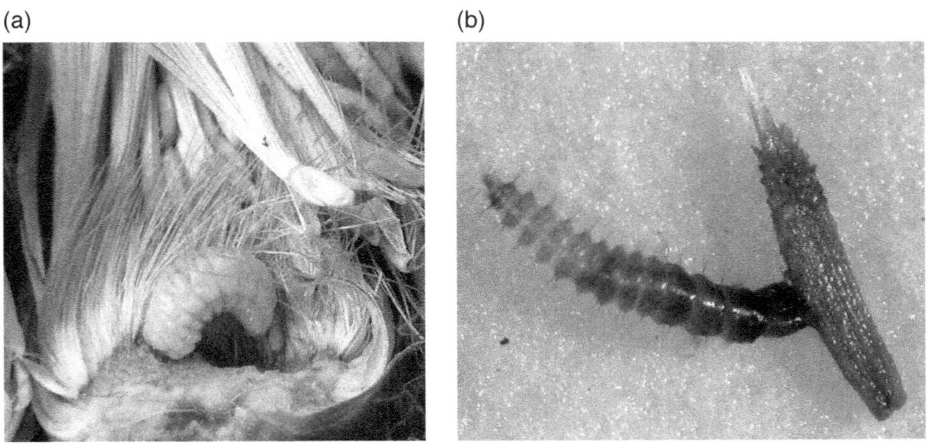

Figure 8.1 Seed predators. (a) Mature larva of *Larinus turbinatus* Gyllenhal (Coleoptera: Curculionidae) feeding in the flower head of *Cirsium heterophyllum* (L.) Hill. (Asteraceae) Czech Republic, Hrabětice. Actual size of the larva: 10 mm. *Source:* Photo credit: Jiří Skuhrovec. (b) First instar larva of *Amara montivaga* Sturm (Coleoptera: Carabidae) feeding on a seed of *Taraxacum* spp. agg. (Asteraceae). Czech Republic, Prague-Ruzyně. Actual size of the larva: 4 mm. *Source:* Photo credit: Pavel Saska.

destroy the seeds they attack (DeSousa et al. 2003). Larvae of Diptera (mainly Tephritidae and Cecidomyiidae) and Hemiptera suck the nutritious fluids from the seeds and attacked and intact seeds within the same inflorescence are almost impossible to distinguish (Koprdova et al. 2015). However, the viability of seeds from infected *Centaurea cyanus* inflorescences was reduced by 40% compared with intact ones (Koprdova et al. 2015). Infestation by pre-dispersal predators varies substantially, as exemplified by *Amaranthus retroflexus* in which the number of destroyed seeds ranged from 0% to 80% per plant, although the seed loss due to predation was below 10% in most plants (DeSousa et al. 2003). While pre-dispersal seed predation per plant is generally low, it can be substantial over a local area (Koçak et al. 2007). Examples from nonagricultural settings indicate that plant population dynamics can still be affected by pre-dispersal seed predation (Kolb et al. 2007), and success stories of practical weed control using pre-dispersal seed predators provide enough support for the importance of pre-dispersal predators in shaping weed population dynamics (see Section 8.5).

8.2.2 Post-dispersal Seed Predation

After dispersal from the mother plant, seeds may stay exposed on the soil surface for some time, before they germinate or become incorporated in the soil by rain and soil disturbance or by the activity of soil fauna. It is generally assumed that seeds are buried within a few weeks after release, if they do not germinate or are not eaten by predators. This assumption was tested by Westerman et al. (2009) using beads of different sizes. They observed that a significant proportion of seeds became inaccessible to predators immediately after release from the plant as they were hidden in the soil cracks or under plant residue. As the rate of burial is inversely related to bead size, they concluded that larger seeds are in general at greater risk of predation than small ones as they stay on the soil surface longer. Post-dispersal predation is, therefore, a periodic event concentrated in a relatively short period of time of seed availability.

The guild of post-dispersal predators of weed seeds includes birds (Holland et al. 2006), rodents (Daedlow et al. 2014), and various invertebrate groups such as carabid beetles (Saska et al. 2008) and their larvae (Figure 8.1b), ants (Minkey and Spafford 2016), crickets (O'Rourke et al. 2006), slugs (Honek et al. 2017), and terrestrial isopods (Saska 2008). However, the range of invertebrate seed predators is probably much wider than we currently know (Lundgren et al. 2013; Frei et al. 2019). Although the literature suggests that post-dispersal weed seed predators are mainly generalists (Bohan et al. 2011; McHugh et al. 2020), specific food requirements in many species indicate tight associations with particular plant species (Talarico et al. 2016 and references therein; Saska et al. 2019b).

The relative importance of particular taxonomic groups of weed seed predators is inconsistent in literature reports. Among the 33 papers that we compared for vertebrate and invertebrate seed predation, 14 studies reported that vertebrates, usually rodents, were important and 16 found that invertebrates, mainly carabids, crickets, or ants, were more important than vertebrates. In three studies, both groups were similarly important or the relative importance of each group of predators varied among seed species, which indicates that seeds are not equally palatable to all predators (see Section 8.3). Climate-related or regional differences may also play a role. For example, ants are the dominant weed seed predators in semiarid and arid zones (Baraibar et al. 2009; Evans and Gleeson 2016) as they are well adapted to warm climates; however, their importance is lesser in irrigated fields (Baraibar et al. 2009). Crickets were considered as top seed predators in studies in the United States and Asia (O'Rourke et al. 2006; Davis and Raghu 2010, Ichihara et al. 2011), but carabid beetles dominated the invertebrate seed predator guild in arable fields in temperate Europe (Saska et al. 2008; Baraibar et al. 2012).

The magnitude of weed seed predation varies among studies. Due to methodological problems associated with estimating seed predation estimates in the field (discussed in detail in Section 8.7), we offer examples using studies that provide estimates of annual weed seed losses. If continuous seed exposure was assumed, the range of estimated annual weed seed loss was 32–70% in a study of Westerman et al. (2003b), depending on the species. Using the data from multiple studies and species of seeds, the range of annual seed loss was estimated to be 2–89% (Davis et al. 2011); however, in this analysis a four-day seed exposure was assumed. This variability may be due to the predator preference for certain seeds (Section 8.3) and the presence of the key predator species (Carbonne et al. 2020). The importance of post-dispersal seed predation for weed population dynamics is further discussed in Section 8.5.

8.2.3 Seed Predation from the Seedbank and After Release from the Bank

Predation of weed seeds that persist in the soil is poorly studied and apparently underestimated. Since the seedbank provides reserves for future weed populations, more attention to this issue should be given. In general, surface-active predators have difficulties finding seeds covered by even a thin layer of soil (Hulme 1998; White et al. 2007; Kulkarni et al. 2015), and progressively deeper burial makes it more difficult to locate the seeds (Kulkarni et al. 2015).

Following burial in soil, a suite of edaphic seed predators may further deplete the seedbank. Earthworms are probably the most important seed predators at this stage. Some seeds may survive passage through the earthworm gut and retain germinability, but earthworms are able to obtain nutrients from many types of seeds (Forey et al. 2011). Earthworms selectively feed on particular seed species and seem to respond to volatiles emanating from the seeds, enabling them to be efficient seed locators and seed patch exploiters (Clause et al. 2017). On the other hand, storage of seeds in earthworm burrows provides safe sites for weed seeds (Schutte et al. 2010) and supports seed persistence at a given site. This applies also to epigeal predators that form underground seed caches, such as carabid larvae of the tribe Harpalini (Talarico et al. 2016), ants (Minkey and Spafford 2016), or rodents (Hulme 1998).

Due to biotic soil perturbation or tillage, seeds from the soil bank may eventually come to the surface and become exposed to predators for a second time. Very little is known about seed predation after exhumation from the seedbank. Following a half-year burial, the acceptance rate of four species of seeds by two carabid species was unchanged, one species of seed was accepted at a higher rate than in the fresh state (*Tripleurospermum inodorum*), while one lost its attractiveness to seed predators (*Taraxacum* spp. agg.) (Martinkova et al. 2006). Extended periods in the seed bank weather seeds by reducing the quality of their seed coat (toughness and elasticity) and erode seed outgrowths such as trichomes,

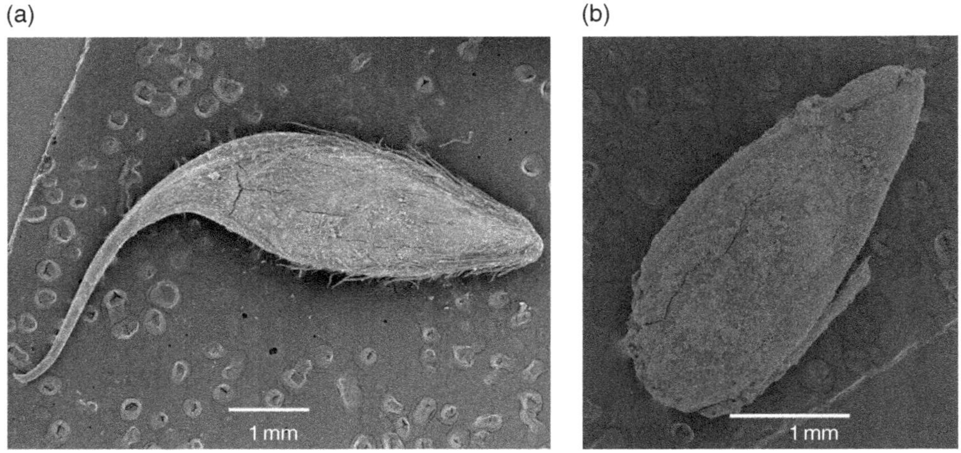

Figure 8.2 Changes in external morphology of *Geum urbanum* L. (Rosaceae) seeds resulting from burial in soil. Czech Republic, Prague-Ruzyně. (a) Fresh seed and (b) seed after seven years of burial. Photo credit: Hana Foffová.

hooks, wings, etc. (Saska et al. 2020; Figure 8.2). These changes may substantially increase or reduce seed acceptability to predators. For example, excavated seeds of *Persicaria lapathifolia* that were buried for six years were eaten more than fresh seeds by three species of carabids; however, the response was opposite in weathered seeds (Saska et al. 2019a). In addition to morphological changes, chemical processes such as leaching and degradation of chemical compounds protecting the seed coat or softening of lignin impregnation also occur (Dalling et al. 2011). These changes in seed properties may be due to, or interact with, the activity of seed surface-colonizing soil microorganisms (Dalling et al. 2011). Microbial communities and metabolic activity most likely influence the attractiveness to predators of seeds from the soil bank, but these processes remain largely unknown and warrant further research.

8.3 Seed Defence Versus Seed Selection by Predators

Seeds constitute a nutritionally rich and abundant food source. During seed formation, the mother plant concentrates reserves in the seeds to provide enough energy for germination even after a prolonged stay in the soil seedbank, thus the amount of reserves is directly linked to plant fitness (Bretagnolle et al. 2016). The storage molecules in seeds are mainly starches and fatty acids, which are compounds of high caloric value for seed predators (Gaba et al. 2014); this is what makes the seeds attractive (Fischer and Turke 2016). Gaba et al. (2019) reported that carabid beetles preferentially fed on seeds rich in lipids.

Seeds are not defenceless; in fact, they are the most heavily defended part of plants (Dalling et al. 2011). The risk of predation and pathogen attack creates significant selection pressure on defensive sets of traits, called seed defence syndromes (Dalling et al. 2011). The strategies of seed defence are diverse and generally related to seed persistence. Transient seeds do not form a seedbank and mechanical and chemical barriers against their consumption are weak (Thompson et al. 1997). Their strategy is to escape from predators by germinating shortly after dispersal from the mother plant. By contrast, seeds that persist in the soil and form a persistent seedbank are usually well protected against predation, either physically or chemically.

Physical defence is usually accomplished by seed morphology. For example, a typical adaptation is a hard seed coat, which is difficult to penetrate. The resistance of the seed coat depends on its toughness, thickness, and the degree of weathering. Lundgren and Rosentrater (2007) measured seed destruction rates by three invertebrate species in relation to seed structural characteristics. They found that seed density and compression strength affected consumption positively and seed mass and volume negatively, but not in all predator species. More research on the mechanical resistance of seeds against predation is needed. It is, however, important to distinguish between the effects of seed structure from differences in attractiveness between species of seeds based on other properties. In order to understand this relationship better, experiments need to be designed utilizing species of weeds that have polymorphic seeds. Only initial steps have been made in this direction. Honek et al. (2017) demonstrated large differences in consumption by carabid beetles of varied seed morphs produced by *Atriplex sagittata*. Polymorphic seeds (the morphs are here referred to as 'black' and 'brown') are dispersed in the autumn and over winter on the ground surface, protected from predation by low temperatures. In the spring, 'black seeds', which are slow to germinate and destined to enter the soil seedbank, are exposed to predation by *Pseudoophonus rufipes*, a medium–large carabid species. 'Brown' seeds are palatable to carabids but escape seed predation by germinating quickly; however, resulting seedlings will later fall victim to predation by slugs.

Seed coat texture is another character that may affect seed acceptance by predators. Bojňanský and Fargašová (2007) analyzed the seed morphology of Central European weed species. Their classification distinguishes 20 types of three-dimensional seed shapes, 24 types of seed surfaces, and at least 14 types of seed protuberances (ribs, hairs, spines, wings, etc.). Little is known of the effects of the various seed types on seed predation. One would expect that globular seeds with smooth testas would be avoided compared with seeds with elongated or angular shape, because the latter may be more easily grasped and squeezed by chewing seed predators. We found several examples in choice experiments where smooth and rounded seeds were rejected by carabid beetles, while seeds of similar size and shape endowed with fine surface structures were accepted (Honek et al. 2007). This may relate to differences observed between the smooth rejected seeds of *Amaranthus retroflexus* (surface smooth or faint verrucose, glossy) and the accepted seeds of *Capsella bursa-pastoris* (surface bare, lustrous, reticulate scalariform).

Physical seed defence may be effective only to a subset of foraging seed predators. The ratio of seed size to predator size is an important determinant of seed predation (Saska et al. 2019b). Within each taxonomic category of the predators, there is a large variation in body size. For instance, the dry body mass of common central European beetle predators varies between 1 mg (*Trechus quadristriatus*) and 30 mg (*Pseudoophonus rufipes*) (Honek et al. 2007). Vertebrate predators (mice and birds) are heavier, and they may weigh up to tens or hundreds of grams. The variation of body size (mass) of weed seed predators thus extends at least over four orders of magnitude. There is considerable evidence to show that larger carabid beetle species prefer larger seeds, while smaller species utilize smaller seeds (Honek et al. 2007; Saska et al. 2019b). In the field, rodents prefer larger species of seeds, while invertebrates preferentially feed on smaller ones (Brust and House 1988). The fundamental differences between seed predators is not limited to their size but can also be influenced by their body structure. There are enormous differences in size, form, and performance of the mouthparts and the alimentary tract. Chewing mouthparts are typical for seed predators but their mechanism is different between vertebrate and invertebrate seed predators. The differences involve the number, position, movement, and shape of the mouthparts. While in some groups of seed predators such as birds, the relationship between shape of the mouthparts (beak) and diet are relatively well known; in other seed predators, taxa studies have only been conducted at a basic level, for example, for invertebrate groups where the mouthpart size and structure are clearly limiting factors for handling seeds. Specialized granivorous carabids (Acorn and Ball 1991) and their larvae (Zetto Brandmayr et al. 1998) have typically broad head and stout mandibles that can create forces strong enough to crush even very hard seeds.

Chemical defence is based on secondary metabolites concentrated in seeds. Although these metabolites are often present in the seeds in higher concentrations than in the plant (Dalling et al. 2011), the effectiveness against seed predators has not been investigated. The increase in consumption of *Tripleurospermum inodorum* after six months burial, compared with fresh seeds, was speculatively attributed to the loss in chemical defence (Martinkova et al. 2006), but empirical support for this conclusion is lacking. Recent data have shown that invertebrates as well as vertebrates are able to respond to volatiles released from seeds (Paulsen et al. 2013; Clause et al. 2017; Kulkarni et al. 2017b), but these semiochemicals may more likely be associated with imbibition (Law and Gallagher 2015) and germination rather than chemical defence. In fact, the amount of volatiles emanated differs between seeds with hard and soft coats. Nondormant and nonpersistent seeds with soft coats are more easily located by rodents than persistent dormant seeds with impermeable coats, because these seeds with impermeable coats do not produce enough volatile cues for rodents to locate them (Hulme 1998; Vander Wall 1998; Paulsen et al. 2013). Thus, producing seeds with hard coats may be a successful strategy for plants to increase seed persistence via reduced predation (Vander Wall 1998; Paulsen et al. 2013). Detection of active substances released from seeds and perceived by seed predators as cues requires chemical analyses, isolation, and identification, followed by an experimental proof of their efficiency.

Seed selection by a predator is further shaped by the risk of competition and predation. This topic has been more thoroughly investigated for rodents (e.g. Reed et al. 2005), but recent findings indicate that invertebrate predators are able to adjust their consumption according to predation and competition risks (Charalabidis et al. 2017, 2019). When the presence of a predator was simulated by providing chemical cues on the surface of the experimental arena, the proportion of *Harpalus affinis* that accepted seeds and individual seed consumption increased significantly (Charalabidis et al. 2017). Interestingly, response to the risk of predation was different between a specialist and a generalist (Charalabidis et al. 2019), suggesting that foraging strategies of carabid beetles for seeds are species specific. Increased density of carabids did not cause increased per capita seed predation, suggesting that seed consumption is influenced by the density of conspecifics (Carbonne et al. 2019). In rodents, it is expected that competition and risk of predation is driven by the efficiency of exploiting the seed patch (Reed et al. 2005). Thus, more seed predation can be expected in patches with higher rodent density, while less efficient patch exploitation generally occurs in patches in which the predation risk for rodents is high, such as habitats with sparse vegetation that do not provide enough shelter. Understanding the behavioural responses of seed predators to the risk of predation or increased competition is important for better linking predator abundance data with actual weed seed predation.

8.4 Spatiotemporal Variation in Seed Predation

8.4.1 Landscape Scale

Intensive agricultural production methods leading to landscape simplification are now considered to be responsible for reduction of biodiversity of functions and species in agroecosystems, which increases the risk of hampering ecosystem services provided by biodiversity. Several studies have attempted to investigate the variation in seed predation at the landscape scale, with equivocal results. Menalled et al. (2000) suggested that seed predation should be higher in more complex landscapes. Trichard et al. (2013) found that seed predation by carabid beetles was positively correlated to landscape diversity and proportion of cover of temporary grassland within a 1000 m buffer around the focal fields. McHugh et al. (2020) report positive influence of woody habitat on seed removal of *Chenopodium album* (but not for *Poa trivialis*). On the other hand, seed predation by carabids was also higher in landscapes with a larger total area of annual crops (Jonason et al. 2013; Petit et al. 2017), and differences in the usual farming practices may complicate the picture further. For example, seed predation increased with landscape complexity in conventional fields but decreased in organic fields (Fischer et al. 2011). Clearly more research is needed in this area.

8.4.2 Field Scale and Density Dependence

Weeds naturally occur in patches that differ in size, density, diversity, and distribution within fields; thus, weed seed availability in crops is highly clustered. Seed predators need to respond to such an uneven spatial distribution of their food in order to be able to exploit this source. Predators can aggregate in patches of naturally occurring weeds (Honek et al. 2005; Kulkarni et al. 2017a) or augmented seed patches (Honek and Martinkova 2001). The expectation that sites with higher abundance or activity of seed predators will experience higher seed predation levels seems to be reasonable, but this is only partially supported by available evidence. Saska et al. (2008) reported a negative relationship between seed predator abundance and seed removal: high seed removal was recorded inside the field where both weeds and carabids were scarce, while at the margins where weeds were abundant, the seed removal was very low despite high seed predator density and diversity. Satiation or other aspects of density dependence might have been responsible for this result.

In terms of seed patch exploitation, there are three basic types of responses the predators can exhibit (Westerman et al. 2008): (i) density dependence, (ii) inverse density dependence, and (iii) density independence. If the response is density dependent, the proportion of seeds predated increases with seed density, which indicates that seeds from a patch with higher seed density will have a lower chance to survive than those from a patch with lower seed density. This can be achieved by predator aggregation in high-density patches and/or by functional response of individual predators to seed density (Baraibar et al. 2012). The inverse density dependence means a declining proportion of seeds consumed with increasing seed density in a patch. This suggests a lack of predator aggregation in the patch, perhaps due to limited mobility relative to the size of the patch (Westerman et al. 2008), increased competition for seeds (Charalabidis et al. 2019; Carbonne et al. 2019), or satiation (Carbonne et al. 2019). Inverse density dependence may also result if there are high-density patches of seeds with seed densities exceeding the ability of seed predators to increase their seed consumption in proportion to seed availability (Daedlow et al. 2014). Consequently, seeds in a high-density patch have a higher probability of survival than those in a low-density patch. When the proportion of consumed seeds remains constant over the range of seed densities, there is a density-independent response (Westerman et al. 2008).

Although the results are inconsistent across studies of density dependence, some trends can be deduced. Rodents tend to exhibit density-dependent responses (Baraibar et al. 2012; but see Daedlow et al. 2014). Carabid beetles showed inverse density dependence or density independence (Westerman et al. 2008) and density dependence at the beginning but inverse density dependence later on (Carbonne et al. 2019). The reason for the opposite responses in the two studies might have arisen from different mobility of individual predators and different per capita voracity or from methodological issues. Rodents are clearly more mobile than carabid beetles (or other surface-active invertebrates); therefore they are able to cover much larger areas when searching for food and may have a higher likelihood of finding and aggregating at a weed seed patch. Also, rodents need to consume much more food on a daily basis to maintain their endothermic metabolism. Daedlow et al. (2014) concluded that the inverse density dependence they observed for rodent predation did not have biological meaning since 88% of seed production was lost due to predation. Carabids are able to aggregate at patches with increased seed densities (Honek and Martinkova 2001; Kulkarni et al. 2017a) and in the short term respond in a density-dependent way, but they get satiated quickly (Carbonne et al. 2019). Thus, carabids are likely to efficiently exploit smaller patches of weed seeds. Since ants are eusocial insects that share information about the location of food sources, their response to seed density and seed patch size is density and size dependent (Torra et al. 2016).

8.4.3 Seasonal Scale

Observed seed removal rates typically vary over time (Westerman et al. 2003a, 2006; Saska et al. 2008), but the patterns observed across studies are diverse, and even within the same study, the variation may be species specific or vary between years (Westerman et al. 2003a; O'Rourke et al. 2006; Saska et al. 2008; Davis et al. 2013). Sometimes temporal trends associated with predator activity can be seen. For example, removal of *Setaria faberi* seeds exposed for 48 hours was about 10% in July and ranged between 30% and 90% in August and September (O'Rourke et al. 2006). However, temporal variation in seed removal cannot always be attributed to the change in predator activity or density. Saska et al. (2008) found week-to-week variation from c. 0–80% of removed seeds exposed for seven days, and only one (*Stellaria media*) out of the four species of seeds used showed a distinct seasonal trend in removal. This brings up the question to what degree these temporal fluctuations in seed predation are resulting from stochastic events, such as fluctuations in temperature, or to what degree these oscillations represent real phenological patterns in seed predation.

8.5 The Significance of Seed Predation for the Population Dynamics of Weeds

Seed predation can only be considered as a promising tool supporting integrated weed management if seed predators are able to reduce seed densities that result in the weed populations not exceeding economical thresholds. A sensitivity analysis by Firbank and Watkinson (1986) shows that an annual loss of 25–50% of newly produced seeds can halt weed population growth. Several empirical and modelling studies (Westerman et al. 2003b, 2005, 2006; Davis et al. 2011) demonstrate that these levels can be regularly achieved in the field conditions via post-dispersal seed predation. Moreover, annual reductions in seed input are accumulated over the years even in highly persistent species (Maron and Gardner 2000). Analysis of the soil seedbank density in cultivated fields across the United Kingdom revealed that seedbanks contained lower densities of seeds in fields with higher abundance of carabid beetles, indicating that seed predators are delivering the desired ecosystem service and depleting weed seedbanks (Bohan et al. 2011).

Tests have also been done to verify the biological control potential of pre-dispersal seed predators against invasive weeds (Myers and Sarfraz 2017). Biological control via pre-dispersal seed predation seems to be effective at low plant population densities but may be less successful at higher plant densities due to plant compensatory mechanisms coping with the presence of pre-dispersal predators (Garren and Strauss 2009). Several species of thistles (Asteraceae) that are abundantly infested by pre-dispersal predators reproduce vegetatively as well as sexually. Pre-dispersal seed predation likely has little impact on the persistence of populations of these weeds once they are established but may rather effectively reduce dispersal and the likelihood of colonization of new habitats. If pre-dispersal seed predation occurs in combination with root feeding, as is the case for many curculionid beetles, biological control may be successful (Myers and Sarfraz 2017).

8.6 Field and Crop Management Effects on Weed Seed Predation

The ultimate goal of weed seed predation research is to develop strategies that can be implemented as a part of integrated weed management (Petit et al. 2018), although the reality is that weed control in arable fields cannot rely solely on seed predators. Contemporary weed control offers an array of tactics that together with field management and crop protection measures may interfere with the ecosystem service of seed predation.

Tillage is a key agronomical tool that modulates the input of weed seeds in the soil bank. During tillage, seeds that are left on the soil surface are incorporated into the soil and become largely inaccessible to predators (see Section 8.2.3). Cropping systems that rely on regular tillage therefore experience the greatest input of weed seeds in the soil seedbank as the predators may not have enough time to find and exploit the seed stock if tillage happens shortly after harvest. Consequently, postponing tillage is suggested as an affordable management option to reduce weed seed input into the soil bank (Westerman et al. 2006; Baraibar et al. 2017). This suggestion is supported by multiple studies that showed greater seed predation in systems with reduced tillage, including no-till systems (Menalled et al. 2007; Baraibar et al. 2009; Trichard et al. 2014; Petit et al. 2017) or zone tillage (Pretorius et al. 2018).

Conservation agriculture, besides reducing tillage intensity, introduces cover crops during the period when fields would normally be left fallow, but the effect of cover crops on seed predation has not been conclusive (Davis et al. 2013; Trichard et al. 2013; Blubaugh et al. 2016). The spatiotemporal patterns in seed predation are crop specific and correlated to light interception (Heggenstaller et al. 2006). They suggested that growing phenologically dissimilar crops may provide an environment that favours seed predators in the form of suitable habitats with enough vegetation cover, in which the activity of the predators and predation were highest. Seed losses due to predation were also greater in more diversified crop rotation (Westerman et al. 2005). Adding mulch also seems to be beneficial for seed predation (Cromar et al. 1999; Pullaro et al. 2006). Not surprisingly, reduced input farming reinforced seed predation service compared with conventional fields (O'Rourke et al. 2006).

8.7 Methodological Aspects of Studying Seed Predation

8.7.1 Estimating Pre-dispersal Seed Predation

Field observations of predation events utilize two major approaches. One is the dissection of the inflorescence or developing pod that provides direct evidence on the composition and abundance of the seed predator community (Honek and Martinkova 2005). However, eggs or early larval stages can be easily overlooked due to their small size, and the proportion of damaged seeds can be underestimated if the predator development is disrupted before its completion. The other method includes collecting ripening heads or pods and isolating them in paper bags or glass vials until the resident predators complete their development and emerge (Koprdova et al. 2015). If rearing conditions are properly set in order to minimize unnatural mortality, the incidence of infestation and proportion of seeds damaged can be directly compared, e.g. between the terminal and lateral inflorescences or between populations.

8.7.2 Removal of Sentinel Seeds

In agricultural settings, the magnitude of pest control, including seed predation, can be estimated using artificially exposed (sentinel) prey or seeds (McHugh et al. 2020). In the case of seed predation, seed cards (Westerman et al. 2003b) have become the most frequently used and usually consist of strips of sandpaper with seeds glued on the abrasive surface. Seeds on cards are accessible to most groups of predators that forage on the ground, although some important taxa, such as ants or carabid larvae, are not able to remove the seeds from the cards. Other commonly used options include seeds glued to the bottom of

Petri dishes (Lundgren et al. 2013), trays filled with plasticine in which the seeds are placed (Honek et al. 2003), or dishes with soil on top of which the seeds have been spread (Marino et al. 1997; Baraibar et al. 2012). Substrates with seeds are usually presented in transects or grids, exposed for short periods of time (1–14 days), and periodically replaced. The advantage of most of these substrates is that they can be prepared in advance so collection or replacement in the field is time effective (Shuler et al. 2008).

Using sentinel seeds allows for comparing the specific removal rates among multiple species of seeds or for a qualitative assessment of the relative contribution of particular groups of seed predators to seed removal. This is usually assured by covering a subset of seed cards with wire netting, which exclude vertebrates from accessing the seeds (Westerman et al. 2003a; Saska et al. 2014). Seed removal by invertebrates is then estimated based on the cards from the exclosure, and removal by vertebrates is based on the difference between the open and covered cards. The downside of using sentinel seeds is that the exact taxonomic identity of the predators cannot be directly ascertained if video imaging is not used (Fischer and Turke 2016), but can be inferred from parallel measurements of abundance or activity of the dominant taxa on the study site or from traces of animal activity (bites, presence of droppings or mucus, etc.).

However, caution must be exercised when sentinel seed removal data are interpreted for several reasons. Each method of seed exposure may give statistically different removal rates, and the suitability of particular methods may be seed species specific (Gallandt 2005; Shuler et al. 2008; Saska et al. 2014). Also, exposure of sentinel seeds may not provide realistic estimates of seed predation. The use of the term 'seed removal' is probably more appropriate for this kind of data, because 'predation' is derived from the difference between the number of exposed and retrieved seeds or expressed as a proportion of removed seeds. It is also impossible to follow the fate of the missing seeds, and these studies fail to account for natural processes like seed washing into soil cracks or under plant residues. Although the number of seeds that are presented often represent realistic densities relative to the usual seed density in seedbanks or per m^2 of soil surface, in most studies the density of sentinel seeds does not reflect the seed density that is present on the soil surface at the moment of exposure ('background seed density'), and seed density dependent response of the predators may obscure the resulting removal rates (see Section 8.4.2). Further bias in the estimates occurs if the background seed density is not even among sampling points or over time, which is usually the case as weed distribution inside the field is patchy (Saska et al. 2008). Without adjusting exposed seed densities to the background seed density, exposing sentinel seeds provides only a measure of encounter frequency under given environmental conditions and presence of predators with particular demands. Caution must be taken before using sentinel seed removal data as a proxy for quantifying the actual loss of seeds prior to input in the soil seedbank through predation (see Section 8.5) as erroneous conclusions can be drawn.

8.7.3 Seedling Emergence Studies

In natural systems, the estimation of seed predation based on seedling emergence has been frequently used (Turnbull et al. 2000). The method is based on sowing a defined seed density in the study system, and the calculation of subsequent seedling recruitment. If exclosures are used, the relative contribution of particular groups of seed predators can be estimated. Turnbull et al. (2000) distinguished two types of sowing experiments – seed augmentation and seed introduction. Under the seed augmentation method, seeds of species already present in the study site are distributed, while under seed introduction, seeds of species not previously present on-site are added. The choice between these two should be carefully considered as the absence of a species in a site may also mean that the site is unsuitable for establishment; therefore no germination cannot necessarily be interpreted as a result of predation (Turnbull et al. 2000).

Seedling emergence studies are much less frequent in arable areas than non-crop habitats. Seeds are either presented in dishes or metal frames with soil and allowed to germinate in the field (Davis et al. 2011, 2013) or sown as described earlier (White et al. 2007; Westerman et al. 2008; Daedlow et al. 2014). In combination with exclosures, the latter approach can be relatively powerful in revealing realistic estimates of seed losses due to predation, but possible seed density dependence in predator response (Section 8.4.2) should also be taken into account when a weed seed sowing experiment is planned.

8.7.4 Gut Content Analysis of Predators

The classical method of gut content analysis is a morphological investigation of the food remains in beetles' gut under microscope. Using this approach, Skuhravý (1959) was able to demonstrate that several species of carabid beetles previously believed as solely carnivorous, such as *Poecilus cupreus*, readily consumed plant material including seeds, and remnants of the seeds could be identified to species level. Hengeveld (1980) found plant material in the gut of all 24 species of carabids inspected.

Immunochemical methods (enzyme-linked immunosorbent assay, ELISA) were recently used to reveal seed-centred trophic linkages (Lundgren et al. 2013; Blubaugh et al. 2016). Lundgren et al. (2013) coated seeds of *Taraxacum* with a rabbit IgG marker before spreading them in the field. After a given exposure time, invertebrates were collected and screened for the presence of the marker in their gut. The uncovered community of seed predators was taxonomically more diverse than expected and included taxa that have been neglected (Coleoptera: Elateridae, Scarabaeidae, and Curculionidae) or poorly known (Diplopoda, Isopoda: Oniscidea) as granivores.

The most recent approaches are based on DNA recognition. Next-generation sequencing (NGS) has the potential to simultaneously identify many types of prey via DNA metabarcoding (Wallinger et al. 2015; Sint et al. 2018; Frei et al. 2019), especially when specific polymerase chain reaction (PCR) primers are available. This technique also allows for non-invasive analysis, utilizing regurgitates instead of the whole body extracts or dissected guts, circumventing the problem of overabundance of nontarget DNA or cross-contamination (Wallinger et al. 2015). Using regurgitates, plant DNA was detectable up to five days post-feeding, but the trajectory of detectability in time varied among species of seeds. In *Capsella bursa-pastoris* the seeds could not be detected for longer than 32 hours post-feeding in the harpaline carabid *Pseudophonus rufipes* (Sint et al. 2018). The NGS-based analyses have been evolving rapidly, providing a unique opportunity to reveal complex trophic interactions centered on weed seeds.

8.7.5 Estimating Consumption and Preferences of Predators in the Laboratory

Decisions by predators whether to accept a seed is not a random process. Predators usually show more or less prominent preferences for seeds (as described in Section 8.3). Laboratory-based estimates of per capita seed consumption (Klimes and Saska 2010) provide an idea on how many seeds the individual predators can eat in the field. Manipulating abiotic conditions, such as temperature (Saska et al. 2010) or seed/predator density, may be very insightful for understanding the variation in observed predation in the field.

Preferences can be studied in choice and no-choice experiments. Choice experiments ('cafeteria experiments') are typically conducted by presenting an array of seed species in an arena at the same time (Honek et al. 2007). These experiments provide preferences for particular species of seed relative to the consumption of other species. Consumption of individual types of diet are not independent of each other and the relative preference of a species may change if presented with a different set of diet. This phenomenon is a possible criticism of this approach. However, the strength of this type of experiment is that many species can be tested simultaneously and if the same array of seeds is offered to multiple species of predators, 'general' attractiveness of particular species of seed to predators can be inferred. Associations of particular species of a predator with a species of seed can be revealed if pronounced preference for a certain seed is consistent (Honek et al. 2007; Talarico et al. 2016; Saska et al. 2019b).

The difference between no-choice experiments from choice experiments is straightforward – in no-choice experiments individual predators are offered one type of diet only. Charalabidis et al. (2017, 2019) introduced novel weed science terms that were otherwise commonly used in behavioural sciences and that are related to the time and energy invested by the predator in seed selection – 'choosiness'. Individuals are choosy if they accept only a few species of seeds and invest a lot of time and energy in finding the preferred ones, and less choosy if they immediately accept a wider array of seeds. The preference is then inferred from the behaviour of the predator and the resulting differences in consumption (Charalabidis et al. 2017). Movement tracking software is very useful in further analyzing the behaviour of predators, such as the relative length of time spent by searching, handling seeds, and feeding (Charalabidis et al. 2017).

In laboratory experiments, seeds can be presented either on substrates similar to those found in the field (Honek et al. 2007; Saska et al. 2014), on filter paper (Klimes and Saska 2010; Charalabidis et al. 2017), or spread on top of soil (Shuler et al. 2008). Placing seeds on soil is closest to the field conditions as predators can hide in the soil but retrieving the remaining seeds is labour intensive (Shuler et al. 2008). Using filter paper assures a rapid evaluation of the number of consumed versus intact seeds but may stress the predators if shelters are not provided (Klimes and Saska 2010), and this may distort their natural behaviour. Using seed-carrying substrates, placed on top of the soil, combines the benefits of the two previously mentioned approaches, i.e. easy seed counting and shelter provisioning; however, it may introduce bias inherent to the substrate (Saska et al. 2014), because the handling time needed to collect a seed can be substrate specific.

8.8 Directions for Future Research and Conclusions

In order to be able to manage seed predation within a system of weed management in cultivated fields, more research is needed to better understand both abiotic and biotic interactions influencing seed predation. This chapter illustrated several areas that require considerable additional research. We identified the following knowledge gaps:

1) *The effect of field and crop management on seed predator populations and the magnitude of seed predation in a landscape context.* Although we know less about seed predation in arable settings than in natural habitats, a methodologically rigorous synthesis should be possible as the number of published papers increase every year. A global meta-analysis might reveal common patterns driving seed predation in different agroecosystems. As making general conclusions is difficult based on individual studies, compiling the results of multiple independent studies may provide strong inference on crop and field management favourable for both the seed predators and weed seed predation service. Surprisingly, no meta-analysis has been published on seed predation in agricultural fields. Meta-analysis should focus on the following aspects that influence seed predation: (i) global patterns, (ii) factors acting at landscape level, and (iii) field and crop management. The analysis should consider differences between particular groups of seed predators and species of weed seeds used.
2) *Behaviour of seed predators.* In order to better understand the spatiotemporal dynamics of seed predation, studies focusing on the behaviour of seed predators at individual and community levels are also needed. Disentangling the influence of aggregative behaviour, density dependence, seed selection process, competition, and intraguild predation on actual predation rates in the field and finding a rigorous methodological approach on how to measure it are essential to making reasonable predictions of seed predation.
3) *Food web studies.* Better understanding of the granivorous community is needed. The work of Davis and Raghu (2010) using structural equation modelling clearly demonstrated that a range of biotic and abiotic processes have an impact on invertebrate as well as vertebrate seed predation and that to fully understand this phenomenon, it is necessary to approach seed predation in the context of the whole food web on arable fields. The NGS provides a potentially useful tool to approach this problem (Frei et al. 2019)
4) *Linking seed predation with seed defence.* More research is needed to learn how mechanical and chemical defences of weed seeds affect the decisions of predators to approach and consume the seeds. Utilizing chemical cues emanating from seeds has been documented, but to what extent it modulates the persistence of seeds needs more information. In the soil seedbank, focusing on multitrophic interactions between seeds, soil-borne microorganisms, and seed predators provides a new direction to research that may reveal complex relationships shaping seed persistence in the soil.

Acknowledgements

This work was supported by the CSF project 17-00043S.

References

Acorn, J.H. and Ball, G.E. (1991). The mandibles of some adult ground beetles: structure, function, and the evolution of herbivory (Coleoptera: Carabidae). *Canadian Journal of Zoology* 69: 638–650.

Baraibar, B., Westerman, P.R., and Recasens, J. (2009). Effects of tillage and irrigation in cereal fields on weed seed removal by seed predators. *Journal of Applied Ecology* 46: 380–387.

Baraibar, B., Daedlow, D., De Mol, F., and Gerowitt, B. (2012). Density dependence of weed seed predation by invertebrates and vertebrates in winter wheat. *Weed Research* 52: 79–87.

Baraibar, B., Canadell, C., Torra, J. et al. (2017). Weed seed fate during summer fallow: the importance of seed predation and seed burial. *Weed Science* 65: 515–524.

Blubaugh, C.K., Hagler, J.R., Machtley, S.A., and Kaplan, I. (2016). Cover crops increase foraging activity of omnivorous predators in seed patches and facilitate weed biological control. *Agriculture, Ecosystems & Environment* 231: 264–270.

Bohan, D.A., Boursault, A., Brooks, D.R., and Petit, S. (2011). National-scale regulation of the weed seedbank by carabid predators. *Journal of Applied Ecology* 48: 888–898.

Bojňanský, V. and Fargašová, A. (2007). *Atlas of Seeds and Fruits of Central and East-European Flora*. The Carpathian Mountains Region. Dordrecht: Springer.

Bretagnolle, F., Matejicek, A., Gregoire, S. et al. (2016). Determination of fatty acids content, global antioxidant activity and energy value of weed seeds from agricultural fields in France. *Weed Research* 56: 78–95.

Brust, G.E. and House, G.J. (1988). Weed seed destruction by arthropods and rodents in low-input soybean agroecosystems. *American Journal of Alternative Agriculture* 3: 19–25.

Carbonne, B., Bohan, D.A., Charalabidis, A., and Petit, S. (2019). Weed seed availability, carabid intraspecific interference and their interaction drive weed seed consumption by carabid beetles. *Agricultural and Forest Entomology* 21: 396–406.

Carbonne, B., Bohan, D.A., and Petit, S. (2020). Key carabid species drive spring weed seed predation of *Viola arvensis*. *Biological Control* 141: 104148.

Charalabidis, A., Dechaume-Moncharmont, F.X., Petit, S., and Bohan, D.A. (2017). Risk of predation makes foragers less choosy about their food. *PLoS One* 12: e0187167.

Charalabidis, A., Dechaume-Moncharmont, F.X., Carbonne, B. et al. (2019). Diversity of foraging strategies and responses to predator interference in seed-eating carabid beetles. *Basic and Applied Ecology* 36: 13–24.

Clause, J., Forey, E., Eisenhauer, N. et al. (2017). Seed selection by earthworms: chemical seed properties matter more than morphological traits. *Plant and Soil* 413: 97–110.

Cromar, H.E., Murphy, S.D., and Swanton, C.J. (1999). Influence of tillage and crop residue on postdispersal predation of weed seeds. *Weed Science* 47: 184–194.

Daedlow, D., Westerman, P.R., Baraibar, B. et al. (2014). Weed seed predation rate in cereals as a function of seed density and patch size, under high predation pressure by rodents. *Weed Research* 54: 186–195.

Dalling, J.W., Davis, A.S., Schutte, B.J., and Arnold, A.E. (2011). Seed survival in soil: interacting effects of predation, dormancy and the soil microbial community. *Journal of Ecology* 99: 89–95.

Davis, A.S. and Raghu, S. (2010). Weighing abiotic and biotic influences on weed seed predation. *Weed Research* 50: 402–412.

Davis, A.S., Daedlow, D., Schutte, B.J., and Westerman, P.R. (2011). Temporal scaling of episodic point estimates of seed predation to long-term predation rates. *Methods in Ecology and Evolution* 2: 682–690.

Davis, A.S., Taylor, E.C., Haramoto, E.R., and Renner, K.A. (2013). Annual postdispersal weed seed predation in contrasting field environments. *Weed Science* 61: 296–302.

DeSousa, N., Griffiths, J.T., and Swanton, C.J. (2003). Predispersal seed predation of redroot pigweed (*Amaranthus retroflexus*). *Weed Science* 51: 60–68.

Evans, T.A. and Gleeson, P.V. (2016). Direct measurement of ant predation of weed seeds in wheat cropping. *Journal of Applied Ecology* 53: 1177–1185.

Firbank, L.G. and Watkinson, A.R. (1986). Modeling the population-dynamics of an arable weed and its effects upon crop yield. *Journal of Applied Ecology* 23: 147–159.

Fischer, C. and Turke, M. (2016). Seed preferences by rodents in the agri-environment and implications for biological weed control. *Ecology and Evolution* 6: 5796–5807.

Fischer, C., Thies, C., and Tscharntke, T. (2011). Mixed effects of landscape complexity and farming practice on weed seed removal. *Perspectives in Plant Ecology, Evolution and Systematics* 13: 297–303.

Forey, E., Barot, S., Decaens, T. et al. (2011). Importance of earthworm-seed interactions for the composition and structure of plant communities: a review. *Acta Oecologica-International Journal of Ecology* 37: 594–603.

Frei, B., Guenay, Y., Bohan, D.A. et al. (2019). Molecular analysis indicates high levels of carabid weed seed consumption in cereal fields across Central Europe. *Journal of Pest Science* 92: 935–942.

Gaba, S., Collas, C., Powolny, T. et al. (2014). Skylarks trade size and energy content in weed seeds to maximize total ingested lipid biomass. *Behavioural Processes* 108: 142–150.

Gaba, S., Deroulers, P., Bretagnolle, F., and Bretagnolle, V. (2019). Lipid content drives weed seed consumption by ground beetles (Coleoptera, Carabidae) within the smallest seeds. *Weed Research* 59: 170–179.

Gallandt, E.R. (2005). Experimental substrate affects rate of seed removal in assays of invertebrate seed predation. *Weed Technology* 19: 481–485.

Garren, J.M. and Strauss, S.Y. (2009). Population-level compensation by an invasive thistle thwarts biological control from seed predators. *Ecological Applications* 19: 709–721.

Heggenstaller, A.H., Menalled, F.D., Liebman, M., and Westerman, P.R. (2006). Seasonal patterns in post-dispersal seed predation of *Abutilon theophrasti* and *Setaria faberi* in three cropping systems. *Journal of Applied Ecology* 43: 999–1010.

Hengeveld, R. (1980). Polyphagy, oligophagy and food specialization in ground beetles (Coleoptera, Carabidae). *Netherlands Journal of Zoology* 30: 564–584.

Holland, J.M., Hutchison, M.A.S., Smith, B., and Aebischer, N.L. (2006). A review of invertebrates and seed-bearing plants as food for farmland birds in Europe. *Annals of Applied Biology* 148: 49–71.

Honek, A. and Martinkova, Z. (2001). Aggregation of ground beetles (Carabidae, Coleoptera) on winter rape seeds dispersed on the ground. *Plant Protection Science* 37: 97–102.

Honek, A. and Martinkova, Z. (2005). Pre-dispersal predation of *Taraxacum officinale* (dandelion) seed. *Journal of Ecology* 93: 335–344.

Honek, A., Martinkova, Z., and Jarosik, V. (2003). Ground beetles (Carabidae) as seed predators. *European Journal of Entomology* 100: 531–544.

Honek, A., Martinkova, Z., and Saska, P. (2005). Post-dispersal predation of *Taraxacum officinale* (dandelion) seed. *Journal of Ecology* 93: 345–352.

Honek, A., Martinkova, Z., Saska, P., and Pekar, S. (2007). Size and taxonomic constraints determine the seed preferences of Carabidae (Coleoptera). *Basic and Applied Ecology* 8: 343–353.

Honek, A., Martinkova, Z., Koprdova, S., and Saska, P. (2017). The role of invertebrates in seedling establishment in a heterocarpic plant, *Atriplex sagittata*. *Annals of Applied Biology* 171: 264–276.

Hulme, P.E. (1998). Post-dispersal seed predation and seed bank persistence. *Seed Science Research* 8: 513–519.

Ichihara, M., Maruyama, K., Yamashita, M. et al. (2011). Quantifying the ecosystem service of non-native weed seed predation provided by invertebrates and vertebrates in upland wheat fields converted from paddy fields. *Agriculture, Ecosystems & Environment* 140: 191–198.

Jonason, D., Smith, H.G., Bengtsson, J., and Birkhofer, K. (2013). Landscape simplification promotes weed seed predation by carabid beetles (Coleoptera: Carabidae). *Landscape Ecology* 28: 487–494.

Klimes, P. and Saska, P. (2010). Larval and adult seed consumption affected by the degree of food specialization in Amara (Coleoptera: Carabidae). *Journal of Applied Entomology* 134: 659–666.

Koçak, E., Kesdek, M., and Yildrim, E. (2007). A new anise (*Pimpinella anisum* L.) pest: *Carterus dama* (Rossi, 1792) (Coleoptera: Carabidae). *S.Ü. Ziraat Fakültesi Dergisi* 21: 1–3.

Kolb, A., Ehrlen, J., and Eriksson, O. (2007). Ecological and evolutionary consequences of spatial and temporal variation in pre-dispersal seed predation. *Perspectives in Plant Ecology, Evolution and Systematics* 9: 79–100.

Koprdova, S., Bellanger, S., Skuhrovec, J., and Darmency, H. (2015). Does gall midge larvae cause pre-dispersal seed mortality and limit cornflower population growth? *Acta Oecologica-International Journal of Ecology* 69: 167–172.

Kulkarni, S.S., Dosdall, L.M., Spence, J.R., and Willenborg, C.J. (2015). Depth of seed burial and gender influence weed seed predation by three species of ground beetle (Coleoptera: Carabidae). *Weed Science* 63: 910–915.

Kulkarni, S.S., Dosdall, L.M., Spence, J.R., and Willenborg, C.J. (2017a). Field density and distribution of weeds are associated with spatial dynamics of omnivorous ground beetles (Coleoptera: Carabidae). *Agriculture, Ecosystems & Environment* 236: 134–141.

Kulkarni, S.S., Dosdall, L.M., Spence, J.R., and Willenborg, C.J. (2017b). Seed detection and discrimination by ground beetles (Coleoptera: Carabidae) are associated with olfactory cues. *PLoS One* 12: e0170593.

Larios, L., Pearson, D.E., and Maron, J.L. (2017). Incorporating the effects of generalist seed predators into plant community theory. *Functional Ecology* 31: 1856–1867.

Law, J.J. and Gallagher, R.S. (2015). The role of imbibition on seed selection by *Harpalus pensylvanicus*. *Applied Soil Ecology* 87: 118–124.

Lundgren, J.G. and Rosentrater, K.A. (2007). The strength of seeds and their destruction by granivorous insects. *Arthropod-Plant Interactions* 1: 93–99.

Lundgren, J.G., Saska, P., and Honek, A. (2013). Molecular approach to describing a seed-based food web: the post-dispersal granivore community of an invasive plant. *Ecology and Evolution* 3: 1642–1652.

Marino, P.C., Gross, K.L., and Landis, D.A. (1997). Weed seed loss due to predation in Michigan maize fields. *Agriculture, Ecosystems & Environment* 66: 189–196.

Maron, J.L. and Gardner, S.N. (2000). Consumer pressure, seed versus safe-site limitation, and plant population dynamics. *Oecologia* 124: 260–269.

Martinkova, Z., Saska, P., and Honek, A. (2006). Consumption of fresh and buried seed by ground beetles (Coleoptera: Carabidae). *European Journal of Entomology* 103: 361–364.

McHugh, N.M., Moreby, S., Lof, M.E. et al. (2020). The contribution of semi-natural habitats to biological control is dependent on sentinel prey type. *Journal of Applied Ecology* 57: 914–925.

Menalled, F.D., Marino, P.C., Renner, K.A., and Landis, D.A. (2000). Post-dispersal weed seed predation in Michigan crop fields as a function of agricultural landscape structure. *Agriculture, Ecosystems & Environment* 77: 193–202.

Menalled, F.D., Smith, R.G., Dauer, J.T., and Fox, T.B. (2007). Impact of agricultural management on carabid communities and weed seed predation. *Agriculture, Ecosystems & Environment* 118: 49–54.

Minkey, D.M. and Spafford, H. (2016). Removal and burial of weed seeds by ants (Hymenoptera: Formicidae) from the soil surface of a cropped area in Western Australia. *Environmental Entomology* 45: 1199–1204.

Myers, J.H. and Sarfraz, R.M. (2017). Impacts of insect herbivores on plant populations. *Annual Review of Entomology* 62: 207–230.

O'Rourke, M.E., Heggenstaller, A.H., Liebman, M., and Rice, M.E. (2006). Post-dispersal weed seed predation by invertebrates in conventional and low-external-input crop rotation systems. *Agriculture, Ecosystems & Environment* 116: 280–288.

Paulsen, T.R., Colville, L., Kranner, I. et al. (2013). Physical dormancy in seeds: a game of hide and seek? *New Phytologist* 198: 496–503.

Petit, S., Trichard, A., Biju-Duval, L. et al. (2017). Interactions between conservation agricultural practice and landscape composition promote weed seed predation by invertebrates. *Agriculture, Ecosystems & Environment* 240: 45–53.

Petit, S., Cordeau, S., Chauvel, B. et al. (2018). Biodiversity-based options for arable weed management. A review. *Agronomy for Sustainable Development* 38: 48.

Pretorius, R.J., Hein, G.L., Blankenship, E.E. et al. (2018). Comparing the effects of two tillage operations on beneficial epigeal arthropod communities and their associated ecosystem services in sugar beets. *Journal of Economic Entomology* 111: 2617–2631.

Pullaro, T.C., Marino, P.C., Jackson, D.M. et al. (2006). Effects of killed cover crop mulch on weeds, weed seeds, and herbivores. *Agriculture, Ecosystems & Environment* 115: 97–104.

Reed, A.W., Kaufman, G.A., and Kaufman, D.W. (2005). Rodent seed predation and GUDs: effect of burning and topography. *Canadian Journal of Zoology* 83: 1279–1285.

Saska, P. (2008). Granivory in terrestrial isopods. *Ecological Entomology* 33: 742–747.

Saska, P., van der Werf, W., de Vries, E., and Westerman, P.R. (2008). Spatial and temporal patterns of carabid activity-density in cereals do not explain levels of predation on weed seeds. *Bulletin of Entomological Research* 98: 169–181.

Saska, P., Martinkova, Z., and Honek, A. (2010). Temperature and rate of seed consumption by ground beetles (Carabidae). *Biological Control* 52: 91–95.

Saska, P., Koprdova, S., Martinkova, Z., and Honek, A. (2014). Comparing methods of weed seed exposure to predators. *Annals of Applied Biology* 164: 301–312.

Saska, P., Honěk, A., Foffová, H., and Martinková, Z. (2019a). Burial-induced changes in the seed preferences of carabid beetles (Coleoptera: Carabidae). *European Journal of Entomology* 116: 133–140.
Saska, P., Honěk, A., and Martinková, Z. (2019b). Preferences of carabid beetles (Coleoptera: Carabidae) for herbaceous seeds. *Acta Zoologica Academiae Scientiarum Hungaricae* 65 (Suppl. 1): 55–74.
Saska, P., Foffová, H., Martinková, Z., and Honěk, A. (2020). Persistence and changes in morphological traits of herbaceous seeds due to burial in soil. *Agronomy* 10: 448. https://doi.org/10.3390/agronomy10030448.
Schutte, B.J., Liu, J.Y., Davis, A.S. et al. (2010). Environmental factors that influence the association of an earthworm (*Lumbricus terrestris* L.) and an annual weed (*Ambrosia trifida* L.) in no-till agricultural fields across the eastern US Corn Belt. *Agriculture, Ecosystems & Environment* 138: 197–205.
Shuler, R.E., DiTommaso, A., Losey, J.E., and Mohler, C.L. (2008). Postdispersal weed seed predation is affected by experimental substrate. *Weed Science* 56: 889–895.
Sint, D., Guenay, Y., Mayer, R. et al. (2018). The effect of plant identity and mixed feeding on the detection of seed DNA in regurgitates of carabid beetles. *Ecology and Evolution* 8: 10834–10846.
Skuhravý, V. (1959). Potrava polnich strevlikovitych [Die Nahrung der Feldcarabiden]. *Casopis Ceskoslovenské spolecnosti entomologické* 56: 1–18.
Talarico, F., Giglio, A., Pizzolotto, R., and Brandmayr, P. (2016). A synthesis of feeding habits and reproduction rhythm in Italian seed-feeding ground beetles (Coleoptera: Carabidae). *European Journal of Entomology* 113: 325–336.
Thompson, K., Bakker, J.P., and Bekker, R.M. (1997). *The Soil Seed Banks of North West Europe: Methodology, Density and Longevity*. Cambridge: Cambridge University Press.
Torra, J., Atanackovic, V., Blanco-Moreno, J.M. et al. (2016). Effect of patch size on seed removal by harvester ants. *Weed Research* 56: 14–21.
Trichard, A., Alignier, A., Biju-Duval, L., and Petit, S. (2013). The relative effects of local management and landscape context on weed seed predation and carabid functional groups. *Basic and Applied Ecology* 14: 235–245.
Trichard, A., Ricci, B., Ducourtieux, C., and Petit, S. (2014). The spatio-temporal distribution of weed seed predation differs between conservation agriculture and conventional tillage. *Agriculture, Ecosystems & Environment* 188: 40–47.
Turnbull, L.A., Crawley, M.J., and Rees, M. (2000). Are plant populations seed-limited? A review of seed sowing experiments. *Oikos* 88: 225–238.
Vander Wall, S.B. (1998). Foraging success of granivorous rodents: effects of variation in seed and soil water on olfaction. *Ecology* 79: 233–241.
Wallinger, C., Sint, D., Baier, F. et al. (2015). Detection of seed DNA in regurgitates of granivorous carabid beetles. *Bulletin of Entomological Research* 105: 728–735.
Westerman, P.R., Hofman, A., Vet, L.E.M., and van der Werf, W. (2003a). Relative importance of vertebrates and invertebrates in epigeaic weed seed predation in organic cereal fields. *Agriculture, Ecosystems & Environment* 95: 417–425.
Westerman, P.R., Wes, J.S., Kropff, M.J., and Van der Werf, W. (2003b). Annual losses of weed seeds due to predation in organic cereal fields. *Journal of Applied Ecology* 40: 824–836.

Westerman, P., Liebman, M., Menalled, F.D. et al. (2005). Are many little hammers effective? – Velvetleaf (*Abutilon theophrasti*) population dynamics in two- and four-year crop rotation systems. *Weed Science* 53: 382–392.

Westerman, P.R., Liebman, M., Heggenstaller, A.H., and Forcella, F. (2006). Integrating measurements of seed availability and removal to estimate weed seed losses due to predation. *Weed Science* 54: 566–574.

Westerman, P.R., Borza, J.K., Andjelkovic, J. et al. (2008). Density-dependent predation of weed seeds in maize fields. *Journal of Applied Ecology* 45: 1612–1620.

Westerman, P.R., Dixon, P.M., and Liebman, M. (2009). Burial rates of surrogate seeds in arable fields. *Weed Research* 49: 142–152.

White, S.S., Renner, K.A., Menalled, F.D., and Landis, D.A. (2007). Feeding preferences of weed seed predators and effect on weed emergence. *Weed Science* 55: 606–612.

Zetto Brandmayr, T., Giglio, A., Marano, I., and Brandmayr, P. (1998). Morphofunctional and ecological features in carabid (Coleoptera) larvae. In: *Phylogeny and Classification of Caraboidea* (eds. G.E. Ball, A. Casale and A. Vigna Taglianti), 449–489. Torino: Museo Regionale di Scienze Naturali.

9

Modelling the Persistence of Weed Populations

Jonathan Storkey[1], Antoine Gardarin[2], Nathalie Colbach[3], Helen Metcalfe[1], and Alice Milne[1]

[1] *Rothamsted Research, Harpenden, Hertfordshire, UK*
[2] *UMR Agronomie, INRAE, AgroParisTech, Thiverval-Grignon, France*
[3] *Agroécologie, AgroSup Dijon, INRAE, Univ. Bourgogne, Univ. Bourgogne Franche-Comté, F-21000, Dijon, France*

9.1 Why Do We Need Models to Predict Weed Persistence?

Weeds are adapted to highly disturbed environments characterised by ephemeral, short-lived opportunities for growth and reproduction. They tend, therefore, to be inherently persistent, with long-lived seedbanks that enable populations to survive periods of unfavourable growing conditions. This means there is a significant time lag in the response of weed populations to a change in management (such as a new herbicide or tillage practice) that presents a significant challenge to research approaches based on short-term field experimentation that seek informed sustainable weed management strategies. While the experimental response of weeds to contrasting management treatments over one or more years will be indicative of trends in community dynamics, in the context of 'real world' farming systems, data of this kind cannot be relied upon to predict long-term community dynamics. This is especially true when seeking to understand the impact of environmental change, which generally occurs over decadal time scales (Varanasi et al. 2016). One way to meet this challenge is to attempt to capture the effects of management and the environment on weeds in a computer model that simulates changes in populations over many years. These so-called population dynamics models all, generally, have a similar structure based on the life cycle of an annual weed (Holst et al. 2007) that can be thought of as having four 'states': seeds in the seedbank, seedlings, mature plants, and fresh seed either still on the maternal plant or on the soil surface (Figure 9.1).

Any individual weed in a field will be in one of these four states. Models attempt to predict the transition of populations between states in response to environmental variables or management options. For example, the proportion of the seedbank that will germinate in a given year will be a product of the moisture and temperature of the soil, the depth it has been buried by previous cultivation and the timing of any soil disturbance. A model consists of a

Persistence Strategies of Weeds, First Edition. Edited by Mahesh K. Upadhyaya, David R. Clements, and Anil Shrestha.

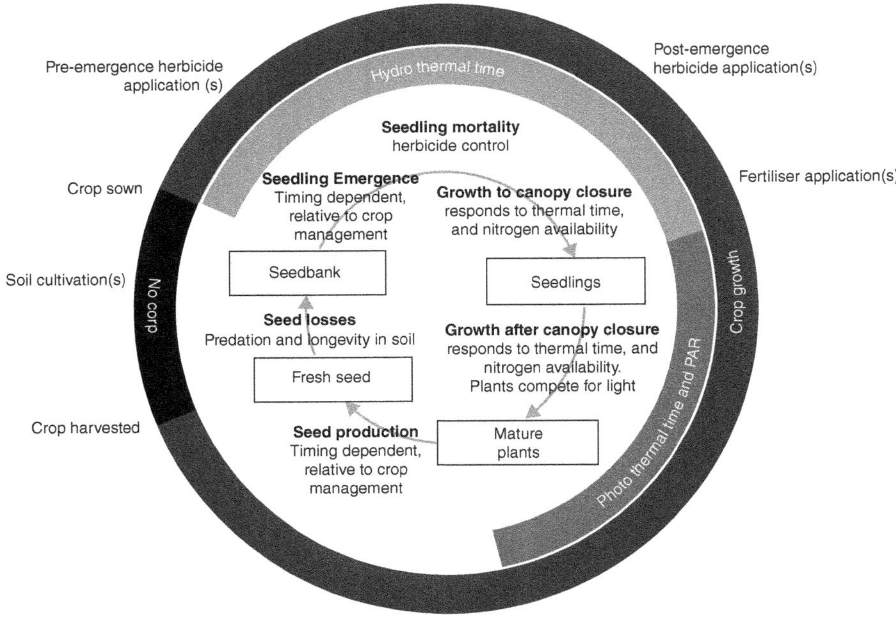

Figure 9.1 Schematic of the system that needs to be captured in models of weed population dynamics and persistence (based on model discussed in Section 9.2.3). Weed species will vary in their response to management through the cropping season (outer ring and text) and environmental conditions (inner ring). Transitions between different states (boxes in inner circle) will be determined by processes that may be regulated by the environment, management, or biological interactions. Models will vary in the complexity with which they simulate the system, from models that derive empirical relationships describing transitions between states in a range of 'static' conditions to 'process-based' approaches that attempt to capture all management × environment × biological interactions dynamically. Environmental variables are usually included in models as some measure of 'biological time' that reflects the amount of temperature, moisture, photosynthetically active radiation (PAR), or daylight hours over the cropping season, e.g. thermal time (temperature), hydrothermal time (temperature and moisture), or photothermal time (temperature and daylength).

series of equations that predict these relationships with 'parameters' that may be specific to individual species, e.g. the base temperature below which seeds do not germinate. Modelling transitions between states can either be based on 'empirical' relationships between management explanatory and weed response variables (e.g. the relationship between depth of seed burial and probability of a seed germinating) or 'process based' (e.g. modelling the growth rate of the hypocotyl in response to soil moisture and temperature). These approaches tend to lie along a 'tractability–prediction' continuum with trade-offs between the amount of data required to parameterize the model and the resolution in the response of the weeds.

9.1.1 The Challenge of Modelling Weed Persistence

Whether or not a weed species can persist in a cultivated field will largely be determined by the properties of the soil seedbank. This chapter will, therefore, focus on the challenge of modelling seedbank dynamics and early growth stages. Whether or not a new seed shed onto the soil surface survives and emerges to produce a following generation will depend on the survival of the embryo against ageing, avoidance of predation by arthropods or vertebrates, survival of pathogen attacks and its dormancy profile in the context of opportunities for germination, and emergence as well as the ability of the emerged plant to capture resources and survive weed control. These processes will be affected differently by the management operations performed in arable fields and be dependent on the local environment. For example, tillage can bury seeds and protect them from predation, but it can also reduce the probability of successful germination and emergence if buried too deep. The challenge facing modellers that aim to explain variation in weed persistence between fields and predict the impact of change is to integrate these factors in sufficient detail to retain the power to discriminate between the response of contrasting species while avoiding the need to derive values for excessive numbers of parameters.

In this chapter, we present contrasting approaches to meet the challenge of modelling weed persistence that lie at different points along this 'tractability–prediction' continuum. Firstly, a 'broad-brush', largely empirical approach is described that applies ecological strategy theory to weed communities to predict the impact of management on broad functional groups based on morphological and phenological traits available in the literature for all weed species. This approach is largely based on quantifying empirical relationships between management and the distributions of these traits. Secondly, we discuss a mechanistic, process-based model that predicts the impact of the local environment and management on the behaviour of the weed seedbank at the level of the species. Traits are, again, used in this example, but this time they are used to derive species-specific parameters in the model and have been measured experimentally for a subset of weeds. We do not present a comprehensive description of the models here but, rather, use them to illustrate different approaches to predicting weed persistence and the central importance of having an underlying detailed understanding of weed biology that is explored in other chapters.

9.2 'Broad-Brush' Ecological Approaches to Modelling Weed Persistence

9.2.1 Ecological Strategies of Weeds

Predicting the persistence of weeds in cultivated fields is a specific example of the broader challenge of understanding the assembly of plant communities in contrasting habitats. One promising approach to meet this challenge is to characterise plant communities based on the distribution of plant traits as opposed to traditional metrics of species diversity. Central to this discipline of functional ecology is the concept of the 'plant ecological strategy scheme' (PESS) that assumes trade-offs and synergies between traits that reflect contrasting evolutionary solutions to environmental constraints. One approach to modelling the impact of the environment and management on the persistence of weed communities would be to quantify the relative response of these different strategies ('functional groups') or the individual traits that underlie them.

The most widely accepted PESS is that introduced by Grime (1974) who identified three contrasting strategies at the extremes of a triangle based on species adaptation to gradients of soil fertility and disturbance. 'Competitors' (C) are characterised by traits that make them effective at resource acquisition, such as a fast growth rate and tall stature, and are adapted to high fertility environments with low disturbance. 'Stress tolerators' (S) are adapted to low fertility environments with low disturbance and have traits that conserve nutrients in the plant such as long leaf life span. Finally, 'ruderals' (R) are typically found in habitats with high disturbance and high fertility with a short life span and fast growth rate. So-called 'regenerative traits', e.g. seed mass, flowering, and seedbank persistence, vary between the strategies and reflect fundamental ecological trade-offs such as that between competition and colonisation (mediated via seed mass) (Moles and Westoby 2004). All species can be assigned to one of these, or an intermediate, functional group (e.g. competitive–ruderals [CR]).

In the context of the C–S–R ecological strategy scheme, cropped fields with regular cultivation and application of nutrients, either as inorganic fertilisers or manure, represent a habitat at the extreme of the soil fertility and disturbance gradient. Consequently, weed floras are dominated by R species; they are predominantly annuals with a short life cycle, rapid growth rate, and high seed production (Gaba et al. 2017). These traits enable weeds to complete their life cycle between disturbance events and rapidly capture applied nutrients. A recent trait-based analysis of two national weed floras (in the United Kingdom and France) compared arable weeds to equivalent floras found in grassland habitats to identify traits that typified adaptation to arable fields (Bourgeois et al. 2019). As a group, weeds occupied an extreme 'ruderal trait space', and this was reflected in significant differences in key traits when compared with the grassland flora. Particularly relevant to this discussion on weed persistence was the finding that weeds tend to flower earlier and for longer, meaning they are able to produce seed opportunistically avoiding the selection pressure of soil disturbance and herbicides.

9.2.2 Ecological Strategies and the Historical Effect of Management on Weed Floras

While the contemporary weed flora is characterised by extreme R strategies reflecting the dominance of ploughed systems and high inputs of inorganic fertilisers and herbicides, this has not always been the case, and one might expect changes in cropping system management to be reflected in shifts in the functional characteristics of the weed flora. That is, traits that meant weed species could persist in past management scenarios may be selected against in current systems. These insights are important for predicting the potential impact of new management options on the weed flora and identifying traits that respond to management change. Several studies of this type exist in the literature.

Storkey et al. (2010) compared the traits of the common weed flora in the United Kingdom to a group of species that had declined since the 1960s, owing to the increase in nitrogen fertiliser and herbicide use. The combination of short stature, late flowering, and large seed emerged as a combination of traits that had been selected against by modern crop management – a so-called 'rare weed trait syndrome'. Examples include *Scandix pecten-veneris* and *Agrostemma githago*. These trends can be related to the large-seeded species being better adapted for competition for nutrients, an advantage in fields with lower fertiliser applications. However, large-seeded species tend to have shorter seedbank persistence (Thompson et al. 1993) and have a limited capacity to buffer against high herbicide mortality (owing to lower seed production per plant). The decline of this functional group can, therefore, be attributed to the increased use of herbicides and inorganic fertiliser (Storkey et al. 2012). A study of changes in the functional characteristics of weed floras in France over a similar period also identified shifts in traits that explain the relative persistence of contrasting weed species (Fried et al. 2012). Again, regenerative traits emerged as being important in explaining changes in community composition. 'Successful' species under current management tended to have a smaller seed size (high seedbank persistence and fecundity) and the ability to germinate and flower all year round – a generalist strategy that enables seed set across a range of crop types and the 'avoidance' of herbicide applications. Although there was also some evidence for a shift in traits in response to reduced disturbance under minimum tillage systems (towards a perennial life cycle and larger seed), crop type appeared to be the dominant management option driving weed functional traits in this and other studies (Gunton et al. 2011). One interesting result from the Gunton et al. (2011) study was the shift in C–S–R strategies in different crop types related to their architecture, with a shift towards more competitive species in taller crops such as sunflower (*Helianthus annuus*) and maize (*Zea mays*).

9.2.3 Modelling Future Changes in the Functional Composition of Weed Seedbanks

Can we use new knowledge on the past effects of management on the change in distribution of 'persistence traits' to develop methodologies for predicting the effect of future management on weed communities? Meeting this challenge will require a life cycle model of weeds (Figure 9.1) that operates at the level of traits or functional groups (as opposed to individual species). This approach represents a departure from traditional weed life cycle models that are developed for individual species (e.g. Freckleton and Watkinson 1998 for *Chenopodium album* in sugar beet [*Beta vulgaris*]) and are underpinned by empirical data measured at the species level. Rather, it identifies weed functional groups or ideotypes that reflect trade-offs between traits and contrasting ecological strategies. Such a model will be positioned at the interface between functional plant ecology and weed science and seeks to integrate data from functional trait databases of national floras with detailed weed life cycle data. The approach assumes that concepts developed in theoretical ecology to do with the processes underpinning community assembly and plant coexistence are useful for understanding the response of weed in cropped fields. It also relies on the existence of data that explicitly link the ecology of weeds to the dynamic response of their populations to management.

Metcalfe et al. (2020) have published a model of this type that is founded on the principle of identifying relationships between functional traits in large ecological databases and life cycle parameters that predict shifts between demographic states (e.g. seedbank to emerged seedlings) under contrasting management scenarios. Here, we illustrate the approach using the example trait of seedbank persistence. In his compilation of the seedbank characteristics of a national flora, Thompson et al. (1997) identified three categories of seedbank persistence for UK plants: transient, short-term persistent and long-term persistent and published these ecological data for the UK flora, including weeds. For a small subset of common weed species, including *Galium aparine* (short-term persistent) and *Papaver rhoeas* (long-term persistent), the decline of the weed seedbank in cultivated fields has been well-characterised (Lutman et al. 1993). By assigning each of the three categories (identified by Thompson et al. 1997) – transient, short-term, and long-term persistent – an average value for weed seedbank decline (according to the data from Lutman et al. 1993), it is possible to estimate the survival rate of all weeds in each of the persistence categories (Figure 9.2a). This parameter will determine the decline of the seedbank under contrasting strategies, and, although an intensive weed seedbank decline study cannot be done for every weed species, the whole flora can be modelled at the level of the seed persistence category. The behaviour of the different persistence categories under contrasting tillage (Figure 9.2b) can be predicted using a whole life cycle model that relies on a similar 'broad-brush' ecological approach to modelling transitions between the remaining life cycle stages (Storkey et al. 2015; Metcalfe et al. 2020).

Using the impact of different tillage regimes on an observed local species pool in a field on a research farm as a worked example, following 30 years of simulation with continuous inversion tillage, the proportion of the seedbank made up of weeds in the 'long-term persistent' seedbank category was predicted to reduce, and there was a predicted increase in 'short-term persistent' seedbank types (Figure 9.2). As germination is prevented when seeds are moved deep into the soil profile, tillage is an important weed management tool, and it can change the functional composition of the weed community according to the seedbank type. While having limited power to predict the change in absolute numbers of individual weed species, this approach is useful for giving an insight into how a change in practice might influence the functional composition and absolute abundance of weeds. By aiming to predict changes in trait distributions across the whole regional species pool, as opposed to the dynamics of individual species, this model is restricted to using functions that can be related to traits widely reported in the literature. The following section shows how a similar approach to relating traits to model parameters can be used in a 'process-based', mechanistic model developed using novel data on traits not routinely measured in the ecological literature.

9.3 A Process-Based Approach to Modelling Weed Persistence

Arable fields often have a wide range of weed species in the seedbank with continuous shifts in weed communities in response to changes in cropping systems in the medium and in the long term (see Section 9.2.2). The approach described earlier to predict these shifts assigns weeds to functional groups based on broad categories of 'seedbank persistence' using online trait databases and predicts contrasting responses using parameters estimated at the level of the functional group. While this pragmatic approach can capture the effect of management on the functional composition of communities, it is less powerful in terms of predicting the response of individual species. Modelling the diversity of behaviours at the species level requires a tailored approach that captures the interactions between management, environment, and biological processes (Figure 9.1). When considering the dynamics of the seedbank, these processes include the induction and release of seed dormancy,

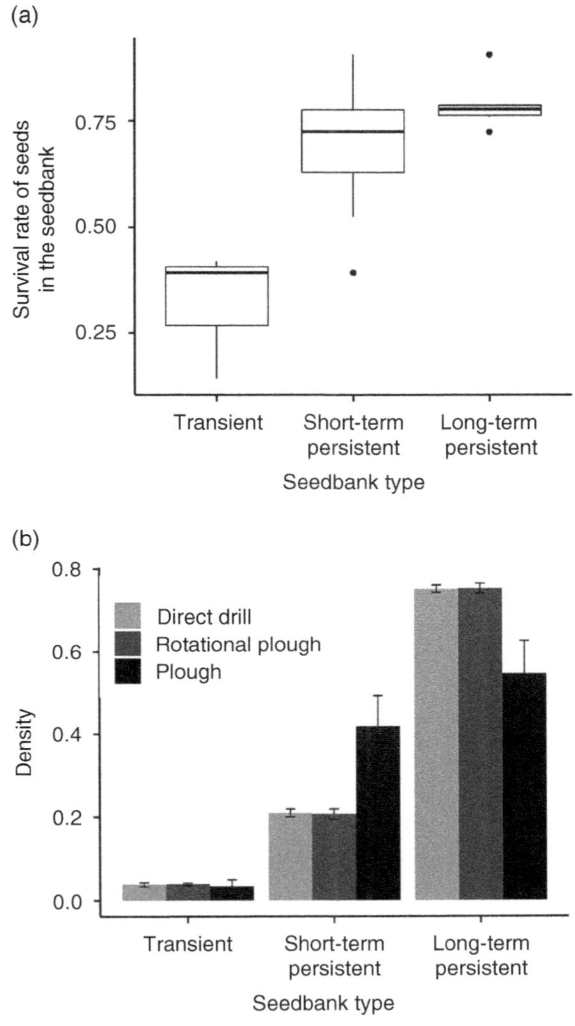

Figure 9.2 (a) Functional group categorization of weeds in terms of seedbank persistence, based on empirical data of seed survival at the level of individual weed species; (b) the proportion of the seedbank in each seedbank category in contrasting tillage scenarios, continuous direct drill (no inversion), rotational ploughing (three seasons direct drill followed by one season plough), and continuous ploughing (full inversion). Results are specific to the functional characteristics of the initial species pool; in this case, the field has low initial abundance of transient species, meaning the contrasting response to management for this group is less clear.

mortality of seeds through predation and rotting, germination, seedling growth, and emergence. This alternative mechanistic, process-based approach to modelling weed communities represents a powerful tool for continuously synthesizing biological knowledge (Colbach 2010). However, in the context of the 'tractability–prediction' trade-off, it also requires large numbers of parameters, which hinders the addition of new species to the model.

Previously, multispecies weed dynamics models of this type relied on the aggregation of independent single-species models, with a species-specific structure and a species-specific set of parameters (with the associated experimental overhead). To simplify the addition of new species, a generic model has been developed using the *same structure for all species* whose parameters are estimated from species traits. This model, parts of which are described in the following sections, is called FLORSYS (weed flora and cropping system) and simulates the effects of cropping systems, in interaction with climatic and soil conditions, on the dynamics of a multispecies weed flora composed of annual species. Following the general structure of weed populations dynamic models (Figure 9.1), FLORSYS is based on the representation of the weed life cycle by a succession of key life stages linked by demographic processes that are regulated by the cropping system, in interaction with the climatic and soil conditions. In the following discussion, the development of the seedbank part of the model and the approaches taken to deriving parameters for multiple species are reviewed with reference to a progression of published studies. To estimate the required parameters for a large number of weeds, they were, where possible, related to easier to measure species traits (*sensu* Violle et al. 2007), e.g. seed mass. However, as opposed to relying on published values in online trait databases (see Section 9.2.3), the relationships between traits and parameters were quantified from a series of bespoke experiments (Gardarin et al. 2012). In the following text, we present this approach for the processes determining seedbank persistence: seed survival in the soil (or mortality), dormancy, and germination.

9.3.1 Modelling Seed Mortality Processes

In situ seed mortality includes any seed disappearance caused by age, diseases, or microscopic predators (Chapter 13) but excludes any seedbank decrease due to germination (Chapter 6). In FLORSYS, predation by macro-organisms (vertebrates and arthropods) is not taken into account because it is assumed to be negligible for buried seeds (Westerman et al. 2006); this assumption would, therefore, be valid for regularly tilled systems. To measure seed mortality, a parameter in the FLORSYS model, an experiment was done in which seeds of 13 weed species were buried 30 cm deep in a field and recovered regularly for two years to measure their viability (germination tests followed by seed crushing tests). Seed viability decreased during the second year of burial, with mortality rates ranging from 0.01 to 0.63 per year, depending on the species. More than 75% of the seeds of *G. aparine* disappeared, whereas more than 80% of *Amaranthus hybridus* or *Polygonum lapathifolium* seeds were still viable. To facilitate including additional species in the model without having to run the field experiment for every species, the mortality parameter was related to seed traits that could be more easily measured. Among the studied trait–mortality relationships, it was found that seed mortality decreased with increasing seed coat thickness (Figure 9.3). This was measured using X-ray images (Figure 9.4). Seed mortality was higher for seeds with thin seed coats (e.g. *Geranium dissectum* and *Digitaria sanguinalis*) and lower for seeds with thick seed coats (e.g. *Fallopia convolvulus* and *Datura stramonium*). This relationship explained half the between-species variability in mortality rate (Gardarin et al. 2010).

Despite its importance in governing several processes in seedbank dynamics, seed coat thickness has rarely been reported in the weed science literature. Seminal envelopes, such as the seed coat or the achene wall, provide a mechanical and chemical barrier, protecting the seed and embryo from parasitic attacks, predation, and wide variations in temperature and humidity. Seeds with thick seminal envelopes (e.g. *Ambrosia artemisiifolia* and *Polygonum* spp.) have lower mortality. The biochemical composition of the seed shells and reserves, which may contain phenolic compounds and alkaloids with antibacterial and antifungal properties, may also contribute to seed protection and may affect predator preferences.

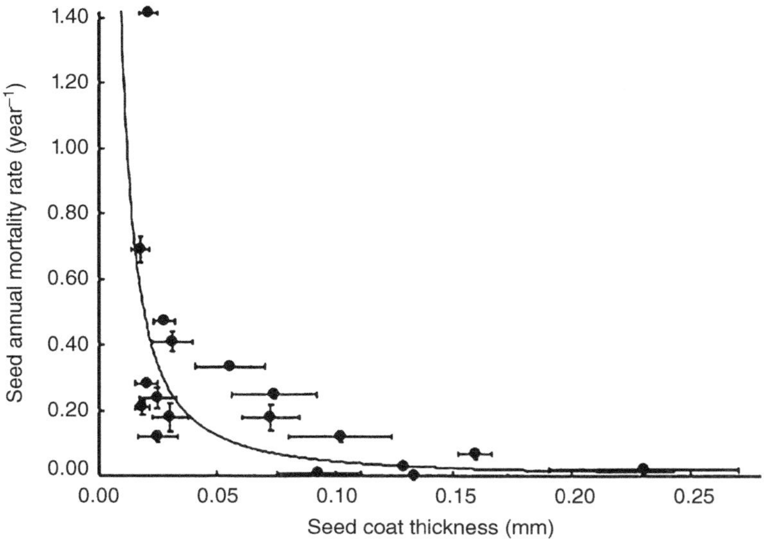

Figure 9.3 Annual seed mortality rate of seeds buried deep in the soil as a function of seed coat thickness (± standard deviation). A mortality rate exceeding 1 means that all the seeds lost their viability in less than one year. *Source:* Data from Gardarin et al. (2010).

9.3.2 Modelling Seed Dormancy Processes

Most weed species display cyclic dormancy, being dormant in one season and nondormant in another, thus optimising their chance to establish and reproduce in a favourable environment (Chapter 3). Variation in seed dormancy can be modelled using a broken-stick linear model, which is essentially a series of regression lines that predict the dynamics of dormancy in time (Figure 9.5). The proportion of nondormant seeds varies within the season, between a minimum and maximum level. Four timing parameters define the dates of dormancy induction (onset and end) and release (onset and end). In the first year after seed production, the dormancy levels may differ from the subsequent ones due to primary dormancy acquired while on the maternal plant. To quantify dormancy level–trait relationships, data were collected on seasonal seed dormancy fluctuations obtained after seed burial experiments under comparable germination conditions, in terms of temperature and light. Together with data from new experiments, the dataset comprised information on dormancy fluctuations for 29 weed species from 32 literature sources. Again, these parameters were related to seed traits to allow species outside of the germination dataset to be included in the model.

Minimum (when dormancy was low) and maximum (when dormancy was high) proportions of nondormant seeds in the first and the second years of burial were almost always explained by the same seed traits. Both minimum and maximum dormancy significantly increased with the seed shape index (the variance of the relative seed dimensions, ranging from nearly zero for spherical seeds to nearly 1 for elongated or flattened seeds), with a lower dormancy level for elongated seeds, even when accounting for phylogenetic relatedness between species (Gardarin and Colbach 2015). From an evolutionary point of view, natural

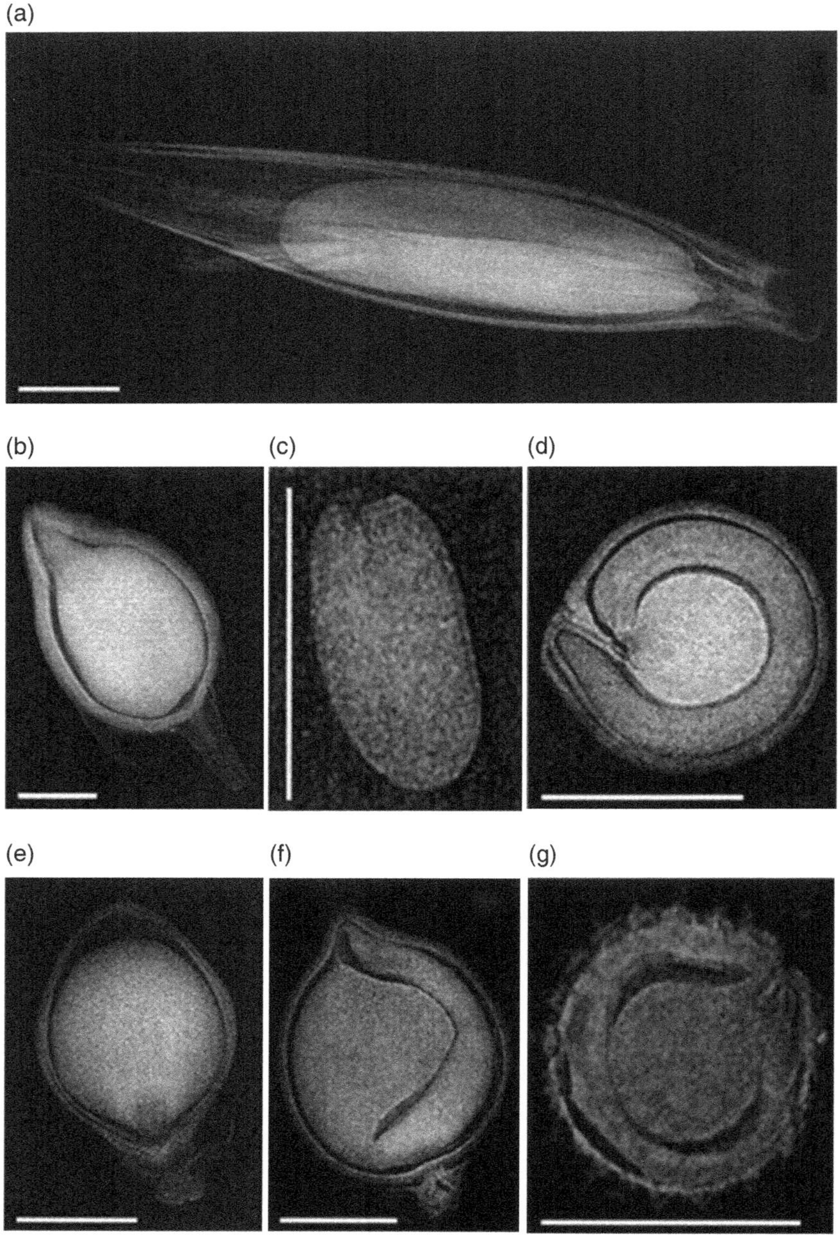

Figure 9.4 X-ray images of seeds of seven weed species with contrasted seed coat thickness: (a) *Avena fatua*, (b) *Ambrosia artemisiifolia*, (c) *Capsella bursa-pastoris*, (d) *Chenopodium album*, (e) *Echinochloa crus-galli*, (f) *Polygonum lapathifolium*, and (g) *Stellaria media*. Scale bars represent 1 mm.

burial of spherical seeds is easier than for flattened seeds, the latter tending to remain on the soil surface (Benvenuti 2007). Seeds located on the surface are more prone to predation than buried ones (Hulme and Borelli 1999) and may have been selected for lack of dormancy and immediate germination to limit such mortality risks (Grime et al. 1981). The seed dormancy

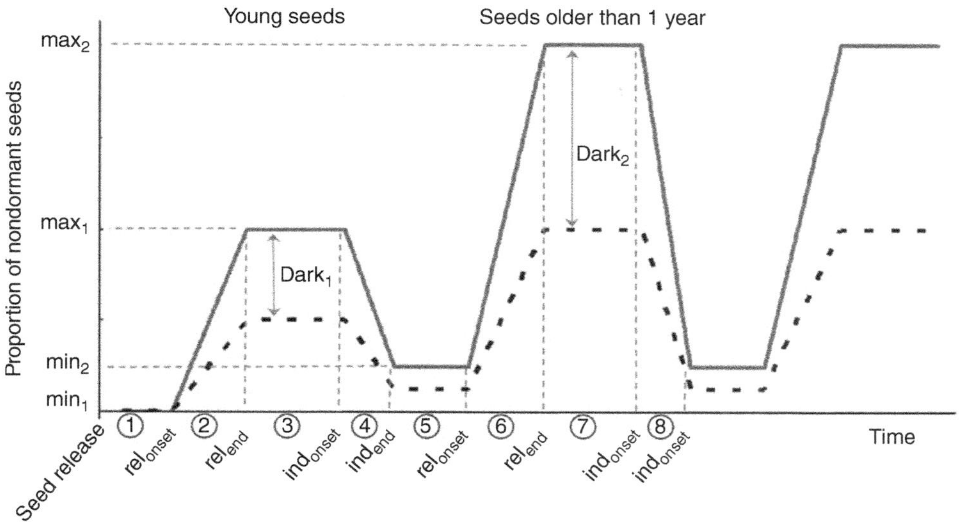

Figure 9.5 Representation of the parameters used in a model of the change in seed dormancy through a season. In the light (solid line), the proportion of nondormant seeds vary between min_1 and max_1 for young seeds and between min_2 and max_2 for seeds aged more than one year. In the dark (dotted line), these proportions were proportionally reduced by a factor of $dark_1$ and $dark_2$, respectively, for young and old seeds. Dormancy release occurs between rel_{onset} and rel_{end} dates, and dormancy is induced from day ind_{onset} to day ind_{end}. *Source:* Gardarin and Colbach (2015).

of each species also increased with increasing coat thickness or dry mass, which were all correlated (Gardarin and Colbach 2015). The seed coat represents a physical and maybe chemical barrier. We observed in our data set that the thicker the coat, the larger this barrier effect, resulting in coat-imposed dormancy and a lower germination proportion.

Seed morphological traits (mass, shape, coat thickness) explained 38–56% of the variability in the level of seed dormancy, while no relation was found with seed lipid or protein content. The unexplained variability was a product of the interaction with environmental factors; a large proportion of the variability in the level of dormancy is not species specific but results from prevailing conditions experienced by the maternal plant during seed production including heat and moisture stress (Chapters 3). As for the dates of dormancy fluctuations, they were not related to any trait but could be related to average weed emergence periods (with a 10-day time scale) using emergence calendars for each species. This kind of information can be obtained by questioning agricultural botanists with a long experience of weed emergence surveys in field conditions or interrogating the literature. It is also possible to group species into categories defined by the periodicity of emergence (Storkey et al. 2015) – this is the approach that was taken in the trait-based approach to modelling weed persistence (Section 9.2.3). The dates of dormancy induction are positively correlated with the date of the end of emergence of each species, that is, the later the end of the average emergence period, the later the dormancy induction period. The dates of dormancy release increased with the date of the onset of emergence in the field. It was therefore possible to estimate dates of dormancy induction and release for a wide range of species by knowing the usual approximate dates of weed emergence onset and end in the field (Figure 9.5).

9.3.3 Modelling Seed Germination Processes

In the field, germination can be triggered when tillage is performed in a moist soil or when precipitations moisten a dry soil layer. Germination only occurs when the soil temperature and soil water potential exceed the base values for a given species. Modelling the date and rate of germination is essential as it determines the date of seedling emergence relative to the crop, crop–weed competitive interactions, and the ability of the weed to replenish the seedbank. In the following text, we again describe the parameters used in the FLORSYS model and present the analysis of the traits associated with these germination dynamics.

A dataset of 185 germination time courses covering 25 weed species under optimal germination conditions was used to quantify the following model parameters: the germination time lag (time from imbibition to the first germinated seed), the time to mid-germination (time to reach 50% of the final germination percentage), and the maximal rate (slope of the germination curve at mid-germination). The correlations between seed germination parameters and morphological, chemical, and physiological seed traits were then analysed, using seed dormancy level as a covariate (Section 9.2.3) since germination occurred earlier and was faster in nondormant seed lots. Seed traits, together with dormancy level (final germination rate) explained 41–59% of the variability in germination speed parameters (Gardarin et al. 2011). Germination was more rapid in species with a high base temperature for germination; a trait that is often available in the literature. This relationship indicates that species requiring high temperature germinate faster than those germinating earlier at low temperatures. Thus, both types of species achieve germination and subsequently grow and develop in a similar time considering the range of temperatures in which they each germinate in field conditions.

Lipid-rich seeds (exceeding 30% lipid content) also had earlier and more rapid germination (Gardarin et al. 2011). This effect was found for species with a wide range of seed lipid contents, varying from 3.5% (*F. convolvulus*) to 43.6% (*P. rhoeas*) and was a novel finding as a reverse correlation was expected between seed lipid content and germination rate. The reactivation of metabolism at seed imbibition is associated with the beginning of the mobilization of reserves. The use of lipid reserves requires a greater quantity of oxygen, which makes the germination of lipid-rich species more sensitive to partial oxygen pressure (Al-Ani et al. 1985; Raymond et al. 1985), which can thus slow the rate of germination. The last trait that was found to affect germination was the surface-area-to-mass ratio that was positively related to the germination lag (Gardarin et al. 2011). It was hypothesised that germination timing partly depends on the time required for seed imbibition. Imbibition should be faster in seeds with a large area for water to enter relative to seed water demand, probably linked to seed mass. However, although logical, the present correlation only explains a small proportion of the variability in the germination rate and remains to be confirmed on a larger number of species.

9.3.4 Integrating Seedbank Processes into the Multi-annual Weed Dynamics Model

Modelling seed germination and pre-emergent growth alone is not enough to predict the persistence of weeds in fields. The processes regulating seedbank dynamics, described in Sections 9.3.1–9.3.3, have been aggregated with other submodels to complete the life cycle of annual weeds (Munier-Jolain et al. 2013; Colbach et al. 2014b,c; Mézière et al. 2015). The resulting model, FLORSYS, can be visualised as a virtual field on which cropping systems can be tested with different pedoclimates and weed floras in terms of crop production and weed (dis)services (Colbach et al. 2014a). The input variables of FLORSYS consist of (i) a description of the simulated field (daily weather, latitude, and soil characteristics); (ii) all the crops and management operations in the field, with dates, tools, and options; and (iii) the initial weed seedbank, which is either measured on soil samples or, more feasible, estimated from regional flora assessments (Colbach et al. 2016). Parameters have been estimated for 25 frequent and contrasting annual weed species and 33 cash and cover crop species, using relationships with traits where necessary. The input variables influence the annual life cycle of annual weeds and crops, with a daily time step. After emergence, the crop–weed canopy is simulated in 3D to account for heterogeneous canopies in terms of species and emergence cohorts, with a simplified representation of each crop and weed plant (Figure 9.6). A series of post-emergence processes (e.g. photosynthesis, respiration, growth, shade response), driven by light availability and air temperature, are modelled at the level of the individual plant. At plant maturity, weed seeds are added to the soil seedbank and crop seeds are harvested to determine crop yield. As for the pre-emergence submodel, functional relationships were established to estimate difficult-to-measure parameters from more easily measured or accessible species traits and characteristics (Colbach et al. 2019).

This thorough simulation of the whole weed life cycle produces very detailed outputs, at a daily time step and in 3D. These virtual measurements are essential to provide a mechanistic understanding of why a given technique or cropping system results in a given weed seedbank community. To simplify the comparison of different cropping systems, the detailed outputs are translated into indicators of weed impact on crop production and the environment (Mézière et al. 2015). For instance, crop yield loss due to weeds is the relative difference in yield of simulations versus without weeds. Other indicators use a trait-based approach (i.e. community-weighted mean of trait) to assess weed services (e.g. potential trophic resources for carabids), weighting weed state variables (e.g. weed seeds on soil surface during insect activity season) by species traits reflecting the contribution to the service (e.g. seed lipid content) (Mézière et al. 2015).

9.3.5 Case Study Using the Process-Based Model

In this section, a simulation with FLORSYS is presented to illustrate the power of the model to simulate the effect of management on weed persistence and show an example of the kind of output the models can produce, capturing the time lag in the response of weeds to management that is inherent in their population dynamics.

Weed seed persistence is the reason why a management decision in a given year can have consequences for many years to come. To illustrate this, we assessed the effect of cover and companion crops in winter wheat (*Triticum aestivum*) on weed persistence in a wheat/barley (*Hordeum vulgare*)/oilseed rape (*Brassica napus*) rotation in three virtual fields whose management only differs during the first year (Figure 9.7). During that year, one field was grown only with wheat, one had a cover crop during summer fallow, and the third one associated wheat with a companion crop. Cover and companion crops are grown to

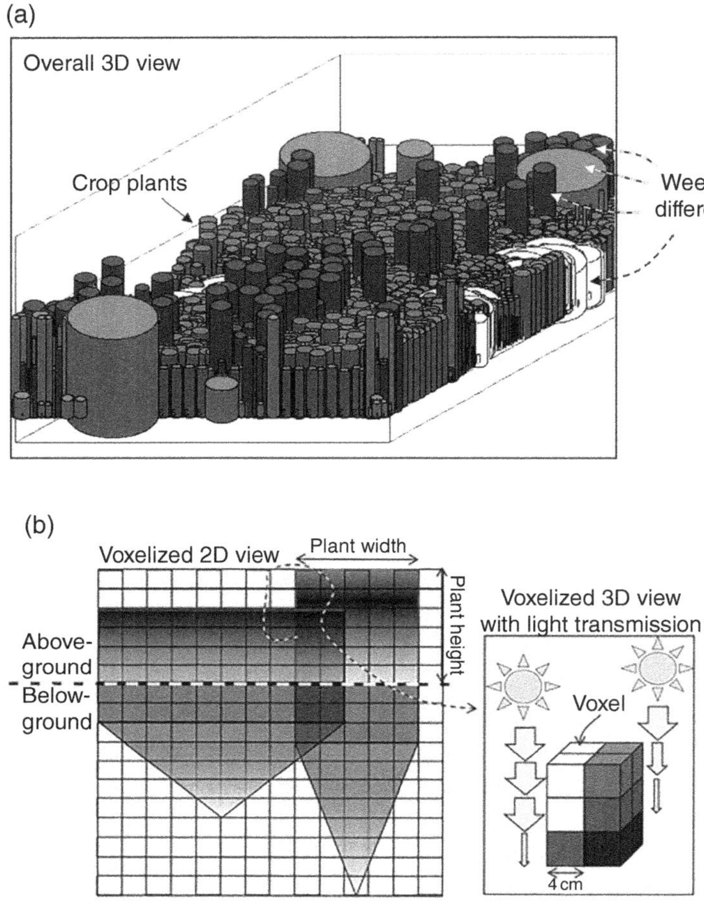

Figure 9.6 3D individual-based representation of multispecies canopy in FLORSYS. (a) 3D aboveground view of a sample of a simulated field of 3×6 m^2 where each crop and weed plant is represented as a cylinder. The different grey shades show plants of different species. (b) 2D view of two plants, detailing above-and belowground plant volume (with leaf area and root biomass distributed along layers illustrated here by grey shades) and space discretized with voxels (3D pixels, here for instance 4x4x4 cm^3) to simulate light trickling down in the aboveground part of the canopy and overlapping root systems competing for soil resources. *Source:* Nathalie Colbach.

occupy the soil surface and leave no available niches for weeds. Simulations using FLORSYS show that these solutions, particularly the companion crop, are indeed efficient in reducing weed densities in the cash crop grown during the year the solutions are tested, compared with a sole wheat crop (Figure 9.7). However, the beneficial effect of the companion crop was also still visible intermittently during the next nine years, even though the three fields were managed identically during that time. More importantly, weed densities increased in the field with the initial cover crop compared to the control. This was a consequence of the cover crop restricting opportunities for controlling weeds in false seed beds, resulting in a denser seedbank than in the control. These seeds emerged during the third year, with a nefarious effect still visible in the tenth year. This case study illustrates the necessity of

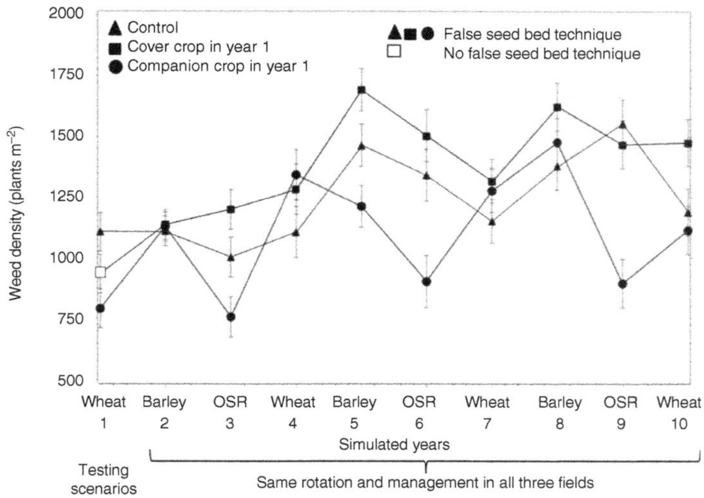

Figure 9.7 Long-term effects of cover (mustard) and companion crops (field bean) and the associated management practices in the wheat grown in year 1 on multispecies weed dynamics over 10 years. OSR is oilseed rape. Mean and standard deviation of the maximum weed density in the cash crop of 10 weather repetitions from Burgundy. *Source:* Nathalie Colbach.

long-term modelling approaches to capture the impact of interacting management effects that would not be possible using conventional field experimentation approaches.

9.3.6 The Weed Seed Traits Selected by Management Practices

A central theme running through this chapter has been the value of understanding how the traits of weed species can be used to predict model parameters and the response of their population dynamics under contrasting management scenarios. Identifying which traits should be targeted in these analyses largely relies on a knowledge of the underlying biology or evidence from the published literature. However, it is also possible to use models to identify the extent to which management discriminates between, or acts on, different traits related to their persistence. Models such as FLORSYS can be used to ask this kind of theoretical question by establishing multiple scenarios covering contrasting pedoclimates and cropping systems (Colbach and Cordeau 2018). The simulated data are analysed similarly to data collected in actual fields (Colbach et al. 2014d) but allow the coverage of considerably more situations and measurements and can be used to better understand underlying processes and trends.

The example of Colbach and Cordeau (2018), Table 9.1, shows an example of such an analysis. In this case, the two weed species traits determining the timing of germination and emergence (Section 9.3.3), i.e. base temperature and base water potential, were those that most interacted with management practices on average over all regions, weather conditions, and cropping systems. Species that require a warmer soil were rarer in longer rotations (negative correlation for base temperature in line [1] in Table 9.1), consisting of many different species or varieties [4] and comprising long cover crops [3] as well as frequent winter crops [5]. They are also disadvantaged by frequent [9–11] and deep tillage [8] but

Table 9.1 Which weed species traits related to seedbank persistence are selected by management practices? Relationships identified by fourth-corner analyses on 272 cropping systems from 7 French and Spanish regions simulated over 28 years and 10 weather repetitions. Pearson correlation coefficients between practices and species traits/parameters, showing only coefficients exceeding 0.20 in absolute values and significant at $p < 0.001$ (positive relationships indicated with bold outlines and negative relationships with shaded boxes). Fourth-corner analyses investigate correlations between three matrixes consisting of management practices in fields, weed species densities in these same fields, and weed species traits/parameters, using permutations to test the null hypothesis that species are distributed independently of scenarios and traits/parameters. Here, practices and weed densities were averaged over 27 years.

		Seed traits					
	Cropping system components	Area/mass	Coat thickness	Shape[a]	Mass	Base temperature	Base water potential
[1]	Rotation length (years)					−0.21	
[2]	% cover crops during fallow		0.20				
[3]	Duration of crop cover (moths/year)					−0.24	
[4]	Number of crops or varieties in rotation					−0.22	
[5]	% winter crops					−0.30	−0.20
[6]	Sowing date of winter crops					−0.24	
[7]	Harvest date of winter crops					−0.29	−0.21
[8]	Mean tillage depth (cm)	−0.27		−0.34		−0.29	−0.27
[9]	Superficial tillage (operations/year)			−0.29		−0.40	−0.34
[10]	Superficial tillage (April–September)			−0.29		−0.39	−0.32
[11]	Superficial tillage (October–March)					−0.28	−0.26
[12]	Days from harvest to first till	−0.24			0.24		
[13]	Years between ploughing			0.20		0.20	0.20
[14]	Days from ploughing to sowing					−0.23	−0.24
[15]	Rolling operations/year					−0.27	−0.22
[16]	Days from harvest to rolling					−0.21	
[17]	Nonsystemic herbicides/year					−0.22	

[a] Elongated and flattened seeds have a high shape index; round seeds have a small index value.
Source: Method from Colbach et al. (2014d), data from Colbach and Cordeau (2018).

favoured by rare mouldboard ploughing [13]. Conversely, seed traits that are related to persistent seeds, i.e. thick seed coats, small seed areas per seed mass, and round seeds were better adapted to long crop covers as well as deep and frequent tillage.

Knowing the weed density or the frequency of key weed traits in a field or a cropping system is not enough to assess whether the weed flora is harmful for crop production or beneficial for the agroecosystem. Thanks to its weed impact indicators, FLORSYS allows the weed traits that drive weed (dis)services to be determined, using the same methods as for investigating the links between management practices and species traits (i.e. running a virtual field network with FLORSYS and analysing the correlations between weed species densities, traits, and (dis)services). For instance, weed contribution to reducing nitrate leaching during summer fallow was shown to increase in species with a high seed lipid content and a large seed area per mass. Both traits result in fast and early germination, which is crucial to ensure that weeds take up nitrogen fast enough to keep it from leaching to groundwater. Similarly, weeds with thinner seed coats protected the soil more against erosion as such seeds are less dormant and can emerge during fallow.

9.4 Conclusions

The need to understand the complex, interacting effects of the environmental, management, and biological components of an arable field on the persistence of weeds (Figure 9.1) means that weed population modelling will continue to make an important contribution to sustainable weed management. However, precisely because of this complexity and associated uncertainty, care should be taken in how models are presented and interpreted (Holst et al. 2007). It is unlikely that any model, regardless of its complexity, will be able to accurately predict the change in absolute numbers of weed seeds of a given species under contrasting management scenarios. However, as illustrated earlier, models do have the power to ask 'what-if' questions to do with the relative response of different weed species or the likelihood of a management change to reduce overall abundance or functional composition of the community. When applied in this 'hypothesis-generation-and-testing way', they have an important role to play in exploring future trajectories of weed communities under future, novel, environmental, and management conditions.

References

Al-Ani, A., Bruzau, F., Raymond, P. et al. (1985). Germination, respiration and adenylate energy charge of seeds at various oxygen partial pressures. *Plant Physiology* 79: 885–890.

Benvenuti, S. (2007). Natural weed seed burial: effect of soil texture, rain and seed characteristics. *Seed Science Research* 17: 211–219.

Bourgeois, B., Munoz, F., Fried, G. et al. (2019). What makes a weed a weed? A large- scale evaluation of arable weeds through a functional lens. *American Journal of Botany* 106: 90–100.

Colbach, N. (2010). Modelling cropping system effects on crop pest dynamics: how to compromise between process analysis and decision aid. *Plant Science* 179: 1–13. https://doi.org/10.1016/j.plantsci.2010.04.009.

Colbach, N. and Cordeau, S. (2018). Reduced herbicide use does not increase crop yield loss if it is compensated by alternative preventive and curative measures. *European Journal of Agronomy* 94: 67–78.

Colbach, N., Biju-Duval, L., Gardarin, A. et al. (2014a). The role of models for multicriteria evaluation and multiobjective design of cropping systems for managing weeds. *Weed Research* 54: 541–555.

Colbach, N., Busset, H., Roger-Estrade, J., and Caneill, J. (2014b). Predictive modelling of weed seed movement in response to superficial tillage tools. *Soil & Tillage Research* 138: 1–8.

Colbach, N., Collard, A., Guyot, S.H.M. et al. (2014c). Assessing innovative sowing patterns for integrated weed management with a 3D crop:weed competition model. *European Journal of Agronomy* 53: 74–89.

Colbach, N., Granger, S., Guyot, S.H.M., and Mézière, D. (2014d). A trait-based approach to explain weed species response to agricultural practices in a simulation study with a cropping system model. *Agriculture, Ecosystems & Environment* 183: 197–204.

Colbach, N., Bertrand, M., Busset, H. et al. (2016). Uncertainty analysis and evaluation of a complex, multi-specific weed dynamics model with diverse and incomplete data sets. *Environmental Modelling & Software* 86: 184–203.

Colbach, N., Gardarin, A., and Moreau, D. (2019). The response of weed and crop species to shading. Which parameters explain weed impacts on crop production? *Field Crops Research* 238: 45–55.

Freckleton, R.P. and Watkinson, A.R. (1998). Predicting the determinants of weed abundance: a model for the population dynamics of *Chenopodium album* in sugar beet. *Journal of Applied Ecology* 35: 904–920.

Fried, G., Kazakou, E., and Gaba, S. (2012). Trajectories of weed communities explained by traits associated with species' response to management practices. *Agriculture Ecosystems & Environment* 158: 147–155.

Gaba, S., Perronne, R., Fried, G. et al. (2017). Response and effect traits of arable weeds in agro-ecosystems: a review of current knowledge. *Weed Research* 57: 123–147.

Gardarin, A. and Colbach, N. (2015). How much of seed dormancy in weeds can be related to seed traits? *Weed Research* 55: 14–25.

Gardarin, A., Dürr, C., Mannino, M.R. et al. (2010). Seed mortality in the soil is related to seed coat thickness. *Seed Science Research* 20: 243–256.

Gardarin, A., Dürr, C., and Colbach, N. (2011). Prediction of germination rates of weed species: relationships between germination parameters and species traits. *Ecological Modelling* 222: 626–636.

Gardarin, A., Dürr, C., and Colbach, N. (2012). Modeling the dynamics and emergence of a multispecies weed seed bank with species traits. *Ecological Modelling* 240: 123–138.

Grime, J.P. (1974). Vegetation classification by reference to strategies. *Nature* 250: 26–31.

Grime, J.P., Mason, G., Curtis, A.V. et al. (1981). A comparative study of germination characteristics in a local flora. *Journal of Ecology* 69: 1017–1059.

Gunton, R.M., Petit, S., and Gaba, S. (2011). Functional traits relating arable weed communities to crop characteristics. *Journal of Vegetation Science* 22: 541–550.

Holst, N., Rasmussen, I.A., and Bastiaans, L. (2007). Field weed population dynamics: a review of model approaches and applications. *Weed Research* 47: 1–14.

Hulme, P.E. and Borelli, T. (1999). Variability in post-dispersal seed predation in deciduous woodland: relative importance of location, seed species, burial and density. *Plant Ecology* 145: 149–156.

Lutman, P.J.W., Cussans, G.W., Wright, K.J. et al. (1993). The persistence of seeds of 16 weed species over six years in two arable fields. *Weed Research* 42: 231–241.

Metcalfe, H., Milne, A.M., Deledale, F., and Storkey, J. (2020). Using functional traits to model plant community dynamics. *Ecology* 101 (11) https://doi.org/10.1002/ecy.3167.

Mézière, D., Petit, S., Granger, S. et al. (2015). Developing a set of simulation-based indicators to assess harmfulness and contribution to biodiversity of weed communities in cropping systems. *Ecological Indicators* 48: 157–170.

Moles, A.T. and Westoby, M. (2004). Seedling survival and seed size: a synthesis of the literature. *Journal of Ecology* 92: 372–383.

Munier-Jolain, N.M., Guyot, S.H.M., and Colbach, N. (2013). A 3D model for light interception in heterogeneous crop:weed canopies. Model structure and evaluation. *Ecological Modelling* 250: 101–110.

Raymond, P., Al-Ani, A., and Pradet, A. (1985). ATP production by respiration and fermentation, and energy charge during aerobiosis and anaerobiosis in twelve fatty and starchy germinating seeds. *Plant Physiology* 79: 879–884.

Storkey, J., Moss, S.R., and Cussans, J.W. (2010). Using assembly theory to explain changes in a weed flora in response to agricultural intensification. *Weed Science* 58: 39–46.

Storkey, J., Meyer, S., Still, K.S., and Leuschner, C. (2012). The impact of agricultural intensification and land-use change on the European arable flora. *Proceedings of the Royal Society B: Biological Sciences* 279: 1421–1429.

Storkey, J., Holst, N., Bojer, O.Q. et al. (2015). Combining a weed traits database with a population dynamics model predicts shifts in weed communities. *Weed Research* 55: 206–218.

Thompson, K., Band, S.R., and Hodgson, J.G. (1993). Seed size and shape predict persistence in the soil. *Functional Ecology* 7: 236–241.

Thompson, K., Bakker, J.P., and Bekker, R.M. (1997). *The Soil Seed Banks of North West Europe: Methodology, Density and Longevity*. Cambridge University Press.

Varanasi, A., Prasad, P.V.V., and Jugulam, M. (2016). Impact of climate change on weeds and herbicide efficacy. *Advances in Agronomy* 135: 107–146.

Violle, C., Navas, M.L., Vile, D. et al. (2007). Let the concept of trait be functional! *Oikos* 116: 882–892.

Westerman, P.R., Liebman, M., Heggenstaller, A.H., and Forcella, F. (2006). Integrating measurements of seed availability and removal to estimate weed seed losses due to predation. *Weed Science* 54: 566–574.

10

Influence of Agronomic Practices on the Persistence of Weed Seedbanks

Clarence Swanton and Saeed Vazan

Department of Plant Agriculture, University of Guelph, Guelph, ON, Canada

10.1 Introduction

Weeds represent the most persistent threat to successful crop production. Uncontrolled populations of weeds have been estimated to result in a 50% yield reduction for corn (*Zea mays*) grown in North America (Soltani et al. 2016). Globally, the annual cost of controlling weeds in agriculture is expected to steadily increase in value from US$ 27.2 billion in 2016 to US$ 39.15 billion by 2022 (Anonymous 2015). This value does not include tillage operations or manual efforts that are also expended in an effort to control weeds. Despite our best efforts with such selection pressures as herbicide technology, tillage, and pure manual muscle power, weeds continue to persist and to befuddle our weed management strategies. The wellspring from which all of these weed problems arise is the weed seedbank.

The weed seedbank is the most likely source of any weed population to be found in agriculture. The seedbank is the "memory" of aboveground plant communities (Cavers and Benoit 1989) and provides insight into past agronomic practices as well as future potential weed problems (Swanton and Booth 2004). Weeds that escape management efforts can quickly replenish the seedbank given the potential for huge seed return. For example, of the eleven weed species reported in Table 10.1, estimates of seed productivity per plant range from a low of 520 to 890 for yellow foxtail (*Setaria viridis*) to a high of 30 000 to 176 000 seeds per plant for common lambsquarters (*Chenopodium album*). Combining this potential seed return on a per plant basis with the estimated number of weed seeds per meter present within arable weed seedbanks (Tables 10.1 and 10.2), one can quickly see the challenge in attempting to reduce the size of the seedbank based on agronomic management practices. Murphy et al. (2006) reported that after six years of a corn, soybean (*Glycine max*), and winter wheat (*Triticum aestivum*) rotation including intensive use of herbicides, the existing seedbank estimates were 8000 compared with 49 000 seeds m^{-2} in a no-tillage versus moldboard plow tillage cropping system. On-farm studies such as this highlight how persistent the weed seedbank remains despite our agronomic practices.

Persistence Strategies of Weeds, First Edition. Edited by Mahesh K. Upadhyaya, David R. Clements, and Anil Shrestha.

Table 10.1 Estimated seed production for common agricultural weeds found in North America.

Author/year	Species	Seeds/plant	Location
Schwartz et al. (2016)	Palmer amaranth (*Amaranthus palmeri*)	13 384–60 221	Arkansas, Tennessee, Illinois, Missouri, and Nebraska
	Tall waterhemp (*Amaranthus tuberculatus*)	11 833–82 811	Nebraska, Missouri, Wisconsin, and Illinois
Weaver (2001)	Fleabane (*Conyza canadensis*)	2 000–230 000	Ontario
Bosnic and Swanton (1997)	Barnyardgrass (*Echinochloa crus-galli*)	120–3 460	Ontario
Harrison (1990)	Common lambsquarters (*Chenopodium album*)	30 000–176 000	Ohio
Defelice et al. (1989)	Giant foxtail (*Setaria faberi*)	460–600	Missouri
Warwick and Black (1986)	Velvetleaf (*Abutilon theophrasti*)	1 466–4 740	Ontario, Michigan, and Ohio
Nadeau and Morrison (1986)	Green foxtail (*Setaria viridis*)	2 720–3 050	Manitoba
	Yellow foxtail cited as (*Setaria lutescens*)	520–890	Manitoba
Weaver and McWilliams (1980)	Redroot pigweed (*Amaranthus retroflexus*)	13 860–34 600	California
Dickerson and Sweet (1971)	Common ragweed (*Ambrosia artemisiifolia*)	3 135–32 485	Washington, Maine, New Hampshire, Ontario, New York, Pennsylvania, Indiana, Missouri, Virginia, and Florida

The ability of weed seeds to persist in the soil is, by far, one of the most striking characteristics of a weed seedbank. One of the best examples of this is the *Papaver* spp. made famous by the World War 1 poem "In Flanders Fields" written in May 1915 by Canadian physician Lieutenant Colonel John McCrae, (McCrae 1915). Speculation arose that seeds of this poppy had been in the soil since the Franco–Prussian War, which occurred from 1870 to 1871. Further support for the persistence of weed seeds within the seedbank also arose from such studies as the Beal experiment (Kivilaan and Bandurski 1981) and the Duvel experiment (Toole and Brown 1946).

The persistence of seeds within an arable seedbank, however, can be influenced by the frequency and intensity of disturbance. In arable fields, disturbance caused by cultural practices such as tillage can impose a strong selection pressure on weed seed persistence. Seed persistence is known to increase in habitats that experience frequent and intense density-independent mortality, i.e. disturbance (Thompson et al. 1998). Thompson et al. (1998) developed a seed longevity index in order to characterize persistence in varying habitats. They found significant differences in seed longevity between a woodland and an

Table 10.2 Estimated number of seeds in the seedbank under different cropping systems and management.

Author/year	Location	Tillage	Cropping system	Soil type	Herbicide use (−/+)	Sampling depth (cm)	Seeds m^{-2}
Légère et al. (2011)	Quebec	Moldboard plow	Spring barley monoculture	Kamouraska clay	+	0–15	7353
		Moldboard plow	Barley–canola–wheat–soybean		+		3964
		Chisel plow	Spring barley monoculture		+		5933
		Chisel plow	Barley–canola–wheat–soybean		+		7920
		No-till	Spring barley monoculture		+		10106
		No-till	Barley–canola–wheat–soybean		+		7187
Murphy et al. (2006)	Ontario	No-till	Corn–soybeans–winter wheat	Conestoga silty loam, Melanic Brunisol and grey–brown podzolic	+	0–15	8000
		No-till	Corn–soybeans		+		11000
		Chisel plow	Corn–soybeans–winter wheat		+		34000
		Chisel plow	Corn–soybeans		+		36000
		Moldboard plow	Corn–soybeans–winter wheat		+		49000
		Moldboard plow	Continuous corn		+		56000

Sosnoskie et al. (2006)	Ohio	Moldboard plow	Six rotations involving corn, soybean, oat, and hay	Silt loam	+	0–10	1 350–1 411
		Minimum till			+		2 726–3 459
		No-till			+		4 308–7 790
Davis et al. (2005)	Michigan	Moldboard plow	Corn–soybean–wheat	Kalamazoo silt loam	+	0–5	22 600
		Moldboard plow and reduced N input	Corn–soybean–wheat		+		29 350
		Moldboard plow and manure	Corn–soybean–wheat		–		18 540
		No-till	Corn–soybean–wheat		+		21 780
Buhler et al. (2001)	Iowa	Moldboard plow, disking, ridge tillage, sweep plow	Legume/grass hay–corn–soybean–corn–oat	Nicollet loam	–	0–20	2 810–70 780
Dorado et al. (1999)	Spain	No-till	Three rotations involving barley, vetch, and sunflower	Calcic Haploxeralf	+	0–25	35 793
		Moldboard plow			+		21 841
Barberi et al. (1998)	Italy	Moldboard plow	Continuous corn	Xerofluvent	+	0–30	27 601
		Organic	Continuous corn		–		100 761
		Reduced energy input	Continuous corn		+		67 687
		Strip cultivation	Continuous corn		+		63 633

+ or – implies herbicide use or no herbicide use, respectively.

arable field, thus suggesting that habitat disturbance selects for increased seed persistence. In addition, the disturbance caused by tillage will also have a direct impact on the vertical distribution of seed within the soil profile.

10.2 Tillage: Vertical Distribution of Seeds Within the Weed Seedbank Influences Weed Seed Persistence

The level of soil disturbance and the vertical positioning of weed seeds within the soil profile will vary with tillage type, depth of tillage, and soil type. In a long-term tillage experiment that consisted of four different tillage systems, Clements et al. (1996) sampled the seedbank after seven and eight years of a corn–soybean rotation (Figure 10.1). The moldboard plow, which disturbs soil to the deepest depth of the tillage systems in the study, had the most uniform vertical placement of seed within the top 15 cm contrasted with no-tillage, which had the highest concentration of seed in the top 5 cm of the soil seedbank. Similar results were reported by Ball (1992), Cardina et al. (1991), Nichols et al. (2015), and Yenish et al. (1992). In general, as tillage is transitioned from moldboard plowing to a no-till system, the majority of the weed seeds migrate close to the soil surface. Variations on this concept, however, have been observed with differing soil types. Cardina et al. (1991) observed that the effect of tillage on seed depth differed with soil type, and Swanton et al. (2000) suggested that soil strength, pore size distribution, and alterations in soil physical conditions may play a role in modifying the vertical placement of seed within the soil profile.

The vertical positioning of the seed within the seedbank has a direct impact on its persistence as it influences both abiotic and biotic variables such as microbial activity and seed predation to which the seed is exposed. Persistence is reduced when seed is placed close to the soil surface compared with seed located deeper within the soil profile. Seed loss near the soil surface will exceed that of deeper placed seed because of higher seed germination rates and subsequent seedling mortality caused by such factors as seedling desiccation, wet–dry cycles, severe temperature fluctuations, and enhanced microbial activity (Kremer 1993; Lonsdale 1993; Mohler and Galford 1997; Fierer et al. 2003; Long et al. 2015). The physical positioning of the microbial community within the soil profile can alter

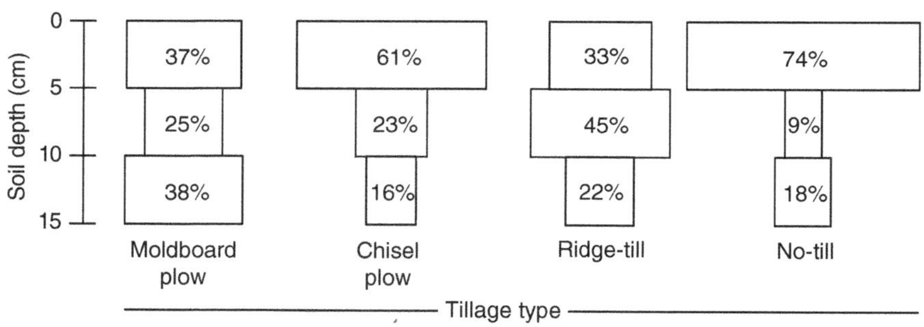

Figure 10.1 Effect of tillage methods on the vertical distribution of weed seeds in the soil after six growing seasons in a corn–soybean rotation. *Source:* Adapted from Clements et al. (1996).

microbial diversity and function. For example, Fierer et al. (2003) reported that microbial densities were one to two orders of magnitude higher at the soil surface than at a depth of 2 m. At this level within the soil profile, microbial communities are well supplied with carbon substrates, which occur from the input of chemical exudates released from plant roots, root decay, and crop surface residue. The fate of weed seeds that have been vertically placed within this zone of enhanced microbial activity may include physiological alterations within the seed, resulting in loss of viability prior to germination, increased rates of seed decay, and losses from post-germination mortality caused by reduced seedling vigor as a result of fungi and bacterial infections of the emerging seedling stems and roots (Kremer 1993; Kennedy 1999). In addition, seed losses resulting from seed predators and pathogens are highest for seed placed near or on the soil surface (Cromar et al. 1999; White et al. 2007). In field experiments designed to determine the effects of selected tillage systems and crop residue type on seed predation, Cromar et al. (1999) reported that ground-dwelling invertebrates consumed 80–90% of the weed seed found on the soil surface. In addition to the high rates of predation, the persistence of seeds at or near the soil surface can also be influenced by light.

10.3 Light Penetration and Soil Disturbance Can Reduce Seed Persistence

The persistence of buried weeds seeds can be reduced by germination caused by direct light penetration into the soil and/or through very brief exposure to sunlight as a result of soil disturbance caused by tillage. The ability of phytochrome (light sensing protein that regulates photomorphogenesis including seed germination), within weed seed to detect millisecond-to-second exposures of light pulses, is thought to be a major evolutionary adaptive trait for species survival (Scopel et al. 1994). Light does not penetrate the soil very well, but is influenced by pore size, soil cracks, color, and moisture levels (Tester and Morris 1987; Scopel et al. 1994). Tester and Morris (1987) suggested that enough light to trigger seed germination may be found at depths of 4–5 mm within the soil profile. The depth to which light can penetrate the soil profile is influenced by tillage.

It is well established that soil disturbance caused by tillage can stimulate seed germination (Chancellor 1964; Wareing 1969; Woolley and Stoller 1978; Roberts and Potter 1980; Bliss and Smith 1985; Suzuki and Waller 1987). Both deep and shallow tillage can expose buried seeds to milliseconds of sunlight, which can trigger seed germination. Warnes and Andersen (1984) reported that wild mustard (*Brassica kaber*) seed populations decreased to half of the original after seven growing seasons in undisturbed soil, whereas seed populations decreased to lower than 3% of initial populations in a treatment that involved plowing three times a year plus additional cultivation to remove emerging weeds. Botto et al. (1998) reported that daytime cultivations with a moldboard plow increased weed seedling emergence by as much as 200% compared with plowing at night. Jensen (1995) reported similar findings and suggested that repeated shallow soil disturbance with a harrow increased the cumulative number of weed seeds exposed to light, thereby, increasing seedling emergence of common lambsquarters during daylight compared with nighttime harrowing.

Despite evidence supporting the concept of light-induced germination, this response was not consistent across studies (Buhler et al. 1997; Juroszek et al. 2003). The degree to which light plays an important role will depend on several factors, such as seedbank species composition, state of seed dormancy and seasonality in sensitivity to light, tillage equipment and depth of operation, and soil micro-environmental conditions within the seedbank such as temperature, aeration, moisture, and compaction. For example, the sensitivity of the weed species to light can be highly variable. Field experiments conducted by Scopel et al. (1994) reported that nighttime cultivation reduced the emergence of seedling dicotyledons compared with cultivation under daylight condition, whereas the response of monocots, predominantly annual bluegrass (*Poa annua*), was not consistent. Additionally, the requirement for light-induced seed germination varied with weed species and season. In field experiments conducted by Botto et al. (1998), the contribution of monocots and dicots to the total seedling emergence pattern was found to vary with season. In late winter tillage trials conducted in Argentina, monocots contributed less than 10% of the total population of emerged seedlings. This contrasted with a monocot contribution of 19–66% to the emerged seedling population when trials were conducted in the spring or summer. Other variables including the degree of seed hydration and thermal cues received at the time of tillage may influence the ability of the seed to integrate light signals (Benech Arnold et al. 1988; Scopel et al. 1994).

10.4 Diverse Crop Rotations Do Not Consistently Reduce Weed Persistence

The use of diverse crop rotations has often been considered as an important management tool that will reduce the persistence of weed seedbanks over time. By definition, a rotation is "the sequence of crops grown in succession on a particular field" (Wibberley 1996). This is in direct contrast to monoculture, where the same crop is planted year after year in the same field. The underlying principle underpinning the role of crop rotations in weed management is that crop diversity will impose differing selection pressures such as planting dates, planting pattern, weed competitive traits, cultural practices including herbicides, nutrients requirements, and differing maturity dates. These differing selection pressures will act as filters on selected traits of the emerging weed species, which may in principle lead to an increased level of seedling mortality or reduced fitness of the surviving weed populations (see Booth and Swanton 2002).

Research conducted on weed seedbank dynamics in response to crop rotation alone is very limited. Most studies are confounded by tillage comparisons and weed management studies, or the size of the seedbank was not sampled at the beginning of the study, with results reported only upon completion of the experiment (see Barberi and Lo Cascio 2001; Cardina et al. 2002). This type of study makes it impossible to access the impact of crop rotation alone. For example, Ball (1992) stated that the cropping sequence influenced species composition in the seedbank; however, this conclusion was confounded by the herbicides used in the study. Schreiber (1992) reported that a crop rotation consisting of corn, soybean, and wheat was particularly effective in reducing giant foxtail (*Setaria faberi*) seed in the seedbank. Again, this result was invariably influenced by the three levels of weed management utilized in this study. The authors state that "there was no way to establish an untreated level of control without complete loss of yield due to weed competition." Liebman and Dyck (1993), however, summarized seven studies that reported 29 rotation sequences and had at least one rotation that did not include the use of herbicides. Of these 29 rotational sequences, 12 studies reported seed densities. Of these 12 studies, 9 reported lower weed seedbank densities in the crop rotational treatments, and no difference in seedbank density in 3 of the studies, when compared with the monocrop treatment. It should be noted, however, that the authors did not report any differences in tillage practices associated with these studies. In a rotational study involving moldboard plowing, Doucet et al. (1999) reported that in the absence of weed control measures, weed seedbank densities were similar among monocultures and 2–3-year rotations. This lack of consistency in the seedbank data reported for crop rotation studies may be attributed to an interaction between crop rotation and tillage.

The interaction of a diverse crop rotation plus the selection pressure imposed by tillage will invariably influence the size of the weed seedbank and the population dynamics of particular species. We identified three rotational studies by tillage type that sampled the seedbank at the beginning and end of the experiments (Table 10.3). There was no consistency in response to the degree of rotational complexity and type of tillage. For example, Murphy et al. (2006) reported that over a six-year rotation study, seedbank densities declined most rapidly in the no-till with either a two or three crop rotation. Seedbank densities remained relatively constant in the moldboard plow treatments despite a continuous corn versus a diverse corn, soybean, and wheat rotation. In addition, Barberi and Lo Cascio (2001) reported that tillage system had a greater influence on seedbank size and weed species composition than did crop rotation. In terms of species composition, Cardina et al. (2002) reported a significant interaction between tillage and crop rotation for fall panicum (*Panicum dichotomiflorum*), giant foxtail, common lambsquarters, and yellow woodsorrel (*Oxalis stricta*). Despite differences in rotation, total seed density of fall panicum was higher in continuous corn compared with the corn–soybean or corn–oats–hay rotations. This response was not observed in the moldboard plowed treatments. Interestingly, the authors also noted several weed species, including redroot pigweed (*Amaranthus retroflexus*), common chickweed (*Stellaria media*), corn speedwell (*Veronica arvensis*), barnyardgrass (*Echinochloa crus-galli*), yellow woodsorrel, and common lambsquarters that had no interaction with tillage and crop rotation. Diverse crop rotations remain an important tool in managing emerged weed populations; however, the response of the weed seedbank to rotational complexity is highly variable.

10.5 Control of Weed Seed at Harvest Has Potential to Reduce Seed Persistence

At harvest, there is a unique opportunity to reduce the amount of new weed seed return to the seedbank. Control of weed seed at harvest is centered on the reproductive biology of weeds that have escaped control during the growing season. This strategy focuses on reducing weed seed input into the seedbank. It is most successful with weeds that are tall enough to be combine harvested and retain a significant portion of their seed within the seed head

Table 10.3 Impact of crop rotation and tillage system on weed seedbank dynamics.

Author/year	Rotation	Length of study in years	Tillage system	Soil type	Soil organic matter (%)	Seedbank dynamics
Murphy et al. (2006)	Continuous corn	6	Moldboard plow	Conestoga silty loam, Melanic Brunisol, and grey–brown podzolic	5	No effect
	Corn–soybean		Chisel plow			(−)
	Corn–soybean		No-till			(−)
	Corn–soybean–winter wheat		Moldboard plow			No effect
	Corn–soybean–winter wheat		Chisel plow			(−)
	Corn–soybean–winter wheat		No-till			(−)
Buhler et al. (2001)	Legume/grass hay–corn–soybean–corn–oat	5	Moldboard plow, disking, ridge tillage, sweep plow	Nicollet loam	N/A	(+)
Ball (1992)	Continuous corn	3	Moldboard plow	N/A	N/A	(−)
	Continuous corn		Chisel plow			No effect
	Continuous pinto bean		Moldboard plow			No effect
	Continuous pinto bean		Chisel plow			(+)

(−) or (+) means the seedbank declined or increased, respectively.

at the time of crop harvest. Although seed retention will vary with species and environmental conditions, studies have identified several annual weed species that are suitable for harvest seed control. For example, species such as Palmer amaranth and tall waterhemp were observed to retain more than 95% of their seed at soybean maturity in several Midwestern states of the United States (Schwartz et al. 2016). Studies conducted in Western Canada found seed retention to be greater than 90% at harvest in species such as common lambsquarters, yellow foxtail, wild mustard (*Sinapis arvensis*), *Galium* spp., volunteer canola (*Brassica napus*), and bassia scoparia (*Kochia scoparia*) (Burton et al. 2016; Beckie et al. 2017; Tidemann et al. 2017). Harvest seed control utilizes mechanical rather than chemical tools such as field impact mills, which mechanically crush weed seeds, or chaff carts, which collect and allow for the removal of weed infested crop chaff from the field. An impact mill was found to be highly effective in destroying seed of 10 annual weed species commonly found in soybean fields in the Midwestern and Mid-Atlantic United States (Shergill et al. 2020). Studies conducted in Manitoba, Canada, reported that chaff collection during wheat harvest contained significant amounts of wild oat (*Avena fatua*) seed and was effective in reducing seed dispersal (Shirtliffe and Entz 2005). This strategy of destroying or removing of weed seed at harvest using mechanical rather than chemical means may prove most effective in reducing the seed persistence of herbicide-resistant weed species.

10.6 Role of Cover Crops and Microbial Populations

There has recently been a revival of interest in the role of cover crops for the associated benefits of both soil health and weed suppression. Cover crops have been shown to reduce wind and water erosion and increase soil water content early in the growing season (Wagner-Riddle et al. 1994), to moderate soil temperatures (Gupta et al. 1983) and enhance soil microbial interactions (Wortman et al. 2013). Several authors have suggested the role of cover crops for weed management as an important component of an integrated weed management strategy (Swanton and Weise 1991; Williams et al. 1998; Teasdale and Mohler 2000). Cover crops act in two capacities, one to suppress weed emergence and growth and the other to encourage microbial populations.

Cover crops are generally thought to provide an alternative selection pressure that may reduce weed seedling emergence, establishment, and fitness (Moore et al. 1994; Galloway and Weston 1996). Studies to date have recognized that the level of weed control provided by cover crops is highly variable and inconsistent (Moore et al. 1994; Teasdale 1996; Hayden et al. 2012). In order to deal with this inconsistency, Osipitan et al. (2018) recently suggested that cover crops may provide commercially acceptable weed control only during the early stages of crop development.

We were not able to locate any long-term studies, which evaluated the role of cover crops and the persistence of weed seeds in the seedbank. In terms of weed seed persistence, however, there has been an increase in research involving the interaction of cover crops, tillage, nitrogen, and soil microbial populations, which may prove to be a fruitful area of study (Swanton et al. 1999; Gallandt et al. 2005; Davis et al. 2006; Mohler et al. 2012, 2018). Mohler et al. (2018) suggested that the nitrate released from decomposing hairy vetch (*Vicia villosa*) cover crop may enhance fatal germination of seeds of common lambsquarters and Powell amaranth (*Amaranthus powellii*). Davis et al. (2006) reported a strong correlation between weed seed decay and soil microbial populations, suggesting a possibility of manipulation of soil microbial populations in order to enhance seed decay. Similarly, Mohler et al. (2012) suggested that the incorporation of fresh cover crop residue by tillage may partially suppress weed seedling emergence as a result of enhanced fungal invasion of seeds and seedlings. In addition, it is well-known that the persistence of weed seeds remaining on the soil surface is very limited because of seed predation by predators. Gallandt et al. (2005) stated that cover crops should be managed to provide late season habitats to encourage surface seed predation by invertebrates. Cromar et al. (1999) in fact postulated that soil surface invertebrate feeding on weed seeds may be the most effective broad-spectrum biological control available to reduce weed seed persistence.

10.7 Livestock, Pasture, and Manure Management Can Reduce Weed Seed Persistence

The persistence of the weed seedbank can be reduced over time with proper pasture and grazing management. In a recent study, Schuster et al. (2016) reported on a 15-year integrated crop–livestock system experiment that explored the effect of grazing intensity on the size of the weed seedbank. The results from this study found that seedbank size was significantly reduced when grazing intensity of cattle was low and when the pasture was managed under no-tillage conditions. Similar results were reported in a separate follow-up study by Schuster et al. (2018), who found reduced grazing under no-tillage management with a sheep grazed winter cover crop dramatically reduced the size of the weed seedbank over the 12 years of this study. Both studies stressed the importance of maintaining a high level of forage vegetative growth available for animals during the grazing phase in order to reduce the size of the weed seedbank.

Manure is known to be a source of viable seed input. Weed seeds are well-known to pass through the digestive tracts of animals (Atkeson et al. 1934). In 1994, Pleasant and Schlather (1994) surveyed 26 dairy farms in New York State. Among the farms surveyed, most had weed seed counts of less than 100 000 seeds per 1000 kg of fresh manure. The authors concluded that although this total seed input was viable, it represented a minor source of seed input when compared with the amount of viable seed found in the seedbank or produced by weed escapes in the field. Alternatively, if there is a requirement to reduce the viable seeds within manure, composting processes that generate temperatures greater than 60 °C have been reported to dramatically reduce the viability of seeds in manure (Eghball and Lesoing 2000). However, it should be noted that the temperature required and the length of exposure to these temperatures is species dependent (Larney and Blackshaw 2003).

10.8 Conclusions

One could argue that agronomic practices have been employed for decades in an effort to reduce weed seedbanks. Of the agronomic practices highlighted in this chapter, tillage and pasture, and grazing management were by far the most significant variables influencing long-term weed seedbank dynamics. Harvest weed seed control has potential; however, further long-term studies are required to confirm the impact of this strategy on seedbank dynamics. In dealing with weed seedbanks, we are confronted with two realities; only a small percentage of the seeds in the seedbank will ever germinate (Zhang et al. 1998) and the spatial variability of seedbanks within a single field is extremely diverse (Luschei 2003). This would suggest that the path forward may be best served by simply accepting the reality that weed seedbanks will be ever present despite our weed management and agronomic practices. In addition, weed seeds represent an important source of energy for food webs. If the goal of weed science remains the elimination of weed seeds from the soil profile, then, intuitively, this may lead to collapse of food webs and a rapid decline in soil health. Regardless, however, research must continue to address the ecological significance of weed seedbanks and the agronomic practices that may contribute to the best management of a persistent weed seedbank.

References

Anonymous (2015). Herbicides market by type (glyphosate, 2, 4-D, diquat), crop type (cereals & grains, oilseeds & pulses, fruits & vegetables), mode of action (non-selective, selective), and region – global forecast to 2022 (Online). https://www.marketsandmarkets.com/Market-Reports/herbicides-357.html (accessed 20 March 2019).

Atkeson, F., Hulbert, H., and Warren, T. (1934). Effect of bovine digestion and of manure storage on the viability of weed seeds. *Agronomy Journal* 26: 390–397.

Ball, D.A. (1992). Weed seedbank response to tillage, herbicides, and crop rotation sequence. *Weed Science* 40: 654–659.

Barberi, P. and Lo Cascio, B. (2001). Long-term tillage and crop rotation effects on weed seedbank size and composition. *Weed Research* 41: 325–340.

Barberi, P., Cozzani, A., Macchia, M., and Bonaire, E. (1998). Size and composition of the weed seedbank under different management systems for continuous maize cropping. *Weed Research* 38: 319–334.

Beckie, H., Blackshaw, R., Harker, K.N., and Tidemann, B. (2017). Weed seed shatter in spring wheat in Alberta. *Canadian Journal of Plant Science* 98: 107–114.

Benech Arnold, R., Ghersa, C., Sanchez, R., and Fernandez, A.G. (1988). The role of fluctuating temperatures in the germination and establishment of *Sorghum halepense* (L.) Pers. regulation of germination under leaf canopies. *Functional Ecology* 2: 311–318.

Bliss, D. and Smith, H. (1985). Penetration of light into soil and its role in the control of seed germination. *Plant, Cell and Environment* 8: 475–483.

Booth, B.D. and Swanton, C.J. (2002). Assembly theory applied to weed communities. *Weed Science* 50: 2–13.

Bosnic, A.C. and Swanton, C.J. (1997). Influence of barnyardgrass (*Echinochloa crus-galli*) time of emergence and density on corn (*Zea mays*). *Weed Science* 45: 276–282.

Botto, J.F., Scopel, A.L., Ballaré, C.L., and Sánchez, R.A. (1998). The effect of light during and after soil cultivation with different tillage implements on weed seedling emergence. *Weed Science* 46: 351–357.

Buhler, D.D., Hartzler, R.G., and Forcella, F. (1997). Implications of weed seedbank dynamics to weed management. *Weed Science* 45: 329–336.

Buhler, D.D., Kohler, K.A., and Thomspon, R.L. (2001). Weed seed bank dynamics during a five-year crop rotation. *Weed Technology* 15: 170–176.

Burton, N.R., Beckie, H.J., Willenborg, C.J. et al. (2016). Evaluating seed shatter of economically important weed species. *Weed Science* 64: 673–682.

Cardina, J., Regnier, E., and Harrison, K. (1991). Long-term tillage effects on seed banks in three Ohio soils. *Weed Science* 39: 186–194.

Cardina, J., Herms, C.P., and Doohan, D.J. (2002). Crop rotation and tillage system effects on weed seedbanks. *Weed Science* 50: 448–460.

Cavers, P. and Benoit, T.D. (1989). Seed banks in arable land. In: *Ecology of Soil Seed Banks* (eds. M. Leck, V. Parker and R. Simpson), 309–328. California; San Diego, CA: Academic Press.

Chancellor, R. (1964). Emergence of weed seedlings in the field and the effects of different frequencies of cultivation. *Proceedings of the 7th British Weed Control Conference* 1964: 599–606.

Clements, D.R., Benoit, D.L., Murphy, S.D., and Swanton, C.J. (1996). Tillage effects on weed seed return and seedbank composition. *Weed Science* 44: 314–322.

Cromar, H.E., Murphy, S.D., and Swanton, C.J. (1999). Influence of tillage and crop residue on postdispersal predation of weed seeds. *Weed Science* 47: 184–194.

Davis, A.S., Renner, K.A., and Gross, K.L. (2005). Weed seedbank and community shifts in a long-term cropping systems experiment. *Weed Science* 53: 296–306.

Davis, A.S., Anderson, K.I., Hallett, S.G., and Renner, K.A. (2006). Weed seed mortality in soils with contrasting agricultural management histories. *Weed Science* 54: 291–297.

Defelice, M.S., Brown, W.B., Aldrich, R.J. et al. (1989). Weed control in soybeans (*Glycine max*) with reduced rates of postemergence herbicides. *Weed Science* 37: 365–374.

Dickerson, C. and Sweet, R. (1971). Common ragweed ecotypes. *Weed Science* 19: 64–66.

Dorado, J., Del Monte, J., and Lopez-Fando, C. (1999). Weed seedbank response to crop rotation and tillage in semiarid agroecosystems. *Weed Science* 47: 67–73.

Doucet, C., Weaver, S.E., Hamill, A.S., and Zhang, J. (1999). Separating the effects of crop rotation from weed management on weed density and diversity. *Weed Science* 47: 729–735.

Eghball, B. and Lesoing, G.W. (2000). Viability of weed seeds following manure windrow composting. *Compost Science & Utilization* 8: 46–53.

Fierer, N., Schimel, J.P., and Holden, P.A. (2003). Variations in microbial community composition through two soil depth profiles. *Soil Biology and Biochemistry* 35: 167–176.

Gallandt, E.R., Molloy, T., Lynch, R.P., and Drummond, F.A. (2005). Effect of cover-cropping systems on invertebrate seed predation. *Weed Science* 53: 69–76.

Galloway, B.A. and Weston, L.A. (1996). Influence of cover crop and herbicide treatment on weed control and yield in no-till sweet corn (*Zea mays* L.) and pumpkin (*Cucurbita maxima* Duch.). *Weed Technology* 10: 341–346.

Gupta, S., Larson, W., and Linden, D. (1983). Tillage and surface residue effects on soil upper boundary temperatures 1. *Soil Science Society of America Journal* 47: 1212–1218.

Harrison, S.K. (1990). Interference and seed production by common lambsquarters (*Chenopodium album*) in soybeans (*Glycine max*). *Weed Science* 38: 113–118.

Hayden, Z.D., Brainard, D.C., Henshaw, B., and Ngouajio, M. (2012). Winter annual weed suppression in rye–vetch cover crop mixtures. *Weed Technology* 26: 818–825.

Jensen, P.K. (1995). Effect of light environment during soil disturbance on germination and emergence pattern of weeds. *Annals of Applied Biology* 127: 561–571.

Juroszek, P., Drews, S., Neuhoff, D., and Kopke, U. (2003). Light dependent germination of weeds in a natural seed bank during a 2-years course of tillage. *Aspects of Applied Biology* 69: 125–130.

Kennedy, A. (1999). Soil microorganisms for weed management. *Journal of Crop Production* 2: 123–138.

Kivilaan, A. and Bandurski, R.S. (1981). The one hundred-year period for Dr. Beal's seed viability experiment. *American Journal of Botany* 68: 1290–1292.

Kremer, R.J. (1993). Management of weed seed banks with microorganisms. *Ecological Applications* 3: 42–52.

Larney, F.J. and Blackshaw, R.E. (2003). Weed seed viability in composted beef cattle feedlot manure. *Journal of Environmental Quality* 32: 1105–1113.

Légère, A., Stevenson, F.C., and Benoit, D.L. (2011). The selective memory of weed seedbanks after 18 years of conservation tillage. *Weed Science* 59: 98–106.

Liebman, M. and Dyck, E. (1993). Crop rotation and intercropping strategies for weed management. *Ecological Applications* 3: 92–122.

Long, R.L., Gorecki, M.J., Renton, M. et al. (2015). The ecophysiology of seed persistence: a mechanistic view of the journey to germination or demise. *Biological Reviews* 90: 31–59.

Lonsdale, W. (1993). Losses from the seed bank of *Mimosa pigra*: soil micro-organisms vs. temperature fluctuations. *Journal of Applied Ecology* 30: 654–660.

Luschei, E. (2003). Comparison of the effectiveness of seedbank sampling to seedling counts in reducing the uncertainty in estimates of weed population size. *Aspects of Applied Biology* 69: 137–142.

McCrae, J. (1915). Flanders fields (Online). http://www.greatwar.co.uk/poems/john-mccrae-in-flanders-fields.htm (accessed 22 March 2019).

Mohler, C.L. and Galford, A. (1997). Weed seedling emergence and seed survival: separating the effects of seed position and soil modification by tillage. *Weed Research* 37: 147–155.

Mohler, C., Dykeman, C., Nelson, E., and Ditommaso, A. (2012). Reduction in weed seedling emergence by pathogens following the incorporation of green crop residue. *Weed Research* 52: 467–477.

Mohler, C.L., Taylor, A.G., Ditommaso, A. et al. (2018). Effects of incorporated rye and hairy vetch cover crop residue on the persistence of weed seeds in the soil. *Weed Science* 66: 379–385.

Moore, M.J., Gillespie, T.J., and Swanton, C.J. (1994). Effect of cover crop mulches on weed emergence, weed biomass, and soybean (*Glycine max*) development. *Weed Technology* 8: 512–518.

Murphy, S.D., Clements, D.R., Belaoussoff, S. et al. (2006). Promotion of weed species diversity and reduction of weed seedbanks with conservation tillage and crop rotation. *Weed Science* 54: 69–77.

Nadeau, L.B. and Morrison, I.N. (1986). Influence of soil moisture on shoot and root growth of green and yellow foxtail (*Setaria viridis* and *S. lutescens*). *Weed Science* 34: 225–232.

Nichols, V., Verhulst, N., Cox, R., and Govaerts, B. (2015). Weed dynamics and conservation agriculture principles: a review. *Field Crops Research* 183: 56–68.

Osipitan, O.A., Dille, J.A., Assefa, Y., and Knezevic, S.Z. (2018). Cover crop for early season weed suppression in crops: systematic review and meta-analysis. *Agronomy Journal* 110: 2211–2221.

Pleasant, J.M. and Schlather, K.J. (1994). Incidence of weed seed in cow (*Bos* sp.) manure and its importance as a weed source for cropland. *Weed Technology* 8: 304–310.
Roberts, H. and Potter, M. (1980). Emergence patterns of weed seedlings in relation to cultivation and rainfall. *Weed Research* 20: 377–386.
Schreiber, M.M. (1992). Influence of tillage, crop rotation, and weed management on giant foxtail (*Setaria faberi*) population dynamics and corn yield. *Weed Science* 40: 645–653.
Schuster, M.Z., Pelissari, A., De Moraes, A. et al. (2016). Grazing intensities affect weed seedling emergence and the seed bank in an integrated crop–livestock system. *Agriculture, Ecosystems and Environment* 232: 232–239.
Schuster, M.Z., Harrison, S., De Moraes, A. et al. (2018). Effects of crop rotation and sheep grazing management on the seedbank and emerged weed flora under a no-tillage integrated crop-livestock system. *The Journal of Agricultural Science* 156: 810–820.
Schwartz, L.M., Norsworthy, J.K., Young, B.G. et al. (2016). Tall waterhemp (*Amaranthus tuberculatus*) and Palmer amaranth (*Amaranthus palmeri*) seed production and retention at soybean maturity. *Weed Technology* 30: 284–290.
Scopel, A., Ballare, C., and Radosevich, S. (1994). Photostimulation of seed germination during soil tillage. *New Phytologist* 126: 145–152.
Shergill, L.S., Bejleri, K., Davis, A., and Mirsky, S.B. (2020). Fate of weed seeds after impact mill processing in midwestern and mid-Atlantic United States. *Weed Science* 68: 92–97.
Shirtliffe, S.J. and Entz, M.H. (2005). Chaff collection reduces seed dispersal of wild oat (*Avena fatua*) by a combine harvester. *Weed Science* 53: 465–470.
Soltani, N., Dille, J.A., Burke, I.C. et al. (2016). Potential corn yield losses from weeds in North America. *Weed Technology* 30: 979–984.
Sosnoskie, L.M., Herms, C.P., and Cardina, J. (2006). Weed seedbank community composition in a 35-yr-old tillage and rotation experiment. *Weed Science* 54: 263–273.
Suzuki, T. and Waller, G.R. (1987). Purine alkaloids in tea seeds during germination. ACS Symposium Series. Oxford University Press: 289–294.
Swanton, C.J. and Booth, B.D. (2004). Management of weed seedbanks in the context of populations and communities. *Weed Technology* 18: 1496–1503.
Swanton, C.J. and Weise, S.F. (1991). Integrated weed management: the rationale and approach. *Weed Technology* 5: 657–663.
Swanton, C.J., Shrestha, A., Roy, R.C. et al. (1999). Effect of tillage systems, N, and cover crop on the composition of weed flora. *Weed Science* 47: 454–461.
Swanton, C.J., Shrestha, A., Knezevic, S.Z. et al. (2000). Influence of tillage type on vertical weed seedbank distribution in a sandy soil. *Canadian Journal of Plant Science* 80: 455–457.
Teasdale, J.R. (1996). Contribution of cover crops to weed management in sustainable agricultural systems. *Journal of Production Agriculture* 9: 475–479.
Teasdale, J.R. and Mohler, C.L. (2000). The quantitative relationship between weed emergence and the physical properties of mulches. *Weed Science* 48: 385–392.
Tester, M. and Morris, C. (1987). The penetration of light through soil. *Plant, Cell and Environment* 10: 281–286.
Thompson, K., Bakker, J.P., Bekker, R.M., and Hodgson, J.G. (1998). Ecological correlates of seed persistence in soil in the north-west European flora. *Journal of Ecology* 86: 163–169.

Tidemann, B.D., Hall, L.M., Harker, K.N. et al. (2017). Suitability of wild oat (*Avena fatua*), false cleavers (*Galium spurium*), and volunteer canola (*Brassica napus*) for harvest weed seed control in western Canada. *Weed Science* 65: 769–777.

Toole, E.H. and Brown, E. (1946). Final results of the Duvel buried seed experiment. *Journal of Agricultural Research* 72: 201–210.

Wagner-Riddle, C., Gillespie, T., and Swanton, C.J. (1994). Rye cover crop management impact on soil water content, soil temperature and soybean growth. *Canadian Journal of Plant Science* 74: 485–495.

Wareing, P. (1969). Germination and dormancy. In: *The Physiology of Plant Growth and Development* (ed. M.B. Wilkins), 605–646. New York, NY: McGraw-Hill Book Co., Inc.

Warnes, D.D. and Andersen, R.N. (1984). Decline of wild mustard (*Brassica kaber*) seeds in soil under various cultural and chemical practices. *Weed Science* 32: 214–217.

Warwick, S. and Black, L. (1986). Genecological variation in recently established populations of *Abutilon theophrasti* (velvetleaf). *Canadian Journal of Botany* 64: 1632–1643.

Weaver, S.E. (2001). The biology of Canadian weeds. 115. *Conyza canadensis*. *Canadian Journal of Plant Science* 81: 867–875.

Weaver, S.E. and McWilliams, E.L. (1980). The biology of Canadian weeds.: 44. *Amaranthus retroflexus* L., *A. powellii* S. Wats. and *A. hybridus* L. *Canadian Journal of Plant Science* 60: 1215–1234.

White, S.S., Renner, K.A., Menalled, F.D., and Landis, D.A. (2007). Feeding preferences of weed seed predators and effect on weed emergence. *Weed Science* 55: 606–612.

Wibberley, J. (1996). A brief history of rotations, economic considerations and future directions. *Aspects of Applied Biology* 47: 1–10.

Williams, M.M., Mortensen, D.A., and Doran, J.W. (1998). Assessment of weed and crop fitness in cover crop residues for integrated weed management. *Weed Science* 46: 595–603.

Woolley, J.T. and Stoller, E.W. (1978). Light penetration and light-induced seed germination in soil. *Plant Physiology* 61: 597–600.

Wortman, S.E., Drijber, R.A., Francis, C.A., and Lindquist, J.L. (2013). Arable weeds, cover crops, and tillage drive soil microbial community composition in organic cropping systems. *Applied Soil Ecology* 72: 232–241.

Yenish, J.P., Doll, J.D., and Buhler, D.D. (1992). Effects of tillage on vertical distribution and viability of weed seed in soil. *Weed Science* 40: 429–433.

Zhang, J., Hamill, A.S., Gardiner, I., and Weaver, S.E. (1998). Dependence of weed flora on the active soil seedbank. *Weed Research* 38: 143–152.

11

Clonal Growth, Resprouting, and Vegetative Propagation of Weeds

Jitka Klimešová[1,2] *and Jana Martínková*[1]

[1]*Department of Experimental and Functional Morphology, Institute of Botany of the Czech Academy of Sciences, Třeboň, Czech Republic*
[2]*Department of Botany, Faculty of Sciences, Charles University, Praha, Czech Republic*

11.1 Introduction

Weeds are unwanted plants growing in various habitats and subjected to different management practices, e.g. mechanical, chemical, or biological control. To survive and persist under these practices, weeds adopt various strategies. The focus of this chapter is to describe tolerance strategies that enable plants to survive weed control measures by placing regenerative organs out of range of disturbance, usually belowground. We introduce the disturbance caused by weeding practices in a framework of disturbance severity and frequency (Section 11.2), describe strategies that weeds use to cope with recurrent disturbance (Section 11.3), list plant traits that help plants tolerate disturbance (Section 11.4), and place the tolerance strategy in an evolutionary perspective (Section 11.5), exploring examples of how plants that tolerate mechanical control respond to changes in weeding practices.

11.2 Weeding as a Disturbance Regime

Traditionally, weeding on arable land has been practiced by manual removal of weeds. This practice was replaced in some parts of the world by tilling, i.e. mechanized plowing that inverts the soil several times per year. This method has been recently supplemented or partly replaced by the use of herbicides. Other methods like cultural (e.g. crop rotation, cover crops) and biological (e.g. application of seed predators) control are less common but are used as well.

Mechanical weed management can be described as a disturbance regime with specific timing, frequency, severity, intensity, predictability, and spatial extent (Miller et al. 2011).

Persistence Strategies of Weeds, First Edition. Edited by Mahesh K. Upadhyaya, David R. Clements, and Anil Shrestha.

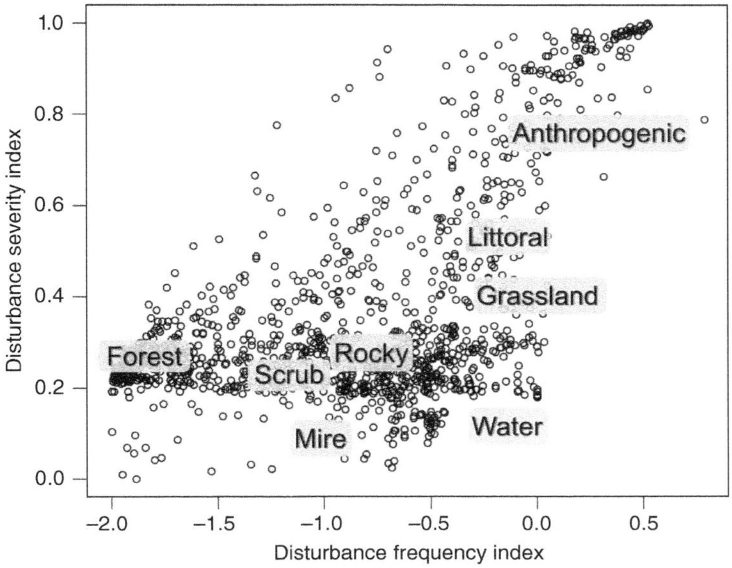

Figure 11.1 Position of habitats like arable land ("Anthropogenic") in the space delimited by preferences for disturbance regime in the flora of the Czech Republic (according to Herben et al. 2016). Disturbance severity index is defined as the extent of disturbance: low values are for a regime where disturbance affects only aboveground plant parts and high values are for disturbance affecting belowground plant parts as well. Disturbance frequency index is disturbance return time (log scale). These indices were calculated from species' occurrence in vegetation with certain disturbance regimes. Source: Herben et al. 2016.

For weeding in agriculture, the timing of disturbance is generally predictable and has high frequency and severity. For example, on arable land, weeding is done several times per year, which removes most aboveground biomass and fragments the belowground biomass to a depth of 10 cm (Stephens 1982). Such intensive disturbance is rare in nature (Figure 11.1), and only a limited number of plant species is capable of dealing with it (Herben et al. 2016).

Since the second half of the twentieth century, unwanted plants have been controlled using herbicides that target metabolic pathways necessary for plant life. With increased herbicide use, the intensity of mechanical control has decreased (Nichols et al. 2015). Reduced tillage or no-till crop production systems are being adapted as a soil protection measure (Ringselle et al. 2015), which have reduced the frequency and severity of soil disturbance. We now have new biological tools for weed control, e.g. bioherbicides, seed predators, or herbivores that usually target certain sensitive life history stages of weeds. These weed control measures are species specific and characterized by a low disturbance severity and generally small spatial extent of disturbance.

With increasing pressure of alien plants and the danger of alien species invading natural communities, weed control is increasingly undertaken outside of agricultural systems. The use of targeted weeding methods includes biological controls, herbicides, mechanical weeding, and fire. The disturbance associated with weed control measures applied to invasive species may often be even more severe and frequent compared to those applied on arable land. The timing of invasive species control methods is more flexible than in arable systems because there is no need to sustain agricultural production and could be applied at any time throughout a season.

11.3 Plant Strategies Under Recurrent Disturbance

Plants adapt different strategies to deal with disturbance. The following three are the most important:

i) *Avoidance strategy*. This involves avoidance in time or space. Time avoidance is usually related to a short life cycle, i.e. if a plant matures promptly and produces seeds before the next disturbance event (annual plants). Space avoidance may be attained by specialized plant architecture. For example, plants that are subjected to herbivory produce flat rosettes laying on the soil surface so that large herbivores cannot feed on them.
ii) *Resistance strategy*. This involves traits that exclude or minimize negative effects of disturbance on a plant. These traits usually represent a very specific adaptation to a particular type of disturbance, e.g. thorns preventing herbivory by large animals.
iii) *Tolerance strategy*. This involves adaptation to biomass loss. Such plants store buds and carbohydrates out of reach of disturbance. Tolerance is a general strategy that allows plants to survive disturbance of various severities and frequencies caused by different agents.

These three strategies are not mutually exclusive, and combinations do occur; however, with more intensive disturbance regimes, the suite of tenable strategies is reduced (Vesk and Westoby 2004a). Along the gradient of disturbance severity and frequency, disturbance tolerance is typically seen for intermediate disturbance severity and frequency (Bellingham and Sparrow 2000). Severe disturbance of plant communities once every several years allows accumulation of belowground reserves of buds and carbohydrates and successful vegetative regeneration from the belowground organs (Iwasa and Kubo 1997). Plant communities adapted to such disturbance regime include natural communities in fire-prone areas of Mediterranean-type climate and tropical grasslands or human-managed pastures and meadows (Hendrickson and Briske 1997; Pausas et al. 2018).

When disturbance is more frequent than once per year and it removes not only aboveground plant parts but also fragments from the belowground biomass, avoidance strategy is more typical, e.g. on arable land. Such plants produce numerous seeds and have a short life span (Grime 1977). Highly and repeatedly disturbed habitats feature a small but significant group of plants with a tolerance strategy. Instead of having a very short life span, they are perennial, and do not restore populations from seed, but regenerate (resprout) from preserved vegetative organs (Herben et al. 2018a). Co-occurrence of annual and perennial weeds with contrasting strategies on arable land resembles situation in fire-prone areas where seeders (usually shrubs regenerating from seeds after disturbance) co-occur with resprouters (usually shrubs regenerating from bud-bearing organs) (Bellingham and Sparrow 2000).

11.4 Plant Traits Typical for Tolerance Strategies and Resprouting Limitations

The tolerance strategy, i.e. resprouting after damage, is facilitated by a plant's ability to store buds and assimilates in specialized, usually belowground, organs. Resprouting may be limited or prevented when bud banks or carbohydrate stores are exhausted due to seasonal regrowth or a previous disturbance. Another important factor that limits resprouting capacity is nutrient availability (Wise and Abrahamson 2008). When low soil nutrient contents create stressful conditions, plants may be absent altogether. Stressful conditions due to a low content of soil nutrients are devoid of plants when they are subjected to frequent and severe disturbance. However, it is not clear whether this absence of plant life is due to harsh conditions or due to a lack of preadapted plants because such habitats are very rare in nature (Grime 1977; Herben et al. 2018b).

11.4.1 Bud Bank

Buds are small meristematic regions capable of growing into a shoot or branch. They are commonly present on all stems on a plant as they are produced as the stems grow. It is difficult to discern whether the presence of buds on a plant is a disturbance adaptation or not (Klimešová and Klimeš 2007). In contrast to this passive production of buds during stem growth, signs of active bud bank formation are (i) the location of buds in places where they are protected from negative impacts of disturbance, (ii) a higher rate of bud protection, and (iii) carbohydrate storage located out of reach of disturbance (Ott et al. 2019). To place the bud bank belowground, a plant needs to develop some specialized adaptations because stem growth is usually geotropically negative. Belowground germination, pulling of stem bases belowground by contractile roots, positive geotropic growth of young stems forming future rhizomes, and reliance on consequential burial by substrate or litter are some adaptations that allow formation of

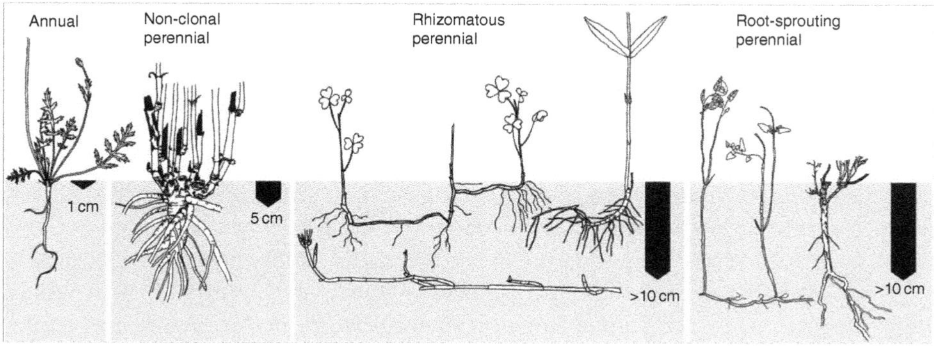

Figure 11.2 Examples of vertical distribution of bud banks. In annual plants, bud bank is at the soil surface. In non-clonal perennial plants, bud bank is at caudex (perennial stem bases) that are usually not deeper than 5 cm. In rhizomatous perennial plants, the bud bank is located from 5 to 10 cm deep in the soil. In the root-sprouting perennials, buds are located deeper than 10 cm in the soil. Sources: Kollmann et al. (2011); West et al. (2012).

bud bank belowground (Ott et al. 2019). A bud bank may also be produced adventitiously out of a plant shoot system, i.e. on roots in soil (Raju et al. 1966; Bartušková et al. 2017).

There are two measures of the bud bank that are relevant to bud bank functioning: vertical distribution of buds and their number (Klimešová et al. 2019). Vertical distribution of buds in relation to soil surface is a good predictor of plant tolerance to damage (Pausas and Paula 2020). The deeper the bud location, the higher the probability of plant survival. Therefore, soil disturbance during weeding and depth of tilling selects for species with certain depth of buds and bud-bearing organ location (Figure 11.2). Good examples of species that have a deep placement of the bud bank are *Equisetum arvense* (rhizomes can be found 70 cm deep in the soil) and *Cirsium arvense* (roots with adventitious buds reach a depth of two meters) (Wehsarg 1954). On the other hand, shallow bud banks on roots of *Rumex acetosella* or buds produced aboveground in *Allium vineale* might also represent successful persistence strategies when fragmented, but not when buried in soil.

The number of buds reported for various perennial weeds of arable land is comparable with the number for grassland species and ranges from few hundreds to tens of thousands per square meter (Korsmo 1930; Ott et al. 2019). However, arable land bud banks, compared with grassland banks, have a higher resprouting rate. The typical disturbance in arable land (e.g. plowing) causes fragmentation of belowground organs, and the resting buds located on rhizomes or roots may be released from correlative inhibition and produce new shoots after plowing (Leakey et al. 1978a).

Another unique aspect of an arable land is that a high number of species are able to form buds on roots and resprout from them (Bartušková et al. 2017). This ability is important because it enables plants to form buds *de novo* on root fragments. Contrary to the stem bud bank, of which the size is defined by the number of stem nodes, the bud bank on the roots of root-sprouting species is difficult to quantify because there is no morphological limitation. Root buds may arise potentially everywhere on the root system. The success of root sprouters as weeds is due to roots generally being located much deeper in the soil profile and therefore less vulnerable to mechanical control operations (Figure 11.2; Kollmann et al. 2011; West et al. 2012). Root sprouters are also often listed among the most troublesome weeds of the world (e.g. *C. arvense* and *Convolvulus arvensis*).

The bud bank develops gradually during a plant ontogeny as a plant increases in size. In early stages, the bud bank is small, and plants may avoid disturbance through their relatively small stature. Plants adapt a tolerance strategy only when they accumulate enough buds and sufficient carbohydrate storage (Martínková et al. 2004). In some special cases, however, plants in early life stages possess a higher sprouting ability than older plants. This happens in annuals resprouting from the hypocotyl, whose ability to do so is confined to the first few weeks of life when the epidermis may produce buds (Rauh 1937). In biennials, resprouting occurs from roots and is the most vigorous when they are young and vegetative, because accumulated carbohydrates stored in a root have not yet been used for flowering (Sosnová et al. 2014). Younger trees resprout more successfully (Del Tredici 2001; Vesk 2006) than older trees, often because basal buds may be overgrown by wood during secondary thickening.

11.4.2 Carbohydrate Storage

Stored carbohydrates are the only source of energy for plants to recover after biomass removal as photosynthetic area is either destroyed or very limited. However, the relationship between stored carbohydrate and vegetative regeneration is not well understood. We do not know whether coexisting species that differ in carbohydrate storage also differ in their resprouting ability. However, there is a clear effect of the size of storage organ on resprouting success, particularly when a plant resprouts from root/shoot fragments of different sizes or from root/shoot fragments buried to different soil depths (Cuthbertson 1972; Bourdôt 1984; Shrestha et al. 2018). It is well known that the resprouting ability of a plant is reduced during certain life history stages, e.g. during spring regrowth, flowering, or fruiting. This is probably because of a low level of stored carbohydrates, exhausted due to either rebuilding of green biomass or diversion to seed production (Raju et al. 1964; Cuthbertson 1972; Taab et al. 2018). Information in this regard is useful in determining the timing of weed management (Verwijst et al. 2018).

11.4.3 Nutrients

Sufficient nutrient concentration is necessary for breaking bud dormancy and for accumulating new biomass after plant damage (Ott et al. 2019). Nutrients are usually plentiful in anthropogenic habitats, but in some stages of plant life, they may be limited if they are exhausted by rapid growth (Leakey et al. 1977). Therefore, together with carbohydrates, soil nutrient level may also be important in determining seasonal fluctuations in resprouting capability and vigor (Wise and Abrahamson 2008).

Soil fertility also affects plant's investments in regenerative organs, growth, and carbohydrate storage (Janeček et al. 2014). Under conditions of nutrient shortage, belowground organs like rhizomes tend to be less branched and have a smaller bud bank (Tomlinson and O'Connor 2004). They also tend to have higher carbohydrate concentrations (Janeček et al. 2014) because the stored carbohydrates cannot be used for resprouting due to the nutrient shortage. In root-sprouting *R. acetosella*, the number of buds per root length is the same or higher in low compared with high nutrient conditions, but the plants growing in soils with high nutrient availability produce more roots, buds, and shoots and resprout more vigorously after damage (Klimeš and Klimešová 1999).

11.4.4 Plant Organs Responsible for the Tolerance Strategy

Belowground plant parts that possess buds for resprouting (bud-bearing organs) usually store carbohydrates and are responsible for clonal growth. However, both of these functions (storage and clonal growth) can also be provided by other organs, which do not serve in regeneration after damage. These "clonal organs" facilitate recolonization of disturbed areas after weeding. "Storage organs" without the autonomous capacity to resprout may serve to supply energy for seasonal regrowth of plants in undisturbed habitats. Belowground "bud-bearing organs" are usually characterized by certain correlated sets of tolerance traits, e.g. bud bank size and depth (Herben and Klimešová 2020), and it is generally

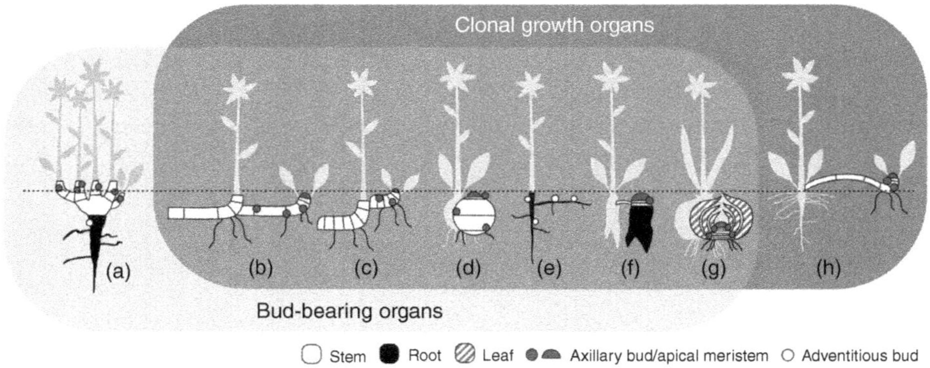

Figure 11.3 Types of bud-bearing/clonal growth organs in disturbed anthropogenic communities: (a) caudex of perennial non-clonal plant with bud bank, (b) hypogeogenous rhizome, (c) epigeogenous rhizome, (d) stem tuber, (e) root with adventitious sprouting, (f) root tuber, (g) bulb, and (h) stolon. Shaded outlines of plants depict seasonal growth, while solid lines depict the part of the biomass that provides perennation during seasonal rest. For a detailed description of organs, see Table 11.1.

expected that plants with similar "bud-bearing organs" and accompanying traits exhibit similar responses to disturbance and hence to weed control efforts (Leakey 1981).

"Bud-bearing organs" can be hypogeogenous and epigeogenous rhizomes, roots with adventitious sprouting, bulbs, and tubers (Figure 11.3, Table 11.1). Important "clonal growth organs" enabling weeds to recolonize disturbed areas are horizontal rooting stems, i.e. stolons of creeping plants, or dispersible vegetative diaspores located aboveground like bulbils.

A richness of bud-bearing organs in a habitat correlates with the severity of disturbance (Figure 11.4). At the most severe end of the disturbance gradient, in terms of perennials, only species with hypogeogenous rhizomes and roots with adventitious sprouting occur due to their deep belowground placement and a large bud bank. In habitats under severe disturbance, the spectrum of bud-bearing organs has been found substantially reduced from the whole pool available or occurring in Central Europe (Klimešová and Klimeš 2008). The most universal type of bud-bearing organ on the disturbance gradient consists of roots with adventitious sprouting. Although plants with this ability also occur in other vegetation types (Figure 11.4), root sprouting is the most vigorous among the species from disturbed habitats (Klimešová et al. 2008, 2017; Bartušková et al. 2021).

Perennial root sprouters like *C. arvense*, *C. arvensis*, *Euphorbia esula*, *Convolvulus repens*, *Sonchus arvensis*, and *Acroptilon repens* were reported as the most important weeds in pre-herbicide agriculture in the United States (Johnson et al. 2009). These species resprout from roots spontaneously, and root fragmentation enhances their resprouting ability substantially. There are other weeds (e.g. biennials *Barbarea vulgaris* and *Barbarea stricta*) in which root-sprouting ability is triggered by injury as well. Fragmentation of *B. vulgaris* or *B. stricta* root systems produces several plants that behave as short-lived perennials (Martínková et al. 2016). Perennial non-clonal weed *Knautia arvensis* may change its growth form after injury, producing horizontal roots and resprouting from them instead of forming a taproot (Klimešová and Martínková 2004).

Table 11.1 Bud-bearing organs in weedy species, their description, and examples according to the CLO-PLA3 database.

Bud-bearing organ	Description	Examples
Caudex with main root	Belowground perennial stems with bud bank (caudex) and primary root system without adventitious roots and buds; senescing taproot of old plants decays from the root center, in some species, causing plant fragmentation; an old individual genet disintegrates into ramets bearing parts of the main root and one or a few shoots; vegetative spreading is poor; the taproot serves as a storage organ and vascular link between shoots	*Symphytum officinale*
Hypogeogenous rhizome	Belowground stems with adventitious roots; the rhizome is formed belowground and grows horizontally at a species-specific depth, and after some time it becomes orthotropic and forms aboveground shoots; the horizontal part of the rhizome bears bracts, a few roots developed at the nodes, and it has long internodes; vegetative spreading is often fast, up to several tens of centimeters per year; persistence of the hypogeogenous rhizome differs considerably among species	*Elytrigia repens*
Epigeogenous rhizome	Belowground stems with adventitious roots; the rhizome is formed aboveground; its distal part is covered by soil and litter or pulled into the soil by the contraction of roots; nodes bear green leaves, and the internodes are usually short; the rhizomes serve as a bud bank and storage organs; vegetative spreading is usually slow (up to a few centimeters per year), and persistence of epigeogenous rhizomes differs considerably among species	*Ranunculus acris*
Bulb	A storage organ consisting of storage leaves and a shortened stem base; the bulb is formed by organs produced within a single season or in the course of several seasons; by itself it represents one renewal bud; in addition, some plants produce smaller bulbs and/or bulbils; plants growing from them morphologically and in size resemble seedlings	*Allium vineale*
Stem tuber	Belowground, usually short-lived storage and regenerative organ of shoot origin; offspring tubers are attached to a parent tuber or produced at the end of a hypogeogenous rhizome; the parent plant dies back in autumn, except for the stem tuber(s), which bear one dominant bud, each utilized for spring regrowth; in summer old tubers decay and new ones are formed; in addition, some plants produce smaller tubers and/or tubercules; plants growing from them resemble in morphology and size to seedlings	*Cyperus esculentus*
Root tuber	Belowground storage organ; in some plants it serves as a regenerative organ of root origin bearing a bud or buds of stem origin; in that case the plant dies back in autumn, except for the root tuber(s) with the buds, later utilized for spring regrowth; in summer old tubers decay and new ones are formed	*Ranunculus ficaria*
Stolon	Horizontal stem growing on soil surface and having adventitious roots, rooting in the soil and providing connection between offspring plants or formed by a creeping axis of a plant; nodes on the stem bear leaves, internodes are usually long, the stem serves as a storage organ and a bud bank; vegetative spreading may be fast, and persistence of the horizontal aboveground stems differs considerably among species	*Potentilla anserina*

(*Continued*)

Table 11.1 (Continued)

Bud-bearing organ	Description	Examples
Root with adventitious buds	Plant roots (main root including the hypocotyl and adventitious roots) form adventitious buds spontaneously or after an injury; when buds are formed on horizontal roots, they may enable an extensive clonal growth; persistence of the roots with adventitious buds considerably differs among species	*Cirsium arvense*
Bulbils aboveground	Small vegetative diaspores produced in axils of leaves on stems aboveground or belowground; depending on the location of storage within the buds, they are called bulbils (storage located in scale leaves), stem tubercules (storage located in the stem part), and root tubers (storage located in an adventitious root); they soon fall down from the parent plant and immediately start to grow, having no dormancy; young plants regenerating from bulbils and tubercules resemble seedlings in their morphology and size	*Allium vineale*

Source: Based on (Klimešová and Klimeš 2006).

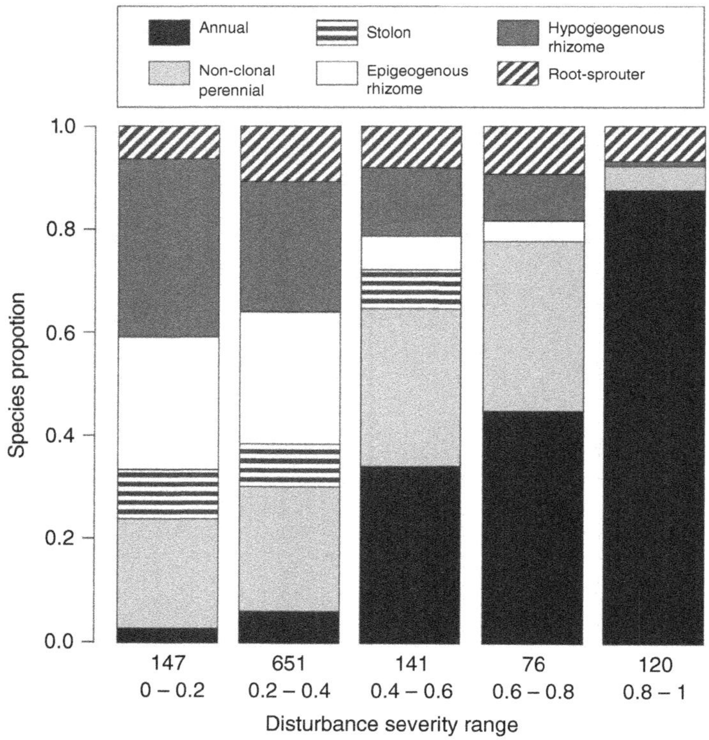

Figure 11.4 Spectra of bud-bearing/clonal growth organs in plants along gradients of disturbance severity with low values on left and high values on right (for the gradient of disturbance severity, see *y*-axis in Figure 11.1). The numbers along the x-axis, i.e., 147, 651, 141, 76, and 120 denote the number of species in individual disturbance categories. The most severely disturbed habitats like arable land have a disturbance severity range 0.8–1.0. A description of bud-bearing organs is in Table 11.1, and their appearance in Figure 11.3.

Some weeds combine several modes of clonal growth and/or bud-bearing organs and exhibit both vegetative regeneration after fragmentation and seed production. This life history strategy ensures that buds on different organs or buds and seeds differ in dormancy, and cannot be easily exhausted by repeated damage. For example, in monocarpic root-sprouting *Rorippa palustris* and *B. vulgaris*, the bud bank on stems is first activated after damage, and only when this bud bank is removed or exhausted, the root buds resprout (Bartušková and Klimešová 2010). Moreover, as typical monocarpic plants, the species produce a high number of seeds (Klimešová et al. 2014). The invasive wetland herb *Alternanthera philoxeroides* does not produce seeds in the invaded range (Australia and Asia), but spreads laterally through long branched stolons and has strong regenerative ability, both from stem and root fragments (Sainty et al. 1997; Wang et al. 2018).

11.4.5 Recolonization and Spread of Vegetative Propagules in Space and Time

Perennial weeds of arable lands typically exhibit extensive lateral spread. This reflects the fact that perennial weeds usually employ long hypogeogenous rhizomes (e.g. *Elytrigia repens* and *E. arvense*) and stolons (e.g. *Cynodon dactylon* and *Potentilla anserina*) or sprout from roots (e.g. *C. arvense* and *E. esula*). Such organs combine a high regenerative ability when fragmented and the ability to rapidly recolonize disturbed areas devoid of plants. The lateral spread by stolons or rhizomes allows the plant to spread several meters per year on bare soil; however, the total length of branched rhizome system may be much longer. For example, *E. repens* produces about 80 m of rhizomes per year (Leakey 1981).

The greatest dispersal is seen in rhizomes and roots fragments and tubers spread by machinery commuting among fields. This phenomenon makes some perennial weeds, which otherwise may not have any adaptations for long distance dispersal, highly capable of expanding their local distributions. The spread via rhizome fragments among fields is well documented for *E. arvense* (Cloutier and Watson 1985), *Solidago canadensis*, (Weber 2011), *Mikania micrantha* (Li et al. 2013), and *Cyperus esculentus* (Follak et al. 2015). After its introduction as an ornamental plant to Europe in the beginning of nineteenth century from South Africa, *Oxalis pes-caprae* has spread on arable land across the whole of Southwestern Europe. The majority of *O. pes-caprae* individuals are sterile pentaploids producing belowground bulbs that are dispersed by agricultural machinery (Papini et al. 2017).

In addition to spreading spatially, weeds also spread temporally. Dormant diaspores (seeds or buds) survive until suitable conditions for germination or sprouting occur. In contrast to seedbanks, where diaspores may survive up to hundreds of years, viability of buds in the bud bank is limited by the longevity of the bud-bearing organ. Survival of bud-bearing organs like rhizomes is usually dependent on the presence of aboveground green shoots. Survival of bud-bearing organs occurs rarely, while aboveground organs are not produced during the whole growing season (Shefferson et al. 2018). Nevertheless, in some species, rhizome or root fragments are capable of resisting decomposition for more than a year (Leakey et al. 1978b), and some tubers or bulbs may remain dormant even longer. For example, small belowground bulbs of *A. vineale* may be dormant for six years, while bulbils of the same species originating in inflorescence are dormant only until the next growing season (Lazebny 1960).

11.5 Tolerance Strategy in an Evolutionary Perspective

Anthropogenic habitats are the youngest type of biome (Mucina 2019) where preadapted plants that are able to withstand severe disturbance are occurring and their regenerative abilities are further selected. Many plants adapted to disturbance prevail in present-day landscapes because anthropogenic disturbance is a very strong and frequent selection factor (Kuneš et al. 2019). Weeds with a high degree of genetic and phenotypic variability have the potential to evolve lineages able to cope with weeding. This potential is also valid for vegetatively regenerating weeds. Understanding the evolution of weeds is an important area of research in weed ecology (Neve et al. 2018), which could provide effective strategies for sustainable weed control in the future.

Weeding practice is a very strong and highly selective force, which has been acting for a relatively short time (in evolutionary terms) of a few thousands of years. Therefore, the majority of persistence traits have been pre-evolved in natural ecosystems, where severe and frequent disturbances have been occurring for millions of years. Changes in weed flora are, however, due to both species turnover (when species better adapted to current disturbance regime outcompete the species that are not able to deal with current disturbance regime) and to selection of better adapted individuals in genetically variable weed populations.

11.5.1 Changes in Weed Flora

For understanding changes in weed flora composition, it is crucial to consider not only benefits but also costs of having tolerance traits. Although production of buds in a bud bank is not very costly for a plant (Vesk and Westoby 2004b), the cost of building bud-bearing organs and formation of carbohydrate storage can greatly affect performance of a resprouter (e.g. a perennial weed) in a community. Resprouters usually have a lower competitive ability compared with seeders, because investments into belowground organs lead to a shorter stature (Midgley 1996). When disturbance pressure decreases, resprouters are outcompeted by seeders (Bellingham and Sparrow 2000; Fidelis et al. 2014). During the first stage of succession on former arable land, there is a replacement of annuals and perennial weeds by perennial herbs that do not invest into deep belowground regenerative structures and, later on, by woody plants (Latzel et al. 2011).

The most noxious weeds of the pre-herbicide era were adapted to deep plowing, severe mechanical damage (see preceding text), and burial of seeds deep in the soil profile. No-till systems, on the other hand, are characterized by accumulation of seeds on the soil surface. They are also characterized by high soil compaction that delays emergence of crops as well as weeds and shifts weed species composition from annual dicots to graminoids, both annual and perennial (Hernández Plaza et al. 2015). Without herbicide application, no-till systems may be heavily infested by perennial weeds with various types of bud-bearing organs (Moyer et al. 1994). Herbicide application, on the other hand, leads to contrasting changes in weed flora as herbicides tend to be more successful in perennial weeds eradication than in eradication of annual weeds (Johnson et al. 2009; Holt et al. 2013). This effect is because perennating belowground organs are destroyed by herbicide, while seeds in the soil remain viable. Perennial weeds rely more on bud banks than on seedbanks. Annuals evolve resistance against herbicides more rapidly than perennial plants due to shorter generation times and the greater likelihood that resistant genotypes will spread (Holt et al. 2013). Despite this prevailing trend, numerous perennial weeds resistant to herbicides are recently reported (e.g. rhizomatous *Agropyron repens*, *Sorghum halepense*, and root-sprouting *E. esula* and *R. acetosella*) (Holt et al. 2013). How herbicide resistance of perennial weeds affects their resprouting capacity is not yet clear.

Recent application of similar weeding practices over vast areas has led to a reduction of native weed diversity, and there is a need to protect this diversity (Lososová and Simonová 2008; Johnson et al. 2009; Storkey et al. 2012; De Cauwer et al. 2017; Hawkins et al. 2019). Protection is needed not only because of biodiversity loss itself but also due to an increasing awareness that diversity of native weeds may have important functions in crop production and soil protection. Species-rich communities, for example, host more pollinators and provide less opportunities for dominance of noxious weed genotypes (Wezel et al. 2014; Neve et al. 2018).

Only 2% of annual plants in Central Europe have adventitious sprouting ability (Bartušková et al. 2017), and several of them are weeds typically infesting root crops in Europe. For example, annual plants, *Anagallis arvensis*, *Euphorbia exigua*, *Kickxia elatine*, *Kickxia spuria*, and *Microrrhinum minus*, survive weeding by regenerating from adventitious buds on the hypocotyl (Latzel et al. 2009). Because of contemporary changes in weeding practices in root crops that do not rely on repeated hand pulling, annual resprouting weeds are becoming rare (Lososová et al. 2004).

11.5.2 Selection of Weed Genotypes

Is it better for sustainability of crop production to have a high diversity of weeds (species as well as genotypes) with multiple strategies adapted to variable weeding techniques or a low diversity of weeds well adapted to uniform weeding practice? This is a crucial question for managers. The answers are dependent not only on biological traits of weeds but also on the economics of weed control strategy. While the economic aspect is beyond the scope of this chapter, we would like to emphasize that there are examples showing how new tolerant weeds may evolve and how we can predict their success. Potential for evolution of new disturbance-tolerating weeds is determined by the variability in resprouting ability among individuals in a population. A survival strategy providing the highest plant performance in a given disturbance regime will be favored and will prevail in a population. A weed with this strategy becomes common and potentially problematic in crop production.

If we want to predict selection of tolerant, i.e. resprouting, weed species, we should identify whether the resprouting after weeding is more advantageous over seed regeneration under various disturbance regimes. We will have to apply weed management to manipulate the spatial extent of the disturbance, its severity, and/or frequency and compare performance of individuals that originated from seeds versus individuals regenerated from bud banks (Figure 11.5). Such complex experimental assessment of strategy (resprouting versus seeding) and its dependence on disturbance regime may be recommended when a new weeding practice is developed either for an agroecosystem or control of invasive plants.

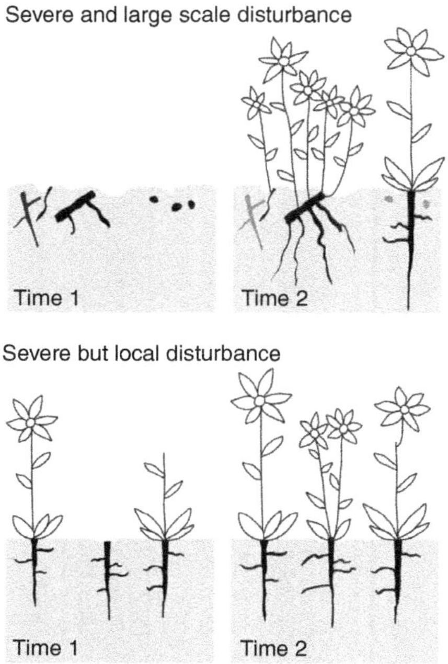

Figure 11.5 Possible disturbance regimes for evaluating successful strategies of weeds with variable life history. In the first scenario (upper pictures), disturbance is on a large scale and affects all plants in a population (Time 1) so that fitness of plants regenerating from seeds and from root or shoot fragments has to be compared (Time 2). In the second scenario (lower pictures), the disturbance is on a small scale and will affect only selected individuals in a population (Time 1) so that fitness of plants regenerating from bud bank and fitness of undisturbed plants has to be compared (Time 2). The plant strategy (regeneration from bud bank versus regeneration from seeds) with a higher performance will be promoted by the disturbance regime and prevail in a community.

The performance of weeds can be measured as fitness, i.e. the number of descendants a genotype is producing, or as biomass accumulation. Perennial and some biennial and annual weeds produce both seeds and clonal offspring. Clonal offspring may be established by spontaneous clonal growth or enforced by fragmentation. Therefore, production of both seeds and clonal offspring has to be recorded as both are directly affected by disturbance (Huhta et al. 2003; Martínková and Klimešová 2016). If several measures of plant performance are used, they often show a different response to weeding reflecting complexity of the relationship (e.g. Cripps et al. 2019). Priority should be given to direct measure of seed production in short-lived plants and to bud bank size in perennial plants.

Deciding which individuals to measure for fitness determination depends on the disturbance regime (Figure 11.5). When a weeding technique is planned to affect all members of a population, the fitness of resprouting individuals has to be compared with individuals established from seeds at the time of the disturbance. When examining weeding techniques that affect only some members of a population, the fitness of individuals subject to disturbance has to be compared with unaffected individuals.

The fitness of resprouting individuals may be higher than individuals established from seeds because the fragment of bud-bearing organ is larger than a seed. Resprouting individuals are also larger than seedlings and may have a competitive advantage and consequentially possibly a higher fitness than plant regenerating from seed. Moreover, in the case of very severe disturbances, the fitness of resprouting plants may be increased by fragmentation of bud-bearing organs and therefore by enforced clonality (Martínková and Klimešová 2016). Different timings of disturbance, however, may result in different performance of resprouters versus seeders, as resprouters would exhaust storage, and will not be able to survive winter, whereas seeders will survive and flower next year (Zhang and Shea 2012; Sosnová et al. 2014).

Individuals subject to disturbance do not necessarily have a lower fitness than undisturbed plants. The fitness of disturbed plants is often greater as disturbance may cause overcompensation, as a greater branching of resprouting individuals causes greater production of flowers and seeds than in undamaged less branched plants (Huhta et al. 2003). However, timing of disturbance plays a role as well. A later timing of disturbance, compared with disturbance earlier in the season, may not result in a higher fitness as plants do not have enough time for seed ripening at the end of growing season (Martínková et al. 2008).

In the examples mentioned previously, there is a clear relationship between the fitness of seeders and resprouters and the timing of disturbance. This is the reason why certain highly predictable weed control regimes select for certain weed strategies, while under variable management different weed strategies coexist. Short-term control of invasive plants will profit from such experiment on timing and frequency of disturbance, and such experiments would elucidate whether a selected weeding technique is really reducing fitness of treated plants. Long-term control of arable weeds will profit from understanding the relationship between disturbance regime and weed performance. This will facilitate adoption of weed management strategies to achieve the most effective weed control.

11.6 Conclusions

Tolerance, i.e. survival of biomass removal, is one of the strategies that unwanted plants have evolved due to mechanical weeding, and the strategy is connected with storing of buds and carbohydrates in the soil where they are protected from weed control measures. This strategy is not restricted to perennial plants and can be found in annuals and biennials as well. The majority of weeds tolerating disturbance are dependent not only on resprouting but also produce seeds, so their life history strategy is variable and provides a large potential for adaptation to new weed management techniques.

Weed managers should be aware that large-scale applications of the same weed management technique may select for very noxious and highly specialized weeds. Conversely, the application of diverse weed and cultural management promotes diversity in plant management strategies.

Acknowledgments

We are indebted to Alena Bartušková for drawing the figures, and to the Grant Agency of the Czech Republic (grant 19-13103S), and the long-term research development project of the Czech Academy of Sciences [No. RVO 67985939] for financial support.

References

Bartušková, A. and Klimešová, J. (2010). Reiteration in the short lived root-sprouting herb *Rorippa palustris*: does the origin of buds matter? *Botany* 88: 630–638.

Bartušková, A., Malíková, L., and Klimešová, J. (2017). Checklist of root-sprouters in the Czech flora: mapping the gaps in our knowledge. *Folia Geobotanica* 52: 337–343.

Bartušková, A., Filartiga, A.L., Herben, T. and Klimešová, J. (2021). Comparative analysis of root sprouting in temperate herbs: anatomical correlates and environmental predictors. Annals of Botany 127: 931–941.

Bellingham, P.J. and Sparrow, A.D. (2000). Resprouting as a life history strategy in woody plant communities. *Oikos* 89: 409–416.

Bourdôt, G.W. (1984). Regeneration of yarrow (*Achillea millefolium* L.) rhizome fragments of different length from various depths in the soil. *Weed Research* 24: 421–429.

Cloutier, D. and Watson, A. (1985). Growth and regeneration of field horsetail (*Equisetum arvense*). *Weed Science* 33: 358–365.

Cripps, M.G., Dowsett, C.A., Jackman, S.D. et al. (2019). Genetic variation in tolerance to defoliation in *Cirsium arvense*. *Weed Research* 60: 78–84.

Cuthbertson, E.G. (1972). *Chondrilla juncea* in Australia. 4. Root morphology and regeneration from root fragments. *Australian Journal of Experimental Agriculture and Animal Husbandry* 12: 528–534.

De Cauwer, B., De Ryck, S., Claerhout, S. et al. (2017). Differences in growth and herbicide sensitivity among *Cyperus esculentus* clones found in Belgian maize fields. *Weed Research* 57: 234–246.

Del Tredici, P. (2001). Sprouting in temperate trees: a morphological and ecological review. *The Botanical Review* 67: 121–140.

Fidelis, A., Appezzato-da-Gloria, B., Pillar, V.D., and Pfadenhauer, J. (2014). Does disturbance affect bud bank size and belowground structures diversity in Brazilian subtropical grasslands? *Flora* 209: 110–116.

Follak, S., Aldrian, U., Moser, D., and Essl, F. (2015). Reconstructing the invasion of *Cyperus esculentus* in Central Europe. *Weed Research* 55: 289–297.

Grime, J.P. (1977). Evidence for the existence of three primary strategies in plants and its relevance to ecological and evolutionary theory. *American Naturalist* 111: 1169–1194.

Hawkins, N.J., Bass, C., Dixon, A., and Neve, P. (2019). The evolutionary origins of pesticide resistance. *Biological Reviews* 94: 135–155.

Hendrickson, J.R. and Briske, D.D. (1997). Axillary bud banks of two semiarid perennial grasses: occurrence, longevity, and contribution to population persistence. *Oecologia* 110: 584–591.

Herben, T. and Klimešová, J. (2020). Evolution of clonal growth forms in angiosperms. *New Phytologist* 225: 999–1010.

Herben, T., Chytrý, M., and Klimešová, J. (2016). A quest for species-level indicator values for disturbance. *Journal of Vegetation Science* 27: 628–636.

Herben, T., Klimešová, J., and Chytrý, M. (2018a). Effects of disturbance frequency and severity on plant traits: an assessment across a temperate flora. *Functional Ecology* 32: 799–808.

Herben, T., Klimešová, J., and Chytrý, M. (2018b). Philip Grime's fourth corner: are there plant species adapted to high disturbance and low productivity? *Oikos* 127: 1125–1131.

Hernández Plaza, E., Navarrete, L., and González-Andújar, J.L. (2015). Intensity of soil disturbance shapes response trait diversity of weed communities: the long-term effects of different tillage systems. *Agriculture, Ecosystems and Environment* 207: 101–108.

Holt, J.S., Welles, S.R., Silvera, K. et al. (2013). Taxonomic and life history bias in herbicide resistant weeds: implications for deployment of resistant crops. *PLoS One* 8 (9): e71916.

Huhta, A.-P., Hellström, K., Rautio, P., and Tuomi, J. (2003). Grazing tolerance of *Gentianella amarella* and other monocarpic herbs: why is tolerance highest at low damage levels? *Plant Ecology* 166: 49–61.

Iwasa, Y. and Kubo, T. (1997). Optimal size of storage for recovery after unpredictable disturbances. *Evolutionary Ecology* 11: 41–65.

Janeček, Š., Patáčová, E., and Klimešová, J. (2014). Effects of fertilization and competition on plant biomass allocation and internal resources: does *Plantago lanceolata* follow the rules of economic theory? *Folia Geobotanica* 49: 49–64.

Johnson, W.G., Davis, V.M., Kruger, G.R., and Weller, S.C. (2009). Influence of glyphosate-resistant cropping systems on weed species shifts and glyphosate-resistant weed populations. *European Journal of Agronomy* 31: 162–172.

Klimeš, L. and Klimešová, J. (1999). Root sprouting in *Rumex acetosella* under different nutrient levels. *Plant Ecology* 141: 33–39.

Klimešová, J. and Klimeš, L. (2006). CLO-PLA3: a database of clonal growth architecture of Central-European plants. http://clopla.butbn.cas.cz (accessed 1 April 2019).

Klimešová, J. and Klimeš, L. (2007). Bud banks and their role in vegetative regeneration – a literature review and proposal for simple classification and assessment. *Perspectives in Plant Ecology, Evolution and Systematics* 8: 115–129.

Klimešová, J. and Klimeš, L. (2008). Clonal growth diversity and bud banks of plants in the Czech flora: an evaluation using the CLO-PLA3 database. *Preslia* 80: 255–275.

Klimešová, J. and Martínková, J. (2004). Intermediate growth forms as a model for the study of plant clonality functioning: an example with root sprouters. *Evolutionary Ecology* 18: 669–681.

Klimešová, J., Kociánová, A., and Martínková, J. (2008). Weeds that can do both tricks: vegetative versus generative regeneration of the short-lived root-sprouting herbs *Rorippa palustris* and *Barbarea vulgaris*. *Weed Research* 48: 131–135.

Klimešová, J., Malíková, L., Rosenthal, J., and Šmilauer, P. (2014). Potential bud bank responses to apical meristem damage and environmental variables: matching or complementing axillary meristems? *PLoS One* 9 (2): e88093.

Klimešová, J., Herben, T., and Martínková, J. (2017). Disturbance is an important factor in the evolution and distribution of root-sprouting species. *Evolutionary Ecology* 31: 387–399.

Klimešová, J., Martínková, J., Pausas, J.G. et al. (2019). Handbook of standardized protocols for collecting plant modularity traits. *Perspectives in Plant Ecology, Evolution and Systematics* 40: 125485.

Kollmann, J., Brink-Jensen, K., Frandsen, S.I., and Hansen, M.K. (2011). Uprooting and burial of invasive alien plants: a new tool in coastal restoration? *Restoration Ecology* 19: 371–378.

Korsmo, E. (1930). *Unkräuter im Ackerbau der Neuzeit*. Berlin: Springer.

Kuneš, P., Abraham, V., and Herben, T. (2019). Changing disturbance-diversity relationships in temperate ecosystems over the past 12000 years. *Journal of Ecology* 107: 1678–1688.

Latzel, V., Dospělová, L., and Klimešová, J. (2009). Annuals sprouting adventitiously from the hypocotyl: their compensatory growth and implications for weed management. *Biologia* 64: 923–929.

Latzel, V., Klimešová, J., Doležal, J. et al. (2011). The association of dispersal and persistence traits of plants with different stages of succession in Central European man-made habitats. *Folia Geobotanica* 46: 289–302.

Lazebny, A. (1960). Adaptability in *Allium vineale*. In: *The Biology of Weeds* (ed. J.L. Harper), 223–236. Blackwell Scientific Publication.

Leakey, R.R.B. (1981). Adaptive biology of vegetatively regenerating weeds. *Advances in Applied Biology* 6: 57–90.

Leakey, R.R.B., Chancellor, R.J., and Vince-Prue, D. (1977). Regeneration of rhizome fragments of *Agropyron repens* (L.) Beauv. I. The seasonality of shoot growth and rhizome reserves in single-node fragments. *Annals of Applied Biology* 87: 423–431.

Leakey, R.R.B., Chancellor, R.J., and Vince-Prue, D. (1978a). Regeneration from rhizome fragments of *Agropyron repens* (L.) Beauv. IV. Effects of light on bud dormancy and the development of dominance among shoots on multi-node fragments. *Annals of Botany* 42: 205–212.

Leakey, R.R.B., Chancellor, R.J., and Vince-Prue, D. (1978b). Regeneration from rhizome fragments of *Agropyron repens* (L.) Beauv. III. Effects of nitrogen and temperature on the development of dominance among shoots on multi-node fragments. *Annals of Botany* 42: 197–204.

Li, X., Shen, Y., Huang, Q. et al. (2013). Regeneration capacity of small clonal fragments of the invasive *Mikania micrantha* H.B.K.: effects of burial depth and stolon internode length. *PLoS One* 8 (12): e84657.

Lososová, Z. and Simonová, D. (2008). Changes during the 20th century in species composition of synanthropic vegetation in Moravia (Czech Republic). *Preslia* 80: 291–305.

Lososová, Z., Chytrý, M., Cimalová, S. et al. (2004). Weed vegetation of arable land in Central Europe: gradients of diversity and species composition. *Journal of Vegetation Science* 15: 415–422.

Martínková, J. and Klimešová, J. (2016). Enforced clonality confers a fitness advantage. *Frontiers in Plant Science* 7: 1–10.

Martínková, J., Kočvarová, M., and Klimešová, J. (2004). Resprouting after disturbance in the short-lived herb *Rorippa palustris* (Brassicaceae): an experiment with juveniles. *Acta Oecologica* 25: 143–150.

Martínková, J., Klimešová, J., and Mihulka, S. (2008). Compensation of seed production after severe injury in the short-lived herb *Barbarea vulgaris*. *Basic and Applied Ecology* 9: 44–54.

Martínková, J., Šmilauer, P., Mihulka, S. et al. (2016). The effect of injury on whole-plant senescence: an experiment with two root-sprouting *Barbarea* species. *Annals of Botany* 117: 667–679.

Midgley, J.J. (1996). Why the world's vegetation is not totally dominated by resprouting plants: because resprouters are shorter than reseeders. *Ecography* 19: 92–95.

Miller, A.D., Roxburgh, S.H., and Shea, K. (2011). How frequency and intensity shape diversity–disturbance relationships. *Proceedings of the National Academy of Sciences of the United States of America* 108: 5643–5648.

Moyer, J., Roman, E., Lindwall, C., and Blackshaw, R. (1994). Weed management in conservation tillage systems for wheat production in North and South America. *Crop Protection* 13: 243–259.

Mucina, L. (2019). Biome: evolution of a crucial ecological and biogeographical concept. *New Phytologist* 222: 97–114.

Neve, P., Barney, J.N., Buckley, Y. et al. (2018). Reviewing research priorities in weed ecology, evolution and management: a horizon scan. *Weed Research* 58: 250–258.

Nichols, V., Verhulst, N., Cox, R., and Govaerts, B. (2015). Weed dynamics and conservation agriculture principles: a review. *Field Crops Research* 183: 56–68.

Ott, J.P., Klimešová, J., and Hartnett, D.C. (2019). The ecology and significance of below-ground bud banks in plants. *Annals of Botany* 123: 1099–1118.

Papini, A., Signorini, M.A., Foggi, B. et al. (2017). History vs. legend: retracing invasion and spread of *Oxalis pes-caprae* L. in Europe and the Mediterranean area. *PLoS One* 12 (12): e0190237.

Pausas, J.G. and Paula, S. (2020). Grasses and fire: the importance of hiding buds. A response to Moore et al. (2019) 'Effects of drought and fire on resprouting capacity of 52 temperate Australian perennial native grasses'. *New Phytologist* 226: 957–959.

Pausas, J.G., Lamont, B.B., Paula, S. et al. (2018). Unearthing belowground bud banks in fire-prone ecosystems. *New Phytologist* 217: 1435–1448.

Raju, M.V.S., Steeves, T.A., and Coupland, R.T. (1964). On the regeneration of root fragments of leafy spurge (*Euphorbia esula* L.). *Weed Research* 4: 2–11.

Raju, M.V.S., Coupland, R.T., and Steeves, T.A. (1966). On the occurrence of root buds on perennial plants in Saskatchewan. *Canadian Journal of Botany* 44: 33–36.

Rauh, W. (1937). Die Bildung von Hypocotyl- und Wurzelsprossen und ihre Bedeutung für die Wuchsformen der Pflanzen. *Nova Acta Leopoldina* 4: 395–553.

Ringselle, B., Bergkvist, G., Aronsson, H., and Andersson, L. (2015). Under-sown cover crops and post-harvest mowing as measures to control *Elymus repens*. *Weed Research* 55: 309–319.

Sainty, G., McCorkelle, G., and Julien, M. (1997). Control and spread of alligator weed *Alternanthera philoxeroides* (Mart.) Griseb., in Australia: lessons for other regions. *Wetlands Ecology and Management* 5: 195–201.

Shefferson, R.P., Kull, T., Hutchings, M.J. et al. (2018). Drivers of vegetative dormancy across herbaceous perennial plant species. *Ecology Letters* 21: 724–733.

Shrestha, U., Rosskopf, E.N., and Butler, D.M. (2018). Effect of anaerobic soil disinfestation amendment type and C:N ratio on *Cyperus esculentus* tuber sprouting, growth and reproduction. *Weed Research* 58: 379–388.

Sosnová, M., Herben, T., Martínková, J. et al. (2014). To resprout or not to resprout? Modeling population dynamics of a root-sprouting monocarpic plant under various disturbance regimes. *Plant Ecology* 215: 1245–1254.

Stephens, R.J. (1982). *Theory and Practice of Weed Control*. Berlin: Springer.

Storkey, J., Meyer, S., Still, K.S., and Leuschner, C. (2012). The impact of agricultural intensification and land-use change on the European arable flora. *Proceedings of the Royal Society B: Biological Sciences* 279: 1421–1429.

Taab, A., Andersson, L., and Bostrøm, U. (2018). Modelling the sprouting capacity from underground buds of the perennial weed *Sonchus arvensis*. *Weed Research* 58: 348–356.

Tomlinson, K.W. and O'Connor, T.G. (2004). Control of tiller recruitment in bunchgrasses: uniting physiology and ecology. *Functional Ecology* 18: 489–496.

Verwijst, T., Tavaziva, V.J., and Lunkvist, A. (2018). Assessment of the compensation point of *Cirsium arvense* and effects of competition, root weight and burial depth on below-ground dry weight – leaf stage trajectories. *Weed Research* 58: 292–303.

Vesk, P.A. (2006). Plant size and resprouting ability: trading tolerance and avoidance of damage? *Journal of Ecology* 94: 1027–1034.

Vesk, P.A. and Westoby, M. (2004a). Sprouting ability across diverse disturbances and vegetation types worldwide. *Journal of Ecology* 92: 310–320.

Vesk, P.A. and Westoby, M. (2004b). Funding the bud bank: a review of the costs of buds. *Oikos* 106: 200–208.

Wang, T., Hu, J.T., Wang, R.Q. et al. (2018). Tolerance and resistance facilitate the invasion success of *Alternanthera philoxeroides* in disturbed habitats: a reconsideration of the disturbance hypothesis in the light of phenotypic variation. *Environmental and Experimental Botany* 153: 135–142.

Weber, E. (2011). Strong regeneration ability from rhizome fragments in two invasive clonal plants (*Solidago canadensis* and *S. gigantea*). *Biological Invasions* 13: 2947–2955.

Wehsarg, O. (1954). *Ackerunkrauter*, 249 pp. Berlin: Akademie Verlag.

West, A., Lewis, D., and Richardson, R. (2012). Fragment size and planting depth affect the regenerative capacity of bushkiller (*Cayratia japonica*). *Invasive Plant Science and Management* 5: 397–401.

Wezel, A., Casagrande, M., Celette, F. et al. (2014). Agroecological practices for sustainable agriculture. A review. *Agronomy for Sustainable Development* 34: 1–20.

Wise, M.J. and Abrahamson, W.G. (2008). Applying the limiting resource model to plant tolerance of apical meristem damage. *American Naturalist* 172: 635–647.

Zhang, R. and Shea, K. (2012). Integrating multiple disturbance aspects: management of an invasive thistle, *Carduus nutans*. *Annals of Botany* 110: 1395–1401.

12

Climate Change and the Persistence of Weeds

David R. Clements[1] *and Antonio DiTommaso*[2]

[1] *Department of Biology, Trinity Western University, Langley, BC, Canada*
[2] *Soil and Crop Sciences, School of Integrative Plant Science, Cornell University, Ithaca, NY, USA*

12.1 Introduction

The effects of climate change on plants include an additional dynamic compared to other organisms since plants utilize CO_2 and higher CO_2 levels generally enhance plant growth. Furthermore, increased temperatures due to global warming can also increase plant growth. However, many of the climate extremes associated with climate change such as increased frequency or intensity of extreme meteorological events such as storms, droughts, fires, or floods can negatively impact plant growth or result in plant mortality (Brown et al. 2018; Crockett and Westerling 2018; Du et al. 2018; Maxwell et al. 2019). For our purposes in understanding potential impacts of climate change on weed persistence, it is important to consider differences in the responses of weedy plants to these various aspects of climate change in comparison with other plants, such as crop plants or native vegetation. Indeed, there is a considerable body of research that demonstrates that weeds are able to take advantage of climate change (Clements and DiTommaso 2011; DiTommaso et al. 2014; Varanasi et al. 2016; Ramesh et al. 2017; Waryszak et al. 2018; Ziska et al. 2019).

As described elsewhere in this book, weeds tend to have particular characteristics that afford them competitive advantages over other plants under certain conditions. This question then arises: Are these advantages potentially increased under climate change? Based on research efforts to identify traits associated with invasiveness (Mack et al. 2000; Kolar and Lodge 2002; Sakai et al. 2001; Rejmánek et al. 2005; Pyšek and Richardson 2007), Whitney and Gabler (2008) identified 10 traits that are key to success for invasive species (including various taxa, not just plants). Clements and DiTommaso (2011), in assessing the potential evolution of invasive plants under climate change, made predictions about the scenarios under which these traits could be selected for when climate conditions change (Table 12.1). All 10 of these traits tend to make weeds more persistent, and all the more persistent under climate change in many cases. For example, the typical 'weedy' trait whereby these plants grow rapidly is favoured by the increased duration in growing seasons

Persistence Strategies of Weeds, First Edition. Edited by Mahesh K. Upadhyaya, David R. Clements, and Anil Shrestha.

Table 12.1 Invasive plant character traits targeted for natural selection under climate change.

Character trait	Predictions of invasiveness due to climate change
High growth rate	Faster-growing plants favoured as growing degree days (GDD) expand towards the poles
Wide environmental tolerance	Plants pre-adapted or evolutionarily labile
Short generation time	Plants capable of maturing rapidly favoured as GDD expand towards the poles
Prolific or consistent reproduction	High reproductive levels enable more rapid adaptation to changing conditions
Small seed size	More readily dispersed under climate change (e.g. if potential dispersal by wind is increased)
Prolific dispersal	Effective dispersal may increase with increasing storms or floods or other extreme conditions
Uniparental reproduction	Able to reproduce well even under disturbed conditions
No specialized germination requirements	Able to invade and become established in a broad array of habitats
Strong competitive ability	Outcompete other plants even in changing climates
Effective defences against natural enemies	Under some conditions, these defences would be enhanced under climate change

Source: Modified list from Clements and DiTommaso (2011) based on character traits in Whitney and Gabler (2008).

towards the poles. Even if other plants also took advantage of the increase in growing degree days (GDD), weeds often have better competitive ability as well; therefore, the list of traits in Table 12.1 are best thought of as working synergistically to favour weed persistence. Many of the traits also involve dispersal ability (e.g. prolific dispersal ability, small seed size, or lack of specialized germination requirements), which allows weedy plants to take advantage of changing climates more rapidly than other plants, as they rapidly move into niches newly made available.

It is impossible for any one species to possess all 10 traits in Table 12.1, and adaptations of invasive plants to climate change are by no means limited to this list, but in a given species, many of these may work synergistically. For example, each of four species successfully invading northwards in North America under climate change profiled by Clements and DiTommaso (2012) exhibits a different combination of adaptive traits. Himalayan balsam (*Impatiens glandulifera*), as a relatively tall annual plant, combines a particularly rapid growth rate, prolific reproduction, and seed production, and traits have been shown to vary along a longitudinal gradient (Kollmann and Bañuelos 2004). Although velvetleaf (*Abutilon theophrasti*) also produces large quantities of seeds, it combines this adaptation with a strong competitive ability and seed dormancy, making it a serious agricultural pest, unlike Himalayan balsam (Clements and DiTommaso 2012). Although mostly selfing, genetic differentiation in velvetleaf has been demonstrated (Anderson et al. 1985), allowing for adaptation to new climatic conditions. As perennial species, seed production by knotweeds (*Reynoutria* spp.) and johnsongrass (*Sorghum halepense*) is relatively lower than for Himalayan balsam and velvetleaf, but other traits possessed by these perennials make them highly invasive such as rapid and extensive rhizomatous growth (Clements and DiTommaso 2012). Interestingly, in response to north–south climate gradients, johnsongrass manages to exhibit multiple strategies, acting as an annual in the northern part of its range where the rhizomes fail to survive the winter, with correspondingly greater allocation of resources to reproduce by seed (Warwick et al. 1986). As with annual plants, there is evidence that evolutionary processes can enhance invasive traits in these perennial species, enabling them to adapt to climate change (Gillies et al. 2016; Sezen et al. 2016).

Just as humans are unwittingly engaged in a global scale experiment on the effects of manipulating the Earth's climate (Calzolari et al. 2018; Lowe and Bernie 2018), we are also carrying out a global experiment on weed evolution via the way we alter our management regimes in forestry, agriculture, and urban settings to produce novel selection pressures favouring particular weeds (Clements et al. 2004; Gaines et al. 2019). These two global experiments converge as we seek to adapt to changing climates. An obvious example of this is the growing of crops in new regions, as facilitated by a warming climate (Barbeau et al. 2015; Yang et al. 2015; Cho and McCarl 2017). As seen throughout history, when crops are grown in new areas, it is often inevitable that the specific weeds associated with the crop will follow, and thus climate change will aid the movement and thereby further the persistence of weeds. Therefore, in addition to more direct effects of climate change on weed persistence, there is the potential for numerous indirect impacts of climate change, perhaps including some impacts that are very difficult to anticipate. Indeed, patterns of greater climatic extremes with increasing global temperature raise the spectre of self-reinforcing feedbacks, whereby once a critical threshold is crossed, there may be no way of halting a runaway climate (Steffen et al. 2018). Through the ages, stochastic forces have very much aided and abetted weeds to thrive, and we have reason to believe that climate change acts in a similar stochastic fashion to promote the proliferation of these disturbance-adapted organisms.

Our goal in this review is to catalogue what we know about actual and predicted effects of climate change on weed persistence, through an examination of available case studies of specific weed species and making general projections. In basic terms, weed persistence denotes properties that enable weeds to sustain populations despite human management efforts. These properties generally consist of innate ecophysiological traits, which for many species may favour their persistence under climate change. In terms of invasion ecology, persistence can also incorporate the major stages of invasion: introduction, naturalization, and invasion (Richardson et al. 2000), with the most persistent invasive species progressing readily through all three stages. Peters et al. (2014) described three scales at which weeds are undergoing shifts in concert with climate change: (i) shifts in distribution at the landscape scale, (ii) shifts in niches at the community scale, and (iii) shifts in traits at the population scale, which include the evolutionary shifts described by Clements and DiTommaso (2011). We will examine four main aspects of anticipated changes under climate change affecting weed persistence: weed ecophysiological responses, distribution changes, changes in interactions with crops or natural communities, and evolutionary changes.

12.2 Weed Ecophysiological Responses to Climate Change

Because the impacts of climate change on plants are multidimensional, predicted ecophysiological impacts are not straightforward. As Ziska (2016) points out, ecophysiological impacts to plants under climate change encompass all of the following: changes in biochemistry, plant leaf structure, whole plant structure, community aspects, and, finally, evolution and fitness. Climate change not only involves increased CO_2 concentration by definition but also may involve warming temperatures, increased drought, flooding, storms, or other extremes, with specific impacts varying according to regional climatic patterns. Thus, although increased plant growth is generally predicted with increased CO_2 concentration, other factors related to climate change may reduce plant growth. For example, Nguyen et al. (2017) evaluated how three different components of climate change (carbon dioxide, drought, and temperature) affect the annual herbaceous weed, parthenium weed (*Parthenium hysterophorus*). They found that elevated CO_2 concentration only promoted growth and reproductive output up to a certain threshold temperature regime. Reproductive output of parthenium weed was also promoted by drier soil conditions, highlighting its adaptation to dry conditions, but this potential advantage came at a cost by reducing seed longevity (Nguyen et al. 2017). The responses measured by Nguyen et al. (2017) were unique to this particular species, and in fact, they tested two different biotypes of parthenium weed that exhibited somewhat divergent responses to the treatments. This suggests that weeds like parthenium weed that can exploit dry conditions predicted under climate change in many regions are expected to become more persistent (Nguyen et al. 2017). Other weeds, such as carpetweed (*Galenia pubescens*) may not be as well adapted to exploit drought conditions, because its growth was only increased at higher CO_2 levels under moist conditions (Mahmood et al. 2016).

Since the widespread use of herbicides for weed management began in the 1940s, herbicide efficacy has been a key factor determining the persistence of weeds in agronomic and forestry systems. Although it is recognized that environmental conditions are important determinants of herbicide efficacy, the general assumption has been that such environmental factors remain relatively constant in a given geographic region. However, it is clear that herbicide efficacy will be affected by shifting conditions predicted under climate change (Ziska et al. 2004; Matzrafi et al. 2016; Varanasi et al. 2016; Ziska 2016; Matzrafi 2019). Herbicide modes of action involve at least 19 different target sites operating via a variety of mechanisms that disrupt the physiology of weeds (Duke 2012). These mechanisms range from impacts on amino acid metabolism, hormone receptors, organic compound synthesis, mitosis, photosynthesis, protein phosphatase, to adenosine triphosphate (ATP) energy production (Duke 2012). Because herbicides target plant enzyme systems, and also because their movement via the translocation stream is also important, herbicide efficacy is impacted by environmental factors that alter these plant physiological components (Matzrafi 2019). Key environmental factors affecting herbicide activity include light, CO_2, temperature, relative humidity, precipitation, soil moisture, and wind (Varanasi et al. 2016).

The amount of light reaching plants and thus possibly affecting herbicide efficacy could change if weed architecture is modified by climate change (e.g. increases in branching due to CO_2 levels). Higher CO_2 levels may reduce herbicide efficacy due to increased weed growth rates and/or increased allocation to belowground storage organs such as rhizomes or tubers (Patterson et al. 1999; Ziska et al. 2004). Waryszak et al. (2018) produced evidence from 14 weed species including C_3 and C_4 weeds and weedy shrubs to suggest that with elevated CO_2 levels management of weeds using herbicides would need to be re-evaluated because of projected losses in efficacy. The effects of higher temperatures on herbicide efficacy are complex, involving direct and indirect effects, and potentially leading to either lower or higher herbicide efficacy depending on the specific conditions (Varanasi et al. 2016). Absorption and translocation of herbicides tend to increase with increasing temperatures, but elevated temperatures may also induce more rapid metabolism or lead to more rapid evaporation of the herbicide from the soil (Kells et al. 1984; Atienza et al. 2001; Johnson and Young 2002). This could be especially problematic for more volatile herbicides such as the auxin-(indoleacetic acid)-mimics dicamba and 2,4-D in recently introduced dicamba- or 2,4-D-resistant crops (Bish et al. 2019; Mueller and Steckel 2019). Drier soil conditions can reduce absorption of herbicides by plant roots and lead to the development of thicker cuticles in the leaves of weeds. Therefore, drought conditions produced or exacerbated by climate change could result in reduced herbicide activity (Olson et al. 2000; Skelton et al. 2016; Kumar and Kumar 2017). All of these impacts are somewhat species specific, and interestingly, the evidence suggests that herbicides that weeds are most likely to develop nontarget site herbicide resistance to are also most likely to feature reduced efficacy under changing environmental conditions (Kleinman et al. 2016; Matzrafi et al. 2016; Matzrafi 2019). Therefore, it is difficult not to conclude that multiple lines of evidence point to climate change, seriously reducing our ability to control weeds effectively using what is arguably the largest hammer in the weed management toolbox.

Across the globe, a clear and important impact of climate change during the past few decades is the change in plant phenology and, in particular, an alteration in reproductive phenology in response to earlier springs and other seasonal changes (Thackeray et al. 2016). These seasonal changes represent both a challenge and opportunity for plants; if plants are unable to adjust the timing of flowering, seedling production, and other reproductive processes to climatic changes, their fitness will be reduced. Alternatively, plants that can modify their phenology either via phenotypic plasticity or evolutionary adaptation will be more successful relative to plant species with more rigid life history schedules (see Chapter 15). There is considerable evidence to suggest that non-native plants have a greater capacity to respond phenologically to seasonal changes than native plants (Willis et al. 2010; Zettlemoyer et al. 2019). For example, when Zettlemoyer et al. (2019) experimentally tested the response of 20 native and 20 non-native plant species to climate warming by 3 °C above ambient temperatures, there was no detectable effect on the reproductive phenology of the native species, whereas the non-native species exhibited accelerated phenologies. Long-term observational records have also shown non-natives to exhibit more plastic flowering times than native plants (Willis et al. 2010; Wolkovich et al. 2013). This difference between native and non-native plants would tend to make populations of non-native species including many troublesome weed species more persistent (Møller et al. 2008; Willis et al. 2010; Donnelly et al. 2011; Cleland et al. 2012; Wolkovich et al. 2013; Thackeray et al. 2016).

12.3 Predicted Changes in Weed Distribution

As mentioned earlier, warming temperatures predict northwards expansion of weed distributions in North America, and examining this trend worldwide, one would predict poleward expansion of weed distributions across the various continents, and indeed that is what is observed across the globe (Clements et al. 2014). Similar northwards expansion of species' ranges has been predicted for many weeds in Europe (Hyvönen et al. 2012). These trends are determined by tracking individual species, but it is also important to consider invasive species diversity, manifested as invasive species hotspots. In the continental United States, Allen and Bradley (2016) examined the potential range changes for 896 invasive plants, predicting that by 2050, 20% of invasive plant hotspots would shift northwards while 80% would remain in place. In the southern hemisphere, weed distributions have been documented as expanding southwards in southern continents as observed in Africa (Vardien et al. 2012) and Oceania (Buswell et al. 2011; Taylor et al. 2012; Gallagher et al. 2013), although in drought-prone regions of Australia, ranges for some weed species may contract under climate change (Gallagher et al. 2013). To date, much of the research on impacts of climate change on weeds has focussed on changes in distribution (Peters et al. 2014), although potential ecophysiological alterations have also been well studied (Ziska 2016). Mapping weed distributional changes necessitates accounting for all aspects of climate change, such as increased droughts, wildfires, floods, and wind storms, all of which can impact plant health, dispersal, and reproduction.

12.3.1 Case Studies of Weed Distribution Changes

The number of documented changes in the distribution of weeds where climate change is implicated is growing rapidly across the world. These are particularly well documented for North America, Europe, and Oceania. Across all three of these temperate regions poleward range extension has been observed in response to warming, but other trends are also apparent in response to changing precipitation patterns and climate extremes.

The weed flora of North America is largely a product of its colonial history with European settlement and the ever-expanding cultivation of land, particularly with many weeds introduced through the seed trade during the nineteenth and twentieth centuries (Mack 1991). Weed managers and researchers have tended to focus efforts on weed species in North America established during the colonial period simply referred to as weeds rather than 'invasive weeds', but careful survey work revealed that new weed problems are emerging due to novel species from other subregions of North America or recent invaders of North America (Thomas and Leeson 2007). Northwards weed distribution expansion has been well documented (Table 12.2) and is of particular concern in areas where traditionally many weeds have been unable to establish because of severe winter conditions, like the Northeastern United States (Wolfe et al. 2008; Coiner et al. 2018) or Canada (Clements and DiTommaso 2012). Meanwhile, some weeds are predicted to decline in certain regions of North America as climate changes (Ziska and Runion 2007; McDonald et al. 2009). For example, Canada thistle (*Cirsium arvense*) and proso millet (*Panicum miliaceum*) are likely to decline in the southern parts of their ranges with climate warming (Ziska and Runion 2007).

Table 12.2 Weed distribution changes related to climate change in North America.

Weed species	Range expansion	Attributed mechanism(s)	References
Bromus tectorum (cheatgrass)	Expanded range in Western North America	Dry conditions with spring precipitation and rain instead of snow	Concilio et al. (2013)
Centaurea stoebe (spotted knapweed)	More northerly latitudes	Lack of natural enemies/adaptation to cooler or drier climates	Broennimann et al. (2007)
Datura stramonium (jimsonweed)	Northwards invasion of cropland since the 1950s	Selection for heavier seeds, earlier growth	Warwick (1990) and Weaver et al. (1985)
Echinochloa crus-galli (barnyard grass)	Northwards invasion of Canada from the United States in the nineteenth century	More rapid maturation at each life cycle stage	Potvin (1986)
Reynoutria spp. (knotweeds)	Northwards range expansion in Canada	Genotypic variation in cold tolerance via hybridization	Bourchier and Van Hezewijk (2010)
Sorghum halepense (johnsongrass)	Northwards expansion in North America	Northern populations annual (southern perennial)	Warwick et al. (1986)

Japanese knotweed (*Reynoutria japonica*) has recently shown rapid range expansion in both Ontario and British Columbia, Canada (Bourchier and Van Hezewijk 2010), paralleling rapid range expansion following its introduction to Europe from Asia (Beerling 1993). Bourchier and Van Hezewijk (2010) identified an increase of 53% in suitable habitats for this invasive plant in Southern Ontario for the period 2000–2008, when temperatures were warmer than for the 1971–2000 period. Similarly, they identified considerable room for northwards expansion in British Columbia (Bourchier and Van Hezewijk 2010). Such climate change scenarios threaten to increase greatly the already considerable economic costs for management of Japanese knotweed (Bourchier and Van Hezewijk 2010). Japanese knotweed is likely to develop frost tolerance (Maruta 1983), potentially enabling increased persistence in the face of occasional frost as it extends its range northwards in response to warming. Since the turn of the twenty-first century, increased genetic diversity has been observed in the *Reynoutria* complex in North America through hybridization with giant knotweed (*Reynoutria sachalinense*) (Gaskin et al. 2014; Gillies et al. 2016). Thus, the range expansions already observed in North America are just the beginning of a much larger-scale expansion of distributions in response to climate change and other factors such as repeated use of glyphosate (Strelau et al. 2018), given the ability of the hybrid species to disperse both clonally and via seed and to produce a diversified gene pool (Gillies et al. 2016). See also Chapter 11 of this book.

One of the world's worst invasive plants is kudzu (*Pueraria lobata*) (Global Invasive Species Database 2007), well known for its populations established over large areas of the Southeastern United States (Lindgren et al. 2013). Although generally restricted to fairly warm environments, Sasek and Strain (1990) noted that physiologically it may withstand winter temperatures as low as −15 °C. Therefore, Sasek and Strain (1990) predicted it could readily move northwards in response to warming temperatures in North America. In 2009, the first recorded kudzu invasive populations were documented in Canada (Lindgren et al. 2013). Coiner et al. (2018) assessed the impact of cold temperatures on cell damage in 14 kudzu populations from eastern North America. Temperatures causing substantial damage to plants averaged −9.6 °C for northern and −14.4 °C for southern populations, indicating kudzu can acclimate to winter cold. These workers also reported that consistent with acclimation, the cold tolerance threshold of aboveground kudzu stems at the coldest time of year for a population in southern Ontario, Canada, was −26 °C, while stems insulated from cold extremes survived to −17 °C, colder than the survival limits indicated by kudzu's biogeographic distribution.

When johnsongrass was initially introduced into North America, as a perennial C_4 grass native to Eurasia, it was adapted to the warm, humid conditions of Mediterranean Europe and Africa (Warwick and Black 1983). New ecotypes developed in North America with increasingly cold tolerant rhizomes (Warwick et al. 1986). Remarkably, although southern populations in North America are perennial, northern populations generally have an annual life history that facilitates their overwinter survival (Warwick et al. 1986). Facilitated by these adaptations, McDonald et al. (2009) predicted that the damage niche in maize (*Zea mays*) could move 200–650 km northwards by 2100, resulting in a much greater impact on US maize-growing regions (e.g. Midwestern USA).

The plasticity between perennial and annual forms of johnsongrass is just one of a number of features that enables it to adapt to the northern edge of its range, facilitating future northwards movement under changing conditions (Paterson et al. 2020). In a recent common garden multisite trial in the Eastern United States (New Mexico, Texas, Georgia, Virginia, and New York) by Fletcher et al. (unpublished), johnsongrass plants from the northern region of its North American distribution (e.g. central New York State) were smaller in stature (grew smaller and shorter) than southern populations (e.g. Virginia, Georgia, Texas), but they also initiated flowering eight days earlier, on average. These patterns were consistent across all five gardens and suggest a possible trade-off between the amount of time to reach reproductive maturity and vegetative size, where plants that flower earlier grow smaller than plants that flower later. Adaptation of earlier flowering at higher latitudes, where the growing season is shorter, may have enabled johnsongrass to colonize these regions by allowing them to complete their life cycle by the end of the growing season. This is of particular importance since 100% winter kill of johnsongrass plants growing in the northernmost region of its distribution was reported in this study – supporting previous observations that northern populations of johnsongrass display an annual life history. It should be noted, however, that although shoots originating from all of the five regions died over the winter at the New York site, plants did produce rhizomes including the New York population. Additionally, the New York population overwintered as rhizomes in all other gardens. This suggests that the New York johnsongrass population used for this study has the potential to overwinter as rhizomes in New York if winter temperatures warm to the point that the rhizomes survive until the spring. Winter survival of rhizomes at the northern edge of its range has important implications for the northern range expansion of johnsongrass and is consistent with the range predictions by McDonald et al. (2009).

Johnsongrass also increasingly threatens native tallgrass prairie ecosystems under climate change (Rout et al. 2013). Its rhizomatous growth and allelopathic exudates result in rates of advance of 0.45 m per year in native tallgrass prairies (Rout et al. 2013). Many other weed species exhibit population differences developed along a north–south gradient in North America, such as the more rapid development of barnyardgrass (*Echinochloa crus-galli*) in the northern part of its range (Potvin 1986).

It is important to recognize that climate change involves more than changes in temperature and that some weed species are well equipped to respond to changes in precipitation patterns predicted under climate change. For example, cheatgrass (*Bromus tectorum*) with its relatively short life cycle grows well in areas with relatively little precipitation. In parts of western North America where summer droughts accompanied by some spring precipitation is forecasted under climate change, the range of cheatgrass is predicted to expand (Bradley 2009). This region of North America includes many mountainous areas, where as climate warming occurs there is a predicted shift from snow to rain in many areas, a shift that is already evident. This predicted shift from snow to rain was shown by Concilio et al. (2013) to favour cheatgrass via a simulated reduction in the snowpack which prompted earlier germination and more vigorous growth (Table 12.2). These changes allowed cheatgrass to outcompete other species as is evident throughout its invaded range in North America (Mack 1981).

In addition to changes in rainfall patterns, more frequent storms may increase weed dispersal and weed persistence, as seen, for example, in Vermont, USA, when Japanese knotweed spread to new sites via fragments from the impact of Tropical Storm Irene in 2011 (Colleran and Goodall 2015). Similarly, populations of the invasive sedge, *Carex kobomugi*, benefitted from Hurricane Sandy in 2012, providing better stabilization of dunes on the New Jersey coast than the native grass American beachgrass (*Ammophila breviligulata*) (Charbonneau et al. 2017).

There are as many as 4139 non-native plant species that occur in Europe (van Kleunen et al. 2015). Although most non-native species do not cause serious management problems, many are expanding their ranges rapidly (Table 12.3). Giant hogweed (*Heracleum mantegazzianum*) and Himalayan balsam have been successful invaders in most of Northern Europe (Kollmann and Bañuelos 2004; Pyšek et al. 2008). Adaptation to northern climate conditions has resulted in northern populations of Himalayan balsam flowering earlier and producing less biomass compared with southern populations (Kollmann and Bañuelos 2004). The distributions of Japanese knotweed and common ragweed (*Ambrosia artemisiifolia*) are currently limited by temperature in Europe and are most abundant in Central Europe (Beerling 1993; Beerling et al. 2009; DAISIE 2009), so it is important to monitor their spread with changes in climate. Certainly, common ragweed has already been expanding its range rapidly in certain areas such as France (Chauvel et al. 2006), Austria (Essl et al. 2009), and Hungary (Novák et al. 2009). Research by Kralemann et al. (2018) using common ragweed populations from a native lower latitude region (Kentucky, USA) and non-native higher latitude region (northern Germany) under controlled environment conditions showed that North European invasion by this annual herb is likely to be associated with early flowering, which ensures some viable seed production during the relatively shorter growing season in northern latitudes. Range expansion of common ragweed has been somewhat facilitated by a niche shift from ruderal to agricultural habitats, whereas for knotweed spp. hybridization has been a key factor as also seen in

Table 12.3 Weed distribution changes related to climate change in Europe.

Weed species	Range expansion	Attributed mechanism(s)	References
Ambrosia artemisiifolia (common ragweed)	Expansion in Central Europe	Niche expansion from ruderal to agricultural habitats	DAISIE (2009) and Essl et al. (2009)
Heracleum mantegazzianum (giant hogweed)	Expansion in North and Central Europe	Niche expansion	DAISIE (2009)
Impatiens glandulifera (Himalayan balsam)	Expansion throughout Europe	Flowering phenology differs among populations	Kollmann and Bañuelos (2004)
Reynoutria spp. (knotweeds)	Expansion in Central Europe	Hybridization	Hollingsworth and Bailey (2000)
Rosa rugosa (rugose rose)	Expansion throughout Europe	Effective dispersal via floating seeds	DAISIE (2009)
Stellaria media (common chickweed)	Expansion throughout Europe in agronomic crops	Ability to germinate under a broader array of conditions	Fried et al. (2010) and Peters et al. (2014)

North America. As climate warms, both knotweed spp. and common ragweed would be expected to extend their ranges northwards in Europe.

Weed range expansions in Europe are not limited to non-native plants. For example, common chickweed (*Stellaria media*) originated in Central Europe where it primarily inhabited riparian areas, but as changes in its traits evolved over time, it was able to develop a modified life cycle, in part due to milder climates, and has now greatly expanded its European distribution, in part due to the positive effects of crop fertilization on this weed species (Fried et al. 2010; Peters et al. 2014).

The number of arable weed species in Europe declines from south to north following a climate gradient (Glemnitz et al. 2006), but this gradient may well be disrupted as climate change reduces the temperature difference from south to north. Indeed, predictions of future changes in suitable climate conditions for weed species have provided evidence for this (Hyvönen et al. 2012). For example, Hyvönen et al. (2012) predicted redroot pigweed (*Amaranthus retroflexus*) could invade 500 km further north under projected future climatic conditions in 2080, although this assumed suitable crop species were available (Hyvönen and Ramula 2014). As discussed for North America, the extreme weather events predicted under climate change would be expected to increase dispersal of many weed species, thereby enhancing their persistence. For example, rugose rose (*Rosa rugosa*) is very effective at increasing its distribution via floating seeds, just as fragments and seeds of knotweed spp., can be readily transported in water (Rouifed et al. 2011; Colleran and Goodall 2015). Increased flooding conditions will increase the persistence of such species.

As with North America and Europe, poleward spread (in this case southwards) has been documented for a number of weed species in Oceania (Table 12.4). Gallagher et al. (2013) demonstrated that the distribution of 11 species of problematic alien perennial grasses in Australia could be modified with climate change. For instance, the range of pampas grass

(*Cortaderia selloana*) was actually predicted to decline by 68% by 2050 in Australia due to the negative effects of droughts, which were expected to increase in frequency with climate change (Gallagher et al. 2013). However, many non-crop species will likely fare better under drought conditions than crops or may evolve in response to climate change stresses. The range of the annual legume, sicklepod (*Senna obtusifolia*), although only found in northern Australia at present, could expand southwards via ecotypes adapted to a broader range of temperatures (Tsutsumi 2011). Using historical herbarium records, Buswell et al. (2011) reported that 70% of the weed species found in Australia showed changes in at least one trait, with the most commonly observed modification being a change in height. The observed changes in height may be linked to selection for shorter stature under arid conditions. Eight of the 21 species showed changes in height, including trampweed (*Facelis retusa*), four-leaf all-seed (*Polycarpon tetraphyllum*), French catchfly (*Silene gallica*), and cluster clover (*Trifolium glomeratum*). The expansion of weed ranges in Oceania is influenced by a variety of other factors, such as competitive ability (which could be modified by climate change), dispersal via livestock or machinery, fire dynamics, or flooding (Table 12.4).

12.3.2 What Do These Distribution Changes Indicate About Weed Persistence?

Although numerous changes in the range of weed species have already been observed over the last century (Tables 12.3–12.4), it is important to consider that spatial distribution is a complex relationship given the heterogeneity of ecosystems. Thus, even if a weed in the northern hemisphere is predicted to spread further north under climate change, the new landscapes it is predicted to occupy may not necessarily allow such a change in distribution. There may be ecological filters that prevent successful dispersal and establishment of the weed in the predicted range (Peters et al. 2014).

Usinowicz and Levine (2018) demonstrated that in order to incorporate competitive interactions (which represent one such ecological filter) into predicted range expansion under climate change, it is important to account for competition among multiple migrating species to determine which species will be the most persistent in the new range. In the case of weeds, crop–weed interactions are often key components of the distribution of weeds across the landscape. Work by McDonald et al. (2009) showed how the 'damage niches' of johnsongrass and velvetleaf to a maize crop will shift when weed populations move northwards due to climate change. In other words, the regions where particular weeds most impact crop production are expected to be altered due to climate change. It is therefore useful to consider this kind of interaction further to understand potential impacts of climate change on weed persistence, as we discuss in Section 12.4.

12.4 Impacts of Climate Change on Weed Interactions with Crops

There is a great deal of concern over the impact of climate change on crop production, and much of this concern stems from predicted changes in the relationships between crops and pests (Ziska and McConnell 2015; Deutsch et al. 2018; Wienhold et al. 2018). Weed–crop

Table 12.4 Weed distribution changes related to climate change in Oceania.

Weed species	Range expansion	Attributed mechanism(s)	References
Aira cupaniana (silvery hairgrass)	Expanding into more arid areas of New South Wales	Decreased plant height	Buswell et al. (2011)
Cenchrus ciliaris	Rapid spread throughout Australia since the 1870s	Tolerant of heavy grazing and drought and ability to re-grow after fire	Martin et al. (2015)
Cortaderia selloana (pampas grass)	Rapid spread in Southern Australia since 1901	Wind dispersal and competitive mechanisms	Gallagher et al. (2013)
Eragrostis curvula (African lovegrass)	Rapid spread in Southern, Eastern, and Western Australia since 1914	Low palatability and drought tolerance	Gallagher et al. (2013)
Hyparrhenia hirta (Coolatai grass)	Rapid spread through Australia since 1890s	Ability to germinate across a broad range of temperatures and moisture levels	Chejara et al. (2010)
Nassella neesiana (Chilean needle grass)	Rapid spread in South Eastern Australia since 1941	Transport in sheep wool, mowing, and water	Bourdöt et al. (2012) and Gallagher et al. (2013)
Silene gallica (French catchfly)	Expansion into more arid parts of New South Wales	Decreased plant height	Buswell et al. (2011)
Sporobolus africanus (Parramatta grass)	Spread widely throughout Australia since 1802	Seed dispersal via wind, water, and machinery	Gallagher et al. (2013)
Trifolium glomeratum (cluster clover)	Expansion into more arid parts of New South Wales	Decreased plant height	Buswell et al. (2011)

competition is one of the most significant biotic interactions in agriculture, and so predictions of how this relationship may change under climate change are highly critical for assessing future agricultural productivity.

12.4.1 C_3 and C_4 Crops Versus C_3 and C_4 Weeds

Although the C_4 pathway is well-adapted for warmer, drier conditions, the innate mechanism of C_4 plants for concentrating CO_2 levels at the RUBISCO site, up to an ambient level of 2000 ppm of CO_2, means that their response to increasing levels of CO_2 will not be as pronounced as for C_3 plants (Korres et al. 2016). C_3 plants are able take advantage of rising CO_2 concentrations because their RUBISCO site is not saturated at current atmospheric levels. Brunce and Ziska (2000) and Walthall et al. (2012) conducted laboratory and field crop–weed competition experiments using crops and weeds possessing different photosynthetic pathways. Their findings showed that increased CO_2 levels favoured C_3 crops over C_4 weeds in all six cases, favoured C_3 weeds over C_3 crops in five of six cases, favoured the C_4 weed over the C_4 crop in the one case studied, and favoured the C_3 weed over the C_4 crop in two cases (Korres et al. 2016). Overall, then, increased CO_2 levels tended to favour the weed in all combinations except for C_4 weeds versus C_3 crops. Potentially the most serious consequence of increased CO_2 levels on these crop–weed relationships is that although historically it was C_4 weeds that tended to cause serious management issues in arid areas, C_3 weeds could now present new management challenges in such areas under these changing conditions (Korres et al. 2016).

12.4.2 General Outcomes of Weed–Crop Competition Under Climate Change

A meta-analysis comparing the performance of 117 native plant species with 74 invasive plant species under a range of climate change parameters found that the invasive plants generally performed better with increased levels of CO_2 and climate warming, while not performing as well with decreased precipitation (Liu et al. 2017). However, crops also tend to perform better under climate change in many situations or scenarios, such as yield increases anticipated in Nordic crops (Wiréhn 2018). The question becomes as follows: Will these yield gains by crops be nullified by increases in weed populations due to some of the same plant physiological factors and reduced sensitivity to herbicides (Wiréhn 2018; Matzrafi 2019)? Changes in cropping systems, such as the introduction of new crops to new latitudes may also play a role through changes in weed–crop relationships (Hyvönen and Ramula 2014). Another important consideration is that weeds are generally more malleable than crop plants, in terms of physiological plasticity and greater genetic variation, giving them a potential advantage when conditions change (Naidu and Murthy 2014).

Increasing drought may shift the balance towards weeds, and weed persistence, because certain weed species, such as cheatgrass, are better adapted for predicted reductions in available moisture in particular regions under climate change, such as the Intermountain West in the United States (Reeves et al. 2018). A key issue in these ecosystems is the impact of weeds on wildfire cycles, with wildfire expected to occur more frequently, killing desirable native shrub species such as big sagebrush (*Artemisia tridentata*). These fires are facilitated by increases in the abundance of cheatgrass, which is a poor-quality forage for livestock. Cheatgrass is well adapted to fire disturbance and is a fuel source for the fires (Reeves et al. 2018). The frequency of wildfires has been increasing in many parts of western North America recently (Holden et al. 2018) and weeds like cheatgrass are well-adapted to take advantage of this increasingly important abiotic stress.

12.4.3 What Do These Impacts on Weed–Crop Interactions Indicate About Weed Persistence?

Crop–weed interactions constitute a vast and complex subject, and here we have only provided a very limited overview. However, it is clear that alterations in the relationship between crops and weeds under climate change could potentially make weeds even more competitive with crops than they are at present. Our knowledge of the ecology of weed–crop interactions is woefully inadequate, given the challenge of climate change and other future challenges. Westwood et al. (2018) called for more research to better understand competition mechanisms, including incorporating new findings on how plants respond to light quality signals. They also advocated for increased research on improving crop competition with weeds through altered crop growth responses, enhanced allelopathy, and other mechanisms. Finally, these workers recommended that more effective outreach programs be established to convey alternative weed management strategies to crop managers that account for dynamic and unpredictable crop–weed interactions anticipated under climate change. Ziska et al. (2019) identified four ways weeds are moving targets for management under climate change: (i) greater amplitude of variation around herbicide resistance dynamics, (ii) potential lowering of constraints on weeds posed by pathogens or predators, (iii) phenological shifts due to climate change leading to changes in weed–crop competition, and (iv) changes arising from epigenetic factors, as another source of potentially increased weed variation under climate change. Managers need to be cognizant of the potential for weeds to become more persistent under climate change, because of various shifts predicted to take place, i.e. shifts in distribution, in ecological niche, or in traits that enable weeds to adapt to the changing conditions (Peters et al. 2014). As mentioned, such shifts in traits may be evolutionarily driven, and it is to this possibility that we now turn our attention.

12.5 Evolutionary Impacts of Climate Change on Weeds

12.5.1 Weed Evolution Under Climate Change

Climate change is acting as a selection pressure on weeds, leading to potential changes in their weedy traits (Clements and DiTommaso 2011). Another way of looking at it is that novel environments are being created for weeds via changing climate conditions. These changes go well beyond changes in weather patterns, as changing climates prompt changes in agricultural practices. In turn these changes in cultural practices also exert selection pressures on the weeds that cohabit cropland, rangeland, or forests. By their nature, weeds (especially annuals) are generally adapted to areas of high disturbance, colonizing and establishing more rapidly than other plants. However, to form a self-sustaining population in a new habitat, they must possess traits facilitating their persistence (Weiher et al. 1999; Smith et al. 2011). Thus, when weeds do invade a new area, the individuals that do persist may have developed novel traits to allow them to do so (Levin 2009; Lavergne et al. 2010; Richardson et al. 2013). As a result, the expectation is that on the leading edge of an invasion, the highest level of evolutionary adaptation is occurring (Angert et al. 2011; Doxford and Freckleton 2012).

Thus, in North America, weed species extending their distribution in a northerly direction due to climate change would be expected to feature the highest levels of evolutionary adaptation. Clements and DiTommaso (2012) profiled four such weeds: Himalayan balsam, velvetleaf, Japanese knotweed, and johnsongrass. Interestingly, these four species exhibit a diversity of reproductive systems, showing how evolutionary changes due to climate change is also influenced by the reproductive biology of weeds. Himalayan balsam, as an obligatory outcrossing annual dicot shows maximum potential to diverge across its range, as seen in diverging flowering times across a range of latitudes in Europe (Kollmann and Bañuelos 2004). By contrast, because the annual dicot, velvetleaf, is mostly selfing, its ability to adapt genetically to different climatic conditions is more limited, and yet such weeds are capable of forming locally adapted ecotypes (Clements et al. 2004). Clonal, rhizomatous perennials like Japanese knotweed are likewise not expected to form locally adapted genotypes, and in fact, in the United Kingdom, Japanese knotweed has been celebrated as one of the world's largest clones (Kidd 2000). Still, because of hybridization of Japanese knotweed with giant knotweed, locally adapted knotweed hybrid swarms are likely to spread much further in both Europe and North America than previously anticipated (Gillies et al. 2016). The fourth species profiled by Clements and DiTommaso (2012) was johnsongrass, and even though this perennial grass is primarily selfing, genetically diverse populations of Johnsongrass have developed in North America due to introgression with the related crop, sorghum (*Sorghum bicolor*) (Paterson et al. 2020). Such a mechanism of evolutionary change could well allow a weed to adapt to climate change very effectively. That is, if a crop is planted in new areas, associated weeds could potentially piggyback on whatever human manipulations are required to grow the crop in the new area.

Much of the weed flora in North America originated in Europe, and as many of the examples we have discussed have shown, there is much to be learned from research spanning the European and North American continents, because evolutionary responses to climate change may be similar in both regions (e.g. Himalayan balsam and knotweed spp.). Likewise, it is becoming increasingly apparent that the same is true for Asia, often involving weeds that are widespread and familiar in either North America, Europe, or both. Wan et al. (2017) investigated the niche divergence and habitat suitability of eight weed species in China that were native to North America in light of current and future climate scenarios. The weed species they studied included redroot pigweed, spiny amaranth (*Amaranthus spinosus*), slender amaranth (*Amaranthus viridis*), hairy beggarticks (*Bidens pilosa*), asthmaweed (*Conyza bonariensis*), Canada fleabane (*Conyza canadensis*), gallant soldier (*Galinsoga parviflora*), and cutleaf groundcherry (*Physalis angulata*). Using a Maxent modelling approach, Wan et al. (2017) found that the climate niche breadths for these weeds were on the whole, broader in China than in their native North American range, suggesting adaptation to the adventive range had occurred. Furthermore, the authors suggested that there were still many habitats that might be invaded under current and future climate scenarios, particularly in higher altitude regions like Qinghai, Tibet, and Mongolia (Wan et al. 2017).

Although changes in niche breadth between native and non-native range are very common, it is important to point out that these changes are typically due to an artefact of distribution. Weedy species do not tend to benefit from widespread dispersal in the native range, especially in contrast to this benefit in the non-native range, whether via accidental or deliberate spread by anthropogenic means. Furthermore, differences in available climate space make it difficult to truly measure an expansion or shift of the climatic niche, so one must be cautious in interpreting niche breadth studies such as the study by Wan et al. (2017).

Still, the work by Wan et al. (2017) suggests that climate niche conservatism has been observed to some degree among invasive plant species (Petitpierre et al. 2012, 2015; Callen and Miller 2015). However, Early and Sax (2014) observed the opposite, showing that naturalized ranges for most invasive species in the United States tended to fall outside their native climate ranges in Europe. It is important to account for potential niche shifts under novel climatic conditions as plants adapt to their new surroundings and the new regimes created by human modifications of habitats and climate (Martínez-Cabrera et al. 2012; González-Moreno et al. 2015). The labile nature of weed traits must be carefully evaluated and accounted for in mitigating the spread of invasive plants, keeping in mind that distribution modelling approaches generally fail to incorporate genetic or phenotypic variation (Clements and DiTommaso 2011; Warren and Seifert 2011).

12.5.2 What Do These Evolutionary Impacts Indicate About Weed Persistence?

The various traits we listed previously as influenced by climate change (Table 12.1) tend to be associated with greater weed persistence. Many other traits may also be involved, such as assimilate partitioning, induction of dormancy or seed germination, herbivore tolerance, propagule production and distribution, variability of plant architecture, photosynthetic rate, and seedbank longevity (Kathiresan and Gualbert 2016). Kathiresan and Gualbert (2016) point out that numerous climatic factors forecasted to change in a changing climate can influence seed dormancy and seed germination. Likewise, Cochrane et al. (2015) argue that variation in seed traits among populations is critically important to track under climate change, and fundamentally so because climate change has, in fact, driven the evolution of seed traits that allow for persistence of weeds in particular environments. Examples of present and future implications of this effect on seed traits provided by Cochrane et al. (2015) include observations that seedling emergence rates decline under harsh climatic conditions in Australia (Hoyle et al. 2013; Cochrane et al. 2014). This decrease in seedling emergence may lead to greater persistence if accompanied by increases in the seedbank. There is also the possibility of reduced weed persistence if exposure to the extreme high soil temperatures predicted in Australia under climate change increases seed mortality (Ooi et al. 2009).

A robust assessment of the various factors affecting seed and seedling persistence under climate change is warranted. Cochrane et al. (2015) discussed the following critical questions to be answered to advance our understanding of climate impacts on seed traits:

1) Are certain traits (e.g. seed mass, germination characteristics, or dormancy status) more sensitive to environmental traits than others?
2) Are there critical threshold values for a seed's ability to survive, germinate, and become established as a seedling under challenging climatic conditions?
3) How are these critical threshold values associated with particular plants (e.g. particular genotypes or growth forms)?
4) Are certain persistence traits under climate change ecosystem-specific?

As Cochrane et al. (2015) discussed, evaluating such questions would ideally involve an extensive, worldwide research program that would provide detailed comparisons among species and populations within species. Such a program would evaluate the potential for invasive plants to evolve persistence traits rapidly under the relatively rapid wave of climate change that is projected to occur in the next several decades. Even apart from the ideal research program envisioned by Cochrane et al. (2015), considerable data on rapid evolution under climate change and its implications are available, so that steps could and should be taken in the immediate term to design systems to counter weed evolution and its role in weed persistence (see Chapter 15, by Neve and Ceicedo, of this book).

12.6 Conclusions

To summarize, there are many signs that weeds are exploiting climate change. As we have reviewed, many shifts in distribution have been observed in response to latitudinal temperature shifts, changes in precipitation patterns, and increased frequency of extreme climate events. Although there are observations and future predictions of certain weed species becoming less persistent in the face of climate change, because of the very nature of weeds as pioneer species capable of exploiting novel and highly disturbed ecosystems, on the whole climate change will inevitably favour greater weed persistence. As we reviewed and as described elsewhere in this book, weeds possess many unique ecophysiological qualities that enable them to adapt, often better than their crop cousins. This is not good news for weed management. Throughout human history, we have struggled to manage weeds in order to protect crops, pastures, forests, and wilderness areas. As the human population continues to increase and demand more food and fibre resources in this early part of the twenty-first century, it is important not to ignore the potential of weeds to adapt and take advantage of vulnerabilities in our agroecosystems as well as take advantage of climate modifications that will result in warmer temperatures especially in high latitudes and more extreme climate events (e.g. floods, droughts) in many regions of the world. As Cochrane et al. (2015) aptly state, although we might ignore the proclivity for weeds to develop specific traits that are well adapted to climate change, "we do so at our own peril".

Acknowledgements

David R. Clements is grateful to Leslie Weston and her Plant Interactions Research Group at Charles Sturt University, Wagga Wagga, Australia, for providing a fruitful atmosphere and helpful discussions that were instrumental in the completion of this chapter.

References

Allen, J.M. and Bradley, B.A. (2016). Out of the weeds? Reduced plant invasion risk with climate change in the continental United States. *Biological Conservation* 203: 306–312.

Anderson, R.N., Menges, R.M., and Conn, J.S. (1985). Variability in velvetleaf (*Abutilon theophrasti*) and reproduction beyond its current range in North America. *Weed Science* 33: 507–512.

Angert, A.L., Crozier, L.G., Rissler, L.J. et al. (2011). Do species' traits predict recent shifts at expanding range edges? *Ecology Letters* 14: 677–689.

Atienza, J., Tabernero, M.T., Álvarez-Benedí, J., and Sanz, M. (2001). Volatilisation of triallate as affected by soil texture and air velocity. *Chemosphere* 42: 257–261.

Barbeau, C., Oelbermann, M., Karagatzides, J., and Tsuji, L. (2015). Sustainable agriculture and climate change: producing potatoes (*Solanum tuberosum* L.) and bush beans (*Phaseolus vulgaris* L.) for improved food security and resilience in a Canadian subarctic first nations community. *Sustainability* 7: 5664–5681.

Beerling, D.J. (1993). The impact of temperature on the northern distribution limits of the introduced species *Fallopia japonica* and *Impatiens glandulifera* in north-west Europe. *Journal of Biogeography* 20: 45–53.

Beerling, D.J., Huntley, B., and Bailey, J.P. (2009). Climate and the distribution of *Fallopia japonica*: use of an introduced species to test the predictive capacity of response surfaces. *Journal of Vegetation Science* 6: 269–282.

Bish, M.D., Farrell, S.T., Lerch, R.N., and Bradley, K.W. (2019). Dicamba losses to air after applications to soybean under stable and nonstable atmospheric conditions. *Journal of Environmental Quality* 48: 1675–1682.

Bourchier, R.S. and Van Hezewijk, B.E. (2010). Distribution and potential spread of Japanese knotweed (*Polygonum cuspidatum*) in Canada relative to climatic thresholds. *Invasive Plant Science and Management* 3: 32–39.

Bourdöt, G.W., Lamoureaux, S.L., Watt, M.S. et al. (2012). The potential global distribution of the invasive weed *Nassella neesiana* under current and future climates. *Biological Invasions* 14: 1545–1556.

Bradley, B.A. (2009). Regional analysis of the impacts of climate change on cheatgrass invasion shows potential risk and opportunity. *Global Change Biology* 15: 196–208.

Broennimann, O., Treier, U.A., Müller-Shärer, H. et al. (2007). Evidence of climatic niche shift during biological invasion. *Ecology Letters* 10: 701–709.

Brown, B.J., McLaughlin, B.C., Blakey, R.V., and Morueta-Holme, N. (2018). Future vulnerability mapping based on response to extreme climate events: dieback thresholds in an endemic California oak. *Diversity and Distributions* 24: 1186–1198.

Brunce, J.A. and Ziska, L.H. (2000). Crop ecosystems responses to climatic change. Crop/weed interactions. In: *Climate Change and Global Crop Productivity* (eds. K.R. Reddy and H.F. Hodges), 333–348. Wallingford: CAB International.

Buswell, J.M., Moles, A.T., and Hartley, S. (2011). Is rapid evolution common in introduced plant species? *Journal of Ecology* 99: 214–224.

Callen, S.T. and Miller, A.J. (2015). Signatures of niche conservatism and niche shift in the North American kudzu (*Pueraria montana*) invasion. *Diversity and Distributions* 21: 853–863.

Calzolari, G., Casari, M., and Ghidoni, R. (2018). Carbon is forever: a climate change experiment on cooperation. *Journal of Environmental Economics and Management* 92: 169–184.

Charbonneau, B.R., Wootton, L.S., Wnek, J.P. et al. (2017). A species effect on storm erosion: invasive sedge stabilized dunes more than native grass during Hurricane Sandy. *Journal of Applied Ecology* 54: 1385–1394.

Chauvel, B., Dessaint, F., Cardinal-Legrand, C., and Bretagnolle, F. (2006). The historical spread of *Ambrosia artemisiifolia* L. in France from herbarium records. *Journal of Biogeography* 33: 665–673.

Chejara, V.K., Kriticos, D.J., Kristiansen, P. et al. (2010). The current and future potential geographical distribution of *Hyparrhenia hirta*. *Weed Research* 50: 174–184.

Cho, S.J. and McCarl, B.A. (2017). Climate change influences on crop mix shifts in the United States. *Scientific Reports* 7: 40845.

Cleland, E.E., Allen, J.M., Crimmins, T.M. et al. (2012). Phenological tracking enables positive species responses to climate change. *Ecology* 93: 1765–1771.

Clements, D.R. and DiTommaso, A. (2011). Climate change and weed adaptation: can evolution of invasive plants lead to greater range expansion than forecasted? *Weed Research* 51: 227–240.

Clements, D.R. and DiTommaso, A. (2012). Predicting weed invasion in Canada under climate change: evaluating evolutionary potential. *Canadian Journal of Plant Science* 92: 1013–1020.

Clements, D.R., DiTommaso, A., Jordan, N. et al. (2004). Adaptability of plants invading North American cropland. *Agriculture, Ecosystems and Environment* 104: 379–398.

Clements, D.R., DiTommaso, A., and Hyvönen, T. (2014). Ecology and management of weeds in a changing climate. In: *Recent Advances in Weed Management*, 411 pp (eds. B.S. Chauhan and G. Mahajan), 13–37. Berlin: Springer.

Cochrane, A., Hoyle, G., Yates, C. et al. (2014). Climate warming delays and decreases seedling emergence in a Mediterranean ecosystem. *Oikos* 124: 150–160.

Cochrane, A., Yates, C.J., Hoyle, G.L., and Nicotra, A.B. (2015). Will among-population variation in seed traits improve the chance of species persistence under climate change? *Global Ecology and Biogeography* 24: 12–24.

Coiner, H.A., Hayhoe, K., Ziska, L.H. et al. (2018). Tolerance to subzero winter cold in kudzu (*Pueraria montana* var. *lobata*). *Oecologia* 187: 839–849.

Colleran, B.P. and Goodall, K.E. (2015). Extending the timeframe for rapid response and best management practices of flood-dispersed Japanese knotweed (*Fallopia japonica*). *Invasive Plant Science and Management* 8: 250–254.

Concilio, A., Loik, M.E., and Belnap, J. (2013). Global change effects on *Bromus tectorum* L. (Poaceae) at its high-elevation range margin. *Global Change Biology* 19: 161–172.

Crockett, J.L. and Westerling, A.L. (2018). Greater temperature and precipitation extremes intensify Western US droughts, wildfire severity, and Sierra Nevada tree mortality. *Journal of Climate* 31: 341–354.

DAISIE (2009). *Handbook of Alien Species in Europe*, 399 pp. Dordrecht: Springer.

Deutsch, C.A., Tewksbury, J.J., Tigchelaar, M. et al. (2018). Increase in crop losses to insect pests in a warming climate. *Science* 361 (6405): 916–919.

DiTommaso, A., Qin, Z., and Clements, D.R. (2014). Identifying climate change as a factor in the establishment and persistence of invasive weeds in agricultural crops. In: *Invasive Species and Global Climate Change* (eds. L. Ziska and J.S. Dukes), 353–370. CAB International.

Donnelly, A., Caffarra, A., and O'Neill, B.F. (2011). A review of climate driven mismatches between interdependent phenophases in terrestrial and aquatic ecosystems. *International Journal of Biometeorology* 55: 805–817.

Doxford, S.W. and Freckleton, R.P. (2012). Changes in the large-scale distribution of plants: extinction, colonisation and the effects of climate. *Journal of Ecology* 100: 519–529.

Du, L., Mikle, N., Zou, Z. et al. (2018). Global patterns of extreme drought-induced loss in land primary production: identifying ecological extremes from rain-use efficiency. *Science of the Total Environment* 628: 611–620.

Duke, S.O. (2012). Why have no new herbicide modes of action appeared in recent years? *Pest Management Science* 68: 505–512.

Early, R. and Sax, D.F. (2014). Climatic niche shifts between species' native and naturalized ranges raise concern for ecological forecasts during invasions and climate change. *Global Ecology and Biogeography* 23: 1356–1365.

Essl, F., Dullinger, S., and Kleinbauer, I. (2009). Changes in the spatio-temporal patterns and habitat preferences of *Ambrosia artemisiifolia* during its invasion of Austria. *Preslia* 81: 119–133.

Fried, G., Petit, S., and Reboud, X. (2010). A specialist-generalist classification of the arable flora and its response to changes in agricultural practices. *BMC Ecology* 10: 1–11.

Gaines, T.A., Patterson, E.L., and Neve, P. (2019). Molecular mechanisms of adaptive evolution revealed by global selection for glyphosate resistance. *New Phytologist* 223: 1770–1775.

Gallagher, R.V., Duursma, D.E., O'Donnell, J. et al. (2013). The grass may not always be greener: projected reductions in climatic suitability for exotic grasses under future climates in Australia. *Biological Invasions* 15: 961–975.

Gaskin, J.F., Schwarzländer, M., Grevstad, F.S. et al. (2014). Extreme differences in population structure and genetic diversity for three invasive congeners: knotweeds in western North America. *Biological Invasions* 16: 2127–2136.

Gillies, S., Clements, D.R., and Grenz, J. (2016). Knotweed (*Fallopia* sp.) invasion of North America utilizes hybridization, epigenetics, seed dispersal (unexpectedly) and an arsenal of physiological tactics. *Invasive Plant Science and Management* 9: 71–80.

Glemnitz, M., Radics, L., Hoffmann, J., and Czimber, G. (2006). Weed species richness and species composition of different arable field types – a comparative analysis along a climate gradient from south to north Europe. *Journal of Plant Diseases and Protection* 20: 577–586.

Global Invasive Species Database (2007). 100 of the World's worst weeds. Invasive species specialist group. I.U.C.N. The World Conservation Union Species Survival Commission. University of Auckland, New Zealand. http://www.issg.org/database/welcome (accessed 29 June 2019).

González-Moreno, P., Diez, J.M., Richardson, D.M., and Vilà, M. (2015). Beyond climate: disturbance niche shifts in invasive species. *Global Ecology and Biogeography* 24: 360–370.

Holden, Z.A., Swanson, A., Luce, C.H. et al. (2018). Decreasing fire season precipitation increased recent western US forest wildfire activity. *Proceedings of the National Academy of Sciences of the United States of America* 115: E8349–E8357.

Hollingsworth, M.L. and Bailey, J.P. (2000). Hybridisation and clonal diversity in some introduced *Fallopia* species (Polygonaceae). *Watsonia* 23: 111–121.

Hoyle, G.L., Venn, S.E., Steadman, K.J. et al. (2013). Soil warming increases plant species richness but decreases germination from the alpine soil seed bank. *Global Change Biology* 19: 1549–1561.

Hyvönen, T. and Ramula, S. (2014). Crop–weed competition rather than temperature limits the population establishment of two annual C_4 weeds at the edge of their northern range. *Weed Research* 54: 245–255.

Hyvönen, T., Luoto, M., and Uotila, P. (2012). Assessment of weed establishment risk in a changing European climate. *Agricultural Food and Science* 21: 348–360.

Johnson, B.C. and Young, B.G. (2002). Influence of temperature and relative humidity on the foliar activity of mesotrione. *Weed Science* 50: 157–161.

Kathiresan, R. and Gualbert, G. (2016). Impact of climate change on the invasive traits of weeds. *Weed Biology and Management* 16: 59–66.

Kells, J.J., Meggitt, W.F., and Penner, D. (1984). Absorption, translocation, and activity of fluazifop-butyl as influenced by plant growth stage and environment. *Weed Science* 32: 143–149.

Kidd, H. (2000). Japanese knotweed-the world's largest female! *Pesticide Outlook* 11: 99–100.
Kleinman, Z., Ben-Ami, G., and Rubin, B. (2016). From sensitivity to resistance – factors affecting the response of *Conyza* spp. to glyphosate. *Pest Management Science* 72: 1681–1688.
van Kleunen, M., Dawson, W., Essl, F. et al. (2015). Global exchange and accumulation of non-native plants. *Nature* 525: 100–103.
Kolar, C.S. and Lodge, D.M. (2002). Ecological predictions and risk assessment for alien fishes in North America. *Science* 298: 1233–1236.
Kollmann, J. and Bañuelos, M.J. (2004). Latitudinal trends in growth and phenology of the invasive alien plant *Impatiens glandulifera* (Balsaminaceace). *Diversity and Distributions* 10: 377–385.
Korres, N.E., Norsworthy, J.K., Tehranchian, P. et al. (2016). Cultivars to face climate change effects on crops and weeds: a review. *Agronomy for Sustainable Development* 36 https://doi.org/10.1007/s13593-016-0350-5.
Kralemann, L.E.M., Scalone, R., Andersson, L., and Hennig, L. (2018). North European invasion by common ragweed is associated with early flowering and dominant changes in *FT/TFL1* expression. *Journal of Experimental Botany* 69: 2647–2658.
Kumar, A. and Kumar, M. (2017). Climate change's impacts on weeds and herbicide efficacy: a review. *International Journal of Current Microbiology and Applied Sciences* 6: 2846–2853.
Lavergne, S., Mouquet, N., Thuiller, W., and Ronce, O. (2010). Biodiversity and climate change: integrating evolutionary and ecological responses of species and communities. *Annual Review of Ecology, Evolution, and Systematics* 41: 321–350.
Levin, D.A. (2009). Flowering-time plasticity facilitates niche shifts in adjacent populations. *New Phytologist* 183: 661–666.
Lindgren, C.J., Castro, K.L., Coiner, H.A. et al. (2013). The biology of invasive alien plants in Canada. 12. *Pueraria montana* var. *lobata* (Willd.) Sanjappa & Predeep. *Canadian Journal of Plant Science* 93: 71–95.
Liu, Y., Oduor, A.M., Zhang, Z. et al. (2017). Do invasive alien plants benefit more from global environmental change than native plants? *Global Change Biology* 23: 3363–3370.
Lowe, J.A. and Bernie, D. (2018). The impact of Earth system feedbacks on carbon budgets and climate response. *Philosophical Transactions of the Royal Society A: Mathematical, Physical and Engineering Sciences* 376 (2119): 20170263.
Mack, R.N. (1981). Invasions of *Bromus tectorum* L. into Western North America: an ecological chronicle. *Agro-Ecosystems* 7: 145–165.
Mack, R.N. (1991). The commercial seed trade: an early disperser of weeds in the United States. *Economic Botany* 45: 257–273.
Mack, R.N., Simberloff, D., Lonsdale, W.M. et al. (2000). Biotic invasions: causes, epidemiology, global consequences, and control. *Ecological Applications* 10: 689–710.
Mahmood, A.H., Florentine, S.K., Fernando, N. et al. (2016). Investigations into the effects of elevated carbon dioxide and drought on the growth and physiology of carpet weed (*Galenia pubescens* Eckl. and Zeyh.). In: *Proceedings of the 20th Australasian Weeds Conference. Perth, WA* (eds. R. Randal, S. Lloyd and C. Borger), 347–350. Perth: Weed Society of Western Australia.
Martin, T.G., Murphy, H., Liedloff, A. et al. (2015). Buffel grass and climate change: a framework for projecting invasive species distributions when data are scarce. *Biological Invasions 17*: 3197–3210.

Martínez-Cabrera, H.I., Schlichting, C.D., Silander, J.A., and Jones, C.S. (2012). Low levels of climate niche conservatism may explain clade diversity patterns in the South African genus *Pelargonium* (Geraniaceae). *American Journal of Botany* 99: 954–960.

Maruta, E. (1983). Growth and survival of current-year seedlings of *Polygonum cuspidatum* at the upper distribution limit on Mt. Fuji. *Oecologia* 60: 316–320.

Matzrafi, M. (2019). Climate change exacerbates pest damage through reduced pesticide efficacy. *Pest Management Science* 75: 9–13.

Matzrafi, M., Seiwert, B., Reemtsma, T. et al. (2016). Climate change increases the risk of herbicide-resistant weeds due to enhanced detoxification. *Planta* 244: 1217–1227.

Maxwell, S.L., Butt, N., Maron, M. et al. (2019). Conservation implications of ecological responses to extreme weather and climate events. *Diversity and Distributions* 25: 613–625.

McDonald, A., Riha, S., DiTommaso, A., and Degaetano, A. (2009). Climate change and geography of weed damage: analysis of US maize systems suggests the potential for significant range transformation. *Agriculture, Ecosystems and Environment* 130: 131–140.

Møller, A.P., Rubolini, D., and Lehikoinen, E. (2008). Populations of migratory bird species that did not show a phenological response to climate change are declining. *Proceedings of the National Academy of Sciences of the United States of America* 105: 16195–16200.

Mueller, T.C. and Steckel, L.E. (2019). Dicamba volatility in humidomes as affected by temperature and herbicide treatment. *Weed Technology* 33: 541–546.

Naidu, V.S.G.R. and Murthy, T.G.K. (2014). Crop-weed interactions under climate change. *Indian Journal of Weed Science* 46: 61–65.

Nguyen, T., Bajwa, A.A., Navie, S. et al. (2017). Parthenium weed (*Parthenium hysterophorus* L.) and climate change: the effect of CO_2 concentration, temperature, and water deficit on growth and reproduction of two biotypes. *Environmental Science and Pollution Research* 24: 10727–10739.

Novák, R., Dancza, I., Szentey, L., and Karamán, J. (2009). Arable weeds of Hungary. *Fifth National Weed Survey* (2007–2008). Ministry of Agriculture and Rural Development, Budapest.

Olson, B.L., Al-Khatib, K., Stahlman, P., and Isakson, P.J. (2000). Efficacy and metabolism of MON 37500 in *Triticum aestivum* and weedy grass species as affected by temperature and soil moisture. *Weed Science* 48: 541–548.

Ooi, M.K.J., Auld, T.D., and Denham, A.J. (2009). Climate change and bet-hedging: interactions between increased soil temperatures and seed bank persistence. *Global Change Biology* 15: 2375–2386.

Paterson, A.H., Kong, W., Johnston, R.M. et al. (2020). The evolution of an invasive plant, *Sorghum halepense* L. ('Johnsongrass'). *Frontiers in Genetics* 11: 317. https://doi.org/10.3389/fgene.2020.00317.

Patterson, D.T., Westbrook, J.K., Joyce, R.J. et al. (1999). Weeds, insects and diseases. *Climate Change* 43: 711–727.

Peters, K., Breitsameter, L., and Gerowitt, B. (2014). Impact of climate change on weeds in agriculture: a review. *Agronomy for Sustainable Development* 34: 707–721.

Petitpierre, B., Kueffer, C., Broennimann, O. et al. (2012). Climatic niche shifts are rare among terrestrial plant invaders. *Science* 335: 1344–1348.

Petitpierre, B., MacDougall, K., Seipel, T. et al. (2015). Will climate change increase the risk of plant invasions into mountains? *Ecological Applications* 26: 530–544.

Potvin, C. (1986). Biomass allocation and phonological differences among southern and northern populations of the C_4 grass *Echinochloa crus-galli*. *Journal of Ecology* 74: 915–923.

Pyšek, P. and Richardson, D.M. (2007). Traits associated with invasiveness in alien plants: where do we stand? In: *Biological Invasions* (ed. W. Nentwig), 97–125. Heidelberg, Germany: Springer.

Pyšek, P., Jarošík, V., Müllerová, J. et al. (2008). Comparing the rate of invasion by *Heracleum mantegazzianum* at continental, regional and local scales. *Diversity and Distributions* 14: 355–363.

Ramesh, K., Matloob, A., Aslam, F. et al. (2017). Weeds in a changing climate: vulnerabilities, consequences, and implications for future weed management. *Frontiers in Plant Science* 8: 95.

Reeves, M.C., Manning, M.E., DiBenedetto, J.P. et al. (2018). Effects of climate change on rangeland vegetation in the northern Rockies. In: *Climate Change and Rocky Mountain Ecosystems* (eds. J.E. Halofsky and D.L. Peterson), 97–114. Cham: Springer.

Rejmánek, M., Richardson, D.M., Higgins, S.I. et al. (2005). Ecology of invasive plants: state of the art. In: *Invasive Alien Species: A New Synthesis* (eds. H.A. Mooney, R.N. Mack, J.A. McNeely, et al.), 104–161. Washington, DC, USA: Island Press.

Richardson, D.M., Pyšek, P., Rejmánek, M. et al. (2000). Naturalization and invasion of alien plants: concepts and definitions. *Diversity and Distributions* 6: 93–107.

Richardson, A.D., Keenan, T.F., Migliavacca, M. et al. (2013). Climate change, phenology, and phenological control of vegetation feedbacks to the climate system. *Agricultural and Forest Meteorology* 169: 156–173.

Rouifed, S., Puijalon, S., Viricel, M., and Piola, F. (2011). Achene buoyancy and germinability of the terrestrial invasive *Fallopia* × *bohemica* in aquatic environment: a new vector of dispersion? *Écoscience* 18: 79–84.

Rout, M.E., Chrzanowski, T.H., Smith, W.K., and Gough, L. (2013). Ecological impacts of the invasive grass *Sorghum halepense* on native tallgrass prairie. *Biological Invasions* 15: 327–339.

Sakai, A.K., Allendorf, F.W., Holt, J.S. et al. (2001). The population biology of invasive species. *Annual Review of Ecology and Systematics* 32: 305–332.

Sasek, T.W. and Strain, B.R. (1990). Implications of atmospheric CO_2 enrichment and climatic change for the geographical distribution of two introduced vines in the USA. *Climate Change* 16: 31–51.

Sezen, U.U., Barney, J.N., Atwater, D.Z. et al. (2016). Multi-phase US spread and habitat switching of a post-Columbian invasive, *Sorghum halepense*. *PLoS One* 11 (10): e0164584. https://doi.org/10.1371/journal.pone.0164584.

Skelton, J.J., Ma, R., and Riechers, D.E. (2016). Waterhemp (*Amaranthus tuberculatus*) control under drought stress with 2,4-dichlorophenoxyacetic acid and glyphosate. *Weed Biology and Management* 16: 34–41.

Smith, C., van Klinken, R.D., Seabrook, L., and McAlpine, C. (2011). Estimating the influence of land management change on weed invasion potential using expert knowledge. *Diversity and Distributions* 1: 1–14.

Steffen, W., Rockström, J., Richardson, K. et al. (2018). Trajectories of the earth system in the Anthropocene. *Proceedings of the National Academy of Sciences of the United States of America* 115: 8252–8259.

Strelau, M., Janse van Rensburg, M., and Clements, D.R. (2018). Dose-response methodology for variant populations of Bohemian knotweed (*Reynoutria ×bohemica* Chrtek & Chrtková). *Canadian Journal of Plant Science* 98: 1380–1383.

Taylor, S., Kumar, L., and Reid, N. (2012). Impacts of climate change and land-use on the potential distribution of an invasive weed: a case study of *Lantana camara* in Australia. *Weed Research* 52: 391–401.

Thackeray, S.J., Henrys, P.A., Hemming, D. et al. (2016). Phenological sensitivity to climate across taxa and trophic levels. *Nature* 535: 241–245.

Thomas, A.G. and Leeson, J.Y. (2007). Tracking long-term changes in the arable weed flora of Canada. In: *Invasive Plants: Inventories, Strategies and Action*, Topics in Canadian Weed Science, 165 pp, vol. 5 (eds. D.R. Clements and S.J. Darbyshire), 43–69. Sainte Anne de Bellevue, Québec: Canadian Weed Science Society – Société Canadienne de Malherbologie.

Tsutsumi, M. (2011). Current and potential distribution of *Senecio madagascariensis* Poir. (fireweed), an invasive alien plant in Japan. *Grassland Science* 57: 150–157.

Usinowicz, J. and Levine, J.M. (2018). Species persistence under climate change: a geographical scale coexistence problem. *Ecology Letters* 21: 1589–1603.

Varanasi, A., Prasad, P.V., and Jugulam, M. (2016). Impact of climate change factors on weeds and herbicide efficacy. *Advances in Agronomy* 135: 107–146.

Vardien, W., Richardson, D.M., Foxcroft, L.C. et al. (2012). Invasion dynamics of *Lantana camara* L.(sensu lato) in South Africa. *South African Journal of Botany 81*: 81–94.

Walthall, C.L., Hatfield, J., Backlund, P. et al. (2012) Climate change and agriculture in the United States: effects and adaptation. *USDA Tech. Bull. 1935*. Washington, DC. p. 186.

Wan, J.Z., Wang, C.J., Tan, J.F., and Yu, F.H. (2017). Climatic niche divergence and habitat suitability of eight alien invasive weeds in China under climate change. *Ecology and Evolution* 7: 1541–1552.

Warren, D.L. and Seifert, S.N. (2011). Ecological niche modeling in Maxent: the importance of model complexity and the performance of model selection criteria. *Ecological Applications* 21: 335–342.

Warwick, S.I. (1990). Allozyme and life history variation in five northwardly colonizing North American weedy species. *Plant Systematics and Evolution* 169: 41–54.

Warwick, S.I. and Black, L.D. (1983). The biology of Canadian weeds. 61. *Sorghum halepense* (L.) Pers. *Canadian Journal of Plant Science* 63: 997–1114.

Warwick, S.I., Phillips, D., and Andrews, C. (1986). Rhizome depth: the critical factor in winter survival of *Sorghum halepense* (L.) Pers. (Johnsongrass). *Weed Research* 26: 381–387.

Waryszak, P., Lenz, T.I., Leishman, M.R., and Downey, P.O. (2018). Herbicide effectiveness in controlling invasive plants under elevated CO_2: sufficient evidence to rethink weeds management. *Journal of Environmental Management* 226: 400–407.

Weaver, S.E., Dirks, V.A., and Warwick, S.I. (1985). Variation and climatic adaptation in northern populations of *Datura stramonium*. *Canadian Journal of Botany* 63: 1303–1308.

Weiher, E., Van Der Werf, A., Thompson, K. et al. (1999). Challenging Theophrastus: a common core list of plant traits for functional ecology. *Journal of Vegetation Science* 10: 609–620.

Westwood, J.H., Charudattan, R., Duke, S.O. et al. (2018). Weed management in 2050: perspectives on the future of weed science. *Weed Science* 66: 275–285.

Whitney, K.D. and Gabler, C.A. (2008). Rapid evolution in introduced species, ‘invasive traits’ and recipient communities: challenges for predicting invasive potential. *Diversity and Distributions* 14: 569–580.

Wienhold, B.J., Vigil, M.F., Hendrickson, J.R., and Derner, J.D. (2018). Vulnerability of crops and croplands in the US Northern Plains to predicted climate change. *Climate Change* 146: 219–230.

Willis, C.G., Ruhfel, B., Primack, R. et al. (2010). Favorable climate change response explains nonnative species’ success in Thoreau’s woods. *PLoS One* 5 (1): e8878.

Wiréhn, L. (2018). Nordic agriculture under climate change: a systematic review of challenges, opportunities and adaptation strategies for crop production. *Land Use Policy* 77: 63–74.

Wolfe, D.W., Ziska, L., Petzoldt, C. et al. (2008). Projected change in climate thresholds in the Northeastern U.S.: implications for crops, pests, livestock, and farmers. *Mitigation and Adaptation Strategies for Global Change* 13: 555–575.

Wolkovich, E.M., Davies, T.J., Schaefer, H. et al. (2013). Temperature-dependent shifts in phenology contribute to the success of exotic species with climate change. *American Journal of Botany* 100: 1407–1421.

Yang, X., Chen, F., Lin, X. et al. (2015). Potential benefits of climate change for crop productivity in China. *Agricultural and Forest Meteorology* 208: 76–84.

Zettlemoyer, M.A., Schultheis, E.H., and Lau, J.A. (2019). Phenology in a warming world: differences between native and non-native plant species. *Ecology Letters* 22: 1253–1263.

Ziska, L.H. (2016). The role of climate change and increasing atmospheric carbon dioxide on weed management: herbicide efficacy. *Agriculture, Ecosystems and Environment* 231: 304–309.

Ziska, L.H. and McConnell, L.L. (2015). Climate change, carbon dioxide, and pest biology: monitor, mitigate, manage. *Journal of Agricultural and Food Chemistry* 64: 6–12.

Ziska, L.H. and Runion, G.B. (2007). Future weed, pest and disease problems for plants. In: *Agroecosystems in a Changing Climate* (eds. P.C.D. Newton, R.A. Carran, G.R. Edwards, et al.), 262–279. Boston, MA: CRC.

Ziska, L.H., Faulkner, S., and Lydon, J. (2004). Changes in biomass and root: shoot ratio of field-grown Canada thistle (*Cirsium arvense*), a noxious, invasive weed, with elevated CO_2: implications for control with glyphosate. *Weed Science* 52: 584–588.

Ziska, L.H., Blumenthal, D.M., and Franks, S.J. (2019). Understanding the nexus of rising CO_2, climate change, and evolution in weed biology. *Invasive Plant Science and Management* 13: 79–88.

13

Soil Microbial Effects on Weed Seedbank Persistence: Current Knowledge and Applications for Weed Management

Markus Wagner and Nadine Mitschunas

UK Centre for Ecology & Hydrology, Wallingford, UK

13.1 Introduction

Arable soil contains a seedbank typically numbering thousands of viable weed seeds m^{-2} in the top layer of soil, sometimes exceeding 100 000 seeds m^{-2} (Cavers and Benoit 1989). These seeds contain significant nutritional stores required for seedling establishment and represent an important carbon and energy source for soil microbes (Chee-Sanford 2008). Arable soil also contains large numbers of such microbes, with a single gram typically containing c. 10^9 bacteria (Girvan et al. 2003) and c. 10^2 m of fungal mycelia (Ritz and Young 2004).

Accordingly, studies investigating the effects of soil fungi on buried seed longevity that involve the simultaneous burial, both of fungicide-treated weed seeds to prevent fungal attack and of untreated seeds, have often found seed survival to be moderately (Lonsdale 1993; Gallandt et al. 2004) or even strongly (Fellows and Roeth 1992; van Mourik et al. 2005) boosted by fungicide application, although such effects are not universal (Frost et al. 2019).

Past reviews have discussed the evidence for microbial effects on buried weed seed persistence and the potential of using such microbe–seed interactions for managing weed seedbanks (Kremer 1993, 1998; Kennedy 1999; Wagner and Mitschunas 2008). In the last decade, more evidence has accumulated, further illustrating the important role of various soil microbial groups in affecting seedbank persistence and weed demography. Due to the deployment of more sophisticated approaches, including scanning electron microscopy, advanced statistical techniques, and especially molecular analytical techniques that have only become available more recently, much progress has been made particularly in our understanding of the underlying processes and mechanisms of microbe–seed interaction and of the role of various external factors. A few model systems are now rather well-understood, and the field is advancing towards practical applications in weed control.

In this chapter, we provide an updated overview of the research on interactions between soil-borne microbes and buried weed seeds, highlighting microbial attack mechanisms, seed defence mechanisms, and the roles of the abiotic environment and of biotic influences

Persistence Strategies of Weeds, First Edition. Edited by Mahesh K. Upadhyaya, David R. Clements, and Anil Shrestha.

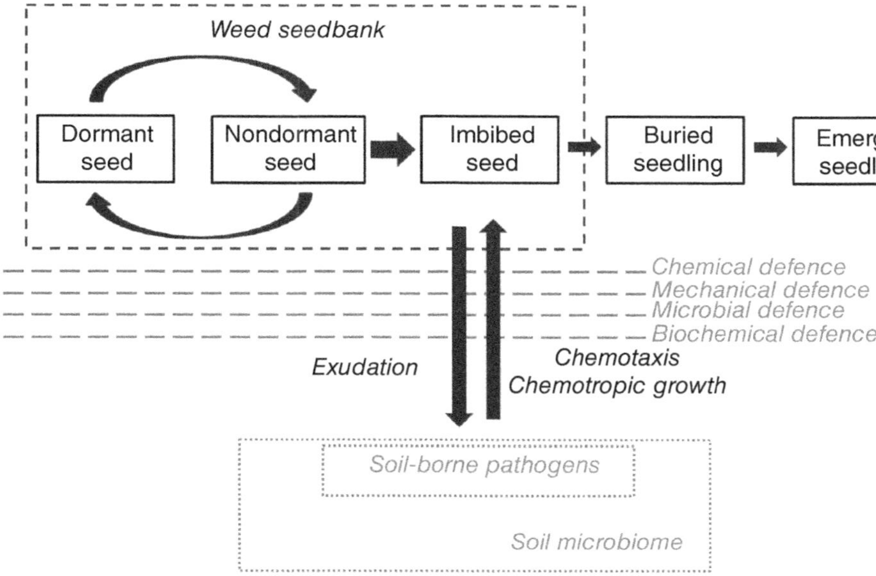

Figure 13.1 Conceptual model of the various components of the weed soil seedbank. The model illustrates seed state transitions within the seedbank and the process of seed germination. It also illustrates seed exudation and microbial movement in the form of chemotaxis and of chemotropic growth that promote the colonization of weed seeds by soil-borne pathogens and seed defence mechanisms with the potential to prevent such colonization.

in modulating these interactions. Figure 13.1 shows a conceptual model illustrating these various interactions and mechanisms. We also give an overview of the range of approaches for utilizing such interactions to manage weed seedbanks in agricultural ecosystems. We conclude this review by highlighting gaps in our knowledge, both in terms of our fundamental understanding and in terms of our ability to develop reliable applications for weed biocontrol.

13.1.1 Focus and Scope of This Chapter

The main focus of this chapter is on direct interactions between buried weed seeds and soil-borne microbes. Seed-borne microorganisms already present at the seed surface and/or internally within the seed before they enter the soil seedbank are discussed in as far as they can affect such interactions, e.g. by affecting seed colonization and attack by soil-borne microbes. Furthermore, a distinction is made between seed and seedling life stages as follows: (i) ungerminated seeds; (ii) seeds undergoing germination, the start of which being defined by the imbibition of water (Bewley et al. 2013) and the end of which being defined by protrusion of the radicle from the seed surface, marking the transition to the seedling stage (Kitajima and Fenner 2000); and (iii) seedlings. A further distinction is made between pathogenic microbes able to attack intact viable seeds and saprophytic microbes incapable of such attack but able to colonize and kill already-damaged seeds. Study systems made up of a given seed pathogen and the seeds of weed species are referred to as pathosystems.

While the focus of this chapter is on microbial effects on the weed seedbank, for two reasons, a strict separation between respective effects on the seed and seedling stages is not always possible. Firstly, some soil-borne pathogens that attack seeds also attack seedlings and more advanced life stages of plants (Thomas et al. 1998). Secondly, there are also practical limitations, e.g. in the context of seed burial experiments, in which a clear distinction between mortality through seed decay and mortality caused by fatal germination often is not possible, due to the rapid degradation of unemerged necrotic seedlings under field conditions. This rapid degradation often makes it impossible to unequivocally attribute the disappearance of a seed to either cause (Davis and Renner 2007). It is also important to note that fatal germination is not always simply a result of germination at a depth too great to permit successful seedling emergence but can also be brought about by pathogen attack, causing the seedling to die before it emerges (Davis and Renner 2007).

13.2 Mechanisms of Microbial Attack

Even before weed seeds enter the soil, they are usually already colonized by microorganisms, both externally (Mishra and Srivastava 1977) and internally (Kirkpatrick and Bazzaz 1979; Kremer 1987). These seed-borne microbial communities (Mishra and Srivastava 1977; Kirkpatrick and Bazzaz 1979; Kremer 1987) differ compositionally between weed species, in relation to both structural and/or physiological seed characteristics (Kremer 1987).

Furthermore, the member species of these seed-borne microbial assemblages are usually nonpathogenic or only weakly pathogenic (Kirkpatrick and Bazzaz 1979; Kremer 1986a; Pitty et al. 1987). Similarly, the majority of soil-borne microbes also lack pathogenicity, i.e. are unable to attack seed with intact seed defences (Pitty et al. 1987; de Luna et al. 2011). For example, de Luna et al. (2011) found that, out of 118 fungal isolates obtained from *Avena fatua* caryopses that had been buried for six months, only 18 (i.e. 15%) elicited fungal decay of caryopses in bioassays, and only six (i.e. 5%) elicited substantial decay. Similarly, Pitty et al. (1987) found that, in their bioassays, only a small number of fungal isolates from *Setaria faberi* caryopses recovered from natural soil seedbanks caused appreciable seed decay. Therefore, only a small number of soil-borne microbes appear to be actively pathogenic, i.e. able to infect intact non-germinating seeds. However, many saprophytic microbes living epiphytically on the outside of seeds might be able to secondarily infect and kill already-damaged seeds (Kremer et al. 1984).

Seed infection by soil pathogens was first investigated for crop plants, with particular focus on the role of water-soluble and volatile seed exudates, e.g. from aged or damaged seeds that have become ‘leaky’ or from germinating seeds. Such exudates can stimulate propagule germination and mycelial growth of seed-rotting fungi (Harman 1983; Harman and Stasz 1986) in the zone immediately surrounding the seed, which in the case of germinating seeds is also referred to as the spermosphere (Nelson 2004; Schiltz et al. 2015). Exudate release during germination typically occurs in three distinct peaks, with the first two associated with membrane damage during initial imbibition and the third occurring with radicle protrusion from the germinating seed (Nelson 2004). As a result, the zone around the emerging radicle represents a major site of exudate release (Schiltz et al. 2015).

An important role of such seed exudates in attracting soil microbes was subsequently also confirmed for weed species by Begonia and Kremer (1994), who demonstrated for velvetleaf (*Abutilon theophrasti*), that deleterious rhizobacteria were chemotactically attracted by seed-exuded sugars and amino acids, initiating bacterial migration towards pre-imbibed seeds across a distance of 2 cm within 24 hours.

For nutrient stores, pathogens have to get past the seed's defences, described in more detail in the following text. Applying the results of earlier research by other authors on crop seeds, Kremer et al. (1984) hypothesized that fungal seed pathogens might enter host seeds mainly through entry points where seed defences are less developed, e.g. in the case of pathogens attacking *A. theophrasti*, by penetrating the seed coat at the location of seed wall discontinuities in the palisade layer in the chalazal region of the seeds. More recently, Pollard (2018) concluded that in the case of fungal attack of ungerminated weed seeds, direct hyphal penetration assisted by cell wall-degrading fungal enzymes appears to be the most common attack strategy. Three fungal enzyme families, including xylanase, protease, and chitinase enzymes, have been implicated in fungal seed attack (Pollard 2018). While xylanases are likely directly involved in seed decay, proteases might help fungal pathogens against seed defence enzymes (Pollard 2018) as described in Section 13.5. With respect to fungal chitinase, it has been hypothesized that this enzyme might help pathogens attack beneficial seed-borne fungi or competing pathogens (Dalling et al. 2011) or increase the pathogen's own cell wall plasticity, thus facilitating infection (Pollard 2018).

Using scanning electron microscopy, hyphal penetration of the seed coat has been observed for a number of fungal species attacking the seeds of crop-parasitic weeds from the Orobanchaceae family. Mycelium of *Fusarium nygamai* entered the seeds of *Striga hermonthica* (Sauerborn et al. 1996a) by means of hyphal perforation of the seed coat and the same applied to *Fusarium oxysporum* f. sp. *orthoceras*, infecting the seeds of *Orobanche cumana* (Thomas et al. 1999), and to *Aspergillus alliaceus*, infecting the seeds of *Orobanche cernua* (Aybeke et al. 2014).

Similarly, *Pyrenophora semeniperda*, a seed pathogen of cool-season grasses (Beckstead et al. 2014, 2016), attacks the seeds of its hosts through hyphal penetration of seed covering structures. This pathogen has been studied in particular detail because of its effectiveness in attacking the seeds of the grass *Bromus tectorum*, an important exotic invasive weed both in agricultural and in natural ecosystems in North America (Upadhyaya et al. 1986). This ability initially made it a plausible candidate for being the causal agent for the observed mass die-off of near-monocultural stands of *B. tectorum* (Baughman and Meyer 2013). Also, *P. semeniperda* represents a good model pathogen for studying seed–pathogen interactions, because it is an obligate seed pathogen, and its reproduction involves the formation of macroscopic fruiting bodies called stromata that allow easy quantification of its prevalence in buried seed populations (Meyer et al. 2018). In addition to its ability to effectively penetrate host seed coverings, two more features are characteristic for its mode of attack, including (i) its apparent non-requirement of chemical signals from the host seed to initiate fungal growth (Meyer et al. 2018) and (ii) its relatively slow development inside the host seed (Finch-Boekweg et al. 2016). Because of these two additional features, it is well adapted to attacking the dormant fraction of the *B. tectorum* seedbank. On the other hand, it appears to be less adapted to attacking nondormant seeds, at least under conditions conducive to quick germination, through which the seeds of both *B. tectorum* (Beckstead et al. 2007) and those of other grasses (Beckstead et al. 2016) have been observed to frequently escape mortality, in what has been described a 'race for survival' (Beckstead et al. 2007). Regarding this 'race' metaphor, it is also interesting that in recent bioassays, metabolites produced by *P. semeniperda* were found to delay germination of *B. tectorum*, illustrating a potential strategy by this fungus to tilt the race in its favour (Lozano-Juste et al. 2019). Due to its specific mode of attack, which is generally suitable for utilizing the nutrient reserves of dormant seeds, *P. semeniperda* had eventually been ruled out as the cause of *B. tectorum* mass die-off events, as during such events, the dormant fraction of the soil seedbank remains unaffected (Baughman and Meyer 2013).

It appears that such die-off events are primarily caused by *Fusarium tricinctum*, another seed pathogen with a different attack pathway (Meyer et al. 2014, 2018). *F. tricinctum* appears to be unable to penetrate the seed coverings of *Bromus* seeds but can cause considerable mortality among the nondormant fraction of the seedbank (Meyer et al. 2018). Time lapse scanning electron microscopy has revealed that its hyphae directionally grow towards the seed's floret attachment scar, which is also the point of radicle emergence during germination, infecting the seed through this entry point (Franke et al. 2014). The most likely explanation according to Meyer et al. (2018) is chemotropic growth in response to exudate release from imbibing *Bromus* seeds, analogous to the already-mentioned exudate-mediated chemotactic migration of deleterious rhizobacteria towards imbibing *Abutilon* seeds (Begonia and Kremer 1994).

In accordance with these findings from North America, an isolate of *F. tricinctum* from root tissue of *Lespedeza davurica*, a legume species native to semiarid grasslands in China, also killed only nondormant but not dormant seeds of either *Lespedeza* or of a co-occurring grass species in inoculation trials under laboratory conditions (Chen et al. 2018). Microscopic observations found the radicles of nondormant seeds killed by *F. tricinctum* in this laboratory trial to be stunted and rotten (Chen et al. 2018). However, a second field-based trial indicated that increased mortality only occurred at the pre-emergence seedling stage rather than during germination (Chen et al. 2018), possibly due to infection occurring later during the germination process and/or proceeding at a slower rate. It could be argued that under such conditions, *F. tricinctum* might act like a seedling pathogen rather than a seed pathogen, leading to fatal germination.

Several other studies have found that some fungi (Kirkpatrick and Bazzaz 1979) and bacteria (Gealy et al. 1996; Miché et al. 2000; Ahonsi et al. 2002; Widmer and Guermache 2004; Banowetz et al. 2008) appear to suppress germination, and some studies involving bacteria showed that cell-free filtrates are sufficient and that microbial metabolites are responsible (Gealy et al. 1996; Widmer and Guermache 2004; Banowetz et al. 2008). In Gealy et al.'s (1996) study, germination of *B. tectorum* was reduced both by cell-containing whole culture broth and by cell-free filtrate, but not by the presence of washed bacterial cells, indicating that phytotoxic metabolites released by the bacteria into the environment are responsible for the observed effect. In one study, the compound deemed responsible was dubbed a 'Germination-Arrest Factor' and described as a natural-occurring herbicide (Banowetz et al. 2008). However, none of these studies appeared to have investigated viability of seeds after the assumed suppression of germination, and in some studies (Gealy et al. 1996; Miché et al. 2000), phytotoxic effects of microbial metabolites manifested themselves after radicle protrusion, with radicles being prevented from further elongation. It

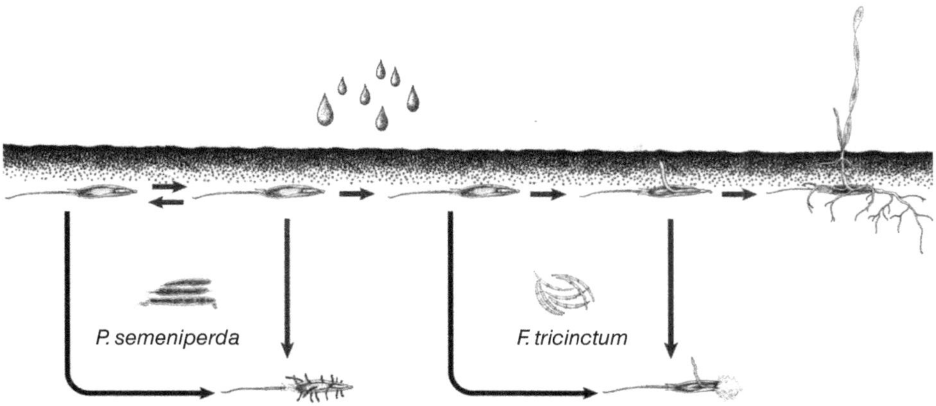

Figure 13.2 Diagram indicating the main niche specialization of *Pyrenophora semeniperda* and of *Fusarium tricinctum*, two seed pathogens well-adapted to attacking the seeds of *Bromus tectorum* (Meyer et al. 2018). The five seeds and seedling stages shown in the diagram are equivalent to those indicated in Figure 13.1. *Source:* Modified from Meyer et al. 2018.

thus appears plausible that in at least some of these studies, the seeds assayed had been killed, i.e. the pathogens used in these studies, as *F. tricinctum* observed in Chen et al.'s (2018) study, primarily attacked nondormant germinating seeds during early-stage germination, in some instances before seedling radicles became visible.

Based on their research with *P. semeniperda* and *F. tricinctum*, Meyer et al. (2018) have concluded that different mechanisms of seed attack as outlined earlier represent a form of niche specialization, with different pathogens either predominantly utilizing dormant seeds within the soil seedbank or nondormant germinating seeds. Figure 13.2 illustrates this niche specialization. However, it is important to note that this specialization is not absolute, as both *P. semeniperda* and *F. tricinctum* can cause some mortality in the other respective group of seeds (Meyer et al. 2018).

Finally, it may also be worth mentioning here that soil microbes can cause fatal germination not only as a result of pathogen infection but also through nonpathogenic microbial stimulation of germination, e.g. via the degradation of seed coverings or via exudation of stimulatory substances, under conditions in which chances of successful seedling establishment are low or non-existent (van Leeuwen 1981). This would, for example, be the case if germination took place at too great a depth for successful plant emergence (van Leeuwen 1981) or in the case of crop-parasitic weeds, in the absence of suitable hosts. For example, Berner et al. (1999) have shown in laboratory experiments that ethylene-producing soil bacteria can stimulate the germination of *S. hermonthica*.

13.3 Abiotic Environmental Factors

Laboratory bioassays are usually carried out under specific sets of abiotic environmental conditions, often involving constant temperature and moisture regimes that facilitate adequate growth and action of the microbial pathogen investigated. Furthermore, in the case of bioassays for pathogen–seed interactions, other biota such as other microbial taxa with the potential to affect the studied pathosystem under field conditions are generally absent. As a result, pathogen–seed interactions occurring under such 'optimal' laboratory conditions are likely more pronounced as would be expected under field conditions. For example, as mentioned earlier, in one study (Chen et al 2018), the killing action of *F. tricinctum* appeared to have been delayed when buried grass seeds were attacked under field conditions, compared with a laboratory assay replicating the same pathosystem in Petri dishes. Discrepancies between laboratory assays and field experiments were also found by other authors. Beckstead et al. (2014) found a much wider potential host range of *P. semeniperda* in laboratory assays, compared with the actual host range they subsequently were able to confirm under field conditions. Similarly, Kennedy et al. (1991) found that, while the application of actinobacteria strains was effective in controlling *B. tectorum* in bioassays and growth chamber studies, outcomes were much less consistent in field experiments.

Differences between laboratory and field studies may be commonplace, as environmental conditions are usually much more spatiotemporally variable in the field than in simplified laboratory environments in which they are generally held constant. Also, seed decay can be expected to be higher under laboratory conditions, where high pathogen inoculum levels are used and weed seeds are surface-sterilized prior to an experiment, than under field conditions. Under the latter, exposure of seeds to the pathogen is effectively lower, and as described in the following text, seeds may also be more resistant due to the antagonistic effects of endogenous microflora towards soil-borne pathogens (Fuerst et al. 2011). Therefore, to more fully understand pathogen–seed interactions under field conditions and to develop reliable weed biocontrol applications, an understanding both of abiotic environmental influences and of biotic interactions affecting a given pathosystem is imperative.

A review of the effects of fungal pathogens on seedbank persistence across ecosystems found that only one abiotic environmental factor, soil moisture, had been explored in more detail (Wagner and Mitschunas 2008). Essentially, this has not changed very much, as our knowledge on the role of other soil environmental factors such as soil pH is still highly deficient. However, due to more recent research on the *B. tectorum–P. semeniperda* pathosystem (Finch et al. 2013; Allen et al. 2018), whose results are described in the following text, our understanding of how soil moisture and its fluctuations can affect the outcome of pathogen–seed interactions has further increased. In the following, we thus focus on discussing the influence of the soil moisture regime on pathogen–seed interactions.

Such an influence has been variously described in the literature. For example, a seed burial study in which seeds of the weed grass *Sorghum bicolor* were buried in two consecutive years with or without fungicide treatment found that added seed mortality due to fungal attack in the no-fungicide treatment was higher in the wetter of the two years (Fellows and Roeth 1992). The results of other studies in non-arable ecosystems, in which seed burial was carried out simultaneously at locations differing in soil moisture regime, also support the notion of buried weed seed mortality being affected by soil moisture (Wagner and Mitschunas 2008). One causal factor for the link between soil moisture and pathogen-related seed mortality may be the role of seed exudate leakage in triggering pathogen spore germination and chemotactic migration and in the case of fungal pathogens also chemotrophic hyphal growth towards seeds releasing exudates. As reviewed in Wagner and

Mitschunas (2008), these processes are dependent on a soil moisture requirement, and seed infection rates can thus be expected to be moisture dependent.

Furthermore, even after a seed has been infected with fungal pathogens, pathogen growth is dependent on moisture, but, crucially, generally less so than seed germination. It has been known for some time that buried crop seeds infected with fungal pathogens through experimental inoculation can be killed by pathogen action occurring at water potentials too low for seed germination (Griffin 1966). More recent research has shown that this also applies to buried weed seeds. One study found that the decay of dormant *B. tectorum* seeds infected by *P. semeniperda* was slowed down at lower water potentials but had nonetheless continued after a transfer to potentials as low as −2 MPa, as indicated by the very rapid development of *Pyrenophora* stromata after subsequent transfer of seeds back to a pure water environment (Finch et al. 2013). In fact, a hydrothermal time model developed by Barth et al. (2015) predicts that under certain conditions, conidial germination of *P. semeniperda* might occur at water potentials as low as −13 MPa, and mycelial growth at water potentials as low as −11 MPa. In contrast, in a study by Meyer and Allen (2009), only 50% of the seeds of *B. tectorum* were typically able to germinate at water potentials of around −1.2 MPa, and seed germination at such water potentials tends to be much slowed down (Thill et al. 1979). In fact, this ability of fungal seed pathogens to continue growing at much lower water potentials than those required for weed seed germination appears to be the case more widely, as shown in Table 13.1, which lists baseline water potentials from the literature and also for other weed and pathogen species discussed in this chapter.

In a study by Kiewnick (1964), experimental burial of *A. fatua* caryopses inoculated with seed-rotting fungi produced the highest levels of fungal-induced seed mortality at soil moisture levels of 50% of water capacity, equating to a water potential of −1.5 MPa, which is insufficient for seed germination (Martinson et al. 2007; Guillemin et al. 2013) but demonstrably allows decay of infected seeds to proceed. At 80% water capacity, seed mortality due to anoxia was more important than mortality through pathogen activity, and at low moisture levels of 20% water capacity, most seeds survived throughout the experiment (Kiewnick 1964), indicating that these moisture levels were too low even for pathogen growth. For nondormant seeds of *B. tectorum*, the previously mentioned experiment by Finch et al. (2013) showed that at reduced water potentials, the 'race for survival' between germinating seeds and attacking pathogens, as described earlier, was markedly tilted in favour of the pathogen. This was because, as seed germination appeared to have been slowed down more strongly than pathogen growth, more seeds ended up being killed before they could complete germination.

In a second experiment, Finch et al. (2013) also demonstrated that longer periods of incubation at reduced water potential lead to an increase in mortality after subsequent transfer of infected seeds to a pure-water environment, as their germination had been prevented for longer periods, while allowing continued pathogen growth. At the same time, for a given incubation water potential, higher incubation temperatures resulted in a marked increase in seed mortality both in dormant seeds as well as in nondormant seeds, as levels of fungal attack were enhanced, indicating a role also of incubation temperature. Furthermore, experimental results suggested that fungal-induced seed mortality of initially nondormant seeds can be further boosted by induction of secondary dormancy,

Table 13.1 Base water potentials for weed seed germination and for mycelial growth and conidial germination of fungal pathogens.

Species	Parameter	Base water potential (MPa)	Sources	Comment
Weeds				
Avena fatua	Seed germination	−1.02	Guillemin et al. (2013)	
Bromus tectorum	Seed germination	−1.27 to −1.11	Meyer and Allen (2009)	Range of model estimates across six seed populations for a 10/20 °C diurnal regime
Chenopodium album	Seed germination	−0.8	Guillemin et al. (2013)	
Digitaria sanguinalis	Seed germination	−0.45	Gardarin et al. (2010a)	
Pathogens				
Fusarium oxysporum	Mycelial growth	−6.5	Brownell and Schneider (1985)	Growth reduction of 50% estimated to occur at −1.6 MPa
Pyrenophora semeniperda	Conidial germination	−13.26	Barth et al. (2015)	
	Mycelial growth	−11.4 to −7.8	Barth et al. (2015)	Lower estimate for a suboptimal-temperature baseline and higher estimate for a supraoptimal-temperature baseline

Values provided for seed germination and conidial germination are model estimates of $\Psi b(50)$, i.e. the base water potential for germination of 50% of nondormant seeds and of conidia. Values provided for mycelial growth are model estimates for Ψb, for 100% growth reduction, i.e. the base water potential at which mycelial growth will cease.

which can take place during incubation at low temperatures and water potentials, tilting the ‘race for survival’ even further in favour of the pathogen as germination is further delayed.

More recent research has led to a revised understanding of the niche of *P. semeniperda* in relation to its relative capability to attack nondormant seeds of *B. tectorum* under field conditions and in relation to the much more variable soil moisture levels that are prevalent under such conditions. Results by Allen et al. (2018) indicated that inoculation with *P. semeniperda* increased the density of killed seeds in the soil seedbank much more than would have been the case if only dormant seeds would have been attacked, strongly implying that nondormant seeds were also killed by the pathogen. In an associated laboratory experiment, after brief initial imbibition in water, seeds were first transferred to very low water potentials between −4 and −150 MPa for dehydration, for periods varying between 1 and 21 days length, before being rehydrated again at the end of the experiment (Allen et al. 2018). With dehydration at −4 MPa, seed mortality was positively correlated with the length of the dehydration period, suggesting cumulative pathogen action, whereas at −150 MPa, no such trend was found. These results show that in spite of previous assumptions of *P. semeniperda* being much better suited to attack dormant seeds of *B. tectorum*, the pathogen can successfully attack nondormant seeds under fluctuating moisture conditions that allow initial infection of the seeds, while subsequently preventing immediate seed germination. Similarly, the attack of nondormant *B. tectorum* seeds by *F. tricinctum* also appears to be modulated by soil moisture. Meyer et al. (2014) found that in a glasshouse experiment with soil moisture maintained at or near field capacity, very rapid seed germination helped such seeds to frequently avoid fatal attack by the pathogen. In contrast, in associated pathogenicity trials in which, following inoculation, seeds were incubated under water stress that suppressed germination, irrespective of fungal isolate used, seed mortality was much higher in a high water stress treatment than in a low water stress treatment (Meyer et al. 2014).

Soil moisture effects may also contribute to the observed effects of placement of *B. tectorum* seeds within the soil profile on the degree of suppression of seedling emergence due to experimental inoculation with *P. semeniperda*, which was most pronounced for seeds buried within the soil profile and less pronounced for seeds placed at the soil surface or within the litter layer (Ehlert et al. 2014). Similarly, an observed positive relationship between the amount of plant litter in a location and the amount of *Pyrenophora*-induced *B. tectorum* seed mortality in the underlying soil seedbank has been hypothesized to be at least partly an indirect effect of litter affecting the soil moisture regime (Beckstead et al. 2012).

13.4 Biotic Interactions

Seed pathogen dynamics and seed infection processes are also affected by biotic interactions, with consequences for the regulation of pathogen–seed interactions in the seedbank.

One biotic interaction not involving other biota is the interaction of uninfected seeds with already infected seeds. The probability of seed-to-seed contamination by a pathogen can be expected to be a function of its prevalence within the soil environment, which in turn should be a function of the population density of buried seeds susceptible to infection. Accordingly, a positive relationship between buried-seed density and pathogen-induced seed mortality was found in field experiments involving the simultaneous burial both of fungicide-treated seeds and of untreated seeds of the tropical weed *S. hermonthica* at a range of seed densities (van Mourik et al. 2005). Moreover, if seed densities of a preferred host plant species are high, pathogen spillover onto the seeds of other, less susceptible, species may result (Beckstead et al. 2010).

Third-party organisms that regulate interactions between pathogens and buried seeds also usually do so by affecting colonization and infection processes. One group whose action can promote infection are seed-feeding insects. Goeden and Ricker (1973) first discussed this possibility, suggesting that buried seed mortality of *Tribulus terrestris* in their study may have been increased by a seed-infesting weevil that facilitated microbial attack of seeds by creating a portal-of-entry for soil microorganisms. The existence of such a mechanism was subsequently experimentally demonstrated for the seed-feeding true bug *Niesthrea louisianica* (Kremer and Spencer 1989a,b). Seed feeding by this bug led to a much higher proportion of mature seeds of *A. theophrasti* being infected by fungi

(Kremer and Spencer 1989a) and to an increased seed mortality during experimental seed burial (Kremer and Spencer 1989b).

Another group of organisms that can affect seed colonization by soil-borne pathogens are nonpathogenic microbes. As shown by several studies, even prior to dispersal from the mother plant, the seeds of many weeds are already internally colonized by nonpathogenic bacteria (Kremer et al. 1984; Kremer 1987) and fungi (Kremer 1986a), and the seeds of particular species are characterized by regular association with specific microbial assemblages (Kremer et al. 1984; Kremer 1986a, 1987). Moreover, seed-borne surface-colonizing microbes can inhibit seed colonization by pathogenic microbes (Kremer 1986a, 1987). The weed species specificity of seed-borne microbial assemblages, along with their antagonistic effects on seed colonization by pathogens, suggests that these microbial assemblages may have been selected by plants specifically to inhibit pathogens (Links et al. 2014), although selection for other benefits, e.g. microbial stimulation of early seedling growth, may also play a role. Based on the selection argument, it can be argued that seed-borne microbial assemblages represent a seed defence mechanism against pathogen attack.

13.5 Seed Defences

The seeds of many weed species are remarkably resistant to microbial decay, due to mechanical, chemical, and biochemical defence features preventing infection by seed-decaying microbes (Halloin 1983). Naturally, the defence features tend to be mainly localized within the seed coat, or its functional equivalent in instances where the ovary wall has become part of the dispersal unit (Mohamed-Yasseen et al. 1994). The intactness of this external seed boundary is crucial not only for preventing pathogens from entering the seed but also for preventing exudate leakage and an associated increase in nutrient availability to microbes in the immediate environment of the seed (Halloin 1983). In the following, mechanical, chemical, and biochemical seed defence mechanisms are described in more detail, along with the already-mentioned microbial seed defence.

Another mechanism considered as a seed defence by some authors is the rapid germination (Dalling et al. 2011; Davis et al. 2016). As illustrated by the example of nondormant *B. tectorum* seeds being attacked by *P. semeniperda* (Beckstead et al. 2007), infected seeds can avoid being killed if capable of rapid germination, thus outrunning the pathogen. The same researchers also showed that interspecific differences in plant susceptibility to attack by this pathogen also depended on germination speed (Beckstead et al. 2010). While the alternative view could be taken that rapid germination, even if evolved as a result of pathogen pressure, represents an avoidance strategy rather than an actual defence mechanism, the capacity of a species for rapid germination might affect how much it would benefit from investment in other seed defences. For this reason, rapid germination is also of relevance for the concept of seed defence syndromes (Dalling et al. 2011; Davis et al. 2016), as discussed below.

An understanding of weed seed resistance against microbial attack is required for utilizing pathogen–seed interactions for weed biocontrol. In the following, each mechanism is discussed individually while also highlighting interrelationships between different types of defence.

13.5.1 Mechanical Defence

The seed coat provides the seed with protection against invasion by deleterious microorganisms (Mohamed-Yasseen et al. 1994) through a combination of serving as a mechanical barrier against invasion and being a seat of chemical and biochemical defences that also inhibit seed colonization and seed coat penetration by pathogens. One important factor affecting the function of the seed coat as a mechanical barrier is its thickness. In a study by Gardarin et al. (2010b) involving 18 weed species, two thirds of interspecies variation in buried-seed mortality could be explained by seed coat thickness alone, and this relationship held up even after phylogenetic relatedness among species was accounted for. One obvious explanation is that thicker seed coats impart better protection against microorganism invasion, but of course, they also shelter seeds better against other adverse environmental conditions such as, e.g. moisture variation (Gardarin et al. 2010b). Also, in this particular study, as seed mortality rates used were the total of both pre-germination seed mortality and mortality through fatal germination, the possibility that lower rates of germination in species with thicker seed coat may also have played a role cannot be discounted. This also applies to a similar finding by Davis et al. (2008) involving five weed species. Thus, while both studies suggest that seed coat thickness might play a role in the defence of seeds against microorganisms, cast-iron evidence for a causal link may yet have to be obtained. However, it is clear that the very existence of an intact seed coat represents an important physical barrier against invasion. This has also been demonstrated in another seed burial experiment by Davis et al. (2008), in which the seeds coats of the seeds of six weed species were pierced or sliced prior to burial, resulting in buried seed mortality markedly higher than in experimental controls consisting of similarly buried undamaged seeds.

13.5.2 Chemical Defence

Chemical seed defence is based on the accumulation of secondary chemical compounds with antimicrobial properties within the seed, primarily within the seed coat (Kremer 1986b; Paszkowski and Kremer 1988; Warr et al. 1992; Mohamed-Yasseen et al. 1994). Many such compounds are phenolic, including *ortho*-dihydroxyphenols (Hendry et al. 1994; Davis et al. 2008), tannins (Kremer et al. 1984; Fellows and Roeth 1992), and flavonoids (Paszkowski and Kremer 1988), although some might be non-phenolic (e.g. Brambl and Gade 1985).

A direct link between seed chemical defence and seed survival in the soil has, for example, been established in a set of 81 woodland plant species in which seed *ortho*-dihydroxyphenol concentration was a good predictor of seed survival (Hendry et al. 1994). On the other hand, in the seed burial experiment by Davis et al. (2008), seed concentrations of *ortho*-dihydroxyphenols and total phenols in six weed species were negatively correlated with seedbank persistence. It appears likely that the contrasting results between these two studies are due to the different habitats of the two sets of species tested. Woodlands tend to be temporally stable ecosystems, and the soil seedbanks of many woodland species are short-lived as a result. In contrast, agricultural habitats are characterized by a much greater frequency of disturbance. Agricultural weeds have adapted to this through generally high seedbank persistence (Bekker et al. 1998). As discussed in the following text in the context of seed defence syndromes, it has been suggested that a strong mechanical defence might be a better guarantor than chemical defence for achieving long-term seedbank persistence and that chemical defence might thus only be a secondary line of defence in weeds (Dalling et al. 2011; Davis et al. 2016).

13.5.3 Biochemical Defence

Seeds also possess biochemical defence mechanisms mediated by the activity of enzymes, also located mainly in the seed coat. Three types of enzymes that have been described as being involved in biochemical seed defence are chitinase, polyphenol oxidase, and peroxidase enzymes. Chitinases can act as fungal growth inhibitors, representing a direct defence against fungal pathogens. Another suggested role is the enhancement of seed relationships with beneficial microbes that are involved in the microbial defence of seeds as discussed in the following text (Pollard 2018). Polyphenol oxidase and peroxidase enzymes appear to play an important role in mobilizing and reinforcing chemical and mechanical defences (Fuerst et al. 2014; Pollard 2018). This includes production of compounds with antimicrobial activity, production of reactive oxygen species, lignification of cell walls to reinforce mechanical defences, and, in the case of polyphenol oxidase, reduction of the bioavailability of microbial substrates (Fuerst et al. 2018).

All three types of enzymes have been explored as part of the biochemical defence of *A. fatua* seeds (Fuerst et al. 2018), with polyphenol oxidase having received the most attention (Anderson et al. 2010; Fuerst et al. 2011, 2014). In the seeds of this grass, polyphenol oxidase is not only located in the caryopsis itself but also in the nonliving hull tissue surrounding the caryopsis, lemma, and palea (Fuerst et al. 2014). The incubation of dormant *A. fatua* seeds with seed-decaying *Fusarium avenaceum* induces polyphenol oxidase activity, and such activated polyphenol oxidase can be readily mobilized into the seed environment (Anderson et al. 2010; Fuerst et al. 2011, 2014), along with peroxidase and chitinase enzymes (Fuerst et al. 2018). Polyphenol oxidase activation in *A. fatua* seeds appears to be highly specific, depending on the attacking pathogen. The strength of its induction has been found to vary among different *Fusarium* isolates, and incubation with an isolate of the genus *Pythium*, a seedling pathogen, even had a negative effect on polyphenol oxidase activity (Fuerst et al. 2011).

13.5.4 Microbial Defence

Microbial defence in weed seeds was first discussed by Kremer et al. (1984) and Kremer (1987) who reported a number of seed-borne bacteria isolated from *A. theophrasti* and from other weed species as antagonistic both to externally seed-borne fungi and to soil-borne fungi. Its actual operation has been demonstrated in assays involving the experimental placement of both unsterilized and surface-sterilized seeds of *A. theophrasti* on the surface of field-collected soil (Kremer 1986a). In these assays, fungal colonization by soil-borne fungi readily occurred on surface-sterilized seeds, but not on unsterilized seeds on which the assemblage of seed-borne fungi persisted, seemingly preventing colonization by soil-borne fungi. Regarding the mechanisms by which microbial defence operates, it has been suggested that this could be either through pre-emption of seed surface space or of seed exudates, or through direct anti-microbial interaction, e.g. via production and release of metabolites inhibiting pathogen growth (Dalling et al. 2011; Schiltz et al. 2015). As pointed out by Nelson (2018), interactions of seed-associated microbes with soil microbiota may be most pronounced during seed germination, when both endophytic and exophytic seed-borne microbes and soil-borne microbes in the spermosphere of the germinating seed are activated through seed imbibition and exudate release.

13.5.5 Complementarity of Seed Defences

Different seed defences can be complementary. For example, once the mechanical barrier against invasion by seed-decaying microorganisms has been breached, other defence mechanisms can provide a secondary line of defence. Davis et al. (2008) found for *Kochia scoparia* that relatively high buried-seed mortality occurred due to more extensive damage from slicing of the seed coat, whereas much lower mortality was observed in response to point-localized damage of the seed coat from piercing. Mortality in response to slicing was negatively correlated in a group of six weed species with seed concentrations of *ortho*-dihydroxyphenols. This indicates that at sufficiently high concentration of chemical defence compounds, such compounds can provide 'back-up' protection against microbial seed decay after a limited breach of the mechanical defence barrier. Similarly, investment in chemical defence can be higher locally within the seed to compensate for functional constraints of mechanical defence as required to ensure orderly germination. This has been shown to be the case for *A. theophrasti*, for which tannin concentration is highest in the chalazal area of the seeds (Kremer et al. 1984), at the prospective point of radicle emergence. Like chemical defence, microbial defence may also act as an additional line of defence after breaching of the mechanical defence. This has been demonstrated by experiments in which scarified maize (*Zea mays*) kernels inoculated with *Pseudomonas* bacteria were more strongly protected against pathogen invasion than scarified but otherwise untreated kernels (Anderson and Liberta 1986).

13.5.6 Seed Defence Syndromes

Davis et al. (2008) observed an inverse relationship across the six weed species tested between concentrations of *ortho*-dihydroxyphenols and seed coat thickness. They also found that seed half-life in the soil was positively related to seed coat thickness and negatively related to seed *ortho*-dihydroxyphenols. This led them to suggest that weed species with short-lived seedbanks may invest more in chemical defence, while species with long-lived seedbanks may invest more in mechanical defence. This suggestion implicitly acknowledges that there may be trade-offs involved between investing in different types of defence. Subsequently, Dalling et al. (2011) implicated concurrent selection on seed dormancy and on antimicrobial defences as a reason for the evolution of specific seed defence syndromes. Accordingly, three main defence syndromes can be distinguished as follows: (i) species with physical seed dormancy, relying mainly on physical defences provided by the hard seed coat that is a prerequisite of physical seed dormancy; (ii) species with physiological seed dormancy, deploying a continuum of physical and chemical defences, depending on pathogen pressure and duration of seed persistence; and (iii) species with very short persistence that show no seed dormancy, i.e. seeds tend to germinate as soon as conditions for emergence are favourable, and lack any obvious chemical and physical defences because the cost of investing in such defences would not be sufficiently offset by any benefits. Dalling et al. (2011) also suggested that the seeds of members of the third group of species may be particularly reliant on microbial defence as provided by beneficial seed-borne microorganisms.

To further quantify underlying trade-offs among defence traits, a five-year seed burial experiment was carried out. The results of this experiment further supported the distinction of two primary seed defence syndromes: a physical–chemical defence for species with longer-lived seedbanks relying more on dormancy and physical defence and a physiological–chemical defence syndrome for species whose seeds are more short-lived in the soil, relying more on chemical defence for persistence (Davis et al. 2016). It has been suggested that the proposed seed defence syndromes could help inform weed management approaches that utilize pathogen–seed interactions for seedbank depletion (Dalling et al. 2011; Davis et al. 2016), which will be discussed next.

13.6 Weed Management Applications

Due to the high environmental cost from large-scale use of chemical herbicides and due to many weed species becoming resistant to a wider range of herbicides, there is now an urgent case for diversification of weed management strategies, including the development and incorporation of biological control options (Abbas et al. 2018). With respect to the latter, decimation of weed seedbanks through manipulation of microbe–seed interactions was first suggested over forty years ago (Stoller and Wax 1974; Putnam and Duke 1978). Since then, interest in weed management using soil microbes has steadily grown (e.g. Kremer 1993, 1998; Kennedy 1999; Wagner and Mitschunas 2008), and it has been suggested that once successfully developed, such approaches might be particularly suitable for managing weeds characterized by large dormant seedbanks that remain unaffected by conventional herbicides (Ehlert et al. 2014). However, for successful implementation of biocontrol targeting weed seedbanks, buried weed seeds, pathogens, and the soil environment must interact in such a manner that seed-decaying microbes can successfully and reliably attack the seedbank (Kennedy and Kremer 1996). Three different approaches have been suggested, including (i) inoculation with seed-pathogenic microbes, (ii) manipulation of soil environmental conditions to boost seed decay, and (iii) weakening of seed defences against microbial attack (Kennedy and Kremer 1996). It could be argued that each of these three approaches deals with one element affecting the success of biocontrol measures targeting the weed seedbank, and for maximum effect all three ingredients need to come together, as shown in the conceptual model in Figure 13.3. Hence, various combinations of individual approaches have been suggested both with each other and with conventional approaches to weed control. In the following, we discuss all three approaches, and how they might be combined.

13.6.1 Targeted Application of Seed Pathogens

Targeted application of pathogenic microbes to reduce weed seedbanks was first suggested by Putnam and Duke (1978). However, four decades later, this approach is still more or less in its infancy (Müller-Stöver et al. 2016). One main reason for this is that, for successful

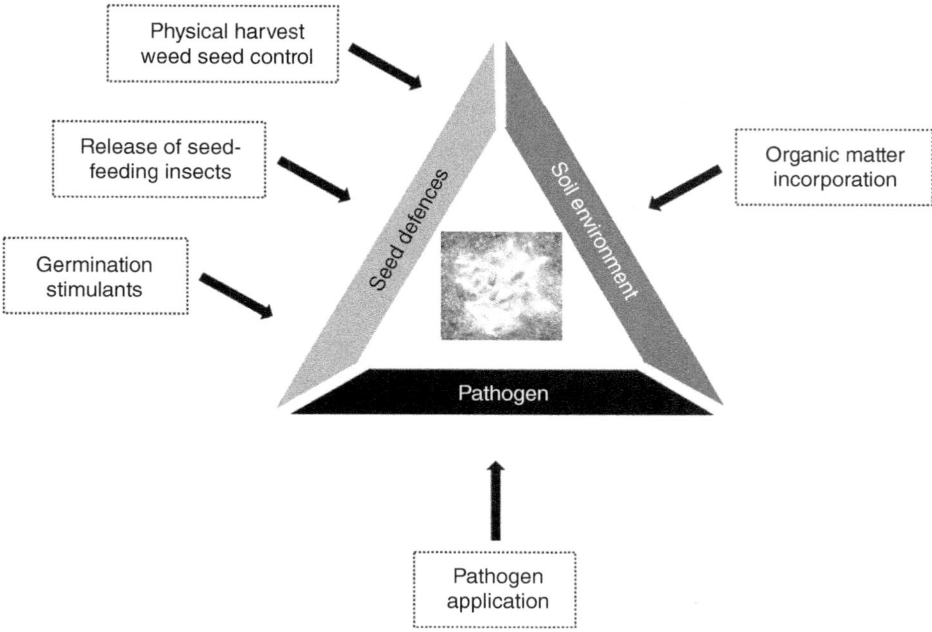

Figure 13.3 Simple conceptual model of the three elements that must be addressed for successful microbial biocontrol of the weed seedbank. Also shown are the main techniques that have so far been explored by different researchers to manipulate these elements for biocontrol purposes.

field-scale application, various obstacles have to be overcome, including a requirement to demonstrate sufficient pathogen activity under less than optimal and often variable field conditions. Most research so far has been limited to bioassays under laboratory conditions and pot experiments in the greenhouse, while very few studies have tested pathogen application in the field (but see Weidemann and Templeton 1988; Kennedy et al. 1991; Sauerborn et al. 1996b; Allen et al. 2018). For reasons already discussed in this chapter, it can be assumed that only a fraction of microbial isolates showing demonstrable seed-pathogenic activity in the laboratory have a potential to yield similar results when applied in the field (Kennedy et al. 1991).

One factor of particular importance for targeted pathogen application is host or target specificity of the prospective control agent. High host specificity is required to avoid unwanted damage both to crops and to wild species, some of which might be closely related to the target weed. For example, a strain of *F. avenaceum* caused decay of *A. fatua* seeds, whereas wheat seeds appeared unaffected (Fuerst et al. 2018). In another example, when a *F. oxysporum* f. sp. *orthoceras* strain isolated from *O. cumana*, a pest of sunflower (*Helianthus annuus*), was tested, it was also able to attack the very closely related *O. cernua*, a tobacco (*Nicotiana tabacum*) pest, but not its less closely related congener *O. aegyptiaca*, a pest of tomatoes (*Lycopersicon esculentum*) (Thomas et al. 1998). As already mentioned, the potential host range for a pathogen as determined in laboratory bioassays is likely wider than the actual host range under more variable field conditions (Beckstead et al. 2014).

Some of the most promising results so far for the targeted application of seed pathogens to control weeds were achieved in experiments on crop-parasitic weeds (Sauerborn et al. 2007), and among these in particular is *S. hermonthica*, a species widespread in northern tropical Africa where it attacks mainly graminaceous crops (Riches and Parker 1995). Studies testing the application of seed-pathogenic *F. nygamai* and of other *Fusarium* species yielded promising results in pot experiments (Abbasher and Sauerborn 1992, 1995; Ciotola et al. 1995; Abbasher et al. 1998; Yonli et al. 2006) and in field trials (Sauerborn et al. 1996b). However, *F. nygamai* is not purely a seed pathogen, and also attacks the very susceptible subterranean seedling stage of *Striga* prior to host attachment as well as older life stages (Abbasher and Sauerborn 1992). Thus, it may be better suited to inflict significant damage in its target weed than, e.g. a purely seed-pathogenic microbe used against a non-parasitic weed. The same caveat might also apply to seed-attacking *Fusarium* spp. and other fungal pathogens successfully tested in pot trials with parasitic *Orobanche* weeds (Thomas et al. 1998; Müller-Stöver et al. 2004; Aybeke et al. 2014).

An alternative to pathogen application might be the application of microbial herbicides in the form of microbial metabolites that are toxic to germinating seeds (Gealy et al. 1996). However, more research will be required to assess the practicability of such an approach.

13.6.2 Manipulation of the Soil Environment

Manipulation of the soil environment to maximize seed depletion through the action of resident microbes was first suggested by Kremer (1993), who highlighted the potential key roles of soil temperature, moisture, and organic matter as determinants of soil microbial activity. It has subsequently been suggested that targeted manipulation of the soil environment may boost activity levels of seed-decaying microorganisms and the natural potential of soils for weed suppression (Kennedy 1999; Kremer and Li 2003). Various ways in which this could be achieved have been proposed, the main one being the addition of organic matter as an alternative substrate for saprophytic microbes and a means to improve soil moisture relations (Pitty et al. 1987; Kremer 1993). Several subsequent studies investigated this approach, but as pointed out by Frost et al. (2019), many early studies did not distinguish between enhanced seed decay and an alternative mechanism of fatal germination, both of which could potentially explain any observed declines in viable seed densities in the soil. In a study involving experimental burial of eight weed species, Davis (2007) made an attempt to make such a distinction using path analysis. Results from that analysis appeared to indicate that seed mortality of *A. theophrasti* may have been increased by a corn stover organic amendment as a result of increased microbial activity, whereas seed persistence of two other species was more affected by effects of soil nitrogen on germination.

Two recent studies compared the effects of cereal and legume cover crops on buried weed seed persistence. Hill et al. (2016) demonstrated that incorporation into soil of red clover residue with narrow C : N ratio decreased seedbank persistence of *Chenopodium album*, whereas incorporation of cereal residue with a high C : N ratio increased seedbank persistence of *Setaria faberi* and *A. theophrasti*. Differential effects of different types of residue on N-limitation of seed-decaying soil microbes was stated as one possible explanation, but the authors carefully qualified their findings by stating that other factors such as seed germination response to N could also have provided an explanation (Hill et al. 2016). In another study, Mohler et al. (2018) found that with the exception of increased persistence of *S. faberi* in one of their two burial experiments, incorporation of rye (*Secale cereale*) cover crop residue did not greatly affect persistence of buried weed seeds. In contrast, incorporation of a leguminous hairy-vetch (*Vicia villosa*) cover crop decreased persistence of buried seeds of two out of four species. With germination of the two species whose seedbank persistence appeared to have declined in the hairy-vetch cover crop treatment, but not that of the other two species, known to be stimulated by raised nitrate levels, Mohler et al. (2018) argued against increased seed decay as an explanation for their results and in favour of fatal germination. To avoid having to rely on such indirect evidence, Frost et al. (2019) combined contrasting treatments of no cover crop versus incorporation of a rye cover crop with fungicide treatment of experimentally buried seeds of two weed species, *Digitaria sanguinalis* and *Amaranthus powellii*. In line with earlier studies, they found a marked increase in the persistence of *D. sanguinalis* seeds. However, this increase was observed regardless of whether seeds had been treated with fungicide or not, indicating that the observed difference could not be explained by differences in decay due to fungal pathogens.

Other more recent studies not included in our previous review (Wagner and Mitschunas 2008) also failed to produce evidence in support of the idea of weed seed decay being positively related to organic matter content. Davis et al. (2006) also found that weed seed mortality was higher in conventionally managed fields than in fields that had received organic amendments and green manure. Similarly, in a comparison of conventional and organic cropping systems, Ullrich et al. (2011) found that differences in buried seed persistence of two weeds were not very consistent, varying between four experimental locations. The same authors also noted lack of consistent associations between soil microbial biomass and weed seed persistence, interpreting this as evidence that soil microorganisms did not make a dominating contribution to seed mortality in their study. One notable exception to these negative findings is a study by Mohler et al. (2012), in which the incorporation of fresh green crop residues appeared to have stimulated pathogen attack of weed seeds and seedlings. However, this appears to have been the case only in the first few days after residue incorporation, i.e. any boost to fungal pathogens appeared to have been very short-lived.

In our previous review (Wagner and Mitschunas 2008), we concluded that the evidence in support of increased weed seed decay through organic matter amendments was not very convincing. Essentially, the more recent studies reviewed here have not fundamentally changed this conclusion. Rather, at least in the case of studies involving incorporation of cover crops, their more critical approach to interpretation of their findings has highlighted altered rates of seed germination as an alternative mechanism by which such management practices might affect the persistence of weed seedbanks.

One other potential approach that should be mentioned here also represents a manipulation of the soil environment, not as a prerequisite for microbial activity but rather a result of such activity. This potential approach is the targeted stimulation of seed germination of parasitic weeds of the genus *Striga* through soil inoculation with ethylene-producing bacteria, which, in the absence of suitable host plants for *Striga*, can be used to deplete its soil seedbank (Berner et al. 1999).

13.6.3 Manipulation of Seed Defences Alone and in Combination

Stoller and Wax (1974) suggested that methods should be developed to rupture weed seed coats, to facilitate microbial entry into the seeds. Such an approach, implemented via physical harvest weed seed control strategies, might be particularly promising in the case of weeds with highly persistent seedbanks and seeds that are characterized by a physical–chemical defence syndrome with strong primary mechanical defence and a weaker secondary chemical defence (Davis et al. 2016). Another way to implement this approach, based on studies by Kremer and Spencer (1989a,b), involves the release of seed-feeding insects that might damage the seeds of target weeds before entering the seedbank, potentially in combination with targeted application of seed-decaying microbes (Kremer 1995, 2000).

Another method to manipulate seed defences might be through application of chemicals. This was essentially tested in bioassays by Kremer and Schulte (1989), who applied two different germination stimulants to the seeds of *A. theophrasti*, both alone and in combination with the fungal pathogen *F. oxysporum*. On their own, ethephon and AC94377 successfully stimulated seed germination. However, when applied in combination with *F. oxysporum*, seed mortality markedly increased. It thus stands to reason that inundative application of pathogens such as *F. tricinctum* that are specialized in attacking seeds during the early stages of germination, and adapted to detect germinating seeds via their exudates, might achieve better control of the weed seedbank if carried out together with a stimulation of target weed seed germination through application of germinations stimulants.

13.6.4 Combination with Conventional Weed Control

The proposition that biocontrol approaches aimed at the weed seedbank can be successfully combined with conventional approaches to weed control was tested in a glasshouse experiment (Ehlert et al. 2014). Specifically, Ehlert et al. (2014) investigated the separate and combined effects of seed inoculation with *P. semeniperda* and subsequent application of the herbicide imazapic on *B. tectorum* seedling emergence and biomass production. Although the added effect of combining both measures on weed biomass were not greater than those of the herbicide on its own, Ehlert et al. (2014) suggested that such a combination could nonetheless be beneficial, as both approaches target different elements of the weed population. While application of the pathogen, due to its specialization on attacking dormant seed, targets the persistent component of the seedbank, herbicide application after seedling emergence can tackle the nondormant component after germination, thus helping to prevent replenishment of the soil seedbank (Ehlert et al. 2014). In their study, inoculation with *P. semeniperda* preceded herbicide application, and thus the pathogen was able to attack *Bromus* seeds already before the herbicide was applied. However, herbicides themselves can affect the soil microbial community (Gu et al. 2019), which means that when both measures are applied closely together, there might be a potential for interaction.

13.7 Future Prospects

In spite of weed biocontrol applications based on seed decay having first been suggested more than four decades ago, very few applications have been successfully tested under field conditions. The reasons for this include high spatiotemporal variability of environmental conditions in field situations, along with complex antagonistic interactions with other organisms, such as the endogenous microflora of the soil. All of these factors can contribute to keep densities and activity of seed-decaying microbes in the soil below those required for effective and reliable management of weed seedbanks. Many of these processes in the field are still only very poorly understood.

On the other hand, remarkable progress has been made with respect to our understanding of seed defences and defence syndromes, and further advances in the field of seed trait functional ecology should facilitate further progress in this area. There is also greater knowledge of seed demographic processes and how these can be better experimentally quantified (Chee-Sanford and Fu 2010) and mechanistically investigated using well-designed field experiments (Frost et al. 2019). More such experiments are needed, along with identification of suitable seed pathogens that could be applied as bioherbicides (Triolet et al. 2020) and field-scale testing at relevant scales of their deployment. Also, more focus may be required on the simultaneous manipulation of environmental conditions and/or of weed seed defences to make such pathogen applications more reliable for practical weed control.

At the other end of the range of experimental scales, i.e. at very small spatial scales, as already illustrated in Section 13.2, modern molecular microbial methods have in recent years enabled more detailed studies of the mechanisms of microbial attack, e.g. enabling characterization of the spatiotemporal dynamics of the release of exudates and of seed defence enzymes into the seed environment (Schiltz et al. 2015; Pollard 2018) and characterization of microbial community dynamics at the seed scale (Chee-Sanford and Fu 2010; Links et al. 2014; Schiltz et al. 2015). More such studies of biochemical and microbial processes at the seed scale are required (Schiltz et al. 2015), and the relevant analytical techniques should also be used to determine more precisely the functional roles of microorganisms interacting with weed seeds (Müller-Stöver et al. 2016), and the interactions of microbes with each other (Links et al. 2014). Furthermore, the effects of environmental variables on such processes will also have to be explored, and all this ideally for a wider range of weed species (Nelson 2018). It has also been suggested that the development and ongoing refinement of 'omics' approaches could facilitate our understanding of changes in the soil microbiome associated with management techniques that might affect seedbank persistence, such as the application of organic amendments to the soil (Vida et al. 2020). Application of this and other molecular microbial methods will further our scientific understanding and the development of a more detailed conceptual framework of how soil-borne seed-decaying microorganisms, weed seeds with their seed-borne microflora, and the environment interact with each other. In turn, such enhanced knowledge will be crucial for the development of reliable biological weed control options based on the action of such seed-decaying microbes.

Acknowledgements

We are indebted to Lucy Hulmes for creating the artwork used in Figure 13.2 and to Patricia Purdy for help in developing the design and layout of this figure.

References

Abbas, T., Zahir, Z.A., Naveed, M., and Kremer, R.J. (2018). Limitations of existing weed control practices necessitate development of alternative techniques based on biological approaches. In: *Advances in Agronomy* (ed. D.L. Sparks), 239–280. Cambridge, MA: Academic Press.

Abbasher, A.A. and Sauerborn, J. (1992). *Fusarium nygamai*, a potential bioherbicide for *Striga hermonthica* control in Sorghum. *Biological Control* 2: 291–296.

Abbasher, A.A. and Sauerborn, J. (1995). Pathogens attacking *Striga hermonthica* and their potential as biological control agents. In: *Proceedings of the Eighth International Symposium on Biological Control of Weeds* (eds. E.S. Delfosse and R.R. Scott), 527–533. Melbourne: DSIR/CSIRO.

Abbasher, A.A., Hess, D.E., and Sauerborn, J. (1998). Fungal pathogens for biological control of *Striga hermonthica* on sorghum and pearl millet in West Africa. *African Crop Science Journal* 6: 179–188.

Ahonsi, M.O., Berner, D.K., Emechebe, A.M., and Lagoke, S.T. (2002). Selection of rhizobacterial strains for suppression of germination of *Striga hermonthica* (Del.) Benth. seeds. *Biological Control* 24: 143–152.

Allen, P.S., Finch-Boekweg, H., and Meyer, S.E. (2018). A proposed mechanism for high pathogen-caused mortality in the seed bank of an invasive annual grass. *Fungal Ecology* 35: 108–115.

Anderson, R.C. and Liberta, A.E. (1986). Occurrence of fungal-inhibiting *Pseudomonas* on caryopses of *Tripsacum dactyloides* L. and its implication for seed survival and agriculture application. *Journal of Applied Bacteriology* 61: 195–199.

Anderson, J.V., Fuerst, E.P., Tedrow, T. et al. (2010). Activation of polyphenol oxidase in dormant wild oat caryopses by a seed-decay isolate of *Fusarium avenaceum*. *Journal of Agricultural and Food Chemistry* 58: 10597–10605.

Aybeke, M., Şen, B., and Ökten, S. (2014). *Aspergillus alliaceus*, a new potential biological control of the root parasitic weed *Orobanche*. *Journal of Basic Microbiology* 54: S93–S101.

Banowetz, G.M., Azevedo, M.D., Armstrong, D.J. et al. (2008). Germination-arrest factor (GAF): biological properties of a novel, naturally-occurring herbicide produced by selected isolates of rhizosphere bacteria. *Biological Control* 46: 380–390.

Barth, C.W., Meyer, S.E., Beckstead, J., and Allen, P.S. (2015). Hydrothermal time models for conidial germination and mycelial growth of the seed pathogen *Pyrenophora semeniperda*. *Fungal Biology* 119: 720–730.

Baughman, O.W. and Meyer, S.E. (2013). Is *Pyrenophora semeniperda* the cause of downy brome (*Bromus tectorum*) die-offs? *Invasive Plant Science and Management* 6: 105–111.

Beckstead, J., Meyer, S.E., Molder, C.J., and Smith, C. (2007). A race for survival: can *Bromus tectorum* seeds escape *Pyrenophora semeniperda*-caused mortality by germinating quickly? *Annals of Botany* 99: 907–914.

Beckstead, J., Meyer, S.E., Connolly, B.M. et al. (2010). Cheatgrass facilitates spillover of a seed bank pathogen onto native grass species. *Journal of Ecology* 98: 168–177.

Beckstead, J., Miller, L.E., and Connolly, B.M. (2012). Direct and indirect effects of plant litter on a seed-pathogen interaction in *Bromus tectorum* seed banks. *Seed Science Research* 22: 135–144.

Beckstead, J., Meyer, E., Reinhart, K.O. et al. (2014). Factors affecting host range in a generalist seed pathogen of semi-arid shrublands. *Plant Ecology* 215: 427–440.

Beckstead, J., Meyer, S.E., Ishizuka, T.S. et al. (2016). Lack of host specialization on winter annual grasses in the fungal seed bank pathogen *Pyrenophora semeniperda*. *PLoS One* 11 (3): e0151058.

Begonia, M.F.T. and Kremer, R.J. (1994). Chemotaxis of deleterious rhizobacteria to velvetleaf (*Abutilon theophrasti* Medik.) seeds and seedlings. *FEMS Microbiology Ecology* 15: 227–236.

Bekker, R.M., Schaminée, J.H.J., Bakker, J.P., and Thompson, K. (1998). Seed bank characteristics of Dutch plant communities. *Acta Botanica Neerlandica* 47: 15–26.

Berner, D.K., Schaad, N.W., and Völksch, B. (1999). Use of ethylene-producing bacteria for stimulation of *Striga* spp. seed germination. *Biological Control* 15: 274–282.

Bewley, J.D., Bradford, K.J., Hilhorst, H.W.M., and Nonogaki, H. (2013). *Seeds: Physiology of Development, Germination and Dormancy*, 3e. New York: Springer.

Brambl, R. and Gade, W. (1985). Plant seed lectins disrupt growth of germinating fungal spores. *Physiologia Plantarum* 64: 402–408.

Brownell, K.H. and Schneider, R.W. (1985). Roles of matric and osmotic components of water potential and their interaction with temperature in the growth of *Fusarium oxysporum* in synthetic media and soil. *Phytopathology* 75: 53–57.

Cavers, P.B. and Benoit, D.L. (1989). Seed banks in arable land. In: *Ecology of Soil Seed Banks* (eds. M.A. Leck, V.T. Parker and R.L. Simpson), 309–328. San Diego: Academic Press.

Chee-Sanford, J.C. (2008). Weed seeds as nutritional resources for soil Ascomycota and characterization of specific associations between plant and fungal species. *Biology and Fertility of Soils* 44: 763–771.

Chee-Sanford, J. and Fu, X. (2010). Investigating the role of microorganisms in soil seed bank management. In: *Current Research, Technology and Education Topics in Applied Microbiology and Microbial Biotechnology* (ed. A. Mendez-Vilas), 257–266. Badajoz: Formatex Research Center.

Chen, T., Nan, Z., Zhang, X. et al. (2018). Does dormancy protect seeds against attack by the pathogenic fungus *Fusarium tricinctum* in a semiarid grassland of Northwest China? *Plant and Soil* 422: 155–168.

Ciotola, M., Watson, A.K., and Hallett, S.G. (1995). Discovery of an isolate of *Fusarium oxysporum* with potential to control *Striga hermonthica* in Africa. *Weed Research* 35: 303–309.

Dalling, J.W., Davis, A.S., Schutte, B.J., and Arnold, A.E. (2011). Seed survival in the soil: interacting effects of predation, dormancy and the soil microbial community. *Journal of Ecology* 99: 89–95.

Davis, A.S. (2007). Nitrogen fertilizer and crop residue effects on seed mortality and germination of eight annual weed species. *Weed Science* 55: 123–128.

Davis, A.S. and Renner, K.A. (2007). Influence of seed depth and pathogens on fatal germination of velvetleaf (*Abutilon theophrasti*) and giant foxtail (*Setaria faberi*). *Weed Science* 55: 30–35.

Davis, A.S., Anderson, K.I., Hallett, S.G., and Renner, K.A. (2006). Weed seed mortality in soils with contrasting agricultural management histories. *Weed Science* 54: 291–297.

Davis, A.S., Schutte, B.J., Iannuzzi, J., and Renner, K.A. (2008). Chemical and physical defense of weed seeds in relation to soil seedbank persistence. *Weed Science* 56: 676–684.

Davis, A.S., Fu, X., Schutte, B.J. et al. (2016). Interspecific variation in persistence of buried weed seeds follows trade-offs among physiological, chemical, and physical seed defenses. *Ecology and Evolution* 6: 6836–6845.

Ehlert, K.A., Mangold, J.M., and Engel, R.E. (2014). Integrating the herbicide imazapic and the fungal pathogen *Pyrenophora semeniperda* to control *Bromus tectorum*. *Weed Research* 54: 418–424.

Fellows, G.M. and Roeth, F.W. (1992). Factors influencing shattercane (*Sorghum bicolor*) seed survival. *Weed Science* 40: 434–440.

Finch, H., Allen, P.S., and Meyer, S.E. (2013). Environmental factors influencing *Pyrenophora semeniperda*-caused seed mortality in *Bromus tectorum*. *Seed Science Research* 23: 57–66.

Finch-Boekweg, H., Gardner, J.S., Allen, P.S., and Geary, B. (2016). Postdispersal infection and disease development of *Pyrenophora semeniperda* in *Bromus tectorum* seeds. *Phytopathology* 106: 236–243.

Franke, J.L., Geary, B., and Meyer, S.E. (2014). Identification of the infection route of a *Fusarium* seed pathogen into nondormant *Bromus tectorum* seeds. *Phytopathology* 104: 1306–1313.

Frost, M.D., Haramoto, E.R., Renner, K.A., and Brainard, D.C. (2019). Tillage and cover crop effects on weed seed persistence: do light exposure and fungal pathogens play a role? *Weed Science* 67: 103–113.

Fuerst, E.P., Anderson, J.V., Kennedy, A.C., and Gallagher, R.S. (2011). Induction of polyphenol oxidase activity in dormant wild oat (*Avena fatua*) seeds and caryopses: a defence response to seed decay fungi. *Weed Science* 59: 137–144.

Fuerst, E.P., Okubara, P.A., Anderson, J.V., and Morris, C.F. (2014). Polyphenol oxidase as a biochemical seed defense mechanism. *Frontiers in Plant Science* 5: 689.

Fuerst, E.P., James, M.S., Pollard, A.T., and Okubara, P.A. (2018). Defense enzyme responses in dormant wild oat and wheat caryopses challenged with a seed decay pathogen. *Frontiers in Plant Science* 8: 2259.

Gallandt, E.R., Fuerst, E.P., and Kennedy, A.C. (2004). Effect of tillage, fungicide seed treatment, and soil fumigation on seed bank dynamics of wild oat (*Avena fatua*). *Weed Science* 52: 597–604.

Gardarin, A., Guillemin, J.-P., Munier-Jolain, N.M., and Colbach, N. (2010a). Estimation of key parameters for weed population dynamics models: base temperature and base water potential for germination. *European Journal of Agronomy* 32: 162–168.

Gardarin, A., Dürr, C., Mannino, M.R. et al. (2010b). Seed mortality in the soil is related to seed coat thickness. *Seed Science Research* 20: 243–256.

Gealy, D.R., Gurusiddaiah, S., and Ogg, A.G. Jr. (1996). Isolation and characterization of metabolites from *Pseudomonas syringae*-strain 3366 and their phytotoxicity against certain weed and crop species. *Weed Science* 44: 383–392.

Girvan, M.S., Bullimore, J., Pretty, J.N. et al. (2003). Soil type is the primary determinant of the composition of the total and active bacterial communities in arable soils. *Applied and Environmental Microbiology* 69: 1800–1809.

Goeden, R.D. and Ricker, D.W. (1973). A soil profile analysis for puncturevine fruit and seed. *Weed Science* 21: 504–507.

Griffin, D.M. (1966). Fungi attacking seeds in dry seed-beds. *Proceedings of the Linnean Society of New South Wales* 91: 84–89.

Gu, X., Cen, Y., Guo, L. et al. (2019). Responses of weed community, soil nutrients, and microbes to different weed management practices in a fallow field in Northern China. *PeerJ* 7: e7650.

Guillemin, J.-P., Gardarin, A., Granger, S. et al. (2013). Assessing potential germination period of weeds with base temperatures and base water potentials. *Weed Research* 53: 76–87.

Halloin, J.M. (1983). Deterioration resistance mechanisms in seeds. *Phytopathology* 73: 335–339.

Harman, G.E. (1983). Mechanisms of seed infection and pathogenesis. *Phytopathology* 73: 326–329.

Harman, G.E. and Stasz, T.E. (1986). Influence of seed quality on soil microbes and seed rots. In: *Physiological-Pathological Interactions Affecting Seed Deterioration* (ed. S.H. West), 11–37. CSSA Special Publication 12. Madison, WI: Crop Science Society of America.

Hendry, G.A.F., Thompson, K., Moss, C.J. et al. (1994). Seed persistence: a correlation between seed longevity in the soil and *ortho*-dihydroxyphenol concentration. *Functional Ecology* 8: 658–664.

Hill, E.C., Renner, K.A., Sprague, C.L., and Davis, A.S. (2016). Cover crop impact on weed dynamics in an organic dry bean system. *Weed Science* 64: 261–275.

Kennedy, A.C. (1999). Soil microorganisms for weed management. *Journal of Crop Production* 2: 123–138.

Kennedy, A.C. and Kremer, R.J. (1996). Microorganisms in weed control strategies. *Journal of Production Agriculture* 9: 480–485.

Kennedy, A.C., Elliott, L.F., Young, F.L., and Douglas, C.L. (1991). Rhizobacteria suppressive to the weed downy brome. *Soil Science Society of America Journal* 55: 722–727.

Kiewnick, L. (1964). Experiments on the influence of seedborne and soilborne microflora on the viability of wild oat seeds (*Avena fatua* L.) II. Experiments on the influence of microflora on the viability of seeds in the soil. *Weed Research* 4: 31–43.

Kirkpatrick, B.L. and Bazzaz, F.A. (1979). Influence of certain fungi on seed germination and seedling survival of four colonizing annuals. *Journal of Applied Ecology* 16: 515–527.

Kitajima, K. and Fenner, M. (2000). Ecology of seedling regeneration. In: *Seeds: The Ecology of Regeneration in Plant Communities*, 2e (ed. M. Fenner), 331–359. Wallingford: CABI.

Kremer, R.J. (1986a). Microorganisms associated with velvetleaf (*Abutilon theophrasti*) seeds on the soil surface. *Weed Science* 34: 233–236.

Kremer, R.J. (1986b). Antimicrobial activity of velvetleaf (*Abutilon theophrasti*) seeds. *Weed Science* 34: 617–622.

Kremer, R.J. (1987). Identity and properties of bacteria inhabiting seeds of selected broadleaf weed species. *Microbial Ecology* 14: 29–37.

Kremer, R.J. (1993). Management of weed seed banks with microorganisms. *Ecological Applications* 3: 42–52.

Kremer, R.J. (1995). Integration of a seed-feeding insect and fungi for management of velvetleaf (*Abutilon theophrasti*) seed production. In: *Proceedings of the Eighth International Symposium on Biological Control of Weeds* (eds. E.S. Delfosse and R.R. Scott), 627–631. Melbourne, Australia: DSIR/CSIRO.

Kremer, R.J. (1998). Microbial interactions with weed seeds and seedlings and its potential for weed management. In: *Integrated Weed and Soil Management* (eds. J.L. Hatfield, D.D. Buhler and B.A. Stewart), 161–179. Chelsea, MI: Ann Arbor Press.

Kremer, R.J. (2000). Combinations of microbial and insect biocontrol agents for management of weed seeds. In: *Proceedings of the X International Symposium on Biological Control of Weeds* (ed. N.R. Spences), 799–806. Bozeman: Montana State University.

Kremer, R.J. and Li, J. (2003). Developing weed-suppressive soils through improved soil quality management. *Soil and Tillage Research* 72: 193–202.

Kremer, R.J. and Schulte, L.K. (1989). Influence of chemical treatment and *Fusarium oxysporum* on velvetleaf (*Abutilon theophrasti*). *Weed Technology* 3: 369–374.

Kremer, R.J. and Spencer, N.R. (1989a). Impact of a seed-feeding insect and microorganisms on velvetleaf (*Abutilon theophrasti*) seed viability. *Weed Science* 37: 211–216.

Kremer, R.J. and Spencer, N.R. (1989b). Interaction of insects, fungi, and burial on velvetleaf (*Abutilon theophrasti*) seed viability. *Weed Technology* 3: 322–328.

Kremer, R.J., Hughes, L.B. Jr., and Aldrich, R.J. (1984). Examination of microorganisms and deterioration resistance mechanisms associated with velvetleaf seed. *Agronomy Journal* 76: 745–749.

van Leeuwen, B.H. (1981). Influence of micro-organisms on the germination of the monocarpic *Cirsium vulgare* in relation to disturbance. *Oecologia* 48: 112–115.

Links, M.G., Demeke, T., Gräfenhan, T. et al. (2014). Simultaneous profiling of seed-associated bacteria and fungi reveals antagonistic interactions between microorganisms within a shared microbiome on *Triticum* and *Brassica* seeds. *New Phytologist* 202: 542–553.

Lonsdale, W.M. (1993). Losses from the seed bank of *Mimosa pigra*: soil micro-organisms vs. temperature fluctuations. *Journal of Applied Ecology* 30: 654–660.

Lozano-Juste, J., Masi, M., Cimmino, A. et al. (2019). The fungal sesquiterpenoid pyrenophoric acid B uses the plant ABA biosynthetic pathway to inhibit seed germination. *Journal of Experimental Botany* 70: 5487–5494.

de Luna, L.Z., Kennedy, A.C., Hansen, J.C. et al. (2011). Mycobiota on wild oat (*Avena fatua* L.) seed and their caryopsis decay potential. *Plant Health Progress* 12 https://doi.org/10.1094/PHP-2011-0210-01-RS.

Martinson, K., Durgan, B., Forcella, F. et al. (2007). An emergence model for wild oat (*Avena fatua*). *Weed Science* 55: 584–591.

Meyer, S.E. and Allen, P.S. (2009). Predicting seed dormancy loss and germination timing for *Bromus tectorum* in a semi-arid environment using hydrothermal time models. *Seed Science Research* 19: 225–239.

Meyer, S.E., Franke, J.L., Baughman, O.W. et al. (2014). Does *Fusarium*-caused seed mortality contribute to *Bromus tectorum* stand failure in the Great Basin? *Weed Research* 54: 511–519.

Meyer, S.E., Beckstead, J., and Allen, P.S. (2018). Niche specialization in *Bromus tectorum* seed bank pathogens. *Seed Science Research* 28: 215–221.

Miché, L., Bouillant, M.-L., Rohr, R. et al. (2000). Physiological and cytological studies on the inhibition of *Striga* seed germination by the plant growth-promoting bacterium *Azospirillum brasilense*. *European Journal of Plant Pathology* 106: 347–351.

Mishra, R.R. and Srivastava, V.B. (1977). Comparison of mycoflora associated with certain crop and weed seeds. *Acta Mycologica* 13: 145–149.

Mohamed-Yasseen, Y., Barringer, S.A., Splittstoesser, W.E., and Costanza, S. (1994). The role of seed coats in seed viability. *Botanical Review* 60: 426–439.

Mohler, C.L., Dykeman, C., Melson, E.B., and DiTommaso, A. (2012). Reduction in weed seedling emergence by pathogens following the incorporation of green crop residue. *Weed Research* 52: 467–477.

Mohler, C.L., Taylor, A.G., DiTommaso, A. et al. (2018). Effects of incorporated rye and hairy vetch cover crop residue on the persistence of weed seeds in the soil. *Weed Science* 66: 379–385.

van Mourik, T.A., Stomph, T.J., and Murdoch, A.J. (2005). Why high seed densities within buried mesh bags may overestimate depletion rates of soil seed banks. *Journal of Applied Ecology* 42: 299–305.

Müller-Stöver, D., Thomas, H., Sauerborn, J., and Kroschel, J. (2004). Two granular formulations of *Fusarium oxysporum* f.sp. *orthoceras* to mitigate sunflower broomrape *Orobanche cumana*. *BioControl* 49: 595–602.

Müller-Stöver, D., Nybroe, O., Baraibar, B. et al. (2016). Contribution of the seed microbiome to weed management. *Weed Research* 56: 335–339.

Nelson, E.B. (2004). Microbial dynamics and interactions in the spermosphere. *Annual Review of Phytopathology* 42: 271–309.

Nelson, E.B. (2018). The seed microbiome: origins, interactions, and impacts. *Plant and Soil* 422: 7–34.

Paszkowski, W.L. and Kremer, R.J. (1988). Biological activity and tentative identification of flavonoid components in velvetleaf (*Abutilon theophrasti* Medik.) seed coats. *Journal of Chemical Ecology* 14: 1573–1582.

Pitty, A., Staniforth, D.W., and Tiffany, L.H. (1987). Fungi associated with caryopses of *Setaria* species from field-harvested seeds and from soil under two tillage systems. *Weed Science* 35: 319–323.

Pollard, A.T. (2018). Seeds *vs* fungi: an enzymatic battle in the soil seedbank. *Seed Science Research* 28: 197–214.

Putnam, A.R. and Duke, W.B. (1978). Allelopathy in agroecosystems. *Annual Review of Phytopathology* 16: 431–451.

Riches, C.R. and Parker, C. (1995). Parasitic plants as weeds. In: *Parasitic Plants* (eds. M.C. Press and J.D. Graves), 226–255. London: Chapman & Hall.

Ritz, K. and Young, I.M. (2004). Interactions between soil structure and fungi. *Mycologist* 18: 52–59.

Sauerborn, J., Dörr, I., Abbasher, A. et al. (1996a). Electron microscopic analysis of the penetration process of *Fusarium nygamai*, a hyperparasite of *Striga hermonthica*. *Biological Control* 7: 53–59.

Sauerborn, J., Abbasher, A.A., Kroschel, J. et al. (1996b). *Striga hermonthica* control with *Fusarium nygamai* in maize. In: *Proceedings of the IX International Symposium on Biological Control of Weeds* (eds. V.C. Moran and J.H. Hoffmann), 19–26. Stellenbosch: University of Cape Town.

Sauerborn, J., Müller-Stöver, D., and Hershenhorn, J. (2007). The role of biological control in managing parasitic weeds. *Crop Protection* 26: 246–254.

Schiltz, S., Gaillard, I., Pawlicki-Jullian, N. et al. (2015). A review: what is the spermosphere and how can it be studied? *Journal of Applied Microbiology* 119: 1467–1481.

Stoller, E.W. and Wax, L.M. (1974). Dormancy changes and fate of some annual weed seeds in the soil. *Weed Science* 22: 151–155.

Thill, D.C., Schirman, R.D., and Appleby, A.P. (1979). Influence of soil moisture, temperature, and compaction and emergence of downy brome (*Bromus tectorum*). *Weed Science* 27: 625–630.

Thomas, H., Heller, A., Sauerborn, J. et al. (1998). The potential of *Fusarium oxysporum* f. sp. *orthoceras* as a biological control agent for *Orobanche cumana* in sunflower. *Biological Control* 13: 41–48.

Thomas, H., Heller, A., Sauerborn, J., and Müller-Stöver, D. (1999). *Fusarium oxysporum* f. sp. *orthoceras*, a potential mycoherbicide, parasitizes seeds of *Orobanche cumana* (sunflower broomrape): a cytological study. *Annals of Botany* 83: 453–458.

Triolet, M., Guillemin, J.-P., Andre, O., and Steinberg, C. (2020). Fungal-based bioherbicides for weed control: a myth or a reality? *Weed Research* 60: 60–77.

Ullrich, S.D., Buyer, J.S., Cavigelli, M.A. et al. (2011). Weed seed persistence and microbial abundance in long-term organic and conventional cropping systems. *Weed Science* 59: 202–209.

Upadhyaya, M.K., Turkington, R., and McIlvride, D. (1986). The biology of Canadian weeds. 75. *Bromus tectorum* L. *Annals of Botany* 83: 453–458.

Vida, C., de Vicente, A., and Cazorla, F.M. (2020). The role of organic amendments to soil for crop protection: induction of suppression of soilborne pathogens. *Annals of Applied Biology* 176: 1–15.

Wagner, M. and Mitschunas, N. (2008). Fungal effects on seed bank persistence and potential applications in weed biocontrol: a review. *Basic and Applied Ecology* 9: 191–203.

Warr, S.J., Thompson, K., and Kent, M. (1992). Antifungal activity in seed coat extracts of woodland plants. *Oecologia* 92: 296–298.

Weidemann, G.J. and Templeton, G.E. (1988). Control of Texas gourd, *Cucurbita texana*, with *Fusarium solani* f. sp. *cucurbitae*. *Weed Technology* 2: 271–274.

Widmer, T.L. and Guermache, F. (2004). Filtrates of rhizosphere bacteria suppressive to *Centaurea solstitialis* seed germination. *Journal of Plant Diseases and Protection* 19: 497–502.

Yonli, D., Traoré, H., Hess, D.E. et al. (2006). Effect of growth medium, *Striga* seed burial distance and depth on efficacy of *Fusarium* isolates to control *Striga hermonthica* in Burkina Faso. *Weed Research* 46: 73–81.

14

The Potential Role of Allelopathy in the Persistence of Invasive Weeds

Sajid Latif[1,2], *Saliya Gurusinghe*[1,2], *and Leslie A. Weston*[2]

[1] *School of Agricultural, Environmental and Veterinary Sciences, Charles Sturt University, Wagga Wagga, NSW, 2678, Australia*
[2] *Graham Centre for Agricultural Innovation, Charles Sturt University, Wagga Wagga, NSW, 2678, Australia*

14.1 Introduction

Crop weeds display a number of physiochemical, physiological, agronomic, and reproductive characteristics, which enable them to successfully establish and persist in ways that differentiate them from typical commercial crops (Cobb and Reade 2011). Weeds interact with other plants in close proximity by maximizing competition for resources as well as by chemical interference through release of bioactive secondary metabolites. The secondary metabolites produced by plants, fungi, algae, and associated microflora frequently serve as natural defenses against potential competitors or predators. This ecological phenomenon is referred to as allelopathy, a term first developed by Molisch (1937), which is broadly defined as the direct or indirect biochemical interactions among plants, insects, and microorganisms living in a certain ecosystem, mediated through the release of allelochemicals (Weston 1996; Rice 2012). While the capacity to interfere with neighboring plants through competition for finite resources such as light, water, nutrients, and space has been studied extensively in the existing weed literature (Einhellig 1996; Niklas and Hammond 2013), there is an exponentially increasing body of evidence showing that chemical interference also impacts plant growth and reproduction in addition to direct competition for resources. Although allelopathy is a broad term referring to the chemical inhibition of germination, growth, survival, and reproduction of another organism by a plant, in this chapter we will limit our focus to the chemical interactions of weed species with other plants and those potentially mediated by the activity of microbes.

Allelopathy as a phenomenon negatively impacting plant growth has been reported in the literature for over 2000 years, mainly through ancient manuscripts (Aliotta et al. 2008). For example, the Greek botanist Theophrastus (300 BCE) first observed that chickpeas cultivated over time inhibited the growth of other crops and weeds by exhausting soil resources. Toxicity of the walnut tree to neighboring plants was later described by the Roman scholar Pliny (1 AD). More recently, De Candolle (1832) performed the first well-described experiment to study the toxicity associated with root exudates of allelopathic species,

Persistence Strategies of Weeds, First Edition. Edited by Mahesh K. Upadhyaya, David R. Clements, and Anil Shrestha.

studying noxious weeds reported to "sicken the soil," thereby adversely affecting the growth and development of other plants (Weston and Duke 2003). It is evident through several historical references in the literature that the inhibitory effect of target species was associated with the release of allelochemicals from donor plants (Weir et al. 2004). Following on from these early studies, research into the mechanisms associated with allelopathic interactions has increased exponentially in the past 25 years, mainly through advancements in analytical chromatography systems providing enhanced detection and quantification of metabolites with improved accuracy and precision. The study of allelopathy, traditionally described in relation to plant–plant interactions, is beginning to reveal complex cross-kingdom interactions involving plants, insects, bacteria, and fungi present in the ecosystem (Zeng 2014; Schandry and Becker 2020).

Past studies have mainly emphasized allelopathic effects on a target plant either in a natural setting or controlled environments. Consequently, this approach generally has unfortunately disregarded the indirect mediation of allelochemicals in plant interactions (Zeng 2014). Such indirect effects include alteration of nutrient availability through changing physicochemical properties of soil and also nutrient mobilization (Stinson et al. 2006; Li et al. 2007; Bakker et al. 2018). These interactions are often cross-kingdom and primarily involve soil microbiota. Interactions between microorganisms and their host plants encompass a repertoire of signal molecules that can be produced from either root cells or microbiome present in close proximity. These interactions could be beneficial, commensal, or pathogenic to the plants.

Plant-produced secondary metabolites facilitate direct and indirect interactions favoring colonization of specific microbial communities. Microbially produced signal molecules binding to surfaces in the rhizosphere often have the ability to regulate plant genes over multiple generations, thus impacting sustainable production. The microbial metagenome is a complex matrix that determines environmental adaptability, resource acquisition, defense responses under biotic and abiotic stressors, and the belowground chemical interactions between plants and associated organisms (Lapsansky et al. 2016; Bakker et al. 2018). Soil microbial communities that coevolve with a particular crop species are dynamically inherited by future crop cycles, sometimes as an adaptive advantage (Lapsansky et al. 2016). However, our understanding of the mechanisms regulating plant and microbial chemical signaling is still in its infancy, but we do know that the rhizosphere microbiome is critical for sustainable crop production and maintenance of ecosystem health. Understanding the role of soil microorganisms in driving allelopathy will be important in determining how allelopathy structures plant communities over time, impacts biodiversity, and ultimately alters agricultural productivity.

Allelopathy, either direct or indirect, has been noted to influence weed invasion and persistence (Hierro and Callaway 2003). Recent studies have been critical in advancing our understanding of plant ecosystems and their drivers and have focused on the impact of allelochemicals on successful invasion by plant invaders in the non-native range, their role in plant succession in dynamic ecosystems, and also in driving rhizosphere interactions with animals, insects, and/or microbial symbionts. This chapter addresses the concepts of allelopathy with regard to weed persistence and invasion success and further investigates allelopathic interactions as key defense strategies of invasive weeds against both crops and native plants.

14.2 Classification of Allelochemicals

Secondary plant products referred to as allelochemicals are metabolites, which do not necessarily play an important role in primary plant metabolism but have traditionally been associated with plant defense. To date, more than 100 000 plant-produced secondary metabolites of low molecular weight have been characterized and initially categorized into classes including phenolics, terpenoids, and alkaloids based on their chemical structure and functional groups (Table 14.1) (Walton and Brown 1999; Ahuja et al. 2012). Some allelochemicals are found ubiquitously in higher plants, while others are synthesized or activated from precursor molecules in response only to specific biotic or abiotic stressors (Bartwal et al. 2013).

Phenolic compounds and their derivatives are the largest and most diverse group of secondary metabolites in plants. They are known to perform a vast array of biological functions and originate from the phenylpropanoid-acetate biosynthetic pathway containing a hydroxylated aromatic hydrocarbon ring with various substitutions (Croteau et al. 2000). Phenolics such as benzoic acids, aldehydes, cinnamic acids, coumarins, tannins, and flavonoids (Zeng et al. 2008) make up to 40% of the structural matrix of plants (Dalton 1999). Simple phenolics are the largest group of allelochemicals reported in the last decade, followed by flavonoids (Weston and Mathesius 2013; Macías et al. 2019). Phenolics and their derivatives also represent the most common class of secondary metabolites playing a critical role in defense mechanisms in higher plants and exhibit varying degrees of phytotoxicity and multiple cellular target sites (Haig 2008). Past studies have identified water-soluble phenolics that were implicated in the phytotoxic effects of chickweed (*Stellaria media*) (Inderjit et al. 1997), *Parthenium hysterophorus*, *Asphodelus tenuifolius*, and *Pluchea lanceolata* on seed germination and seedling growth in wheat (*Triticum aestivum*) and weedy species (Yadav and Chauhan 1998).

Terpenoids, a term adopted from the German word "terpentin" due to the isolation of earliest terpenoids from turpentine (Herz 1963), are derived from both the mevalonic acid and isopentyl pyrophosphate pathways (Haig 2008). Terpenoids are comprised of five carbon isoprene subunits linked together to constitute hemiterpenes, monoterpenes, sesquiterpenes, diterpenes, triterpenes, tetraterpenes, and polyterpenes depending upon the number of isoprene subunits. More than 36 000 terpenoids have been characterized to date (Buckingham et al. 2016), many of which are volatile and are known to be involved in a variety of biological functions in plants including signaling, photo-protection, production of reproductive hormones, and also allelopathy (Croteau et al. 2000). For example, monoterpenes found in many essential oils, including those present in eucalyptus, are reported to exhibit phytotoxic effects on seed germination and seedling growth (Haig 2008). Terpenoids (9-hydroxy-4,7-megastigmadien-9-one, 6,9,10-trihydroxy-4,7-megastigmadien-3-one) were shown among the major chemical constituents of barnyard grass (*Echinochloa crus-galli*) and important in its allelopathic interactions with rice (*Oryza sativa*) in both greenhouse and field experiments (Khanh et al. 2018; Li et al. 2019).

Sesquiterpene lactones released into the soil from "mile-a-minute" weed (*Mikania micrantha*) leaves and flowers by decomposing litter were shown to be associated with the allelopathic effects of this invasive vine on high value crops and also the microbial community (Shao et al. 2005; Liu et al. 2020). In addition, their study suggested that

Table 14.1 Major classes of allelopathic phytochemicals synthesized by invasive plant species.

Class of compound	Weed species	Structure
Phenolic compounds	*Bidens pilosa* (Campbell et al. 1982)	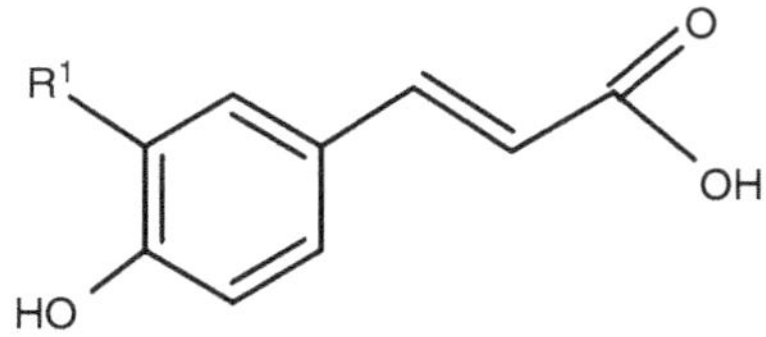 *p*-Coumaric acid $R_1 = H$ Caffeic acid $R_1 = OH$
	Imperata cylindrica (Eussen 1978)	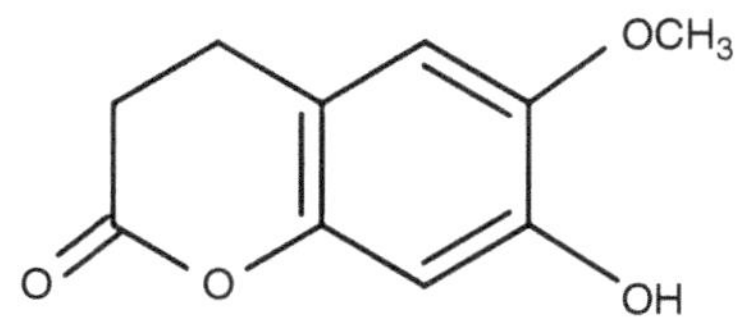 Scopoletin $R_1 = H$ Scopolin $R_1 = glu$
Terpenoids	*Parthenium hysterophorus* (Reinhardt et al. 2009)	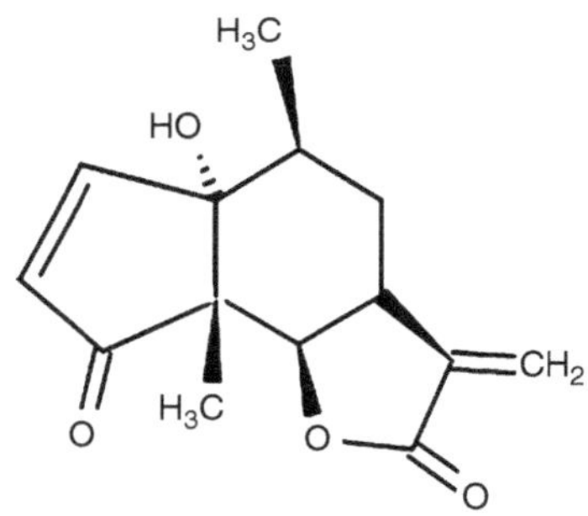 Parthenin
	Pueraria thunbergiana (Kato-Noguchi 2003)	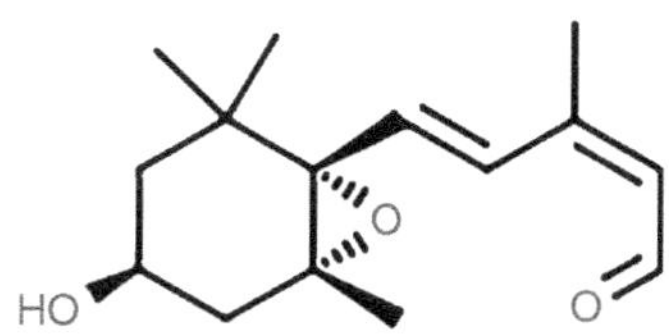 Xanthoxin
	Lantana camara (Ghisalberti 2000)	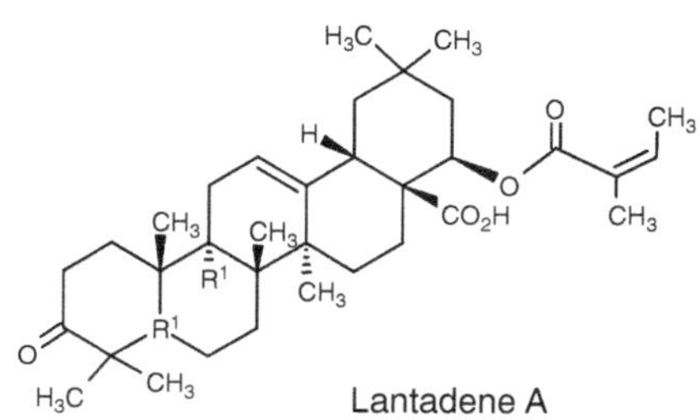 Lantadene A

Table 14.1 (Continued)

Class of compound	Weed species	Structure
Alkaloids	*Datura stramonium* (Lovett and Potts 1987)	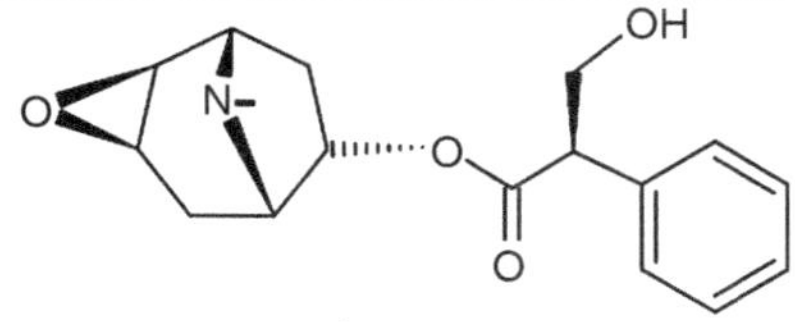 Scoplamine
Benzoxazinoids	*Elymus repens* (Korhammer and Haslinger 1994)	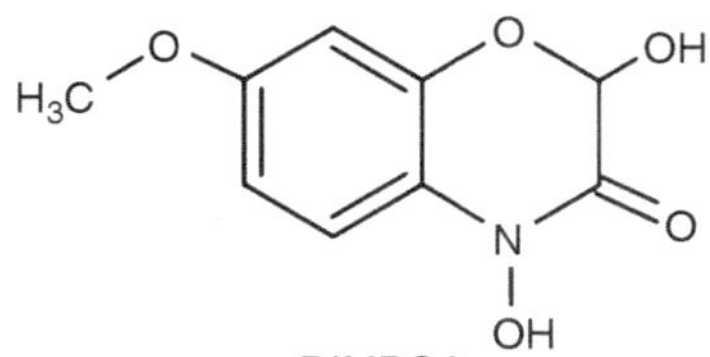 DIMBOA
Quinones	*Sorghum bicolor*	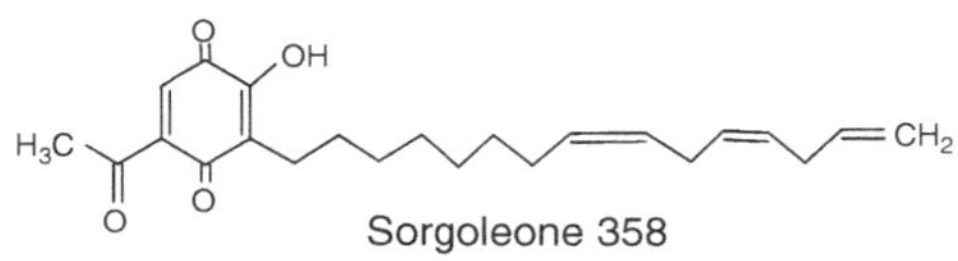 Sorgoleone 358

Source: Walton and Brown 1999; Ahuja et al. 2012.

sesquiterpene lactones produced by *M. micrantha* directly impacted soil microbial communities, increasing nitrogen mineralization and facilitating nitrogen cycling, resulting in the promotion of more rapid growth of this vine than that of neighboring plants.

Alkaloids are nitrogen-containing heterocyclic compounds of plant origin, named due to their alkaline nature. Historically, they are known for their biological and pharmacological properties, including plant defense, and represent great structural diversity, with over 21 000 alkaloids characterized in plants (Yang and Stöckigt 2010; Wink 2011). Biosynthetically, they can typically be categorized into different classes based on the original amino acid from which they were derived, while others are derived from terpenoids (i.e. steroidal alkaloids) or polyketide pathways (Macías et al. 2019). As an example, allelopathic indole alkaloids originate from tryptophan, pyrrolizidine alkaloids from ornithine or arginine, and quinolizidine alkaloids from lysine (Seigler 1998). Pyrrolizidine alkaloids produced by *Senecio angustifolius*, commonly known as "bitter bush," were found to be the main source of soil contamination in Rooibos tea (*Aspalathus linearis*) crop causing yield loss and toxicity (Van Wyk et al. 2016; Van Wyk et al. 2017). Pyrrolizidine alkaloids in Paterson's curse (*Echium plantagineum*) have been shown to be critical to insect defense and plant toxicity to grazing livestock. These are produced constitutively in aboveground plant parts (Skoneczny et al. 2015; Zhu et al. 2017).

Benzoxazinoids (BXs) are another interesting group of secondary metabolites studied for their potential use as natural herbicides and insect deterrents or as lead molecules for the development of herbicides with new modes of action (Frey et al. 2009; Rice et al. 2012; Macías et al. 2019). They are degradation products of hydroxamic acids, mainly synthesized in the *Poaceae* family (Schulz et al. 2013). Numerous studies have described the role of these metabolites in suppressing weeds, their release, microbial degradation, mode of action, and quantification in above- and below-ground parts of the donor and target plants. BXs are biochemically one of the best understood allelochemicals. However, the mechanism of exudation into the rhizosphere by cereal crops requires further investigation (Schandry and Becker 2020; Mwendwa et al. 2021). As there are many other diverse families of allelochemicals that interfere with plant growth and development, we will not describe all in this chapter and refer the reader to numerous reviews and recent books on the subject (Table 14.1).

14.3 Allelochemical Modes of Action

The diverse nature of allelochemicals reported in literature presents a challenge to those who study their specific mode(s) of action as their target site(s) can be numerous and variable and are frequently unknown (Latif et al. 2017). Bioassays designed to study specific modes of actions of allelochemicals often do not provide sufficient evidence for determination, as prior knowledge of associated enzymes, catalysts, and target sites are frequently required for confirmation. This is particularly challenging when an allelochemical acts at multiple sites and impacts multiple biochemical processes in the target plant.

Characterizing the mode of action of potential allelochemicals is important in that it may not only result in the potential for the development of novel herbicides with unique modes of action but may also allow improved management of invasive weeds producing such allelochemicals. Numerous mechanisms of action have been confirmed for herbicides and allelochemical plant growth inhibitors (Table 14.2) using specific bioassays for the identification of specific site of action. These include target sites that limit the functionality of fundamental physiological processes including membrane permeability, water and nutrient uptake, respiration, photosynthesis, protein and nucleic acid synthesis, and growth regulation in susceptible plants. Transcriptomic technologies are now frequently applied to such investigations to determine key genes involved in biosynthesis and provide clues on allelochemical mode of action in targeted plant tissues, by suggesting pathways that may be impacted by specific herbicides. Improved instrumentation has enhanced fundamental knowledge and ability to isolate and identify allelochemicals along with their potential mode(s) of action (Duke et al. 2002). A vast assortment of chemically diverse allelochemicals and untapped opportunities for their discovery in plants, soils, and aquatic ecosystems presents unique opportunities for herbicide discovery, particularly with respect to the identification of novel modes of actions to counter ever-increasing herbicide resistance issues (Dayan and Duke 2014; Duke et al. 2019).

It is important to note that phytotoxicity of any allelochemical is a complex functional process modulated by various factors including concentration, flux rate, activity in soil, physiological stage, climatic conditions, and microbial interactions in the rhizosphere (Weidenhamer 1996; Weston 1996; Rice 2012). Further standardization of research efforts in allelopathy has been suggested to provide evidence of the specific

Table 14.2 Potential mode of action of various allelochemicals associated with invasive weeds.

Mode of action	Allelochemicals	References
Membrane permeability	α-Pinene, mimosine, benzoic acid, cinnamic acid, sarmentine, limonene, and camphor	Yoon et al. (2000), Chai et al. (2013), Baziramakenga et al. (1995, 1997), Huang et al. (2010), and Young and Bush (2009).
Nutrient and water intake	*p*-Coumaric, ferullic, vanillic, salicylic and caffeic acids, 2,3-benzoxazolinones, and *p*-hydroxybenzoic acid	Barkosky and Einhellig (1993); Sánchez-Moreiras and Reigosa (2005); Barkosky and Einhellig (2003); and Abenavoli et al. (2010)
Respiration	Quercetin, kaempferol, camphor, α-pinene and limonene, sorgoleone, juglone, lawsone, naphthotectone, and anthratectone	Latif et al. (2017), Abrahim et al. (2003), Czarnota et al. (2001), Weir et al. (2004), Soderquist, (1973), and Durán et al. (2018)
Photosynthesis	*p*-Hydroxy phenyl acetic acid, *p*-coumaric, ferulic acids, sorgoleone, citral, and sarmentine	Aslam et al. (2017); Graña et al. (2013); and Dayan et al. (2015)
Protein and nucleic acid synthesis	*m*-Tyrosine, vanillic, *p*-coumaric, ferulic acid, momilactones, camptothecin, alangiside, palmatine, and parthenin	Bertin et al. (2007), Mersie and Singh (1993), Li et al. (2010), Kato-Noguchi (2011), Kumar and Barthwal (2018), and Belz (2016)

activity of known allelochemical degradation and/or transformation products and possible antagonistic or synergistic effects (Duke 2015) often attributed to an interplay at the microbial level.

14.4 Synthesis, Localization, and Release of Allelochemicals from Donor Plants

Secondary metabolites, including allelochemicals, are synthesized and localized in almost all plant parts and tissues including stems, leaves, roots, flowers, pollen, bark, fruit, seeds, and rhizomes (Putnam and Tang 1986; Rice 2012). However, many studies have noted that the diversity and concentration are frequently higher in the shoots in contrast to roots (Latif et al. 2020). This is typically explained by the diversity of biosynthetic activity in shoot tissues as opposed to roots. Numerous studies have also reported high concentrations of allelochemicals in seeds, particularly in seed coats, contributing to the longevity of certain weed seeds in the soil by preventing decomposition by microorganisms (Qasem and Foy 2001).

However, allelochemicals must occur in sufficient concentration in leaves, stem, and roots to allow further accumulation in soil or surrounding environment upon release. This is also the case with invasive plants utilizing allelochemicals in effective defense strategies against other plants. Allelochemicals are generally synthesized in cytoplasm of the cell or in specific organelles, such as chloroplast, and are often localized in vacuoles away from the synthetic machinery, or deposited in specialized storage structures such as glandular trichomes to prevent autotoxicity or immediate release into the environment (Duke et al. 1999a,b; Eom et al. 2005). Water-soluble allelochemicals are often stored in vacuoles (Kutchan 2005), while lipophilic compounds are localized in resin ducts, laticifers, glandular trichomes, or cuticle members (Yazaki 2006).

Biologically active concentrations of allelochemicals are required to elicit defense or signaling against potential threats. Significant allocations of carbon and energy resources in terms of adenosine triphosphate (ATP) and equivalent, i.e. nicotinamide adenine dinucleotide phosphate (NADPH2), are required for the synthesis of secondary metabolites, storage, transport, and exudation (Wink 2011). The concentration and diversity of allelochemicals synthesized in plants vary with physiological stage and are impacted by various intrinsic and extrinsic factors including genetics, tissue type, age and environment, and their interactions. Frequently the abundance of allelochemicals can be enhanced in response to exposure to various stimuli including insects, pathogens, ultraviolet light, ionizing radiations, temperature and moisture stress, and herbivory (Albuquerque et al. 2011; Wink 2011). Secondary metabolites are often released in mixtures with other related metabolites, further complicating the interactions with one secondary metabolite often performing more than one biological functions that may include defending the plant against several, often unrelated threats (Macias et al. 2007).

Plants generally possess specialized tissues for the release of allelochemicals into the environment including stomata and glandular trichomes in aboveground tissue and root hairs, border cells, epidermis, and periderm in belowground tissues. In the shoot, glandular trichomes are often the primary tissues for the exudation of secondary volatile metabolites, either synthesized within trichomes or translocated from other plant parts. Volatile compounds can diffuse directly from cell walls on the plant surface, while nonvolatile polar compounds require specialized protein transporters (Weston et al. 2012).

Volatilization and leaching from foliar plant parts, decomposition of nonliving plant parts, and root exudation have each been characterized with respect to the release of allelochemicals over time in the environment in numerous plant species. Root exudation is the most common mechanism of release of allelochemicals into the belowground environment (rhizosphere), through exudation by primary and/or secondary roots and root hairs (Bertin et al. 2003). Root hairs, a unicellular extension of the epidermis, are frequently the interface for chemical interactions in the rhizosphere as they exude the majority of secondary metabolites at varying concentrations, from both living and decomposing roots. Some allelochemicals are synthesized within root hairs and directly released into the soil; examples include sorgoleone and certain naphthoquinones (shikonins), while others, i.e. flavonoids, are translocated to root hairs prior to exudation (Weston et al. 2012). Mechanisms of root exudation including simple diffusion, vesicle transport, and specific ion channels have been previously described in greater detail (Weston et al. 2012). Extraction of plant-produced allelochemicals from rhizosphere soil is challenging due to the complex physical and chemical nature of soil matrices, but relatively recent developments in solid-phase extraction and specialized silicone microprobe tubing to trap soil allelochemicals have facilitated their recovery considering the broad range of both nonpolar to moderately polar metabolites (Weston et al. 2012).

14.5 Factors Affecting Biosynthesis and Release of Allelochemicals

The release of allelochemicals from donor plants into the natural environment is particularly challenging to document when conducting field studies. The release of allelochemicals depends on several intrinsic and extrinsic factors as presented in Figure 14.1. Intrinsic factors include species of plant, cultivar, age, physiological stage, and tissue type, while extrinsic factors typically include pathogen infestation, physical injury, nutrient stress, high temperature, irradiation exposure, soil characteristics, and interference with competitors.

The physicochemical and biotic characteristics of soil affect allelochemical production and fate following release and, hence, the overall fitness of the plant (Cheng 1995; Einhellig 1996; Blum 2011). Soil textural properties affect the degree of adsorption of allelochemicals, with activity of allelochemicals in loose and sandy soil generally being greater compared with heavy clays or loamy soils or those high in organic matter content (Oleszek and Jurzysta 1987), while metabolites leach more quickly from sandy soil (Goslee et al. 2001). Other soil characteristics include soil pH, organic matter, available carbon, and nitrogen affecting allelopathic activity by uptake, immobilization, microbial activity, and accumulation of inorganic ions and nutrients (Dalton et al. 1983; Facelli and Pickett 1991). For example, nutrient stress can result in the upregulation of allelochemical synthesis, particularly phenolic acids, which can interfere further with mineral uptake (Einhellig 1996).

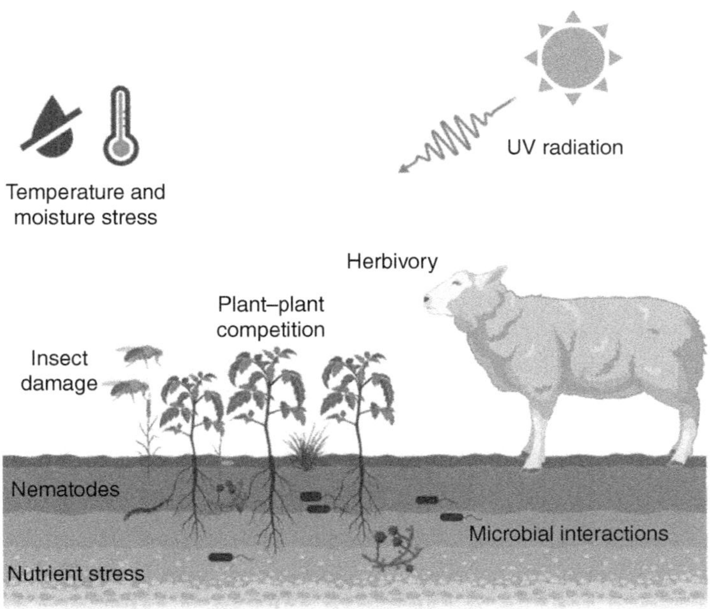

Figure 14.1 Invasive plants are clearly impacted by their environment, and such interactions may lead to enhanced allelopathy, competition for resources, and adaptation to environmental stressors and herbivory.

14.6 The Role of Soil Microorganisms in the Release and Transformation of Allelochemicals

Soil microbiota are important modulators of plant community structure. They can limit or promote growth and establishment of plant species by maintaining a balance between antagonistic and mutualistic interactions. While certain invasive species induce niche construction of microbial communities for nutrient enrichment, others are recruited to facilitate bioactivation and release of defense-related secondary metabolites as well as nitrogen fixation (Dassonville et al. 2011). As the appreciation for soil microbiota and their ecological functions has recently increased, their impact on promoting the persistence of certain weeds through microbially mediated bioactivation of secondary products is gaining traction in the field of allelopathy (Schandry and Becker 2020).

The majority of plants including invasive weeds are known to maintain associations with mycorrhizal fungi and may therefore be dependent on mycorrhizal association for their survival. However, invasive plants that are mycorrhizal independent may actually exploit native plant dependence on mycorrhizal fungi through inhibition of the mycorrhizae (Cipollini et al. 2012). The Eurasian weed species, *Alliaria petiolata* (a member of Brassicaceae), has been shown to produce allyl isothiocyanate (AITC) and benzyl isothiocyanate (BzITC), both degradation products of glucosinolates, which inhibit spore germination of mycorrhizal fungi, thereby suppressing the establishment of mycorrhiza-dependent native plants (Roberts and Anderson 2001).

Microbial interactions facilitate and catalyze several reactions in the benzoxazinoid transformations of plants in the *Poaceae* family. The defensive properties of benzoxazinoids were first identified in maize (*Zea mays*) and were associated with the presence of 2,4-dihydroxy-7-methoxy-1,4-benzoxazin-3-one (DIMBOA) and 2,4-dihydroxy-2H-1,4-ben zoxazin-3(4H)-one (DIBOA) (known inhibitors of herbivorous insects) and later in cereal rye (*Secale cereale*), when the crop was noted to display increased resistance to fungal pathogens (Schulz et al. 2013). Interestingly, benzoxazinoids, released by plants through decomposition or root exudation, are subsequently converted to benzoxazolinones, aminophenoxazinones, acetamides, and malonamic acids through bacterial and fungal degradation (Friebe et al. 1996). The degradation products, classified as aminophenoxazinones (2-amino-3*H*-phenoxazin-3-one [APO] and 2-amino-7-methoxy-3*H*-phenoxazin-3-one [AMPO] produced from benzoxazolin-2-one (BOA) and 6-methoxy-benzoxazolin-2-one (MBOA), respectively), are transformed by oxidation of aminophenols and form exceptionally active microbial metabolites with the ability to persist for extended periods in soil (Mwendwa 2019). Venturelli et al. (2015) demonstrated that both APO and AMPO inhibit a key enzyme, histone deacetylase, required for cell division and growth, hence negatively impacting several aspects of plant growth at a concentration comparable to commercial herbicides.

14.7 Metabolic Profiling of Allelochemicals

The soil rhizosphere presents a complex medium for biochemical interactions between roots, soil, and microbiota and remains relatively uncharacterized with regard to terrestrial ecology. Advances in high precision analytical instrumentation, including high-pressure liquid chromatography interfaced with mass spectrometry, have enabled accurate and robust methodologies for screening, annotation, and identifications of allelochemicals. Due to the chemical diversity of secondary metabolites, multiple, often complementary, analytical platforms are employed to capture the entirety of metabolome of a biological system. A detailed description of metabolic profiling of allelochemicals using metabolomics approaches has been presented by Weston et al. (2015). By performing efficient extractions followed by metabolic profiling of complex matrices including rhizosphere soils, one can describe the functional state of a biological system at a particular point in time with respect to allelopathic interference (Latif et al. 2020). Recent advancements in analytical chemistry have resulted in the utilization of untargeted metabolomics approaches to generate profiles of metabolites, both known and unknown, effectively generating a "fingerprint" of its chemical constituents (Weston et al. 2015). However, this analytical tool has not yet been well utilized for the characterization of allelochemicals in invasive weed species, although it has been applied frequently for identification on novel products and in their isolation from traditional medicinal plants and agronomically important crops (Ha et al. 2015; Zhou et al. 2016). Application of metabolomics and associated bioinformatics is therefore potentially useful for researchers to distinguish the role of allelopathy in the establishment and persistence of exotic weeds from competition for resources, particularly when studying invasive weeds and their persistence. Metabolomics, in conjunction with proteomics and transcriptomics, is emerging as a systematic approach for studying a plant's chemical fingerprint (Weston et al. 2015) and has only recently been successfully employed in the study of the invasive perennial vine *M. micrantha* (Liu et al. 2020). Following profiling, processing of a substantial amount of spectral data is performed by the use of computational tools including those for quality control, binning, and alignment of molecular features across samples, feature extractions, conversion, and normalization of data (Smith et al. 2005). However, as it is always difficult to characterize all plant metabolites in a complex mixture, particularly those in trace quantities, unknown natural products can present challenges with respect to structural elucidation.

14.8 Case Studies of Invasive Plant Species Exhibiting Allelopathic Interactions

Invasive weeds have frequently been redistributed either accidentally or deliberately when established as crops or ornamental plants in non-native ranges, and many have successfully established over time in various geographically distinct locations. Several hypotheses have been proposed to account for the factors that have contributed to the establishment and persistence of invasive plant species. Firstly, the enemy release hypothesis predicting the release from natural enemies in the non-native range provides an advantage to the exotic species (Keane and Crawley 2002). Secondly, the evolution of increased competitive ability hypothesis predicts a reduction of the synthesis and release of chemical compounds and increased physical competitive traits (Blossey and Notzold 1995). Finally, in direct contrast, the novel weapons hypothesis predicts increased production of unique allelochemicals in the exotic species not found in native plants. Such allelochemicals are generally phytotoxic to natives which lack defenses against these novel chemicals (Callaway and Ridenour 2004).

In support of the application of analytical chemistry methods for the study of invasive weeds, Cappuccino and Arnason (2006) showed that the majority of invasive weeds in North America were likely to produce unique phytochemicals when compared with less invasive species. Similar observations were made by Macel et al. (2014) where six exotic species and their native congeners were compared by subjecting plant extracts prepared from them to untargeted metabolic profiling. Results revealed that, in general, the exotic species contained both a higher number of and more unique metabolites compared to their native congeners. Results from these studies clearly support the investigative capacity of metabolic profiling for further identification of the role of allelopathy in the persistence of exotic weed species.

The case studies presented herein provide examples of diverse invasive plants that populate various continents and follow the general trend of a poleward shift due to changing climatic conditions (Parmesan and Yohe 2003), including both monocots and dicots, which interact and sometimes compete with native flora through the production of allelochemicals. The species described in the following text include *E. plantagineum*, Johnsongrass (*Sorghum halepense*), *Reynoutria japonica*, *A. petiolata*, and *P. hysterophorus*, and some of the examples of direct effects of allelochemicals impacting target plant species are presented (Figure 14.2). In contrast, *R. japonica* and *M. micrantha* (Figure 14.2) are presented as examples of indirect allelopathy. These case studies represent excellent models with which to study evolution and adaptation in plant secondary product synthesis and release under the diverse microclimates frequently experienced in the invasive range.

14.8.1 *Echium plantagineum*

Commonly known as Paterson's curse, *E. plantagineum* was introduced to Australia in the 1800s as an ornamental, with recent findings suggesting multiple introduction events, based on genetic evidence (Zhu et al. 2019). Following establishment in New South Wales (NSW), it has become widespread across 30 M ha in Southern Australia and is today considered highly invasive, whereas in its native range in the Iberian Peninsula, infestation is typically more limited and frequently associated with degraded soils caused by movement of livestock (Piggin 1982).

In contrast to the less invasive but related perennial species *Echium vulgare*, *E. plantagineum* has been shown to possess a smaller genome, has an annual or biennial lifecycle, and exhibits higher genetic diversity compared with the perennial Viper's bugloss or *E. vulgare* (Zhu et al. 2017). Recent investigations have suggested that the majority of invasive plants have smaller genome sizes compared with noninvasive co-generic counterparts from the same species (Lavergne et al. 2009). Small genome size has also been associated with enhanced competitive ability and specific traits such as larger leaf area and higher photosynthetic rates associated with reduced allocation of carbon resources for cell cycling (Bennett et al. 1998). *E. plantagineum* is self-incompatible in the native range, but self-compatible in its invaded territory of Australia (Petanidou et al. 2012), potentially enhancing its colonizing ability. It has been found in areas of higher temperature and low rainfall range, while *E. vulgare* occurs more frequently at higher altitudes in areas with cold winter temperatures and high rainfall (Zhu et al. 2017).

Figure 14.2 Representative images of invasive plant species presented as case studies. (a) *Echium plantagineum* in invaded territory, Canberra, Australia. *Source:* Image courtesy of Australian Native Plant Society. (b) *Sorghum halepense* in its invaded range in New South Wales, Australia. *Source:* Image courtesy of Jeff Dahlberg, University of California Cooperative Extension Service. (c) *Parthenium hysterophorus* infestation in Queensland, Australia. *Source:* Image courtesy of Dr. Ali Bajwa, Australia. (d) *Reynoutria japonica* established in Nova Scotia, Canada. *Source:* Image courtesy of David Clements, Canada; (e) *Alliaria petiolata Source:* Image retrieved from https://skaakdotorg.files.wordpress.com/2015/01/garlic-mustard-4863.jpg. (f) *Mikania micrantha*. *Source:* Image courtesy of Navaneeth Krishnan, Creative Commons Zero, Public Domain Dedication.

Foliar tissues of the plant contain high concentrations of pyrrolizidine alkaloids and their N-oxides (Skoneczny et al. 2019); specifically, these metabolites are known to deter insect herbivory and are also associated with livestock toxicity (Molyneux et al. 2011), resulting in acute liver failure (Stegelmeier 2011). *E. vulgare*, which is not considered particularly invasive, also contains pyrrolizidine alkaloids but at comparatively lower concentrations assessed under controlled environmental conditions (Skoneczny 2017). *E. plantagineum* also produces and releases phytotoxic naphthoquinones from its roots, including acetylshikonin and shikonin, both inhibitory to germinating weed seedlings and mammalian cells, *in vitro* (Durán et al. 2017). Furthermore, recent evidence shows clearly that the biosynthesis of phytotoxic naphthoquinones (shikonins) in *E. plantagineum* was upregulated in roots exposed to elevated temperatures and under short-term drought conditions (Skoneczny et al. 2019).

Collectively, evidence from previous work suggests that ecological traits favoring persistence under extreme conditions, genetic diversity, and unique plant secondary chemistry present in *E. plantagineum* have potentially facilitated establishment and persistence of this plant in Australia. Shikonins are naphthoquinones, known to exhibit strong antimicrobial activity and allelopathic effects (Dresler et al. 2015), and they are present in high concentrations in roots and soils of Paterson's curse. It is highly plausible that the elevated expression of shikonins in exotic *E. plantagineum* could impact the rhizosphere microbiome, given the existing knowledge on their antimicrobial activity. Localization of shikonins in root periderm tissue may also provide a barrier against root herbivory in *E. plantagineum*, a mechanism of resistance previously discussed in other exotic plant introductions (Hierro and Callaway 2003).

14.8.2 *Sorghum halepense*

Commonly known as Johnsongrass, *S. halepense* is a perennial grass species reaching up to three meters in height, with fleshy rhizomes. It is a native of the Mediterranean regions of Africa, Asia, and Europe and is globally distributed and cited as one of the world's top 10 worst weeds due to its high vigor and ability to adapt to diverse environments (Holm et al. 1977). Johnsongrass spreads mainly by rhizomes, but is also capable of sexual reproduction by producing selfed or outcrossed seeds. In the 1800s it was introduced from Turkey to North America as a potential forage crop and pasture species; however, it has since expanded its range and established as an invasive weed across the Southern United States (Hatch 2001). *S. halepense* easily hybridizes with other sorghum species, particularly with *Sorghum bicolor* (Essl et al. 2009). This attribute may potentially contribute to its invasion success. Furthermore, its utilization of the C_4 cycle of CO_2 metabolism has resulted in greater overall CO_2 assimilation rates compared with C_3 photosynthesis, particularly in warmer climates with temperatures exceeding 30 °C, providing a physiological advantage over photosynthetically less-efficient C_3 counterparts (Brown 1999).

Sorghum halepense is difficult to control by herbicides in all cropping systems due to its deep and extensive rhizomes that make herbicide translocation difficult (Vila-Aiub et al. 2012). It is also difficult to distinguish it from other related species including Sudan grass, Sorghum × Sudan grass hybrids, or annual grain sorghum at the same developmental stage (Guertin 2001), and at times hybridization is thought to occur in natural settings except maize. In recent years, certain *S. halepense* biotypes have evolved resistance to numerous herbicide families (Powles and Yu 2010; Werle et al. 2016). *S. halepense* roots have also been shown to produce and exude the well-characterized allelochemical sorgoleone, a photosystem II and respiration inhibitor (Czarnota et al. 2001) as well as an inhibitor hydroxyphenylpyruvate dioxygenase (HPPD), an important site for herbicidal mode of action in existing commercial herbicides (Weston and Mathesius 2013).

Roots of *S. halepense* exude significantly higher quantities of sorgoleone than commercial sorghum species, *S. bicolor* (Czarnota et al. 2003). Sorgoleone is produced and exuded by living sorghum root hairs, with exudation occurring shortly after seedling radicle emergence (Czarnota et al. 2003). Sorgoleone has also been shown to be persistent in soil following soil incorporation, with several metabolites detectable up to two months. Later studies found that mineralization of sorgoleone in soils by microorganisms varied according to the functional groups attached to sorgoleone (Gimsing et al. 2009). These observations suggest that *S. halepense* may utilize sorgoleone as a potent allelochemical against neighboring competitors during early growth stages, but its ability to continually release sorgoleone over time along with its persistence could impact its specific activity under field conditions.

More recent studies have also identified that the production of numerous phenolic compounds by *S. halepense* differed based on developmental stages of plants (Huang et al. 2015), with *p*-hydroxybenzoic acid being exuded in high concentration by seedlings and three other phenolic compounds (*p*-hydroxybenzaldehyde, *p*-hydroxybenzoic acid, and ethyl *p*-hydroxybenzoate) being produced at the jointing stage and in high abundance at flowering and reproduction. This is similar to the finding of Weston and colleagues in studies of Sorghum × Sudan grass hybrids (Weston et al. 1989). Results with both genotypes suggested that allelochemical exudation at early growth stages may enhance interference with neighboring plants followed by competitive advantages at canopy and rhizome formation. The production of phenolics and bioactive quinones constitute an arsenal of secondary metabolites that could contribute to the plant's allelopathic potential from seedling stages to maturity.

14.8.3 *Parthenium hysterophorus*

The invasive herbaceous plant *P. hysterophorus*, commonly known as parthenium weed, is native to tropical and subtropical regions of America (Dale 1981). This plant was introduced to over 40 countries worldwide including India, Pakistan, Nepal, Bangladesh, Australia, Ethiopia, Tanzania, and South Africa (Shi et al. 2015). Parthenium weed is a tall (up to 2.5 m) annual or short-lived perennial with an erect, highly branched growth habit (Adkins and Shabbir 2014). Rapid germination and vigorous early growth, along with production of allelopathic compounds, provides the plant with a competitive advantage over neighboring plants (Navie et al. 1998).

Interestingly, this highly invasive weed has exhibited numerous adverse effects on biodiversity of native ecosystems, agricultural crop production, and also on human and animal health (Adkins and Shabbir 2014; Allan et al. 2018). Apart from competition for resources, it has been proposed that allelopathic interference plays a major role in its invasiveness. Both above- and below-ground litter of *P. hysterophorus* have been shown to inhibit germination and growth of various plants in a dose-responsive manner. Litter collected from the invasive regions exhibits a greater level of phytotoxicity compared with litter from the native range (Shi and Adkins 2018). The secondary chemistry in *P. hysterophorus* is unique in a way that parthenin, a sesquiterpene lactone, is produced in abundance in the aerial portions of the plant (Shi et al. 2015; Allan et al. 2018). Parthenin has been shown to inhibit germination and growth of weed seedlings at low concentrations ranging from μM to mM (Belz et al. 2007).

Several other phytotoxic sesquiterpene lactones have also been described and are known to be produced by *P. hysterophorus*. These include ambrosin, hysterin, coronopilin, damsin, and hymenin (Rodríguez et al. 1977; de la Fuente et al. 2000). A comparison between two biotypes of *P. hysterophorus*, Clermont (high invasive potential) and Toogoolawah (low invasive potential), in Eastern Australia showed that polyphenol content was significantly higher in the more invasive biotype Clermont, exhibiting enhanced vegetative growth and higher seed production compared with the less invasive Toogoolawah biotype (Bajwa et al. 2017). Furthermore, the presence of various phytotoxic flavonoids, including kaempferol and quercetin 3-*o*-glycoside, has been noted in foliar tissues of the plant, suggesting upregulation of the phenylpropanoid biosynthetic pathway in *P. hysterophorus*. A recent study demonstrated that the allelopathic potential of its plant residues was significantly reduced after vermicomposting, providing a potential management strategy for reducing its phytotoxicity (Hussain et al. 2016).

14.8.4 *Alliaria petiolata*

Commonly known as garlic mustard, *A. petiolata* is a biennial flowering plant of the Brassicaceae family originating from Eurasia. It is now considered a highly invasive plant impacting forest ecosystems in the Eastern United States and Canada (Stinson et al. 2007; Van Riper et al. 2010). A previous study on the invasion history of the plant in North America showed that multiple introductions occurred from six European regions (Durka et al. 2005). The authors hypothesized that both genetic diversity established by multiple introductions and lack of inbreeding depression due to high selfing rates contributed to its invasion success.

The persistence of *A. petiolata* in invaded territories has also been attributed to its ability to compete for resources, often forming dense monotypic stands (Meekins and McCarthy 1999; Wixted and McGraw 2010). However, the suppression of other plants in complex forest ecosystems is also evidently impacted by the production of allelochemicals impacting native plants, their mycorrhizal fungal associations, and soil bacterial communities (Cipollini and Cipollini 2016). Interestingly, *A. petiolata* has been used for over 6000 years as a spice in Northern Europe, because of the pungency characteristic of the mustard family due to the presence of glucosinolates (Vaughn and Berhow 1999). Glucosinolates have been investigated in the past for their phytotoxicity and weed suppressive ability (Bialy et al. 1990; Vaughn 1999; Shauffer 2017; Rahman et al. 2018). Early studies established that the plant contained high levels of AITC and BzITC, phytotoxic degradation products of the two major glucosinolates produced by this plant (Vaughn and Berhow 1999). Both compounds were found to strongly inhibit radicle elongation of wheat and garden cress.

Allelopathic effects of *A. petiolata* from soil conditioning was demonstrated by Prati and Bossdorf (2004), where soil previously used for growing *A. petiolata* was subsequently found to be inhibitory to the growth of the test species, *Geum laciniatum*. Growth inhibition of the test species recovered significantly when the same soil was treated with activated charcoal, which presumably absorbed the allelopathic root exudates released by *A. petiolata* into the soil matrix. Under field conditions, it has been established that soil conditioning by *A. petiolata* was reduced with increasing age of the plant, correlated with reduced concentration of glucosinolates in roots. A recent study showed that while the presence of *A. petiolata* was associated with an overall reduction in diversity of other plant species, the impact was greater toward alien species than for native species, with which the plant is more likely to coexist (O'Sullivan et al. 2019). However, the underlying mechanisms for the differential suppression of alien species relative to that of native species remain unknown.

As with many species of *Brassicaceae*, *Alliaria* spp. also generally lacks strong associations with mycorrhizal fungi. An early laboratory study showed that water leachates of the plant inhibited the growth of test species tomato (*Lycopersicon esculentum*) and sorghum and also reduced mycorrhizal colonization in those plants (Roberts and Anderson 2001). The authors further demonstrated that leachates reduced the arbuscular mycorrhizal inoculum potential in the same test species. This study provided the first insights into the reduction of the competitive ability of other plants by garlic mustard, through interference with the formation of mycorrhizal associations.

Several studies have also focused on evaluating the impact of *A. petiolata* on the soil bacterial community composition. Burke and Chan (2010) observed that soil obtained from forest floors containing *A. petiolata* litter did not significantly alter soil bacterial community structure with respect to species richness, evenness, or diversity. However, a recent study conducted under controlled conditions demonstrated that following growth of *A. petiolata* in field collected soil, the metabolic potential of microbial communities was significantly reduced (Price and Blair 2019). The impact of aqueous plant extracts and synthetically produced BzITC and AITC was evaluated by Portales-Reyes et al. (2015). Results showed that both the aqueous extract and synthetic BzITC had no significant impact on legume growth or nodulation, while AITC reduced nodulation. However, field validation of this study is yet to be performed. Overall, over 20 years of studies evaluating the direct or indirect allelopathic potential of *A. petiolata* have highlighted several plausible mechanisms utilized by this species to persist in invaded territories, including allelopathic interference. It also highlights the need for stringent experimental design, both under controlled conditions and field settings, using a multidisciplinary approach to clearly distinguish the contribution of allelopathy and soil microbial communities to invasion success of weeds.

14.8.5 *Reynoutria japonica*

R. japonica, commonly known as Japanese knotweed, originated in Asia, specifically Japan, Taiwan, and China (Child and Wade 2000). It was introduced to North America, Canada, and Europe in the 1800s, typically as an ornamental plant (Weston et al. 2005). Various synonyms exist when naming this plant, including *Fallopia japonica*, *Polygonum japonicum*, *Polygonum zuccarinii*, and *R. japonica*. A dwarf variant of *R. japonica*, previously identified as *Polygonum cuspidatum* (var. compactum), was also introduced to the United Kingdom between 1850 and 1880 as an ornamental garden plant. Interestingly, many of these introductions were male sterile and exhibited dark red stem pigmentation (Bailey 1994; Child and Wade 2000). An additional *Reynoutria* species, *R. sachalinense*, was male fertile and exhibited greater genetic diversity than *R. japonica*, which was concurrently introduced to the United Kingdom in the mid-1800s from the Japanese island of Sakhalin. Hybridization of *R. japonica* and *R. sachalinense* resulted in the hybrid *R.* × *bohemica* (syn. *Fallopia bohemicum*), which has established successfully in England and throughout North America (Weston et al. 2005; Gillies et al. 2016).

The incredible capacity for vegetative regeneration and high biomass accumulation has enabled Japanese knotweed to physically outcompete native plants in its invaded ranges and resulted in it being named as one of the world's worst weeds (Barney et al. 2006). Rhizomes can be found at depths of 5–6 m, rendering herbicide translocation difficult for effective control (Weston et al. 2005; Bashtanova et al. 2009). In areas where herbicides can be applied, several application methods have been investigated, including direct stem injection, injection following cutting of stems, wick application of herbicides or foliage, and foliar application by spraying, further demonstrating the complications associated with chemical control of the plant (Weston et al. 2005). Non-herbicidal control of the weed includes removal of the above- and below-ground components by excavation or hand pulling and management by repeated mowing, smothering, or grazing (Seiger and Merchant 1997; Weston et al. 2005; Delbart et al. 2012). Introduction of an insect as biological control agents in the European and American continents may possibly further facilitate reduction of Japanese knotweed, but this has not yet been adopted as an effective control strategy (Grevstad et al. 2013).

Japanese knotweed is also known to produce unique secondary metabolites, including polyphenols, which may act as antioxidants or allelochemicals suppressing plant growth (Vastano et al. 2000; Dommanget et al. 2014). Studies involving the hybrid *R.* × *bohemica* have shown that the interaction with native plants occurred by delaying the time to maturity for natives, most notably, time to flowering, thereby effectively interfering with pollination timing by native pollinators and limited production (Parepa et al. 2012). Further studies have also pointed to the impact of soil microbiota on exotic knotweed (*R.* × *bohemica*) invasion success. In this case, the availability of active soil microbial communities favored the growth and development of the hybrid knotweed, whereas no significant impact was observed on the growth of native plants through enrichment of soil microbiota including both bacteria and fungi (Parepa et al. 2013).

The growth advantage offered by the presence of soil microbiota was diminished in the presence of activated charcoal, suggesting the existence of a plant–microbiome interaction through involvement of plant or microbially produced natural products or allelochemicals. Conversely, the results also suggest that soil microorganisms from invaded sites do not exert biotic resistance toward the invasive knotweed species at a community level further supporting, suggesting, and confirming plant–soil microbial chemical interactions. However, further investigations of such complex interactions are warranted.

Dassonville et al. (2011) further demonstrated that Asian knotweeds significantly alter the community structure of microbial nitrifiers and denitrifiers in the rhizosphere, leading to an altered niche microbiome. This may thus enhance persistence of *Reynoutria* spp. through improved utilization of N and thus competition for resources. As rhizosphere microorganism recruitment and enrichment is associated with root exudation, it is plausible that chemical cues produced by *Reynoutria* spp. may modulate soil microbial communities through synthesis and exudation of specific plant secondary metabolites. However, the structural elucidation of many secondary metabolites associated with allelopathic interference and microbial recruitment will require further experimentation.

14.8.6 *Mikania micrantha*

M. micrantha (commonly referred to as "mile-a-minute" vine) is a member of the *Asteraceae* and is a rapidly growing vine (Day et al. 2016). This fast-growing perennial is native to tropical Central South America and has previously been reported as one of the world's worst weeds (Holm et al. 1977). Today it is listed as one of the 100 worst invaders by the International Union for Conservation of Nature (IUCN). It is a major problem in tropical Asia and the United States, South Pacific and Indian Ocean islands, and Papua New Guinea, resulting in loss of genetic and species diversity and a decline in soil productivity (Day et al. 2016).

Several biological attributes contribute to the invasiveness of *M. micrantha*. Mature plants produce approximately 170 000 seeds m^{-2} that are widely dispersed (Kuo et al. 2002). The climbing growth habit of the plant results in smothering of the canopies of other plants, outcompeting them for sunlight (Shen et al. 2015). Interestingly, the plant exhibits a high level of plasticity to differing nutrient and light conditions, with seed set found to be greatest under relatively low light conditions and low levels of soil nutrients (Yang et al. 2003). A groundbreaking study by Liu et al. (2020) discovered that the genome of this plant possessed many areas of segmental duplication, which may also be responsible for its rapid environmental adaptation and fast growth rate. Gene analysis showed the plant could achieve more efficient photosynthesis, potentially through C-fixation during the day and also enhanced CO_2 absorption at night. The net photosynthetic rate of leaves was found to be higher than other C_3 plants and similar to that of C_4 plants. Further investigation revealed that in *M. micrantha*, the enhanced rate of photosynthesis was enhanced through the activation of Crassulacean acid metabolism (CAM) photosynthesis, with transcription generally silenced in most plants, but activated only in some in response to environmental changes. The increase in primary metabolism driven by enhanced photosynthetic output may also result in an increase in the synthesis of novel secondary plant products, in this case sesquiterpene lactones.

Early experiments investigating the mechanism of invasion and persistence of *M. micrantha* pointed to allelopathy as a contributing factor (Ismail and Mah 1993). Exposure of various fractions of plant extracts prepared from aerial parts of the plant was found to be allelopathic, with three sesquiterpene lactones found to be contributing to the observed phytotoxicity (Shao et al. 2005). Dihydromikanolide, deoxymikanolide, and 2,3-epoxy-1-hydroxy-4,9-germacradiene-12,8: 15,6-diolide were all found to inhibit both germination and seedling growth in the tested plant species under laboratory and field conditions. Liu et al. (2020) further confirmed the localization of the sesquiterpene lactones in leaves and flowers and observed that these compounds were released into the soil by litter fall. Interestingly, the increase in sesquiterpene lactones was shown to be correlated with an increase in ammonium nitrogen in the soil, following recruitment of nitrogen fixing bacteria. Interestingly, the same compounds were shown to decrease the number of denitrifying bacteria in the soil. When established in low-nutrient soils, *M. micrantha* may be able to outcompete neighboring plants incapable of modulating the soil microbiome to increase nutrient availability. Active promotion of nitrogen cycling by the plant through the release of such sesquiterpene lactones provides compelling evidence for the role of secondary metabolites as inducers of indirect allelopathy in *M. micrantha*.

14.9 Conclusions

Invasive plants employ various strategies to persist in invaded territories, and allelopathic interference through the production of secondary metabolites is now understood to play a significant role in successful invasion strategies of certain weeds. The synthesis, translocation, and activation of allelochemicals in plants are always dynamic and vary significantly with respect to the metabolites produced, the target species encountered, and the presence/absence of soil microbiota, resulting in transformation plant metabolites, degradation of residues, and/or activation of root exudation.

Given the complexity of weed invasion and the frequent employment of multiple strategies by successful weed invaders, a multidisciplinary approach is required in future research to further investigate cross-kingdom interactions between invasive plants, herbivores, and soil microbiota. This must also involve exploration of key plant biosynthetic pathways and the microbial metagenome coevolving with associated weed species to facilitate defense and nutrient uptake for successful invasion and persistence of weed species. Plant root exudates produced in response to biotic and abiotic stressors are likely critical in the selection and proliferation of specific microbiomes in the rhizosphere (Berendsen et al. 2018; Sasse et al. 2018). Therefore, we believe that future opportunities await the exploitation of plant–microbial interactions facilitated by organic molecules for sustainable crop production.

In contrast to the study of crop plants used for forages or cereals, the secondary chemistry of invasive plant species has been largely unexplored, likely due to the significant amount of time and funding required to identify and isolate putative allelochemical or phytotoxic constituents. However, the scope of current research performed in plant invasion biology is gradually expanding to include more fundamental aspects of chemical ecology that may contribute to the identification of active allelochemical constituents and evaluation of the physiology of exotic plant invaders.

Advanced technologies for the characterization and identification of phytotoxic plant secondary products in complex matrices such as soil and plant extracts have now enabled rapid advancements in the field of phytochemistry, including the field of allelopathy. The case studies presented have provided several clear examples of the role that allelochemicals may potentially play in the successful invasion strategies and also point to the need for additional opportunities for the application of cutting-edge analytical techniques, such as metabolic profiling, to elucidate the allelochemicals contributing exotic weed interference. In some examples, invasive plants were shown to modulate the biosynthesis and release of phytotoxic secondary metabolites in their invaded range(s), in direct contrast to their native habitat, thereby providing a competitive advantage. Given the continuing challenges associated with weed invasion and range expansion, the use of cutting-edge approaches to study allelochemical biosynthesis, release, and transformation in invaded ranges will undoubtedly provide further evidence for the role of allelochemicals in plant invasion and their impacts on the transformation of native ecosystems post-invasion. A major challenge will thus be how to use this rapidly expanding body of knowledge to more effectively manage plant invasion while protecting biodiversity in native ecosystems.

References

Abenavoli, M., Lupini, A., Oliva, S., and Sorgonà, A. (2010). Allelochemical effects on net nitrate uptake and plasma membrane H+-ATPase activity in maize seedlings. *Biologia Plantarum* 54: 149–153.

Abrahim, D., Takahashi, L., Kelmer-Bracht, A., and Ishii-Iwamoto, E. (2003). Effects of phenolic acids and monoterpenes on the mitochondrial respiration of soybean hypocotyl axes. *Allelopathy Journal* 11: 21–30.

Adkins, S. and Shabbir, A. (2014). Biology, ecology and management of the invasive parthenium weed (*Parthenium hysterophorus* L.). *Pest Management Science* 70: 1023–1029.

Ahuja, I., Kissen, R., and Bones, A.M. (2012). Phytoalexins in defense against pathogens. *Trends in Plant Science* 17: 73–90.

Albuquerque, M.B., Cavalcanti dos Santos, R., Lima, L.M. et al. (2011). Allelopathy, an alternative tool to improve cropping systems. A review. *Agronomy for Sustainable Development* 31: 379–395.

Aliotta, G., Mallik, A.U., and Pollio, A. (2008). Historical examples of allelopathy and ethnobotany from the Mediterranean region. In: *Allelopathy in Sustainable Agriculture and Forestry* (eds. R.-S. Zeng, A.U. Mallik and S.M. Luo), 11–24. New York, NY: Springer.

Allan, S., Shi, B., and Adkins, S.W. (2018). Impact of parthenium weed on human and animal health. In: *Parthenium Weed: Biology, Ecology and Management*, vol. 7, 105–128. Oxfordshire, UK: *CABI*.

Aslam, F., Khaliq, A., Matloob, A. et al. (2017). Allelopathy in agro-ecosystems: a critical review of wheat allelopathy-concepts and implications. *Chemoecology* 27: 1–24.

Bailey, J.P. (1994). *Reproductive Biology and Fertility of Fallopia japonica (Japanese Knotweed) and its Hybrids in the British Isles*, 141–158. Chichester, UK: Wiley.

Bajwa, A.A., Chauhan, B.S., and Adkins, S. (2017). Morphological, physiological and biochemical responses of two Australian biotypes of *Parthenium hysterophorus* to different soil moisture regimes. *Environmental Science and Pollution Research* 24: 16186–16194.

Bakker, P.A., Pieterse, C.M., de Jonge, R., and Berendsen, R.L. (2018). The soil-borne legacy. *Cell* 172: 1178–1180.

Barkosky, R.R. and Einhellig, F.A. (1993). Effects of salicylic acid on plant-water relationships. *Journal of Chemical Ecology* 19: 237–247.

Barkosky, R.R. and Einhellig, F.A. (2003). Allelopathic interference of plant-water relationships by para-hydroxybenzoic acid. *Botanical Bulletin of Academia Sinica* 44: 53–58.

Barney, J.N., Tharayil, N., DiTommaso, A., and Bhowmik, P.C. (2006). The biology of invasive alien plants in Canada. 5. *Polygonum cuspidatum* Sieb. & Zucc.[*Fallopia japonica* (Houtt.) Ronse Decr.]. *Canadian Journal of Plant Science* 86: 887–906.

Bartwal, A., Mall, R., Lohani, P. et al. (2013). Role of secondary metabolites and brassinosteroids in plant defense against environmental stresses. *Journal of Plant Growth Regulation* 32: 216–232.

Bashtanova, U.B., Beckett, K.P., and Flowers, T.J. (2009). Physiological approaches to the improvement of chemical control of Japanese knotweed (*Fallopia japonica*). *Weed Science* 57: 584–592.

Baziramakenga, R., Leroux, G., and Simard, R. (1995). Effects of benzoic and cinnamic acids on membrane permeability of soybean roots. *Journal of Chemical Ecology* 21: 1271–1285.

Baziramakenga, R., Leroux, G.D., Simard, R.R., and Nadeau, P. (1997). Allelopathic effects of phenolic acids on nucleic acid and protein levels in soybean seedlings. *Canadian Journal of Botany* 75: 445–450.

Belz, R.G. (2016). Investigating a potential auxin-related mode of hormetic/inhibitory action of the phytotoxin parthenin. *Journal of Chemical Ecology* 42: 71–83.

Belz, R.G., Reinhardt, C.F., Foxcroft, L.C., and Hurle, K. (2007). Residue allelopathy in *Parthenium hysterophorus* L. – does parthenin play a leading role? *Crop Protection* 26: 237–245.

Bennett, M.D., Leitch, I.J., and Hanson, L. (1998). DNA amounts in two samples of angiosperm weeds. *Annals of Botany* 82: 121–134.

Berendsen, R.L., Vismans, G., Yu, K. et al. (2018). Disease-induced assemblage of a plant-beneficial bacterial consortium. *The ISME Journal* 12: 1496–1507.

Bertin, C., Yang, X., and Weston, L.A. (2003). The role of root exudates and allelochemicals in the rhizosphere. *Plant and Soil* 256: 67–83. https://doi.org/10.1023/A:1026290508166.

Bertin, C., Weston, L.A., Huang, T. et al. (2007). Grass roots chemistry: meta-tyrosine, an herbicidal nonprotein amino acid. *Proceedings of the National Academy of Sciences* 104: 16964–16969.

Bialy, Z., Oleszek, W., Lewis, J., and Fenwick, G. (1990). Allelopathic potential of glucosinolates (mustard oil glycosides) and their degradation products against wheat. *Plant and Soil* 129: 277–281.

Blossey, B. and Notzold, R. (1995). Evolution of increased competitive ability in invasive nonindigenous plants: a hypothesis. *Journal of Ecology* 83: 887–889.

Blum, U. (2011). Plant–plant allelopathic interactions. In: *Plant-Plant Allelopathic Interactions: Phenolic Acids, Cover Crops and Weed Emergence* (ed. U. Blum), 1–7. Dordrecht: Springer.

Brown, R.H. (1999). Agronomic implications of C4 photosynthesis. *C4 Plant Biology* 1: 473–507.

Buckingham, J., Cooper, C., and Purchase, R. (2016). *Natural Products Desk Reference*. Boca Raton, FL, USA: CRC Press, Taylor & Francis Group.

Burke, D.J. and Chan, C.R. (2010). Effects of the invasive plant garlic mustard (*Alliaria petiolata*) on bacterial communities in a northern hardwood forest soil. *Canadian Journal of Microbiology* 56: 81–86.

Callaway, R.M. and Ridenour, W.M. (2004). Novel weapons: invasive success and the evolution of increased competitive ability. *Frontiers in Ecology and the Environment* 2: 436–443.

Campbell, G., Lambert, J.D., Arnason, T., and Towers, G.N. (1982). Allelopathic properties of α-terthienyl and phenylheptatriyne, naturally occurring compounds from species of *Asteraceae*. *Journal of Chemical Ecology* 8: 961–972.

Cappuccino, N. and Arnason, J.T. (2006). Novel chemistry of invasive exotic plants. *Biology Letters* 2: 189–193.

Chai, T.-T., Ooh, K.-F., Ooi, P.-W. et al. (2013). *Leucaena leucocephala* leachate compromised membrane integrity, respiration and antioxidative defence of water hyacinth leaf tissues. *Botanical Studies* 54: 1–7.

Cheng, H.H. (1995). Characterization of the mechanisms of allelopathy: modeling and experimental approaches. In: *Allelopathy*, vol. 582 (eds. Inderjit, K.M.M. Dakshini and F.A. Einhellig), 132–141. Washington, DC, USA: American Chemical Society.

Child, L. and Wade, M. (2000). *The Japanese Knotweed Manual: The Management and Control of an Invasive Alien Weed*. Chichester, England, UK: Packard Publishing Ltd.

Cipollini, D. and Cipollini, K. (2016). A review of garlic mustard (*Alliaria petiolata*, Brassicaceae) as an allelopathic plant. *The Journal of the Torrey Botanical Society* 143: 339–348.

Cipollini, D., Rigsby, C.M., and Barto, E.K. (2012). Microbes as targets and mediators of allelopathy in plants. *Journal of Chemical Ecology* 38: 714–727.

Cobb, A.H. and Reade, J.P. (2011). *Herbicides and Plant Physiology*. Wiley.

Croteau, R., Kutchan, T.M., and Lewis, N.G. (2000). Natural products (secondary metabolites). In: *Biochemistry and Molecular Biology of Plants*, vol. *24* (eds. B. Buchanan, W. Gruissem and R. Jones), 1250–1319. American Society of Plant Physiologists.

Czarnota, M.A., Paul, R.N., Dayan, F.E. et al. (2001). Mode of action, localization of production, chemical nature, and activity of sorgoleone: a potent PSII inhibitor in *Sorghum* spp. root exudates. *Weed Technology* 15: 813–825.

Czarnota, M.A., Paul, R.N., Weston, L.A., and Duke, S.O. (2003). Anatomy of sorgoleone-secreting root hairs of *Sorghum* species. *International Journal of Plant Sciences* 164: 861–866.

Dale, I. (1981). Parthenium weed in the Americas. *Australian Weeds* 1: 8–14.

Dalton, B. (1999). The occurrence and behavior of plant phenolic acids in soil environments and their potential involvement in allelochemical interference interactions: methodological limitations in establishing conclusive proof of allelopathy. In: *Principles and Practices in Plant Ecology: Allelochemical Interactions* (eds. Inderjit, K.M.M. Dakshini and C.L. Foy), 57–74. Boca Raton, FL, USA.: CRC Press.

Dalton, B., Blum, U., and Weed, S.B. (1983). Allelopathic substances in ecosystems. *Journal of Chemical Ecology* 9: 1185–1201.

Dassonville, N., Guillaumaud, N., Piola, F. et al. (2011). Niche construction by the invasive Asian knotweeds (species complex *Fallopia*): impact on activity, abundance and community structure of denitrifiers and nitrifiers. *Biological Invasions* 13: 1115–1133.

Day, M.D., Clements, D.R., Gile, C. et al. (2016). Biology and impacts of pacific islands invasive species *Mikania micrantha* Kunth (Asteraceae). *Pacific Science* 70: 257–285.

Dayan, F.E. and Duke, S.O. (2014). Natural compounds as next generation herbicides. *Plant Physiology* 166: 1090–1105.

Dayan, F.E., Owens, D.K., Watson, S.B. et al. (2015). Sarmentine, a natural herbicide from *Piper* species with multiple herbicide mechanisms of action. *Frontiers in Plant Science* 6: 1–11.

De Candolle, M.A.P. (1832). Physiologie Vegetale Bechet Jeune Lib. *Facultie Medecine Paris* 3: 1474–1475.

De la Fuente, J.R., Uriburu, M., Burton, G., and Sosa, V.E. (2000). Sesquiterpene lactone variability in *Parthenium hysterophorus* L. *Phytochemistry* 55: 769–772.

Delbart, E., Mahy, G., Weickmans, B. et al. (2012). Can land managers control Japanese knotweed? Lessons from control tests in Belgium. *Environmental Management* 50: 1089–1097.

Dommanget, F., Evette, A., Spiegelberger, T. et al. (2014). Differential allelopathic effects of Japanese knotweed on willow and cottonwood cuttings used in riverbank restoration techniques. *Journal of Environmental Management* 132: 71–78.

Dresler, S., Kubrak, T., Bogucka-Kocka, A., and Szymczak, G. (2015). Determination of shikonin and rosmarinic acid in *Echium vulgare* L. and *Echium russicum* J.F. Gmel. by capillary electrophoresis. *Journal of Liquid Chromatography & Related Technologies* 38: 698–701.

Duke, S.O. (2015). Proving allelopathy in crop–weed interactions. *Weed Science* 63: 121–132.

Duke, S.O., Duke, M.V., Paul, R.N. et al. (1999a). Tissue localization and potential uses of phytochemicals with biological activity. In: *Recent Advances in Allelopathy*, A Science for the Future, vol. I (eds. F.A. Macias, J.C.G. Galindo, J.M.G. Molinillo and H.G. Cutler), 237–244. Cadiz, Spain: University of Cadiz Press.

Duke, S.O., Rimando, A.M., Duke, M.V. et al. (1999b). Sequestration of phytotoxins by plants: implications for biosynthetic production. In: *Biologically Active Natural Products: Agrochemicals* (eds. H.G. Cutler and S.J. Cutler), 127–136.

Duke, S.O., Dayan, F.E., Rimando, A.M. et al. (2002). Chemicals from nature for weed management. *Weed Science* 50: 138–151.

Duke, S.O., Stidham, M.A., and Dayan, F.E. (2019). A novel genomic approach to herbicide and herbicide mode of action discovery. *Pest Management Science* 75: 314–317.

Durán, A.G., Gutiérrez, M.T., Rial, C. et al. (2017). Bioactivity and quantitative analysis of isohexenylnaphthazarins in root periderm of two *Echium* spp.: *E. plantagineum* and *E. gaditanum*. *Phytochemistry* 141: 162–170.

Durán, A.G., Chinchilla, N., Molinillo, J.M., and Macías, F.A. (2018). Influence of lipophilicity in O-acyl and O-alkyl derivatives of juglone and lawsone: a structure–activity relationship study in the search for natural herbicide models. *Pest Management Science* 74: 682–694.

Durka, W., Bossdorf, O., Prati, D., and Auge, H. (2005). Molecular evidence for multiple introductions of garlic mustard (*Alliaria petiolata, Brassicaceae*) to North America. *Molecular Ecology* 14: 1697–1706.

Einhellig, F.A. (1996). Interactions involving allelopathy in cropping systems. *Agronomy Journal* 88: 886–893.

Eom, S.H., Senesac, A.F., Tsontakis-Bradley, I., and Weston, L.A. (2005). Evaluation of herbaceous perennials as weed suppressive groundcovers for use along roadsides or in landscapes. *Journal of Environmental Horticulture* 23: 198–203.

Essl, F., Dullinger, S., and Kleinbauer, I. (2009). Changes in the spatio-temporal patterns and habitat preferences of *Ambrosia artemisiifolia* during its invasion of Austria. *Preslia* 81: 119–133.

Eussen, J. (1978). Isolation of growth inhibiting substances from *Alang-alang* (*Imperata cylindrica* (*L*). Bueauv. var. mayor). *The Asian-Pacific Weed Science Society Conference,* Jakarta, Indonesia, (11–17 July 1977).

Facelli, J.M. and Pickett, S.T. (1991). Plant litter: its dynamics and effects on plant community structure. *The Botanical Review* 57: 1–32.

Frey, M., Schullehner, K., Dick, R. et al. (2009). Benzoxazinoid biosynthesis, a model for evolution of secondary metabolic pathways in plants. *Phytochemistry* 70: 1645–1651.

Friebe, A., Wieland, I., and Schulz, M. (1996). Tolerance of *Avena sativa* to the allelochemical benzoxazolinone. Degradation of BOA by root-colonizing bacteria. *Angewandte Botanik* 70: 150–154.

Ghisalberti, E. (2000). *Lantana camara* L. (Verbenaceae). *Fitoterapia* 71: 467–486.

Gillies, S., Clements, D.R., and Grenz, J. (2016). Knotweed (*Fallopia* sp.) invasion of North America utilizes hybridization, epigenetics, seed dispersal (unexpectedly) and an arsenal of physiological tactics. *Invasive Plant Science and Management* 9: 71–80.

Gimsing, A.L., Bælum, J., Dayan, F.E. et al. (2009). Mineralization of the allelochemical sorgoleone in soil. *Chemosphere* 76: 1041–1047.

Goslee, S., Peters, D., and Beck, K. (2001). Modeling invasive weeds in grasslands: the role of allelopathy in *Acroptilon repens* invasion. *Ecological Modelling* 139: 31–45.

Graña, E., Sotelo, T., Díaz-Tielas, C. et al. (2013). The phytotoxic potential of the terpenoid citral on seedlings and adult plants. *Weed Science* 61: 469–481.

Grevstad, F., Shaw, R., Bourchier, R. et al. (2013). Efficacy and host specificity compared between two populations of the psyllid *Aphalara itadori*, candidates for biological control of invasive knotweeds in North America. *Biological Control* 65: 53–62.

Guertin, P. (2001). Observations made during the duration of weed distribution mapping for the USGS Weeds in the West project occurring in the southern Arizona National Park Service management areas. May 1999–June 2001. USGS/BRD, The University of Arizona, Tucson, Arizona, USA.

Ha, I.J., Lee, M.Y., Kwon, Y.K. et al. (2015). Metabolite profiling to discriminate different species and genus from thistles in Korea using liquid chromatography with quadrupole time-of-flight mass spectrometry. *Journal of Separation Science* 38: 502–510.

Haig, T. (2008). Allelochemicals in plants. In: *Allelopathy in Sustainable Agriculture and Forestry* (eds. R.S. Zeng, A.U. Mallik and S.M. Luo), 63–104. New York, USA: Springer.

Hatch, S.L. (2001). Perennial weeds: characteristics and identification of selected herbaceous species. *Great Plains Research* 11: 392–393.

Herz, W. (1963). The organic constituents of higher plants, their chemistry and interrelationships. *Journal of the American Chemical Society* 85: 2876–2876.

Hierro, J.L. and Callaway, R.M. (2003). Allelopathy and exotic plant invasion. *Plant and Soil* 256: 29–39.

Holm, L.G., Plucknett, D.L., Pancho, J.V., and Herberger, J.P. (1977). *The World's Worst Weeds: Distribution and Biology*, 610. Honolulu, HI, USA: University of Hawaii.

Huang, H., Morgan, C.M., Asolkar, R.N. et al. (2010). Phytotoxicity of sarmentine isolated from long pepper (*Piper longum*) fruit. *Journal of Agricultural and Food Chemistry* 58: 9994–10000.

Huang, H., Liu, C., Wei, S. et al. (2015). Dynamic root exudation of phenolic allelochemicals from Johnson grass (*Sorghum halepense*). *Weed Biology and Management* 15: 133–137.

Hussain, N., Abbasi, T., and Abbasi, S. (2016). Vermicomposting transforms allelopathic parthenium into a benign organic fertilizer. *Journal of Environmental Management* 180: 180–189.

Inderjit, Muramatsu, M., and Nishimura, H. (1997). On the allelopathic potential of certain terpenoids, phenolics, and their mixtures, and their recovery from soil. *Canadian Journal of Botany* 75: 888–891.

Ismail, B. and Mah, L. (1993). Effects of *Mikania micrantha* HBK on germination and growth of weed species. *Plant and Soil* 157: 107–113.

Kato-Noguchi, H. (2003). Allelopathic substances in *Pueraria thunbergiana*. Phytochemistry 63: 577–580.

Kato-Noguchi, H. (2011). Barnyard grass-induced rice allelopathy and momilactone B. *Journal of Plant Physiology* 168: 1016–1020.

Keane, R.M. and Crawley, M.J. (2002). Exotic plant invasions and the enemy release hypothesis. *Trends in Ecology & Evolution* 17: 164–170.

Khanh, D.T., Trung, H.K., Anh, H.L., and Xuan, D.T. (2018). Allelopathy of barnyard grass (*Echinochloa crus-galli*) weed: an allelopathic interaction with rice (*Oryza sativa*). *Vietnam Journal of Agricultural Sciences* 1: 97–116.

Korhammer, S.A. and Haslinger, E. (1994). Isolation of a biologically active substance from rhizomes of quack grass [*Elymus repens* (L.) Gould]. *Journal of Agricultural and Food Chemistry* 42: 2048–2050.

Kumar, P. and Barthwal, R. (2018). Structural and biophysical insight into dual site binding of the protoberberine alkaloid palmatine to parallel G-quadruplex DNA using NMR, fluorescence and Circular Dichroism spectroscopy. *Biochimie* 147: 153–169.

Kuo, Y., Chen, T., and Lin, C. (2002). Using a consecutive-cutting method and allelopathy to control the invasive vine, *Mikania micrantha* HBK. *Taiwan Journal of Forest Science* 17: 171–181.

Kutchan, T.M. (2005). A role for intra-and intercellular translocation in natural product biosynthesis. *Current Opinion in Plant Biology* 8: 292–300.

Lapsansky, E.R., Milroy, A.M., Andales, M.J., and Vivanco, J.M. (2016). Soil memory as a potential mechanism for encouraging sustainable plant health and productivity. *Current Opinion in Biotechnology* 38: 137–142.

Latif, S., Chiapusio, G., and Weston, L.A. (2017). Allelopathy and the role of allelochemicals in plant defence. *Advances in Botanical Research* 82: 19–54.

Latif, S., Gurusinghe, S., Weston, P.A. et al. (2020). Metabolomic approaches for the identification of flavonoids associated with weed suppression in selected hardseeded annual pasture legumes. *Plant and Soil* 447: 199–218.

Lavergne, S., Muenke, N.J., and Molofsky, J. (2009). Genome size reduction can trigger rapid phenotypic evolution in invasive plants. *Annals of Botany* 105: 109–116.

Li, L., Li, S.M., Su, J.H. et al. (2007). Diversity enhances agricultural productivity via rhizosphere phosphorus facilitation on phosphorus-deficient soils. *Proceedings of the National Academy of Sciences of the United States of America* 104: 11192–11196.

Li, Z.-H., Wang, Q., Ruan, X. et al. (2010). Phenolics and plant allelopathy. *Molecules* 15: 8933–8952.

Li, L.-L., Zhao, H.-H., and Kong, C.-H. (2019). (−)-Loliolide, the most ubiquitous lactone, is involved in barnyard grass-induced rice allelopathy. *Journal of Experimental Botany* 71: 1540–1550.

Liu, B., Yan, J., Li, W. et al. (2020). *Mikania micrantha* genome provides insights into the molecular mechanism of rapid growth. *Nature Communications* 11: 340.

Lovett, J. and Potts, W.C. (1987). Primary effects of allelochemicals of *Datura stramonium* L. *Plant and Soil* 98: 137–144.

Macel, M., de Vos, R.C., Jansen, J.J. et al. (2014). Novel chemistry of invasive plants: exotic species have more unique metabolomic profiles than native congeners. *Ecology and Evolution* 4: 2777–2786.

Macias, F.A., Molinillo, J.M., Varela, R.M., and Galindo, J.C. (2007). Allelopathy-a natural alternative for weed control. *Pest Management Science* 63: 327–348.

Macías, F.A., Mejías, F.J., and Molinillo, J.M. (2019). Recent advances in allelopathy for weed control: from knowledge to applications. *Pest Management Science* 75: 2413–2436.

Meekins, J.F. and McCarthy, B.C. (1999). Competitive ability of *Alliaria petiolata* (garlic mustard, *Brassicaceae*), an invasive, nonindigenous forest herb. *International Journal of Plant Sciences* 160: 743–752.

Mersie, W. and Singh, M. (1993). Phenolic acids affect photosynthesis and protein synthesis by isolated leaf cells of velvet-leaf. *Journal of Chemical Ecology* 19: 1293–1301.

Molisch, H. (1937). *Der Einfluss einer Pflanze aufdie andere-Allelopathie*, 158–160. Jena: Fischer.

Molyneux, R., Gardner, D., Colegate, S., and Edgar, J. (2011). Pyrrolizidine alkaloid toxicity in livestock: a paradigm for human poisoning? *Food Additives & Contaminants: Part A* 28: 293–307.

Mwendwa, J.M. (2019). Mechanisms of weed suppression in wheat (*Triticum aestivum* L.) and canola (*Brassica napus* L.) in southeastern Australia. Doctoral dissertation. Charles Sturt University, Wagga Wagga, Australia.

Mwendwa, JM, Weston PA, Weidenhamer, JD,Fomsgaard IS, Wu, H, Gurusinghe, S and Weston, LA (2021). Metabolic profiling of benzoxazinoids in the roots and rhizosphere of commercial winter wheat genotypes. Plant and Soil. DOI: 10.1007/s11104-021-04996-9.

Navie, S.C., Panetta, F.D., McFadyen, R.E., and Adkins, S.W. (1998). Behaviour of buried and surface-sown seeds of *Parthenium hysterophorus*. *Weed Research (Oxford)* 38: 335–341.

Niklas, K.J. and Hammond, S.T. (2013). Biophysical effects on plant competition and coexistence. *Functional Ecology* 27: 854–864.

Oleszek, W. and Jurzysta, M. (1987). The allelopathic potential of alfalfa root medicagenic acid glycosides and their fate in soil environments. *Plant and Soil* 98: 67–80.

O'Sullivan, M., Dorken, M.E., and Freeland, J.R. (2019). Garlic mustard (*Alliaria petiolata*) is associated with an overall reduction in plant diversity, but is more likely to co-exist with native than alien species. *Plant Ecology and Diversity* 12: 427–439.

Parepa, M., Schaffner, U., and Bossdorf, O. (2012). Sources and modes of action of invasive knotweed allelopathy: the effects of leaf litter and trained soil on the germination and growth of native plants. *NeoBiota* 13: 15–30.

Parepa, M., Schaffner, U., and Bossdorf, O. (2013). Help from underground: soil biota facilitate knotweed invasion. *Ecosphere* 4: 1–11.

Parmesan, C. and Yohe, G. (2003). A globally coherent fingerprint of climate change impacts across natural systems. *Nature* 421 (6918): 37–42.

Petanidou, T., Godfree, R.C., Song, D. et al. (2012). Self-compatibility and plant invasiveness: comparing species in native and invasive ranges. *Perspectives in Plant Ecology, Evolution and Systematics* 14: 3–12.

Piggin, C. (1982). The biology of Australian weeds. 8. *Echium plantagineum* L. *Journal of the Australian Institute of Agricultural Science* 48: 3–16.

Portales-Reyes, C., Van Doornik, T., Schultheis, E.H., and Suwa, T. (2015). A novel impact of a novel weapon: allelochemicals in *Alliaria petiolata* disrupt the legume-rhizobia mutualism. *Biological Invasions* 17: 2779–2791.

Powles, S.B. and Yu, Q. (2010). Evolution in action: plants resistant to herbicides. *Annual Review of Plant Biology* 61: 317–347.

Prati, D. and Bossdorf, O. (2004). Allelopathic inhibition of germination by *Alliaria petiolata* (Brassicaceae). *American Journal of Botany* 91: 285–288.

Price, G.J. and Blair, A.C. (2019). Garlic mustard as a potential driver of change in soil microbial community metabolism under controlled conditions. *Bios* 90: 218–226.

Putnam, A.R. and Tang, C.-S. (1986). *The Science of Allelopathy*. Ithaca, NY, USA: Wiley.

Qasem, J. and Foy, C. (2001). Weed allelopathy, its ecological impacts and future prospects: a review. *Journal of Crop Production* 4: 43–119.

Rahman, M., Khatun, A., Liu, L., and Barkla, B.J. (2018). Brassicaceae mustards: traditional and agronomic uses in Australia and New Zealand. *Molecules* 23: 231–239.

Reinhardt, C., Belz, R., and Hurle, K. (2009). Role of the allelochemical parthenin in the invasive strategy of the alien plant *Parthenium hysterophorus* L. *South African Journal of Botany* 2: 417–418.

Rice, E.L. (2012). *Allelopathy*. Academic Press.

Rice, C.P., Cai, G., and Teasdale, J.R. (2012). Concentrations and allelopathic effects of benzoxazinoid compounds in soil treated with rye (*Secale cereale*) cover crop. *Journal of Agricultural and Food Chemistry* 60: 4471–4479.

Roberts, K.J. and Anderson, R.C. (2001). Effect of garlic mustard [*Alliaria petiolata* (*Beib. Cavara & Grande*)] extracts on plants and arbuscular mycorrhizal (AM) fungi. *The American Midland Naturalist* 146: 146–152.

Rodríguez, E., Epstein, W., and Mitchell, J. (1977). The role of sesquiterpene lactones in contact hypersensitivity to some North and South American species of feverfew (*Parthenium-Compositae*). *Contact Dermatitis* 3: 155–162.

Sánchez-Moreiras, A. and Reigosa, M.J. (2005). Whole plant response of lettuce after root exposure to BOA (2 (3H)-benzoxazolinone). *Journal of Chemical Ecology* 31: 2689–2703.

Sasse, J., Martinoia, E., and Northen, T. (2018). Feed your friends: do plant exudates shape the root microbiome? *Trends in Plant Science* 23: 25–41.

Schandry, N. and Becker, C. (2020). Allelopathic plants: models for studying plant-interkingdom interactions. *Trends in Plant Science* 25: 176–185.

Schulz, M., Marocco, A., Tabaglio, V. et al. (2013). Benzoxazinoids in rye allelopathy-from discovery to application in sustainable weed control and organic farming. *Journal of Chemical Ecology* 39: 154–174.

Seiger, L.A. and Merchant, H.C. (1997). Mechanical control of Japanese knotweed (*Fallopia japonica* [Houtt.] Ronse Decraene): effects of cutting regime on rhizomatous reserves. *Natural Areas Journal* 17: 341–345.

Seigler, D.S. (1998). Pyrrolizidine, quinolizidine, and indolizidine alkaloids. In: *Plant Secondary Metabolism*, 546–567. New York, USA: Springer.

Shao, H., Peng, S., Wei, X. et al. (2005). Potential allelochemicals from an invasive weed *Mikania micrantha* HBK. *Journal of Chemical Ecology* 31: 1657–1668.

Shauffer, N. (2017). *Investigating the Relationship Between Invasive Garlic Mustard Alliaria petiolata and Blacklegged Ticks Ixodes scapularis*. Honors Thesis, Schenectady, NY: Union College.

Shen, S., Xu, G., Clements, D.R. et al. (2015). Effects of invasive plant *Mikania micrantha* on plant community and diversity in farming systems. *Asian Journal of Plant Sciences* 14: 27–33.

Shi, B. and Adkins, S. (2018). Relative phytotoxicity of parthenium weed (*Parthenium hysterophorus* L.) residues on the seedling growth of a range of Australian native and introduced species. *Crop & Pasture Science* 69: 837–845.

Shi, B., Aslam, Z., and Adkins, S. (2015). The invasive potential of *Parthenium* weed: a role for allelopathy. In: *New Development in Allelopathy Research* (ed. J.E. Price), 135–174. New York, USA: Nova Science Publishers.

Skoneczny, D. (2017). Role of environment and genetics in production of secondary metabolites in the invasive Australian weed *Echium plantagineum* L. Doctoral thesis. Charles Sturt University, Wagga Wagga, Australia. pp. 310.

Skoneczny, D., Weston, P.A., Zhu, X. et al. (2015). Metabolic profiling of pyrrolizidine alkaloids in foliage of two *Echium spp.* invaders in Australia – a case of novel weapons? *International Journal of Molecular Sciences* 16: 26721–26737.

Skoneczny, D., Zhu, X., Weston, P.A. et al. (2019). Production of pyrrolizidine alkaloids and shikonins in *Echium plantagineum* L. in response to various plant stressors. *Pest Management Science* 75: 2530–2541.

Smith, C.A., Maille, G.O., Want, E.J. et al. (2005). METLIN: a metabolite mass spectral database. *Therapeutic Drug Monitoring* 27: 747–751.

Soderquist, C.J. (1973). Juglone and allelopathy. *Journal of Chemical Education* 50: 782.

Stegelmeier, B.L. (2011). Pyrrolizidine alkaloid–containing toxic plants (*Senecio*, *Crotalaria*, *Cynoglossum*, *Amsinckia*, *Heliotropium*, and *Echium* spp.). *Veterinary Clinics: Food Animal Practice* 27: 419–428.

Stinson, K.A., Campbell, S.A., Powell, J.R. et al. (2006). Invasive plant suppresses the growth of native tree seedlings by disrupting belowground mutualisms. *PLoS Biology* 4: 727–731.

Stinson, K., Kaufman, S., Durbin, L., and Lowenstein, F. (2007). Impacts of garlic mustard invasion on a forest understory community. *Northeastern Naturalist* 14: 73–88.

Van Riper, L.C., Becker, R.L., and Skinner, L.C. (2010). Population biology of garlic mustard (*Alliaria petiolata*) in Minnesota hardwood forests. *Invasive Plant Science and Management* 3: 48–59.

Van Wyk, B., Stander, M., and Long, H. (2016). Pyrrolizidine alkaloid contamination of rooibos tea (*Aspalathus linearis*). *Planta Medica* 82: (S 01) p 1106.

Van Wyk, B.-E., Stander, M., and Long, H. (2017). *Senecio angustifolius* as the major source of pyrrolizidine alkaloid contamination of rooibos tea (*Aspalathus linearis*). *South African Journal of Botany* 110: 124–131.

Vastano, B.C., Chen, Y., Zhu, N. et al. (2000). Isolation and identification of stilbenes in two varieties of *Polygonum cuspidatum*. *Journal of Agricultural and Food Chemistry* 48: 253–256.

Vaughn, S.F. (1999). Glucosinolates as natural pesticides. In: *Biologically Active Natural Products: Agrochemicals* (eds. H.G. Cutler and S.J. Cutler), 81–94. Boca Raton, FL, USA: CRC Press.

Vaughn, S.F. and Berhow, M.A. (1999). Allelochemicals isolated from tissues of the invasive weed garlic mustard (*Alliaria petiolata*). *Journal of Chemical Ecology* 25: 2495–2504.

Venturelli, S., Belz, R.G., Kämper, A. et al. (2015). Plants release precursors of histone deacetylase inhibitors to suppress growth of competitors. *The Plant Cell* 27: 3175–3189.

Vila-Aiub, M.M., Balbi, M.C., Distéfano, A.J. et al. (2012). Glyphosate resistance in perennial *Sorghum halepense* (Johnsongrass), endowed by reduced glyphosate translocation and leaf uptake. *Pest Management Science* 68: 430–436.

Walton, N.J. and Brown, D.E. (eds.) (1999). *Chemicals from Plants: Perspectives on Plant Secondary Products*. Singapore: Imperial College Press.

Weidenhamer, J.D. (1996). Distinguishing resource competition and chemical interference: overcoming the methodological impasse. *Agronomy Journal* 88: 866–875.

Weir, T.L., Park, S.-W., and Vivanco, J.M. (2004). Biochemical and physiological mechanisms mediated by allelochemicals. *Current Opinion in Plant Biology* 7: 472–479.

Werle, R., Jhala, A.J., Yerka, M.K. et al. (2016). Distribution of herbicide-resistant shattercane and johnsongrass populations in sorghum production areas of Nebraska and northern Kansas. *Agronomy Journal* 108: 321–328.

Weston, L.A. (1996). Utilization of allelopathy for weed management in agroecosystems. *Agronomy Journal* 88: 860–866.

Weston, L.A. and Duke, S.O. (2003). Weed and crop allelopathy. *Critical Reviews in Plant Sciences* 22: 367–389.

Weston, L.A. and Mathesius, U. (2013). Flavonoids: their structure, biosynthesis and role in the rhizosphere, including allelopathy. *Journal of Chemical Ecology* 39: 283–297.

Weston, L.A., Harmon, R., and Mueller, S. (1989). Allelopathic potential of sorghum-sudan grass hybrid (sudex). *Journal of Chemical Ecology* 15: 1855–1865.

Weston, L.A., Barney, J.N., and DiTommaso, A. (2005). A review of the biology and ecology of three invasive perennials in New York state: Japanese knotweed (*Polygonum cuspidatum*), mugwort (*Artemisia vulgaris*) and pale swallow-wort (*Vincetoxicum rossicum*). *Plant and Soil* 277: 53–69.

Weston, L.A., Ryan, P.R., and Watt, M. (2012). Mechanisms for cellular transport and release of allelochemicals from plant roots into the rhizosphere. *Journal of Experimental Botany* 63: 3445–3454.

Weston, L.A., Skoneczny, D., Weston, P.A., and Weidenhamer, J.D. (2015). Metabolic profiling: an overview – new approaches for the detection and functional analysis of biologically active secondary plant products. *Journal of Allelochemical Interactions* 1: 15–27.

Wink, M. (2011). *Annual Plant Reviews, Biochemistry of Plant Secondary Metabolism*, vol. 40. Chichester, UK: Wiley.

Wixted, K.L. and McGraw, J.B. (2010). Competitive and allelopathic effects of garlic mustard (*Alliaria petiolata*) on American ginseng (*Panax quinquefolius*). *Plant Ecology* 208: 347–357.

Yadav, A. and Chauhan, S. (1998). Studies on allelopathic effect of some weeds. *Journal of Phytological Research* 11: 15–18.

Yang, L. and Stöckigt, J. (2010). Trends for diverse production strategies of plant medicinal alkaloids. *Natural Product Reports* 27: 1469–1479.

Yang, Q., Feng, H., Ye, W. et al. (2003). An investigation of the effects of environmental factors on the flowering and seed setting of *Mikania micrantha* HBK (Compositae). *Journal of Tropical and Subtropical Botany* 11: 123–126.

Yazaki, K. (2006). ABC transporters involved in the transport of plant secondary metabolites. *FEBS Letters* 580: 1183–1191.

Yoon, H.S., Moon, S.C., Kim, N.D. et al. (2000). Genistein induces apoptosis of RPE-J cells by opening mitochondrial PTP. *Biochemical and Biophysical Research Communications* 276: 151–156.

Young, G. and Bush, J. (2009). Assessment of the allelopathic potential of *Juniperus ashei* on germination and growth of *Bouteloua curtipendula*. *Journal of Chemical Ecology* 35: 74–80.

Zeng, R.S. (2014). Allelopathy – the solution is indirect. *Journal of Chemical Ecology* 40: 515–516.

Zeng, R.S., Mallik, A.U., and Luo, S.M. (2008). *Allelopathy in Sustainable Agriculture and Forestry*. New York, USA: Springer.

Zhou, L., Xu, J.-D., Zhou, S.-S. et al. (2016). Integrating targeted glycomics and untargeted metabolomics to investigate the processing chemistry of herbal medicines, a case study on *Rehmanniae radix*. *Journal of Chromatography A* 1472: 74–87.

Zhu, X., Weston, P.A., Skoneczny, D. et al. (2017). Ecology and genetics affect relative invasion success of two *Echium* species in southern Australia. *Scientific Reports* 7: 42792.

Zhu, X., Gopurenko, D., Serrano, M. et al. (2019). Genetic evidence for plural introduction pathways of the invasive weed Paterson's curse (*Echium plantagineum* L.) to southern Australia. *PLoS One* 14 (9): e0222696.

15

Weed Adaptation as a Driving Force for Weed Persistence in Agroecosystems

Paul Neve[1] *and Ana L. Caicedo*[2]

[1] *University of Copenhagen, Department of Plant & Environmental Sciences, Højbakkegård Allé 13, Tåstrup, Denmark*
[2] *Biology Department, University of Massachusetts, Amherst, MA, USA*

> *"Nothing in biology makes sense except in the light of evolution"*
> (Dobzhansky 1973).

15.1 Introduction

In 1956, when speculating on the potential for weed populations to evolve resistance to herbicides, John L. Harper commented that 'arable weeds constitute an ecological group selected and maintained in association by their fitness for existence under conditions of crop cultivation. They comprise species that have been selected by the very cultural practices that were originally designed to suppress them' (Harper 1956). Over time, many definitions have been offered to address the question 'what is a weed?', but none have so eloquently captured the ecoevolutionary basis for the existence, persistence, and ongoing adaptation of weeds in agroecosystems.

While eradication of weeds from crop fields will always be a challenge due to the ongoing migration of weed seeds into agricultural fields, large soil seedbank reservoirs, and incomplete efficacy of weed control technology, it might reasonably be expected that concerted efforts to control weed populations would have, over time, minimized their presence and impacts in agroecosystems (Jordan and Jannink 1997). Though there have been notable successes in reducing the impacts of weeds on crop yield loss, we are in fact engaged in a 'weed management arms race' (Neve et al. 2009), which is characterized by the ongoing coevolution of weed control technologies and weed populations (Figure 15.1). This arms race has been at play ever since the dawn of human crop cultivation, some 12 000 years ago (see Snir et al. 2015) but has intensified in the last 50 years as cropping system diversity has been reduced and dependence on herbicides has increased (Harker 2013; Storkey and Neve 2018).

The intended (e.g. via crop breeding, domestication of animals) and unintended (e.g. evolution of pesticide resistance, pathogen virulence) consequences of human-driven evolution in agriculture are now well-recognized (Thrall et al. 2011; Denison 2012), leading

Persistence Strategies of Weeds, First Edition. Edited by Mahesh K. Upadhyaya, David R. Clements, and Anil Shrestha.

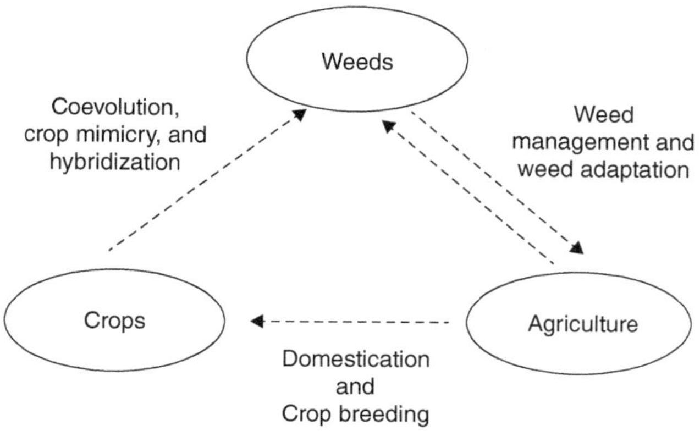

Figure 15.1 The 'weed management arms race' is represented here by the coevolutionary dynamics of human agricultural activity, crop breeding and selection, and weed adaptation. The domestication of wild plants during the early development of agriculture (unconscious selection) and subsequent crop breeding (conscious selection) has led to the development of modern crop cultivars. Widespread cultivation of these crops has created 'opportunity space' for the invasion of agricultural land by ruderal plant species, and subsequent crop–weed coevolution has resulted in the evolution of highly adapted weed ecotypes that mimic the crop lifecycle and morphological characteristics. The evolution of highly adapted weeds has stimulated the development of sophisticated weed control tools, and these tools (for example, herbicides) have exerted extreme selection pressure for weed adaptation. The continuing coevolution of crops, weeds, and crop/weed management systems underpins the establishment, adaptation, and persistence of weeds in agroecosystems. *Source:* Neve et al. (2009).

Thrall et al. (2010) to describe agriculture as a 'crucible of evolutionary change'. Understanding the processes of weed evolution will provide fundamental insights into plant adaptation and the origins of weediness; it will also help us address practical questions relating to ongoing evolution of weeds in the face of changing climates and changing weed management practices (Vigueira et al. 2013; Neve et al. 2018). Fundamentally, efforts to design integrated weed management systems are driven by desires to reduce environmental impact and to reduce weed adaptation, thereby slowing the weed management arms race (Jordan and Jannink 1997; Neve et al. 2009; Harker 2013).

In this chapter, we review historical and contemporary modes of weed evolution, considering the potential for agricultural weeds to evolve from wild plant species that coexist with crops, through hybridization of wild weedy relatives with crops or as a result of crop ferality. We consider the nature and extent of genetic variation in weed populations, this being the raw material for ongoing weed evolution, and we go on to provide some case studies for weed adaptation in modern agroecosystems. We finish by reiterating our central thesis that the capacity of weed populations for rapid and ongoing adaptation to changing environments and management practices is a key facet of weed persistence in agroecosystems.

15.2 Modes of Weed Evolution

Reducing weed adaptation requires a solid understanding of how weeds can arise, that is, from what ancestral backgrounds weeds are capable of evolving. The reason for this is twofold. First, if potential sources of new or continuously arising weeds could be identified, strategies could be set in place to impede further weed origins. Second, if the ancestral background of a weed species determines its potential to evolve specific weedy adaptive traits, more targeted management strategies can be designed to impede the appearance of these adaptations.

Three possible origins for agricultural weed species were first proposed by De Wet and Harlan (1975). Briefly, these are (i) origins from purely wild species, (ii) origins as a product of hybridization between wild and cultivated plants, and (iii) origins from cultivated species. More recently, Gressel (2005) focused on weeds with some degree of cultivated ancestry, which he termed 'feral' plants, distinguishing between exoferal, plants that have evolved through hybridization of crops and wild plants, and endoferal, plants that have evolved through de-domestication of a cultivated crop. These categories overlap with the hybrid and cultivated origins of De Wet and Harlan, with the distinction that 'feral' plants are a much more general category that need not be limited to agricultural weeds because they can also invade wild habitats (Gressel 2005). The most recent treatments of agricultural weed origins (e.g. Vigueira et al. 2013) have largely agreed on the framework established by De Wet and extended by Gressel for potential weed ancestries while expanding on the genetic consequences of each on gene pools of weed species.

Despite the potential benefits of determining where agricultural weeds come from, surprisingly few research efforts have focused on discerning the origins of weed species. For example, more attention has been paid to the origins of weeds with crop relatives because of the availability of genetic resources and the ability to establish ancestor–descendant relationships (Vigueira et al. 2013). Though, as detailed in the following text, weeds representing every type of origin have been documented, the unevenness in research efforts has left us with an incomplete understanding of the prevalence and importance of different weed origins. This lack of comprehensive assessment of weedy origins also impacts our ability to understand the genetic source of weedy adaptive traits that permit weed persistence in our agricultural systems (see Section 15.3).

15.2.1 Weed Origins from Wild Species

De Wet and Harlan (1975) considered weeds evolving from wild plants to be the most common case. This is a reasonable assumption, given the nearly 400 000 species of (wild) vascular plants that are known to exist (Willis 2017), all of which are potential sources of weeds. Originally described as 'colonizer' species capable of invading habitats disturbed by man (De Wet and Harlan 1975), we now recognize that weedy plants with wild origins may either be pre-adapted (i.e. 'exaptation' *sensu* Gould and Vrba (1982)) to agricultural conditions or may be generalists that can tolerate a wide range of environments (Vigueira et al. 2013). More recently, a third route for wild to weed origins has been recognized, which invokes initial adaptive changes as a condition for weed establishment and is discussed further herein.

Despite limited studies, some conclusive examples of pre-adaptation and/or generalist ancestry in wild-origin weeds have been documented. For example, weedy sunflower (*Helianthus annuus*) populations in the United States are most closely related to locally occurring wild sunflowers than domesticated ones, and generally show no signs of bottlenecks (i.e. a reduction in genetic diversity due to a decrease in population size), suggesting that the wild gene pool contains the genetic potential to invade cultivated fields (Kane and Rieseberg 2008). Populations of witchweed (*Striga hermonthica*), a parasitic plant of cereal crops, show genetic differentiation correlated with geography rather than host specificity across Ethiopia (Welsh and Mohamed 2011), suggesting that the species is a generalist that can easily invade new crops. Similarly, no genetic differentiation has been observed for broomrape (*Orobanche foetida*), parasitizing wild and cultivated hosts (Vaz Patto et al. 2008), and within the *Orobanche* genus only species with wide host ranges and annual life histories have evolved to be agricultural weeds, suggesting preadaptation (Schneeweiss 2007). The lack of genetic differentiation between wild and weedy forms of waterhemp (*Amaranthus tuberculatus*), a problematic weed of maize (*Zea mays*) and soybean (*Glycine max*) in the United States, and the presence of traits presumed to be adaptive in agricultural environments in both groups (such as small seeds, wind pollination, and discontinuous germination) suggest that wild waterhemp was preadapted to become an agricultural weed without the need for genetic or phenotypic changes (Waselkov and Olsen 2014). It should be noted that neither preadaptation nor a generalist phenotype precludes later adaptation to selective pressures experienced by weeds in agricultural fields.

Evidence that adaptation specific to agricultural conditions may be necessary in some weeds that have evolved from wild plants also exists. Weedy rice (*Oryza* spp.) has arisen multiple times independently from various different backgrounds (Ziska et al. 2015), but some lineages arising from wild rice (*Oryza rufipogon*) have been detected in South Asia (Huang et al. 2017); these weeds flower significantly earlier under common garden conditions and have a more compact architecture than their wild ancestors (Huang et al. 2017, 2018), suggesting that some adaptive changes have occurred. Similarly, weedy radish (*Raphanus raphanistrum* ssp. *raphanistrum*), a worldwide invasive weed of small grain fields, originated from wild (rather than cultivated) radish populations but is genetically and phenotypically distinct, and is observed to be flowering much earlier and more uniformly than wild radish (Charbonneau et al. 2018).

Adaptive evolution in wild-derived weeds is also implicated in cases where such weeds have evolved herbicide tolerance or resistance (Vigueira et al. 2013), as these traits are not generally expected to be present in wild ancestors (but see Délye et al. 2013 for possible occurrence of herbicide resistance mutation prior to herbicide use). This is also true in the case of crop mimicry – that is, the selection for the weed to resemble its co-occurring crop at some life stages (reviewed in Barrett 1983). Both specific cases of contemporary weed evolution under selection in agroecosystems are discussed in more detail in Section 15.4.

15.2.2 Weed Origins from Crop–Wild Hybrids

More than 2000 plant species worldwide are believed to have been domesticated (Zeven and De Wet 1982; Dirzo and Raven 2003; Milla et al. 2018), although approximately only 90 provide most of the world's food today (Khoury et al. 2014), making crop species an important potential source of weeds. Most weeds with crop ancestors are thought to have exoferal origins, that is, they have arisen through hybridization of crops and their wild relatives (Ellstrand et al. 2010). Such hybridization events may confer two types of advantages: crop parents can supply variants that are adaptive in the agricultural environment, while wild ancestors can provide variants for weediness traits such as seed persistence and proliferation (Vigueira et al. 2013). These trait combinations may make the evolution of crop–wild hybrid weeds occur more rapidly, especially when compared with crop-derived weeds (Gressel 2005). However, dissenting views suggest that linkage between alleles coding for domestication traits, some of which would be deleterious for weeds, should hinder weed evolution in crop–wild crosses (Warwick and Stewart 2005). To our knowledge, no studies have systematically examined the speed of evolution of different weed types.

In cases where the crop or wild relative coexist with the weed, ongoing gene flow (exchange of alleles) is a possibility, often making it difficult to establish the extent to which the initial hybridization event was necessary for weed establishment. However, weed examples with considerable crop–wild contributions abound. Weedy beets (*Beta vulgaris*) in Europe arose through hybridization of domesticated beets that were grown for seed production. Wild beets contain a mix of advantageous cultivated and weedy characters, such as self-fertilization and faster flowering, respectively (Viard et al. 2002; Fénart et al. 2008; Arnaud et al. 2010). Weedy sunflowers in Europe, Australia, and Israel stem from hybridization of cultivated and wild sunflowers in the United States, where wild sunflowers are native (Muller et al. 2011; Lai et al. 2012); presumed contamination of crop seed has spread these weedy strains around the world, where they can no longer undergo gene flow with their wild relatives. Populations of weedy rice with crop and wild contributions also seem to have arisen several times, such as some lineages in South (Huang et al. 2017) and Southeast Asia (Vigueira et al. 2019; Wedger et al. 2019). Furthermore, weeds related to cultivated sorghum (*Sorghum bicolor*), such as shattercane (*S. bicolor*) and Johnsongrass (*S. halepense*), have long been notorious examples of contemporary crop–weed gene flow (Ejeta and Grenier 2005; Morrell et al. 2005; Defelice 2006), though the role of hybridization in the origin of these weeds is confusing, with wild origins also posited (Ohadi et al. 2017).

15.2.3 Weed Origins from Crop Species

The idea of 'crops going bad' has a whiff of comeuppance that both the public and scientists tend to find appealing, judging by the attention paid to weeds falling in this category. Also known as endoferal weeds, De Wet and Harlan (1975) considered these to evolve specifically from abandoned crops through a process of de-domestication. However, the low levels of variation in domesticated crops have been considered a possible hindrance to de-domestication and the adaptability of resulting weeds (Vigueira et al. 2013). Likewise, there has been discussion on the genetic mechanisms for evolving weediness traits from crop ancestors, including the types of mutations involved (Gressel 2005) and the feasibility of reversing multigenic traits (Warwick and Stewart 2005). Given the large number of historically domesticated crops that are not actively cultivated today, there would seem to be a sizeable potential source of crop-derived weeds. However, convincing evidence of weeds originating strictly from crop ancestors requires extensive genetic data to help distinguish the extent to which shared alleles among crop and weed groups are due to a common ancestry versus introgression (Vigueira et al. 2013); to our knowledge, only three well-documented cases of weed origins from crop plants exist. Though limited, these examples show that de-domestication can occur from crop species that are still actively cultivated and provide conflicting evidence about the ease with which de-domestication can occur.

Weedy rye (*Secale cereale*), a problem in many wheat fields, is one such case, originating from cultivated rye fields in California (Burger et al. 2006). Differing from the crop in numerous phenotypes such as seed shattering and flowering time, the evolution of weedy rye is startling in that it may have occurred very quickly, in the span of 60 generations or so (Burger et al. 2007), challenging the notion that weeds can only evolve very slowly or with great difficulty from crop ancestors. A second case, weedy rice, perhaps one of the most studied weeds due to the genetic resources available and its worldwide impact on a major crop, has been shown to have evolved in many cases directly from various rice cultivar backgrounds (e.g. Reagon et al. 2010; Kanapeckas et al. 2016; Huang et al. 2017; Li et al. 2017; Vigueira et al. 2019). The sheer number of independent origins of weedy rice suggests that this transition from crop to weed is one that can occur with ease, belying suggestions that weed origins from crops need to be rare. The third case involves Tibetan semiwild wheat (*Triticum aestivum* ssp. *tibetanum*), a weedy lineage descended from bread wheat, for which a simple genetic change has been shown to underlie the evolution of seed shattering, enabling an easy transition from crop to weed (Jiang et al. 2019). Though the case of Tibetan semiwild wheat again supports an easy transition from crop to weed, the weed has a limited geographic occurrence (known only in Tibet), and no other lineages of weedy descendant of bread wheat have been reported, suggesting that crop-to-weed transitions are not ubiquitous for wheat.

These few examples leave us with several open questions regarding the prevalence and importance of weeds evolving directly from crop ancestors. For example, how easily can such weeds evolve? The weedy rice case suggests that the crop-to-weed transition is simple enough to occur multiple times independently from the same species. The weedy rye and semiwild wheat cases, however, suggest that weeds evolve from crops only in limited circumstances. Ellstrand et al. (2010) and Ellstrand (2019) contend that weed origins from crops (including crop–wild hybrids) are likely under-reported, as noted by the increase in weeds with known crop ancestry in the last 10 years; in contrast, Warwick and Stewart (2005) suggest that very few of the world's worst weeds are sexually compatible with crops and consider 'ferality' a minor concern. Besides determining their actual prevalence, other unresolved questions relevant to controlling weed evolution from crops are as follows: Are crops with certain traits or certain types of genetic variation more susceptible to evolve weedy populations and subspecies? Can weeds evolve from volunteers of crops that are actively in use? Are weeds with crop origins less adaptable to environmental changes and less likely to evolve the ability to invade multiple crop types, given their narrower genetic basis?

15.3 The Genetic Basis of Phenotypic Variation in Weedy and Fitness-Related Traits

Given the economic significance of weedy plants (Pimentel et al. 2001; Oerke 2006), there have been surprisingly few studies to determine the basis of genetic variation in weedy traits (quantity, origin, structure, and architecture) both within and among weed populations. Whereas the last section (Section 15.2.3) focused on the origin and evolution of weed species, here we turn our attention to the weedy traits that underpin adaptation and persistence of weeds in agroecosystems.

Genetic variation in key weedy and fitness-related traits is the raw material on which contemporary weed evolution depends. We propose four key questions that are fundamental to the understanding of genetic variation in weed populations and which will also determine the optimal weed management strategies to mitigate ongoing weed adaptation: (i) What is the role of phenotypic plasticity versus genetic variation in facilitating rapid weed adaptation to new and changing environments? (ii) Does the majority of phenotypic variation in fitness-related traits arise from variation at single major genes or are most traits under poly- or oligogenic control? (iii) Is there standing genetic variation within populations upon which selection can act or is variation created *de novo* via novel mutations, gene flow from surrounding populations, and/or hybridization and introgression with related species? (iv) Do weedy traits evolve once or a few times and spread between populations or are there multiple origins of weedy traits? These questions are fundamental to understanding plant adaptation, in general, leading many authors to advocate for weedy plants as excellent models for contemporary evolution in plant populations (Baucom and Holt 2009; Neve et al. 2009; Vigueira et al. 2013).

15.3.1 Phenotypic Plasticity in Key Weedy Traits

Early views on the genetics of weeds posited that a 'general-purpose genotype' enabled some plant species to rapidly colonize and persist in disturbed agricultural environments via phenotypic plasticity in key weedy traits (Baker 1965). Though undoubtedly a contributing factor, in some cases, to the 'weed syndrome', the primacy of this mode of weed evolution is now questioned with a greater emphasis being placed on the contemporary evolution of weedy traits in genetically diverse plant populations (Clements et al. 2004). These contrasting views of weed adaptation place differing emphases on efforts to manage weed populations. If some plant species are inherently weedy, efforts should focus on reducing the establishment of populations of these species on agricultural land. If, on the other hand, weedy traits evolve in response to selection, the onus is on limiting selection pressure *in situ* via the design and practice of diverse weed management practices.

15.3.2 Origins and Architecture of Genetic Variation

In their review of the practical application of evolutionary principles in agriculture, natural resource management and species conservation, Hendry et al. (2011) commented that most adaptive phenotypic traits are not controlled by a single or just a few major genes. They also concluded that most contemporary evolution will be based on selection from standing genetic variation, with new mutations being more likely to underpin longer-term evolutionary change. In annual plants, where there is a single generation per year and where effective population sizes may be relatively small (compared to fast-reproducing microbes and insect species with multiple generations per year), the rate of generation and fixation of new mutations is likely to be too slow to be a major contributor to the very rapid evolution of weedy traits that is often observed. The widespread and remarkably rapid evolution of resistance to herbicides (see later case study) provides a relevant lens through which to view the importance of standing versus *de novo* and major gene versus polygenic variation in an adaptive trait in weed populations.

Resistance to herbicides has evolved via major gene mechanisms, often endowed by single nucleotide polymorphisms in the genes encoding herbicide target proteins, and via complex, polygenic mechanisms based on altered herbicide metabolism, uptake, and transport (nontarget site resistance [NTSR]; see Beckie et al., this book). While, to our knowledge, no studies have definitively established if single gene mutations in evolved weed populations arose prior to (standing variation) or subsequent to (*de novo*) herbicide selection, Délye et al. (2013) do report the presence of known herbicide resistance conferring mutations in herbarium samples of the grass weed, *Alopecurus myosuroides,* which pre-date herbicide selection. In the case of the putatively polygenic NTSR mechanisms, studies have identified a high frequency of phenotypic herbicide resistance in unselected *Lolium rigidum* populations from Australia (Neve and Powles 2005a), while selection experiments have demonstrated that herbicide resistance can be rapidly selected from standing genetic variation (Neve and Powles 2005b; Busi and Powles 2009). In weedy rice, variation in some weedy traits has been shown to be based on variation at individual genetic loci (awn length, hull pigmentation, and pericarp colour), while other traits (seed shattering, panicle length) appeared to be more quantitative (Qi et al. 2015).

15.3.3 Quantity and Structure of Genetic Variation

Are weed populations inherently genetically variable? Does most variation exist within or between weed populations? Do weedy traits originate from single or multiple origins? Some have argued that invading and establishing weed populations would not harbour a high degree of genetic variability as founder effects in newly invaded habitats, and the passage of populations through genetic bottlenecks and the existence of weeds in homogenous and predictable agricultural environments would all act to limit extent and maintenance of genetic variation. In a study based on allozyme diversity, Hamrick et al. (1979) found that weed species did, in general, exhibit lower diversity than other similar plant species. However, subsequent empirical support for these observations has been limited and, in general, the relatively few studies of weed genetic diversity that have been conducted indicate the maintenance of high levels of variation.

Several studies have contrasted levels of genetic variation in populations of the same species collected from agricultural and 'natural' environments to test the hypothesis that adaptation to agroecosystems results in reduced genetic diversity. In all cases, weedy populations exhibited equal (e.g. *Chenopodium album* and *Amaranthus retroflexus*, Warwick and Black 1986) or greater (e.g. *Capsella bursa-pastoris*, Bosbach and Hurka 1981; *Senecio vulgaris*, Leiss and Müller-Schärer 2001) levels of genetic polymorphism compared with populations from non-cultivated land. Sahli et al. (2008) compared Q_{ST} and F_{ST} values among naturalized (weedy) and native populations of *R. raphanistrum* and concluded that the loss of a rosette and a transition to early flowering had been key adaptations allowing wild radish to become a major agricultural weed.

In particular, population genetics approaches have been used to determine how the rapid evolution of herbicide resistance impacts, and is impacted by, population structure. For example, does intense selection by herbicides force populations through genetic bottlenecks so that resistant populations exhibit lower genetic diversity than their sensitive counterparts? Do patterns of genetic diversity and gene flow within and among populations dictate that resistance evolves rarely and disperses across the landscape, or does it evolve locally on multiple, independent occasions?

In their studies with triazine-resistant *Brassica rapa*, Warwick and Black (1993) observed no difference in the number of polymorphic sites between R (resistant) and S (susceptible) populations, but a slightly reduced allelic diversity in R populations. In populations of *A. myosuroides* from France, Menchari et al. (2007) found no correlation between genetic diversity and the frequency of herbicide resistance. On the other hand, Kuester et al. (2016) used a resurrection experiment approach to demonstrate that adaptation to herbicides over a ca. 10-year period was associated with a loss of genetic diversity in *Ipomoea purpurea*. While these studies may have identified contrasting impacts of weed adaptation on patterns of genetic variation, most studies of outcrossing, herbicide resistance-prone weed species have identified little genetic differentiation (low F_{ST}) between weed populations (Chauvel and Gasquez 1994; Kercher and Conner 1996; Menchari et al. 2007; Sahli et al. 2008; Karn and Jasieniuk 2017). Taken together, these observations of weed population genetics conform to the model proposed by Clements et al. (2004) whereby there is a 'dynamic tension between processes that reduce and restore genetic variation' – founder effects, together with the genetic bottlenecking caused by extreme selection that may reduce variation, but this is rapidly restored by high levels of gene flow (pollen and seed dispersal) in outcrossing and mixed mating species in highly connected agricultural landscapes.

Another fundamental challenge in weed adaptation is to understand the extent to which novel weedy traits evolve independently in multiple populations (convergent evolution) versus cases where the trait evolves once (or rarely) and spreads via seed and pollen dispersal. Weedy rice occurs in many areas of the world, including in agroecosystems where wild rice populations do not naturally occur (Londo and Schaal 2007), and evidence suggests that weedy populations have evolved independently in multiple locations (Huang et al. 2018) through all three of the major modes of weed evolution (Vigueira et al. 2013; see Section 15.2). A high degree of heterogeneity between weedy rice populations is seen in South Asia, with three separate origins of weedy rice posited, two resulting from de-domestication of crop cultivars and one from wild rice. In all cases, despite independent origins there has been convergence for seed shattering, red pericarp pigmentation, and compact plant architecture, suggesting these are essential traits for the evolution of weediness in rice (Huang et al. 2018).

In general, even when considering a single mechanism of evolved herbicide resistance, multiple genetic origins have been observed (Menchari et al. 2007; Délye et al. 2010; Fernández et al. 2013; Okada et al. 2015). However, in the case of glyphosate resistance in *Amaranthus palmeri*, where resistance mutations are considered to arise less frequently than for other classes of herbicide resistance, Molin et al. (2017) concluded that their observations were consistent with a single origin of glyphosate resistance and subsequent spread throughout the United States. In Canadian populations of waterhemp, Kreiner et al. (2019) found that in one location, glyphosate resistance seemed to have evolved through the introduction of preadapted genotypes from the United States; conversely, in a second region, resistance appeared to have evolved locally and independently in nonagricultural genetic backgrounds.

15.4 The Contemporary Evolution of Weeds in Agroecosystems: Evidence and Case Studies

We have established some broad frameworks for the adaptation of weeds in agroecosystems: (i) weeds may arise and evolve from a variety of wild species and crop backgrounds, and (ii) weed evolution is pervasive, contemporary, and ongoing, (iii) meaning that weed adaptation is a major force in enabling weed persistence, with (iv) the notable adaptability of weedy plants being underpinned by dynamic genomes and frequent processes of genetic bottlenecking, gene flow and dispersal, hybridization, introgression, and polyploidization. In this section, we consider the types of selection pressure that are driving weed adaptation and the types of traits that have evolved to increase the fitness of weedy plants in agroecosystems.

In the broad sense, all agricultural weed species arise and evolve because of human-mediated selection in agroecosystems, though we may distinguish between more and less intense selection pressures. Crop choices and crop rotation operate over longer timescales with somewhat less intense selection pressures to select the weed species and communities that will invade and persist in agroecosystems. The most intense selection, which leads to the most rapid, widespread, and notable examples of weed adaptation is imposed by the predominant weed management practices – herbicide application, tillage, cultivation, and hand weeding. This type of adaptation leads to the failure of weed management systems, facilitating the persistence of weed populations and fostering the design of integrated weed management approaches (Neve et al. 2018). We shall focus on more intense and rapid selection for weediness in the examples given here. Over the longer term, less direct selection for weediness is also driven by the concerted impacts of human-mediated climate and environmental change in combination with resulting changes in patterns of crop production. These climatic, environmental, and agronomic drivers lead to ecotypic differentiation between weed populations, which can drive weed range expansions (Clements et al. 2004; McDonald et al. 2009; Clements and Ditommaso 2011, Chapter 12, this book).

In simple terms, weed adaptation may act on morphological (e.g. plant habit, plant architecture, seed size, and shape); phenological (time to flowering, germination timing, seed shattering, and dispersal timing); and physiological (herbicide sensitivity and response to abiotic stress) characteristics. Case studies for morphological, phenological, and physiological adaptation are provided in the following sections (Sections 15.4.1 and 15.4.2).

15.4.1 Crop Mimicry as an Adaptive Strategy for Promoting Weed Persistence in Agroecosystems

Many early studies of weed adaptation focused on crop mimicry (see Barrett 1983), sometimes referred to as 'Vavilovian mimicry' in homage to the Russian plant geneticist, Nikolai Vavilov who, in the 1920s and 1930s, developed ideas about the importance of evolutionary forces driving crop breeding and weed adaptation, including the potential for weeds and crops to co-evolve through mimicry (McElroy 2014). Where weed species can synchronize their life cycles with crops (phenological mimicry) and/or evolve to resemble the crops they infest (morphological mimicry), they become more difficult to distinguish and remove through cultural and physical weed management practices. Two categories of crop mimics have been identified: 'seed mimics' and 'vegetative mimics' (Barrett 1983).

For annual plants that propagate by seed, persistence of populations is promoted by mechanisms that ensure seed survival, dispersal, longevity, and timely germination. For weeds infesting crops, this can be achieved in two ways. Weed populations may evolve to early seed shattering, dispersing weed seeds prior to crop harvest. These seeds are returned to the soil seedbank (where conditions are more benign) by cultivations. Evolved mechanisms of seed dormancy and dormancy cycling may then ensure that seeds germinate, and seedlings emerge from the soil to coincide with favourable environmental and climatic conditions and/or with crop establishment (see Chapters 2–6 of this book). An alternative strategy is for weeds to evolve a non-shattering habit so that their seeds are 'harvested' at the same time as the crop seed, meaning that some weed seeds are stored during unfavourable conditions with the crop seed and are reseeded with the crop at the optimal time of year. Both strategies have pros and cons; early shattering and survival in the seedbank exposes seeds to predation and to microbial attack, whereas as being harvested with the crop means that weed seeds can potentially be removed from stored crop seed via seed cleaning. Indeed, improved seed cleaning has been suggested as a major contributor to the decline in some weed populations in the first part of the twentieth century (Salisbury 1961). However, like any successful weed management strategy, seed cleaning establishes a selection pressure for weed populations to circumvent this control. Although they were carried out some time ago, there have been several notable studies that have demonstrated the evolution of morphological seed mimicry. Perhaps the best-known example concerns the startling mimicry of lentil (*Lens culinaris*) seeds by the coexisting weed, common vetch (*Vicia sativa*) (see Gould 1991 and Figure 15.2). Various explanations were put forward for this phenomenon, including potential crop–weed hybridization. However, Rowlands (1959) was able to establish that the lentil-like seed shape was due to a single recessive mutation, suggesting it had evolved via random mutations in common vetch populations and subsequent unconscious selection by seed cleaning processes. Further examples of seed mimicry are discussed in Barrett (1983) and Gould (1991).

Figure 15.2 Selection for seed mimicry in *Vicia sativa* (common vetch) cultivated in lentil fields in Albion, Washington, USA. The seeds shown on the left show the typical seed colour, shape, and size for common vetch seeds. The seeds in the centre were collected from the field in Albion and had a distinctly different shape and colour to normal common vetch seeds. The seeds on the right are lentil seeds. Clearly, in this case following years of being grown in lentil crops, this common vetch population has evolved seed mimicry. *Source:* Gould (1991).

The best documented case of vegetative mimicry is exhibited by barnyardgrass (*Echinochloa crus-galli*) populations associated with rice cultivation (Barrett 1983). Though not closely related phylogenetically, rice and barnyardgrass are quite similar in morphology. In the case of the subspecies *E. crus-galli* var. *oryzicola*, the morphological similarities to rice are so striking that it is very difficult to tell seedlings of the two species apart (Figure 15.3). The *oryzicola* subspecies is native to Asia, though it has now invaded rice cropping systems in many other parts of the world. It is believed that this variety evolved as a response to hand weeding, where visually similar seedlings were indistinguishable from the rice crop and were preferentially selected in a very specific case of human-directed evolution.

Even though the mimetic forms of barnyardgrass evolved in the non-mechanized agroecosystems of Asia, they persisted as the dominant subspecies in California, USA, long after hand weeding was replaced by other forms of weed control. Indeed, the unique adaptation of *E. crus-galli* var. *oryzicola* to growing with rice was demonstrated by Barrett (1983) who showed that the subspecies rarely exists outside of rice crops. In a comparison of life history traits between populations of *E. crus-galli* and *E. crus-galli* var. *oryzicola*, Barrett and Wilson (1983) found that the mimetic form had seeds that were three to four times heavier, had much lower levels of innate dormancy, and exhibited more emergence in flooded soils than seeds from the non-mimetic form. In addition, the mimic displayed greater seedling vigour, delayed flowering, and produced fewer seeds than barnyardgrass (Barrett and Wilson 1981). Taken together, these data indicate that barnyardgrass rice

Figure 15.3 Adaptation to hand weeding in *Echinochloa crus-galli* (barnyardgrass). From left to right, the plants shown here are a young cultivated rice seedling, a seedling of the *oryzicola* variety of barnyardgrass collected from a rice cultivation field, and a more typical barnyardgrass seedling. The plant in the centre has been unconsciously selected to mimic the morphology of the crop by hand weeding. *Source:* Gould (1991).

mimics have evolved an array of life history characters that have enabled them to evolve a highly adapted 'weed syndrome' for establishing and persisting under rice cultivation in multiple global rice cropping systems.

15.4.2 Herbicide Resistance: Weed Adaptation on Steroids

The evolution of resistance to herbicides in the face of intense and concerted selection pressure is a global phenomenon, and the role of resistance in enabling the persistence of weed populations is discussed by Beckie et al. (Chapter 7, this book). Nevertheless, the subject is worth revisiting here as it represents a striking example of the impacts of human-directed evolution in agroecosystems that, for some mechanisms of resistance, is also based on a form of crop mimicry.

Broadly speaking, we can distinguish between two 'types' of evolved herbicide resistance (Powles and Yu 2010). Target site resistance (TSR) is based on changes in the coding sequence, expression, and/or copy number of the genes that code for herbicide targets, whereas NTSR represents a range of mechanisms that prevent the herbicide from reaching its target in sufficient concentrations to cause plant mortality. These mechanisms include alterations in herbicide metabolism and uptake and transport within and between plant cells and organs (Ghanizadeh and Harrington 2017).

The basis for successful herbicidal weed control in crops is to achieve selectivity so that herbicides are lethal to weeds but not to the crop. This selectivity can be achieved via inherent differences in crop and weed sensitivity, by using safeners that protect the crop, but not the unwanted weeds from phytotoxic herbicide effects or by genetic modification of crops. In many cases, crop safety is conferred by the up-regulation of a suite of metabolism- and defence-related gene families: the glutathione-S-transferases (Cummins et al. 2011), cytochrome P450s (Robineau et al. 1998), glucosyltransferases (Brazier-Hicks and Edwards 2005), and ATP bending cassette (ABC) transporters (Lane et al. 2016). Notably, as we begin to learn more about the biochemical and molecular genetic changes that confer evolved NTSR in weed species, these same families of genes and gene families are implicated (Yuan et al. 2007; Gaines et al. 2014). Essentially, weed species are co-opting the same defence- and metabolism-based pathways that provide inherent protection against herbicides in crop species to evolve resistance under selection – another striking example of weeds mimicking crops to ensure their survival and persistence despite concerted efforts to control them.

We will close this section with a brief consideration of the evolution of resistance to the herbicide glyphosate. This too is a notable example of the adaptive potential of weeds. Until the late 1990s, 25 years after the commercial release of glyphosate, no evolved resistance was reported, leading some to speculate that the risks of evolved glyphosate resistance may be uniquely low (Bradshaw et al. 1997). While it is certainly true that, compared to many other herbicide modes of action resistance risks are lower, the introduction of genetically modified glyphosate-resistant crops in the mid- to late 1990s and increasing use of glyphosate in conventional cropping systems has driven an epidemic of glyphosate resistance.

As of 2020, glyphosate resistance had been reported in 45 weed species and in six continents (Heap 2020). The combination of a paucity of standing genetic variation for glyphosate (inferred from the relatively low risks of resistance; but see Davies and Neve 2017;

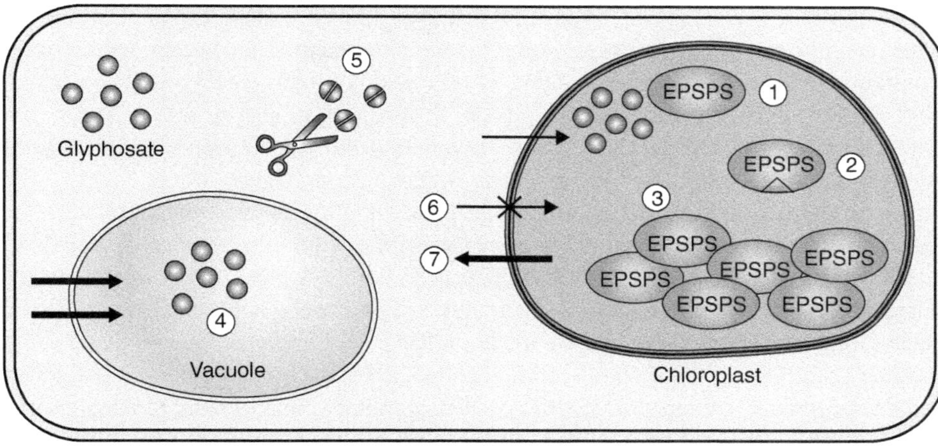

Figure 15.4 The diversity of documented evolved glyphosate resistance mechanisms in weedy plants. In sensitive plants, glyphosate (small gray circles) enters the cytoplasm and is transported into the chloroplast to the target site enzyme, 5-enolpyruvylshikimate 3-phosphate synthase (EPSPS) (1). Mutations changing one or two amino acids in EPSPS can confer target site resistance, first reported in 2002 (2). Extra gene copies of EPSPS can produce extra-sensitive EPSPS, requiring proportionally more glyphosate to cause complete inhibition, first reported in 2010 (3). Glyphosate can be transported and sequestered in the vacuole, with the mechanism first reported in 2010 and the observation of reduced glyphosate translocation first reported in 1999 (4). Some evidence to suggest glyphosate metabolism has been reported, but no specific genes have been identified in weeds to date (5). We predict that mechanisms might exist to reduce glyphosate import into the chloroplast (6) and/or rapidly export glyphosate out of the chloroplast (7), but neither mechanism has been observed to date. *Source:* Gaines et al. (2019).

Comont et al. 2019) and a uniquely intense selection pressure have, however, led weed species to evolve a diverse array of glyphosate resistance mechanisms (Figure 15.4), some of which are extremely rare and are providing novel insights into the nature of plant adaptation in the face of novel environmental stresses (Koo et al. 2018; Gaines et al. 2019). Any notion that glyphosate may be 'resistance-proof' has been emphatically dispelled. Genetically modified glyphosate-resistant crops provided unprecedented and highly efficacious weed control for 5–10 years in the cropping systems to which they were introduced. However, once again we have been reminded that evolution will find a way and that weeds will persist and often thrive in all cropping systems where integrated and diverse weed management is not practiced.

15.5 Applying Evolutionary Thinking to Weed Biology and Management

By their nature, and as recognized by Harper (1956), agricultural weeds are adaptable, via phenotypic plasticity or through the possession of, and ongoing selection for, heritable genetic variation in key weedy traits. In this chapter, we have argued that this adaptability is an important facet of weed persistence in agroecosystems (maybe the most important facet?). More broadly, therefore, weedy and invasive plants provide a cogent example for applied evolutionary biologists to reinforce the notable potential for human agricultural, industrial, and cultural activities to drive rapid rates of contemporary evolution in the world's biota (Palumbi 2001; Carroll et al. 2011), resulting in significant economic and environmental consequences. Clearly, weed adaptation must be an important consideration in our efforts to design sustainable weed management strategies.

It can be argued that the advent and widespread adoption of synthetic herbicides for weed control since the 1950s resulted in an *evolution* of the discipline of weed biology and management (weed science). Before the discovery of herbicides, weed management focused on the application of ecological principles to limit weed abundance and impacts; however, in herbicide-dominated systems, there has been a more explicit focus on weed physiology and the control of weeds through chemical means. Calls for a more integrated weed management (Mortensen et al. 2012), which combines chemical and physical means of weed control with agronomic and agroecological practices to limit weed establishment, spread, and persistence, are founded on the desire to limit weed adaptation. A greater diversity of weed management will make it more difficult for weed species and populations to rapidly adapt to any single management intervention by diversifying selection pressures (see Hendry et al. 2011). Several authors have called for a greater integration of evolutionary principles into weed management (Harper 1956; Jordan and Jannink 1997; Mortimer 1997; Neve et al. 2009; Clements and DiTommaso 2012; Harker 2013), but evolutionary biology remains peripheral to many efforts, and collaborations between weed scientists and evolutionary biologists are notably rare (Baucom and Holt 2009).

There are several specific examples where greater attention to evolutionary potential in weeds may facilitate the design of more sustainable weed management strategies and approaches; here, we will briefly consider three of those – the management and mitigation of herbicide resistance, the modelling of weed population dynamics and range expansions, and the breeding and selection of crop varieties to maximize their competitive potential versus weeds.

15.5.1 The Evolutionary Ecology of Herbicide Resistance

While there have been numerous studies to explore the evolutionary dynamics of herbicide resistance in weeds (experimental evolution, modelling, fitness costs, gene flow and dispersal, population genetics), the major focus has been and remains on establishing the molecular genetic and biochemical basis of resistance traits, often in one or a few 'archetype' weed accessions. These mechanistic studies are important for unravelling the underlying basis of resistance, for developing resistance diagnostics, and for facilitating follow-on studies that seek to determine the agronomic, ecological, and evolutionary drivers for resistance. They do, however, maintain a focus on the *proximate* causes for the evolution of herbicide resistance, whereas concerted progress towards designing strategies to mitigate the evolution of resistance require a deeper understanding of the *ultimate* causes underpinning the origins, selection, spread, and persistence of resistance traits at field to landscape scales (Neve et al. 2014). Such studies can address new concepts in resistance evolution relating to evolutionary origins and modes of evolution (Hawkins et al. 2019), evolutionary rescue (Délye et al. 2013), epidemiology (Hicks et al. 2018; Comont et al. 2019), and non-Mendelian inheritance of resistance traits (Gressel 2015; Dyer 2018) with potential benefits for the design of integrated weed management strategies that reduce evolution of resistance and other weed adaptive traits.

15.5.2 Modelling Weed Population Dynamics

The premise for many models of weed population dynamics is that transitions between various life history stages can be defined by a series of static, species-specific (sometimes population-specific) parameters. In the case of models that define the niche envelope and range limits for weeds, it is usually assumed that ecological and physiological limits are fixed and invariant. The limits of these assumptions and approaches were discussed by Clements and DiTommaso (2012) when considering the potential for weed invasion under climate change. Where these models consider weed populations and the parameters that dictate their dynamics as fixed entities, it is not possible to consider the potential for ongoing evolution to facilitate the invasion of weeds into new environments. Models that dynamically updated range limits to account for ongoing evolution were used to predict the ability of the cane toad (*Chaunus marinus*) to invade beyond its current predicted range in Australia (Urban et al. 2007). These same caveats about incorporating evolutionary potential into models apply to all efforts to model weed populations under new environments and/or management scenarios. Where ongoing evolutionary potential is not considered in models, the persistence of weed populations will be underestimated as models will assume the maintenance of weed control efficacy in management strategies.

15.5.3 Crop Competitiveness

Most attempts to breed modern crop varieties and to screen these breeding lines for optimal yield potential are carried out without competition from weeds. Consequently, traits that maximize crop competitiveness are unconsciously selected against these trade-offs with harvest index and grain yield potential (Denison 2012). The lack of evolutionary thinking in the development and optimization of agricultural systems has led to calls for a more 'Darwinian agriculture' (Denison 2012). In this vein, Weiner et al. (2010) coined the phrase 'evolutionary agroecology' and used this concept to advocate for breeding programmes that would select for highly cooperative individuals that would minimize intraspecific competition between crop plants, while maximizing the crop's ability to suppress weeds.

15.6 Weed Adaptation: A Key Determinant of Weed Persistence in Agroecosystems

The global and repeated evolution of resistance to herbicides provides a salutary example of the impacts of the ongoing weed management arms race in agroecosystems. Some of the earliest noted cases of weed adaptation relate to crop mimicry, which reduced the effectiveness of hand weeding or made it more difficult to clean weed seeds out of harvested grain samples. It is, however, likely that ongoing weed evolution impacts most of the aspects of weed biology and management discussed in this book: seed production and dispersal, seed dormancy, seed longevity, responses to climate change, allelopathy, crop–weed interference and biotic interactions, and herbicide resistance as well as having implications for attempts to model weed population and community dynamics (see Storkey et al., this book).

Harker (2013) noted that while weeds have been evolving for millennia to overcome human attempts to control them and to account for changing environments and cropping practices, the gradual reduction in cropping system diversity and over-reliance on a limited armoury of weed management tools has undoubtedly accelerated rates of weed adaptation in recent decades. If we are to develop integrated weed management systems that slow weed evolution to limit weed persistence, it will be important to integrate applied weed science with evolutionary ecology to answer questions about the potential and mode of weed adaptation (Baucom and Holt 2009). These efforts can be facilitated in the future by a greater access to weed genomics tools to better understand the genetic architecture of weed adaptive traits (Ravet et al. 2018), thus addressing the current observation that the genetic basis of weed adaptation and persistence in agroecosystems remains relatively poorly studied and understood. Access to genomic tools and insights, however, will not be an end in itself. In the future, greater access to high-throughput and remote plant phenotyping technology will enable efforts to link genotype to phenotype. Increasingly this data will also be supplemented by access to contextual environmental and field management metadata, and it will be possible to conduct large-scale evolutionary epidemiology-type studies to unravel the ultimate ecological and evolutionary drivers of weed evolution and design management approaches that limit weed adaptation in agroecosystems (see Hicks et al. 2018; Comont et al. 2020; Comont and Neve 2020).

References

Arnaud, J.F., Fénart, S., Cordellier, M., and Cuguen, J. (2010). Populations of weedy crop–wild hybrid beets show contrasting variation in mating system and population genetic structure. *Evolutionary Applications* 3: 305–318.

Baker, H.G. and Stebbins, G.L. (1965). Genetics of colonizing species. In: *International Union of Biological Sciences Symposia on General Biology 1964*, 147–172. Asilomar, Calif: Academic Press.

Barrett, S.H. (1983). Crop mimicry in weeds. *Economic Botany* 37: 255–282.

Barrett, S.C. and Wilson, B.F. (1981). Colonizing ability in the *Echinochloa crus-galli* complex (barnyardgrass). I. Variation in life history. *Canadian Journal of Botany* 59: 1844–1860.

Barrett, S.C. and Wilson, B.F. (1983). Colonizing ability in the *Echinochloa crus-galli* complex (barnyardgrass). II. Seed biology. *Canadian Journal of Botany* 61: 556–562.

Baucom, R.S. and Holt, J.S. (2009). Weeds of agricultural importance: bridging the gap between evolutionary ecology and crop and weed science. *New Phytologist* 184: 741–743.

Bosbach, K. and Hurka, H. (1981). Biosystematic studies on *Capsella bursa-pastoris* (Brassicaceae): enzyme polymorphism in natural populations. *Plant Systematics and Evolution* 137: 73–94.

Bradshaw, L.D., Padgette, S.R., Kimball, S.L., and Wells, B.H. (1997). Perspectives on glyphosate resistance. *Weed Technology* 11: 189–198.

Brazier-Hicks, M. and Edwards, R. (2005). Functional importance of the family 1 glucosyltransferase UGT72B1 in the metabolism of xenobiotics in *Arabidopsis thaliana*. *The Plant Journal* 42: 556–566.

Burger, J.C., Lee, S.K.Y., and Ellstrand, N.C. (2006). Origin and genetic structure of feral rye in the western United States. *Molecular Ecology* 15: 2527–2539.

Burger, J.C., Holt, J.M., and Ellstrand, N.C. (2007). Rapid phenotypic divergence of feral rye from domesticated cereal rye. *Weed Science* 55: 204–211.

Busi, R. and Powles, S.B. (2009). Evolution of glyphosate resistance in a *Lolium rigidum* population by glyphosate selection at sublethal doses. *Heredity* 103: 318–325.

Carroll, S., Kinnison, M.T., and Bernatchez, L. (2011). In light of evolution: interdisciplinary challenges in food, health, and the environment. *Evolutionary Applications* 4: 155–156.

Charbonneau, A., Tack, D., Lale, A. et al. (2018). Weed evolution: genetic differentiation among wild, weedy, and crop radish. *Evolutionary Applications* 11: 1964–1974.

Chauvel, B. and Gasquez, J. (1994). Relationships between genetic polymorphism and herbicide resistance within *Alopecurus myosuroides* Huds. *Heredity* 72: 336–344.

Clements, D.R. and Ditommaso, A. (2011). Climate change and weed adaptation: can evolution of invasive plants lead to greater range expansion than forecasted? *Weed Research* 51: 227–240.

Clements, D.R. and DiTommaso, A. (2012). Predicting weed invasion in Canada under climate change: evaluating evolutionary potential. *Canadian Journal of Plant Science* 92: 1013–1020.

Clements, D.R., DiTommaso, A., Jordan, N. et al. (2004). Adaptability of plants invading North American cropland. *Agriculture, Ecosystems & Environment* 104: 379–398.

Comont, D. and Neve, P. (2020). Adopting epidemiological approaches for herbicide resistance monitoring and management. *Weed Research* 61: 81–87.

Comont, D., Hicks, H., Crook, L. et al. (2019). Evolutionary epidemiology predicts the emergence of glyphosate resistance in a major agricultural weed. *New Phytologist* 223: 1584–1594.

Comont, D., Lowe, C., Hull, R. et al. (2020). Evolution of generalist resistance to herbicide mixtures reveals a trade-off in resistance management. *Nature Communications* 11: 1–9.

Cummins, I., Dixon, D.P., Freitag-Pohl, S. et al. (2011). Multiple roles for plant glutathione transferases in xenobiotic detoxification. *Drug Metabolism Reviews* 43: 266–280.

Davies, L.R. and Neve, P. (2017). Interpopulation variability and adaptive potential for reduced glyphosate sensitivity in *Alopecurus myosuroides*. *Weed Research* 57: 323–332.

De Wet, J.M. and Harlan, J.R. (1975). Weeds and domesticates: evolution in the man-made habitat. *Economic Botany* 29: 99–108.

Defelice, M.S. (2006). Shattercane, *Sorghum bicolor* (L.) Moench ssp. drummondii (nees ex steud.) de wet ex davidse – black sheep of the family. *Weed Technology* 20: 1076–1083.

Délye, C., Michel, S., Bérard, A. et al. (2010). Geographical variation in resistance to acetyl-coenzyme A carboxylase-inhibiting herbicides across the range of the arable weed *Alopecurus myosuroides* (black-grass). *New Phytologist* 186: 1005–1017.

Délye, C., Jasieniuk, M., and Le Corre, V. (2013). Deciphering the evolution of herbicide resistance in weeds. *Trends in Genetics* 29: 649–658.

Denison, R.F. (2012). *Darwinian Agriculture: How Understanding Evolution Can Improve Agriculture*. USA: Princeton University Press.

Dirzo, R. and Raven, P.H. (2003). Global state of biodiversity and loss. *Annual Review of Environment and Resources* 28: 137–167.

Dobzhansky, T. (1973). Nothing in biology makes sense except in the light of evolution. *American Biology Teacher* **35**: 125–129.

Dyer, W.E. (2018). Stress-induced evolution of herbicide resistance and related pleiotropic effects. *Pest Management Science* 74: 1759–1768.

Ejeta, G. and Grenier, C. (2005). Sorghum and its weedy hybrids. In: *Crop Ferality and Volunteerism*, 123–135. Boca Raton, FL: CRC Press.

Ellstrand, N.C. (2019). The evolution of crops that do not need us anymore. *New Phytologist* 224: 550–551.

Ellstrand, N.C., Heredia, S.M., Leak-Garcia, J.A. et al. (2010). Crops gone wild: evolution of weeds and invasives from domesticated ancestors. *Evolutionary Applications* 3: 494–504.

Fénart, S., Arnaud, J.F., De Cauwer, I., and Cuguen, J. (2008). Nuclear and cytoplasmic genetic diversity in weed beet and sugar beet accessions compared to wild relatives: new insights into the genetic relationships within the *Beta vulgaris* complex species. *Theoretical and Applied Genetics* 116: 1063–1077.

Fernández, L., De Haro, L.A., Distefano, A.J. et al. (2013). Population genetics structure of glyphosate-resistant Johnsongrass (*Sorghum halepense* L. Pers) does not support a single origin of the resistance. *Ecology and Evolution* 3: 3388–3400.

Gaines, T.A., Lorentz, L., Figge, A. et al. (2014). RNA-Seq transcriptome analysis to identify genes involved in metabolism-based diclofop resistance in *Lolium rigidum*. *The Plant Journal* 78: 865–876.

Gaines, T.A., Patterson, E.L., and Neve, P. (2019). Molecular mechanisms of adaptive evolution revealed by global selection for glyphosate resistance. *New Phytologist* 223: 1770–1775.

Ghanizadeh, H. and Harrington, K.C. (2017). Non-target site mechanisms of resistance to herbicides. *Critical Reviews in Plant Sciences* 36: 24–34.

Gould, F. (1991). The evolutionary potential of crop pests. *American Scientist* 79: 496–507.

Gould, S.J. and Vrba, E.S. (1982). Exaptation – a missing term in the science of form. *Paleobiology* 8: 4–15.

Gressel, J. (ed.) (2005). *Crop Ferality and Volunteerism*. CRC Press.

Gressel, J. (2015). Perspective: consider removing 'Inherited' from definitions of pesticide resistance. *Outlooks on Pest Management* 26: 220–222.

Hamrick, J.L., Linhart, Y.B., and Mitton, J.B. (1979). Relationships between life history characteristics and electrophoretically detectable genetic variation in plants. *Annual Review of Ecology and Systematics* 10: 173–200.

Harker, K.N. (2013). Slowing weed evolution with integrated weed management. *Canadian Journal of Plant Science* 93: 759–764.

Harper, J.L. (1956). The evolution of weeds in relation to resistance to herbicides. In: *Proceedings of the Third British Weed Control Conference*, 179–188. United Kingdom: The British Crop Protection Council.

Hawkins, N.J., Bass, C., Dixon, A., and Neve, P. (2019). The evolutionary origins of pesticide resistance. *Biological Reviews* 94: 135–155.

Heap, I (2020). The international herbicide-resistant weed database. www.weedscience.org (accessed January 2020).

Hendry, A.P., Kinnison, M.T., Heino, M. et al. (2011). Evolutionary principles and their practical application. *Evolutionary Applications* 4: 159–183.

Hicks, H.L., Comont, D., Coutts, S.R. et al. (2018). The factors driving evolved herbicide resistance at a national scale. *Nature Ecology and Evolution* 2: 529.

Huang, Z., Young, N.D., Reagon, M. et al. (2017). All roads lead to weediness: patterns of genomic divergence reveal extensive recurrent weedy rice origins from South Asian *Oryza*. *Molecular Ecology* 26: 3151–3167.

Huang, Z., Kelly, S., Matsuo, R. et al. (2018). The role of standing variation in the evolution of weediness traits in South Asian weedy rice (*Oryza* spp.). *G3: Genes, Genomes, Genetics* 8: 3679–3690.

Jiang, Y.F., Chen, Q., Wang, Y. et al. (2019). Re-acquisition of the brittle rachis trait via a transposon insertion in domestication gene Q during wheat de-domestication. *New Phytologist* 223: 961–973.

Jordan, N.R. and Jannink, J.L. (1997). Assessing the practical importance of weed evolution: a research agenda. *Weed Research* 37: 237–246.

Kanapeckas, K.L., Vigueira, C.C., Ortiz, A. et al. (2016). Escape to ferality: the endoferal origin of weedy rice from crop rice through de-domestication. *PLoS One* 11: e0162676.

Kane, N.C. and Rieseberg, L.H. (2008). Genetics and evolution of weedy *Helianthus annuus* populations: adaptation of an agricultural weed. *Molecular Ecology* 17: 384–394.

Karn, E. and Jasieniuk, M. (2017). Genetic diversity and structure of *Lolium perenne* ssp. *multiflorum* in California vineyards and orchards indicate potential for spread of herbicide resistance via gene flow. *Evolutionary Applications* 10: 616–629.

Kercher, S. and Conner, J.K. (1996). Patterns of genetic variability within and among populations of wild radish, *Raphanus raphanistrum* (Brassicaceae). *American Journal of Botany* 83: 1416–1421.

Khoury, C.K., Bjorkman, A.D., Dempewolf, H. et al. (2014). Increasing homogeneity in global food supplies and the implications for food security. *Proceedings of the National Academy of Sciences of the United States of America* 111: 4001–4006.

Koo, D.H., Molin, W.T., Saski, C.A. et al. (2018). Extrachromosomal circular DNA-based amplification and transmission of herbicide resistance in crop weed *Amaranthus palmeri*. *Proceedings of the National Academy of Sciences of the United States of America* 115: 3332–3337.

Kreiner, J.M., Giacomini, D.A., Bemm, F. et al. (2019). Multiple modes of convergent adaptation in the spread of glyphosate-resistant *Amaranthus tuberculatus*. *Proceedings of the National Academy of Sciences of the United States of America* 116: 21076–21084.

Kuester, A., Wilson, A., Chang, S.M., and Baucom, R.S. (2016). A resurrection experiment finds evidence of both reduced genetic diversity and potential adaptive evolution in the agricultural weed *Ipomoea purpurea*. *Molecular Ecology* 25: 4508–4520.

Lai, Z., Kane, N.C., Kozik, A. et al. (2012). Genomics of compositae weeds: EST libraries, microarrays, and evidence of introgression. *American Journal of Botany* 99: 209–218.

Lane, T.S., Rempe, C.S., Davitt, J. et al. (2016). Diversity of ABC transporter genes across the plant kingdom and their potential utility in biotechnology. *BMC Biotechnology* 16: 47.

Leiss, K.A. and Müller-Schärer, H. (2001). Adaptation of *Senecio vulgaris* (Asteraceae) to ruderal and agricultural habitats. *American Journal of Botany* 88: 1593–1599.

Li, L.-F., Li, Y.-L., Jia, Y. et al. (2017). Signatures of adaptation in the weedy rice genome. *Nature Genetics* 49: 811–814.

Londo, J.P. and Schaal, B.A. (2007). Origins and population genetics of weedy red rice in the USA. *Molecular Ecology* 16: 4523–4535.

McDonald, A., Riha, S., DiTommaso, A., and DeGaetano, A. (2009). Climate change and the geography of weed damage: analysis of US maize systems suggests the potential for significant range transformations. *Agriculture, Ecosystems and Environment* 130: 131–140.

McElroy, J.S. (2014). Vavilovian mimicry: Nikolai Vavilov and his little-known impact on weed science. *Weed Science* 62: 207–216.

Menchari, Y., Délye, C., and Le Corre, V. (2007). Genetic variation and population structure in black-grass (*Alopecurus myosuroides* Huds.), a successful, herbicide-resistant, annual grass weed of winter cereal fields. *Molecular Ecology* 16: 3161–3172.

Milla, R., Bastida, J.M., Turcotte, M.M. et al. (2018). Phylogenetic patterns and phenotypic profiles of the species of plants and mammals farmed for food. *Nature Ecology and Evolution* 2: 1808–1817.

Molin, W.T., Wright, A.A., Lawton-Rauh, A., and Saski, C.A. (2017). The unique genomic landscape surrounding the EPSPS gene in glyphosate resistant *Amaranthus palmeri*: a repetitive path to resistance. *BMC Genomics* 18: 91.

Morrell, P.L., Williams-Coplin, T.D., Lattu, A.L. et al. (2005). Crop-to-weed introgression has impacted allelic composition of johnsongrass populations with and without recent exposure to cultivated sorghum. *Molecular Ecology* 14: 2143–2154.

Mortensen, D.A., Egan, J.F., Maxwell, B.D. et al. (2012). Navigating a critical juncture for sustainable weed management. *BioScience* 62: 75–84.

Mortimer, A.M. (1997). Phenological adaptation in weeds – an evolutionary response to the use of herbicides? *Pesticide Science* 51: 299–304.

Muller, M.H., Latreille, M., and Tollon, C. (2011). The origin and evolution of a recent agricultural weed: population genetic diversity of weedy populations of sunflower (*Helianthus annuus* L.) in Spain and France. *Evolutionary Applications* 4: 499–514.

Neve, P. and Powles, S. (2005a). High survival frequencies at low herbicide use rates in populations of *Lolium rigidum* result in rapid evolution of herbicide resistance. *Heredity* 95: 485–492.

Neve, P. and Powles, S. (2005b). Recurrent selection with reduced herbicide rates results in the rapid evolution of herbicide resistance in *Lolium rigidum*. *Theoretical and Applied Genetics* 110: 1154–1166.

Neve, P., Vila-Aiub, M., and Roux, F. (2009). Evolutionary-thinking in agricultural weed management. *New Phytologist* 184: 783–793.

Neve, P., Busi, R., Renton, M., and Vila-Aiub, M.M. (2014). Expanding the eco-evolutionary context of herbicide resistance research. *Pest Management Science* 70: 1385–1393.

Neve, P., Barney, J.N., Buckley, Y. et al. (2018). Reviewing research priorities in weed ecology, evolution and management: a horizon scan. *Weed Research* 58: 250–258.

Oerke, E.C. (2006). Crop losses to pests. *The Journal of Agricultural Sciences* 144: 31.

Ohadi, S., Hodnett, G., Rooney, W., and Bagavathiannan, M. (2017). Gene flow and its consequences in *Sorghum* spp. *Critical Reviews in Plant Sciences* 36: 367–385.

Okada, M., Hanson, B.D., Hembree, K.J. et al. (2015). Evolution and spread of glyphosate resistance in *Conyza bonariensis* in California and a comparison with closely related *Conyza canadensis*. *Weed Research* 55: 173–184.

Palumbi, S.R. (2001). Humans as the world's greatest evolutionary force. *Science* 293: 1786–1790.

Pimentel, D., McNair, S., Janecka, J. et al. (2001). Economic and environmental threats of alien plant, animal, and microbe invasions. *Agriculture Ecosystems and Environment* 84: 1–20.

Powles, S.B. and Yu, Q. (2010). Evolution in action: plants resistant to herbicides. *Annual Review of Plant Biology* 61: 317–347.

Qi, X., Liu, Y., Vigueira, C.C. et al. (2015). More than one way to evolve a weed: parallel evolution of US weedy rice through independent genetic mechanisms. *Molecular Ecology* 24: 3329–3344.

Ravet, K., Patterson, E.L., Krähmer, H. et al. (2018). The power and potential of genomics in weed biology and management. *Pest Management Science* 74: 2216–2225.

Reagon, M., Thurber, C.S., Gross, B.L. et al. (2010). Genomic patterns of nucleotide diversity in divergent populations of US weedy rice. *BMC Evolutionary Biology* 10: 180.

Robineau, T., Batard, Y., Nedelkina, S. et al. (1998). The chemically inducible plant cytochrome P450 CYP76B1 actively metabolizes phenylureas and other xenobiotics. *Plant Physiology* 118: 1049–1056.

Rowlands, D.G. (1959). A case of mimicry in plants – *Vicia sativa* L. in lentil crops. *Genetica* 30: 435–446.

Sahli, H.F., Conner, J.K., Shaw, F.H. et al. (2008). Adaptive differentiation of quantitative traits in the globally distributed weed, wild radish (*Raphanus raphanistrum*). *Genetics* 180: 945–955.

Salisbury, E.J. (1961). *Weeds and Aliens*. London: Collins.

Schneeweiss, G.M. (2007). Correlated evolution of life history and host range in the nonphotosynthetic parasitic flowering plants *Orobanche* and *Phelipanche* (Orobanchaceae). *Journal of Evolutionary Biology 20* (2): 471–478.

Snir, A., Nadel, D., Groman-Yaroslavski, I. et al. (2015). The origin of cultivation and proto-weeds, long before neolithic farming. *PLoS One* 10: e0131422.

Storkey, J. and Neve, P. (2018). What good is weed diversity? *Weed Research* 58: 239–243.

Thrall, P.H., Bever, J.D., and Burdon, J.J. (2010). Evolutionary change in agriculture: the past, present and future. *Evolutionary Applications* 3: 405–408.

Thrall, P.H., Oakeshott, J.G., Fitt, G. et al. (2011). Evolution in agriculture: the application of evolutionary approaches to the management of biotic interactions in agro-ecosystems. *Evolutionary Applications* 4: 200–215.

Urban, M.C., Phillips, B.L., Skelly, D.K., and Shine, R. (2007). The cane toad's (*Chaunus [Bufo] marinus*) increasing ability to invade Australia is revealed by a dynamically updated range model. *Proceedings of the Royal Society B: Biological Sciences* 274: 1413–1419.

Vaz Patto, M.C., Díaz-Ruiz, R., Satovic, Z. et al. (2008). Genetic diversity of Moroccan populations of *Orobanche foetida*: evolving from parasitising wild hosts to crop plants. *Weed Research* 48: 179–186.

Viard, F., Bernard, J., and Desplanque, B. (2002). Crop-weed interactions in the *Beta vulgaris* complex at a local scale: allelic diversity and gene flow within sugar beet fields. *Theoretical and Applied Genetics* 104: 688–697.

Vigueira, C.C., Olsen, K.M., and Caicedo, A.L. (2013). The red queen in the corn: agricultural weeds as models of rapid adaptive evolution. *Heredity* 110: 303–311.

Vigueira, C.C., Qi, X., Song, B.-K. et al. (2019). Call of the wild rice: *Oryza rufipogon* shapes weedy rice evolution in Southeast Asia. *Evolutionary Applications* 12: 93–104.

Warwick, S.I. and Black, L.D. (1986). Electrophoretic variation in triazine-resistant and susceptible populations of *Amaranthus retroflexus* L. *New Phytologist* 104: 661–670.

Warwick, S.I. and Black, L.D. (1993). Electrophoretic variation in triazine-resistant and-susceptible populations of the allogamous weed *Brassica rapa*. *Weed Research* 33: 105–114.

Warwick, S.I. and Stewart, C.N. (2005). Crops come from wild plants: how domestication, transgenes, and linkage together shape ferality. In: *Crop Ferality and Volunteerism* (ed. J. Gressel), 9–30. Boca Raton, FL, USA: CRC Press.

Waselkov, K.E. and Olsen, K.M. (2014). Population genetics and origin of the native North American agricultural weed waterhemp (*Amaranthus tuberculatus*; *Amaranthaceae*). *American Journal of Botany* 101: 1726–1736.

Wedger, M.J., Topp, C.N., and Olsen, K.M. (2019). Convergent evolution of root system architecture in two independently evolved lineages of weedy rice. *New Phytologist* 110: 1031–1042.

Weiner, J., Andersen, S.B., Wille, W.K.M. et al. (2010). Evolutionary agroecology: the potential for cooperative, high density, weed-suppressing cereals. *Evolutionary Applications* 3: 473–479.

Welsh, A.B. and Mohamed, K.I. (2011). Genetic diversity of *Striga hermonthica* populations in Ethiopia: evaluating the role of geography and host specificity in shaping population structure. *International Journal of Plant Sciences* 172: 773–782.

Willis, K.J. (2017). *State of the World's Plants (2017) Report*. UK: Royal Botanic Gardens Kew.

Yuan, J.S., Tranel, P.J., and Stewart, C.N. Jr. (2007). Non-target-site herbicide resistance: a family business. *Trends in Plant Science* 12: 6–13.

Zeven, A.C. and De Wet, J.M. (1982). *Dictionary of Cultivated Plants and their Regions of Diversity: Excluding Most Ornamentals, Forest Trees and Lower Plants*. Wageningen, The Netherlands: Pudoc.

Ziska, L.H., Gealy, D.R., Burgos, N.R. et al. (2015). Weedy (red) rice: an emerging constraint to global Rice production. *Advances in Agronomy* 129: 181–228.

16

Persistence Strategies of Weeds: Synopsis and the Future

Mahesh K. Upadhyaya[1], David R. Clements[2], and Anil Shrestha[3]

[1] *Faculty of Land and Food Systems, University of British Columbia, Vancouver, BC, Canada*
[2] *Department of Biology, Trinity Western University, Langley, BC, Canada*
[3] *Department of Viticulture and Enology, California State University, Fresno, CA, USA*

16.1 Introduction

Weeds are a serious problem in agricultural and many nonagricultural ecosystems. In agriculture, they reduce crop yields significantly by competing for resources necessary for plant growth and development. For example, in corn (*Zea mays*) production in North America, uncontrolled weed populations have been estimated to reduce crop yield by 50% (Soltani et al. 2016). In addition to crop yield loss, weeds also cause many other problems. They can lower the quality of agricultural products, interfere with transportation and recreational activities, poison livestock, lower land values, interfere with harvesting, reduce property values, and cause allergies. Furthermore, many weed control practices are not environmentally sound and drain producer resources (both finances and time).

With limited cultivable land and increasing human population, globally, there is an urgent need to increase agricultural productivity. Globally weeds are responsible for reducing yields of major crops by approximately 34% (Oerke 2006). Therefore, effective weed control is necessary to ensure future food security. In the past six decades, the availability of effective and relatively inexpensive herbicides has led to a heavy reliance on the use of synthetic chemicals for weed control, yet weeds have persisted. During the past six decades, we seem to have forgotten to some extent the principles of sustainability and time-tested cultural weed control options and the fact that weeds are a component in agroecosystems, where many biotic and abiotic factors interact in a complex web. For the development of economically viable, environmentally friendly, and socially responsible weed management programs, it is important to understand why weeds are so persistent. Selective forces of agronomic practices have helped weeds evolve adaptive traits to persist in agricultural systems. A sound knowledge of persistence strategies of weeds at the population, individual plant, and the cellular level could potentially identify targets for controlling weeds in an

Persistence Strategies of Weeds, First Edition. Edited by Mahesh K. Upadhyaya, David R. Clements, and Anil Shrestha.

ecologically responsible way. This could also aid in development of safer and more effective chemicals to target weeds, including chemicals to directly deplete soil seedbanks.

In this chapter, we provide a synopsis of the information, ideas, and future directions provided by the authors of this book on topics related to persistence strategies of weeds. Readers should consult relevant chapters for more in-depth discussion on these topics.

16.2 Weed Propagation, Dissemination, and Seed and Vegetative Propagule Banks

A sound understanding of species diversity, seed germination, persistence of seeds in the soil, and seedling emergence is essential for developing effective weed management strategies. In this section we discuss the gist of Chapters 2 (Netto et al.) and 11 (Klimešová and Martínková) from this book. Annual, biennial, and most perennial weeds generally invade an area as seeds. Since weed seeds can tolerate extreme environmental conditions, they can persist in the soil for long periods and continue to infest an area with new seedlings or, in the case of perennial weeds, vegetative propagules. Information on weed seeds and vegetative propagules, therefore, is very important for understanding persistence of weeds.

Weeds generally produce large number of seeds per plant. For example, redroot pigweed (*Amaranthus retroflexus*) can produce up to 229 175 seeds per plant and common mullein (*Verbascum thapsus*) up to 223 200 seeds per plant (Zimdahl 2018). Depending on the plant species and abundance, seed production per unit area can be enormous. Not all seeds produced, however, are capable of germinating. While prolific seed production is one of the strategies that some weeds use for their persistence, some species produce a relatively moderate number of seeds (e.g. wild oats [*Avena fatua*], only 20–1070 seeds per plant [Beckie et al. 2012]) but still may be very persistent.

Many environmental factors (e.g. temperature, soil moisture, and nutrient status) can significantly affect the size and the number of seeds produced by a plant. The time of weed seedling emergence relative to crop emergence also impacts weed seed production. Weed seedlings that emerge prior to emergence of an associated crop generally produce more seeds compared with those that emerge after the crop. Controlling weeds before they set seeds is important for any weed management strategy.

Weeds can disseminate seeds with (allochory) and without (autochory) help of external vectors. Effective dissemination is important for maintaining seed abundance in an area and for invading new areas. How far a seed disseminates from the parent plant depends on how high the seed is located on a plant, seed characteristics, and the type and effectiveness of the vector involved. Some weeds can eject seeds by force (dehiscence), which can move seeds to short distances from the parent plant. Water, wind, animals (including humans), farm machinery, soil amendments, hay, and crop seeds are some of the vectors for weed seed dissemination. Weeds can disseminate seeds to very long distances with the help of these vectors. A sound understanding of dissemination of weed seeds is very important for developing strategies to prevent their introduction to new areas and spread.

Weeds may be adapted to shattering their seeds prior to crop harvest. These seeds, depending on their dormancy and longevity, can form a persistent seedbank in soil. Seeds that fall on the soil surface can move passively a few millimeters deep by moving through soil cracks, pores, and cavities. The downward movement of water can aid to this transportation. Coarse soil allows this movement more than soils rich in clay due to differences in porosity. Interestingly, some weeds have developed mechanisms that aid their seed burial. For example, wild oats have awns, which rotate in opposite directions upon wetting and drying, and this movement can facilitate their burial through soil cracks and cavities to deeper layers of the soil profile (Somody et al. 1985). Agricultural practices, particularly some types of tillage (e.g. moldboard plowing), can bury weed seeds much deeper in the soil profile. Deeply buried seeds can persist much longer because of a more stable environment and lack of factors (e.g. predators and soil microbes) that either physically damage seeds or make them nonviable. Tillage and crop rotation can also influence weed diversity in the soil seedbanks. A sound understanding of weed seedbanks (diversity, size, distribution in soil profile, seed dormancy cycling, etc.) is essential for development of effective management strategies.

Weed seedbanks are studied either by counting seeds of different species in soil samples or by monitoring seedling emergence. While the former approach is very tedious, labor-intensive, and time-consuming, the latter suffers from a possible lack of correlation between changes in seedbanks and the number of emerged seedlings. Artificial seedbanks created by burying a known number of seeds in either bags or containers to different depths have been used to study the soil seedbank dynamics. The burial process itself, however, may affect soil characteristics and influence seedbank dynamics. Placement of nondormant seeds in an area and monitoring the spread of seedling emergence have been used to study effects of agricultural practices (e.g. tillage) on seed movement in soil banks. Ceramic and plastic beads have also been used to track seed movement in soil in response to different agricultural disturbances.

Models describing dynamics of weed seedbanks with a focus on one species in one crop (Korres and Norsworthy 2017), one species in multiple crops (Osipitan et al. 2019), and multiple species in multiple crops (Colbach et al. 2016) have been developed. Involvement of several factors in regulation of seedbank dynamics and microsite differences makes this work very complex.

Weeds have developed adaptive strategies to deal with disturbances in agroecosystems. These include avoidance strategy (a short life cycle, rosette growth habit to avoid mowing and grazing), resistance strategy (e.g. thorns to protect from herbivores), and tolerance strategy (e.g. storage of buds and reserves safely belowground) to survive after a disturbance. With frequent disturbance, perennial species capable of vegetative reproduction may become significant components of weed populations.

Buds with potential to propagate are collectively referred to as a bud bank, and this reservoir of sprouting potential can play a significant role in persistence of some weed species. Species adapted to disturbances form buds (on stems or roots) as a part of their tolerance strategy. These weeds produce buds at a higher rate and locate these buds and carbohydrate reserves strategically to optimize regeneration potential. Plants can also utilize adaptations such as belowground germination, contractile roots that pull stem bases belowground, and formation of underground rhizomes (Putz and Sukkau 2002; Ott et al. 2019). The presence of buds deeper in the soil, as in horsetail (*Equisetum arvense*) and Canada thistle (*Cirsium arvense*), indicates a higher probability of survival in response to a disturbance.

The soil bud bank can produce new plants following fragmentation of the belowground parts of a plant by tillage. Resprouting of these weeds is lower when food storage, bud bank, and the soil nutrient status are depleted by either a disturbance or a seasonal cycle. Species, plant age, and carbohydrate storage amounts can influence the sprouting ability of weeds (Martínková et al. 2004; Sosnová et al. 2014). A plant's ability to resprout is reduced when stored carbohydrate reserves are depleted in supporting plant growth and seed production (Cuthbertson 1972; Taab et al. 2018). Knowledge of life history stages and resprouting ability is important for the timing of weed management (Verwijst et al. 2018).

In some species, injury can cause root sprouting (e.g. in biennial weeds *Barbarea vulgaris* and *B. stricta*) and produce several plants behaving as short-lived perennials. After damage in *Rorippa palustris* and *B. vulgaris*, bud banks are activated first on stems, and root buds sprout only after these buds are exhausted or removed (Bartušková and Klimešová 2010).

Many perennial weeds normally spread laterally no more than a few meters per year. However, when tillage operations fragment their underground rhizomes, long-distance dispersal is possible. Unlike soil seedbanks, which can be very persistent, longevity of bud-bearing organs and therefore of the buds is limited. These organs generally do not survive when aboveground parts of a plant are not produced for the whole season (Shefferson et al. 2018).

In no-till systems without herbicide use, perennial weeds are favored. In such situations, systemic herbicides may be useful for controlling belowground parts of perennial weeds. Herbicide resistance also develops more rapidly in annual than perennial weeds because of their shorter life cycle and greater potential for dissemination of resistant genotypes.

16.3 Weed Seed Dormancy and Longevity

16.3.1 Seed Dormancy

As in the previous section, here we provide the gist of seed dormancy issues presented by Upadhyaya et al. and Gu in Chapters 3 and 4 of this book. Seed dormancy and its modification by the environment, both during seed development and post-dispersal, play an important role in persistence of weeds (Batlla and Benech-Arnold 2014; Oreja et al. 2017). The environment during seed development can affect dormancy of mature seeds by influencing biochemical and structural changes during development. The position of seeds on a plant, in the inflorescence or in the fruit, is important in determining these influences. Polymorphism for seed dormancy, where seeds produced from one plant or from plants of a population of a species exhibit different dormancy characteristics, has been demonstrated in many weed species. The variability in seed dormancy distributes seed germination over time and leads to the development of persistent soil seedbanks, which makes weed management a challenge.

Dormancy at the time of seed separation from its parent plant is called primary dormancy. This dormancy can be due to several mechanisms including seed coat permeability to water and O_2, a mechanically hard seed coat, presence of germination inhibitors in seed or its coverings, immature embryo, plant hormones, particularly gibberellins and abscisic acid, and possibly some metabolic blocks. Primary dormant seeds undergo after-ripening with time, become nondormant, and can germinate if conditions are right. However, under certain unfavorable conditions, they can become dormant again (secondary or induced dormancy). Secondary dormant seeds become nondormant again with time. Several factors can stimulate germination of primary or secondary dormant seeds in soil seedbanks. While there are many similarities and differences in primary and secondary seed dormancies, it is not clear to what extent their regulations are similar (Hilhorst and Toorop 1997; Chahtane et al. 2017).

The cycle of release and induction of dormancy (dormancy cycling) can occur many times in seedbanks until conditions are optimum for germination so that the resulting plant can grow and complete its life cycle. Seeds tune their dormancy and germination cycle in response to environmental cues. Genetic, physiological, and biochemical makeup of seeds and influence of biotic and abiotic factors are important in regulating dormancy cycling.

Seed longevity, seed dormancy, and influence of many endogenous and environmental signals play important role in persistence of weeds (Martel et al. 2018). Environmental factors can influence dormancy cycling in weed seedbanks. Since the ability to predict the timing of seedling emergence is very important to make weed management decisions, it is necessary to understand the regulation of dormancy cycling in soil seedbanks. Unfortunately, most research on environmental effects on seed dormancy employs only one or two factors, and we therefore know very little about complex interactions among the various factors that regulate dormancy cycling in agroecosystems. A sound understanding of how these factors affect dormancy cycling, their interactions, and their physiological underpinnings could potentially provide ways to manipulate dormancy cycling and improve weed management. For example, manipulation of light conditions during cultivation has been reported to stimulate germination of photoblastic seeds. The light environment can be modified by modifying plant architecture, planting density, row spacing, and the timing of cultivation (day or night) (Batlla and Benech-Arnold 2014).

Although many molecular mechanisms are involved in regulation of seed dormancy, how these work in regulating a complex dormancy cycling in soil seedbanks is not very clear (Finch-Savage and Footitt 2017). Several genes have been identified in regulation of dormancy and its modification by environmental factors during seed development and dormancy cycling. Since tissues of both zygotic and maternal origin are involved in dormancy regulation, genetics of both these tissues is important in this regard (Bentsink and Koornneef 2008). Most of the work on the molecular basis of seed dormancy has been done with crop plants, and has only been studied for a select few weed species, primarily Arabidopsis and weedy rice (*Oryza sativa*). Little information is available for most common agricultural weeds. However, this is an important and interesting area of research, which could potentially provide information that may allow us to manipulate dormancy cycling to improve weed management efficacy.

Weedy rice is a serious problem in rice production systems. Genetic diversity in seed dormancy and influence of temperature during seed development on dormancy have been reported in this species (Gu et al. 2003). Early shattering, seed dormancy, and seed longevity are important traits in persistence of this weed. Many genes control seed dormancy, with each having a small effect on germinability. At present, wild rice is the only common weed in agricultural production systems where quantitative trait loci (QTL) controlling seed dormancy and some associated traits have been mapped. Phenotypic associations suggest coadaptive relations between seed dormancy and some other traits. Genes involved in primary dormancy imposed through embryo, endosperm, or maternal tissues, QTL epistatic and genotype x environment interactions, and association of seed dormancy with wild-like (e.g. seed shattering, pericarp and hull color, awn characters, and seedbank longevity) and crop-mimic (e.g. plant height and flowering time) traits have been studied in this weed. These trait associations could have synergistic effects on success of this species as a weed. More research is needed to improve our understanding of the genetics of seed dormancy and its association with other traits in this and other weed species.

Most weed control methods in current use aim to control or kill seedlings. Effective and practical options to directly deplete soil seedbanks by killing weed seeds with chemicals or chemically stimulating their germination and then killing seedlings are not available. Several chemicals have been reported to stimulate germination of weed seeds under laboratory conditions. However, with some exceptions (e.g. nitrates), they are either ineffective in the soil or their use under field conditions is not practical. More research is needed to identify chemicals that can be used in the field to deplete seedbanks by stimulating seed germination.

16.3.2 Seed Longevity

Seed longevity is discussed in detail by Bajwa et al. in Chapter 6 of this book. Here we highlight the major points of that chapter. Seed longevity, in association with seed dormancy, plays an important role in maintenance of soil seedbanks and therefore in persistence of weeds. An understanding of the mechanisms that regulate seed longevity is important in prioritizing which weed species to control and for how long to exhaust its soil seedbank. In many studies, the time course of germinability of buried seeds has been monitored. These studies show seeds of many weed species can survive in soil seedbanks for very long periods, the number of viable buried seeds decline exponentially, and species differ significantly in their seed longevity.

Conditions during seed development, maturation and storage, and environmental factors in the soil, most importantly temperature and moisture, can influence seed longevity by affecting the biochemistry of seed "aging" (Hay et al. 2006; Long et al. 2015; Leprince et al. 2017). While mechanisms involved in regulation of seed dormancy are poorly understood, several internal and external factors are known to influence seed longevity. The internal factors include genetics, seed characteristics, seed vigor and viability, seed morphology and anatomy, seed dormancy, and seed aging. The external factors that influence seed longevity include moisture and temperature, soil characteristics, land use and management operations, and seed predation and deterioration.

High genetic diversity of wild relatives of domesticated plants has been found to be linked to high seed longevity. Presotto et al. (2020) showed that high genetic diversity of sunflower (*Helianthus annuus*) was correlated with extended longevity of seeds in seedbanks. In another study, dormancy genes in weedy rice seeds were found to be responsible for regulating longevity in seedbanks (Pipatpongpinyo et al. 2020).

Seed vigor decreases as seeds age in soil banks, and reactive oxygen species (ROS) is an important cause for this reduction (Long et al. 2015). The ability of seeds to produce antioxidants to protect themselves from ROS declines with age, which results in greater damage by this highly reactive form of oxygen (Long et al. 2015). The lost seed vigor can, however, be reestablished by repair reactions during seed wetting and drying.

Seed morphology and anatomy can influence seed longevity. Changes in seed quality during long-term burial can change insect feeding preference (Martinková et al. 2006). Relationships between loss of seed viability and seed mass, volume loss, seed shape, and density (Saska et al. 2020) and between seed mortality in soil and seed coat thickness (Gardarin et al. 2010) have been reported. Seed coat chemistry, surface structure, presence of a hard seed coat, burial depth, and mechanisms for deeper burial in soil (e.g. by awns in wild oats) can also influence longevity. Furthermore, relationships between seed longevity and the absence of endosperm, embryo-to-seed ratio, embryo shape, and dormancy have been reported (Probert et al. 2009; Merritt et al. 2014).

Burial depth can influence seed longevity by affecting light, soil moisture and temperature, composition of soil air, decay, and predation. Seed longevity has been reported to be greater at higher pH, lower soil moisture content, and lower soil C : N ratio (Pakeman et al. 2012). It is also affected by soil nutrient status, frequency, intensity and depth of tillage, crop rotation, residual cover, nutritional value of seed, allelochemicals, and soil biota.

Information on longevity of seeds of different species in soil seedbanks could provide estimates of the effort required to deplete seedbanks to an acceptable level. A sound knowledge of factors affecting seed dormancy and longevity could allow us to manipulate these factors to help deplete weed seedbanks.

16.4 Agronomic Practices

A soil seedbank is a "memory" of abundance and diversity of weed infestations in an area (Cavers 1995). It is affected in part by past and present cultural practices and could provide information about future weed problems. Despite serious efforts to control weeds for centuries, weed presence continues because of the soil seedbanks. Every growing season some seeds are withdrawn from, and a new crop of seeds is deposited to these banks. The agronomic practices that impact this phenomenon have been discussed in greater detail by Swanton and Vazan in Chapter 10, and we reemphasize the gist of their discussion and provide a summary in this section.

The persistence of weeds is affected by the intensity and frequency of cultural practices. Significant differences observed between arable fields and woodland suggest that habitat disturbance selects for increased persistence of weeds (Thompson et al. 1998). The type and depth of tillage and soil characteristics affect dormancy cycling and the position of seeds in the soil profile. For example, seeds are distributed more uniformly within the top 15 cm of the soil profile with tillage using a moldboard plow, whereas in shallower tillage regimes they are present mostly in the upper soil profile (Clements et al. 1996; Nichols et al. 2015). The soil type (Cardina et al. 1991), shear strength, and pore size and how soil condition is affected by tillage (Swanton et al. 2000) are important factors in this regard.

The exposure of seeds to biotic and abiotic variables is strongly influenced by the depth of their burial. For example, temperature fluctuations, soil moisture content, wet–dry cycle frequency, soil organic matter content, and microbial and predator activities vary more in the upper, compared with the deeper soil profile (Mohler and Galford 1997; Fierer et al. 2003; Long et al. 2015). The light environment and the gaseous composition of the soil air also fluctuates more in the upper soil profile. These differences have a strong impact on seed germination, dormancy cycling, and persistence of weeds. Weed seeds stand a better chance of persisting at deeper depths than on or near the soil surface (Long et al. 2015).

Light can penetrate only to little depths in the soil and this is affected by the size of soil pores, cracks in soil, soil color, and the soil moisture content. Tillage during daytime can stimulate germination of some weed species (Scopel et al. 1994). In the case of photoblastic seeds, this stimulation is due to a very short exposure of seeds to light during soil disturbance with tillage. Weed species, state of seed dormancy, seasonality in sensitivity to light, depth of tillage, soil moisture content, thermal cues, and some other factors can influence the effect of light on seed germination (Benech-Arnold et al. 1988; Scopel et al. 1994; Benech-Arnold et al. 2000).

Crop rotation is a common practice in many agricultural systems. Growing crops differing in seeding time, planting pattern, cultural practices, plant canopy and root architecture, competitive ability, harvest time, crop nutrient requirement, herbicide application, etc. are expected to affect weed infestation and therefore the soil seedbank. While rotation involving diverse crops is known to influence weed infestation, research on its effects on the size and composition of soil seedbanks is limited, with variable results, which does not allow us to develop a clear conclusion about its effects on soil seedbanks. More research is needed in this important area of weed science.

Seed shattering prior to crop harvest is important for many weeds to ensure deposition of their seeds into the soil banks. Weeds differ with regard to the percentage of seeds that are retained by the plant at the time of harvest of a specific crop. Some weeds can retain up to 90% to 95% of their seeds on plants when the associated crop is harvested (Burton et al. 2016; Schwartz et al. 2016; Tidemann et al. 2017; Beckie et al. 2018a). Mechanical destruction or removal of weed seeds at the time of harvest provides an excellent opportunity to reduce their deposition into seed banks. Field impact mills (Shergill et al. 2020) to mechanically crush seeds or remove harvested seeds with the crop chaff using chaff carts can be effective in this regard.

Cover crops can reduce weed infestation by reducing the germination of photoblastic seeds and seedling emergence and by offering strong competition, which in turn could reduce weed seed production. The presence of cover crops also affects soil microbial and predator activity (Gallandt et al. 2005; Sarrantonio and Gallandt 2008), soil temperature, and the soil nitrate level (Mohler et al. 2018), which can influence seed germination (Adkins et al. 1984). The weed control offered by cover crops, however, is highly variable.

In rangelands, grazing can have a strong impact on weed dissemination, the abundance and diversity of soil seedbanks, and ultimately the persistence of weeds. In a 15-year study involving crop–livestock systems, it was found that when grazing intensity of cattle was low and a reduced tillage management was used, seedbank size was significantly reduced (Schuster et al. 2016). Similarly, in a 12-year study with reduced grazing, a sheep-grazed cover crop, and no-tillage management, the seedbank was reduced (Schuster et al. 2018).

Animal manure can deposit weed seeds to soil seedbanks and could potentially introduce new weed species to a farm, which could be a serious problem. It is therefore important to allow composting of manure for long enough, at sufficient temperatures, to ensure the loss of seed viability due to heat generated during composting (Eghball and Lesoing 2000). Many weed seeds can pass through the digestive tract of animals without losing their viability. The loss of seed viability during passage through an animal's digestive tract depends on the weed species and the type of the animal (e.g. bird, cattle, pig, etc.). Animals can disseminate seeds by excreting viable seeds as they graze or by moving seeds attached to their fur [e.g. hound's tongue (*Cynoglossum officinale*) nutlets].

While reducing the size of seedbanks should be an objective for any weed management strategy, some argue that the goal should not be eliminating the seedbank completely. Weed seeds are an important source of food for some soil-borne fauna, and their removal could eliminate these seed predators and reduce soil health adversely. A proper balance facilitating sustainability of the entire agroecosystem should be the goal.

16.5 Predation, Microbial Effects, and Allelopathy

16.5.1 Seed Predation

Seed predation is an important factor in the loss of weed seeds from soil seedbanks, as detailed by Saska and Honek in Chapter 8 of this book. It is a significant selection pressure on the seed's defense system (Dalling et al. 2011) and involves protective adaptation of weeds and their seeds and of predators for locating, handling, and digesting seeds. Seeds can be subjected to predation when they are still attached to the parent plant, on the soil surface after dispersal, and while buried in the soil seedbanks. Most invertebrate predators belong to the orders Coleoptera, Diptera, Hymenoptera, and Lepidoptera (Hulme and Benkman 2002). Predation of seeds in the soil is not well studied, and its extent may be underestimated. Predators that are active on the soil surface have difficulty in locating seeds that are buried deeper (Kulkarni et al. 2015). Earthworms are important consumers of weed seeds in the soil, and some seeds can pass through their gut and still retain the ability to germinate. Volatiles from seeds can help earthworms locate seeds in the soil (Clause et al. 2017).

Seed quality and surface/coat characteristics can change with time in soil seedbanks, which could influence susceptibility of seeds to predation (Saska et al. 2020). Toughness, thickness, and stability of the seed coat over time, seed texture, the presence of seed surface structures (e.g. hairs, ribs, or spines), seed shape, and the ratio of seed size to the predator size (Saska et al. 2019) are some adaptations that help weed seeds defend themselves against predators. Hard seed coats aid in the persistence of seeds in soil banks (Paulsen et al. 2013). Larger carabid beetles prefer larger seeds whereas smaller species generally prey on smaller seeds (Saska et al. 2019). In addition to predator size, the body structure (e.g. mouth part characteristics) of seed predators is also important in this regard.

Rodents are important seed consumers, with higher predation at high predator density. In patches where predation risk is higher for rodents, less effective predation occurs. More research is needed to link weed seed predation and the abundance of predators and on the relationship between seed predation and landscape characteristics.

Patchy weed infestations make seeds to be available in clusters. Predators respond to this patchy availability of seeds by aggregating in patches. Temporal trends associated with activity of predators have been observed; however, variations in seed predation with time cannot always be related to predator activity or density. Spatiotemporal seed predation patterns are crop specific (Heggenstaller et al. 2006).

Whether weed seed predators could control weeds in crop systems to an acceptable level is an interesting question. If seed predation could reduce weed density to or below economic threshold, seed predation could be a useful tool for weed management. For example, an annual loss of more than 50% of a new crop of seeds has been suggested to reduce the rate of increase of common corn cockle (*Agrostemma githago*) density (Firbank and Watkinson 1986). Lower density of weeds in fields with high abundance of seed predators has been reported (Bohan et al. 2011). Effectiveness of pre-dispersal seed predation may be reduced if a weed responds by increasing its vegetative propagation. However, if pre-dispersal predation occurs in combination with root feeding, effective weed control is possible (Myers and Sarfraz 2017).

If a field is tilled shortly after harvest, seed burial may hamper predator's ability to locate the seeds. Delaying tillage to allow predators to find and attack seeds could reduce seed deposition into soil seedbanks (Baraibar et al. 2017). Greater seed predation in agroecosystems with reduced or no-tillage supports this possibility (Baraibar et al. 2009; Petit et al. 2017). Mulching (Pullaro et al. 2006) and diverse crop rotations (Westerman et al. 2005) have been reported to increase seed predation.

A variety of methodologies have been used in seed predation research. Pre-dispersal seed predation can be assessed by dissecting inflorescence or developing pods, collecting isolated seed heads or pods in bags or glass vials to allow predators to complete their development, and subsequently measuring the seed damage. Post-dispersal seed predation assessment may involve estimating seed predation using sentinel (artificially exposed) seeds or estimating predation by studying seedling emergence, gut content analysis of predators, and estimating consumption and preferences of predators.

More research on effects of agroecosystem management on predator population and predation in a landscape context, behavior of predators at individual and community levels, seed predation in context of food web in fields, and the relationship between predation and seed defense is needed to improve our understanding of weed seed predation in cropping systems.

16.5.2 Microbial Effects

Soil microorganisms can have a significant effect on persistence of weed seeds in the soil seedbanks, as expounded on by Wagner and Mitschunas in Chapter 13 and summarized here. A sound understanding of weed seed–soil microbe interactions can have important implications for the management of weed seedbanks (Wagner and Mitschunas 2008). The use of molecular techniques, scanning electron microscopy, and advanced statistical procedures have advanced our understanding of this interaction and of the factors affecting it.

Some seed-borne microorganisms are already present in or on seeds prior to their entering the soil seedbank. Pathogenic microorganisms can attack viable seeds, whereas saprophytic microorganisms attack damaged or dead seeds. Only a small number of soil-borne microbes are pathogenic. The majority of soil-borne microorganisms are nonpathogenic or weakly pathogenic, and cannot attack weed seeds with intact defenses (Pitty et al. 1987; de Luna et al. 2011). Damaged or aged seeds leak exudates, which can stimulate spore germination and mycelial growth of seed decaying fungi. In velvetleaf (*Abutilon theophrasti*), sugars, and amino acids exuded by seeds chemotactically attract rhizobacteria toward pre-imbibed seed (Begonia and Kremer 1994). Fungal pathogens may enter a seed where its defenses are weak. Chitinases, proteases, and xylanase enzymes have been implicated in attacking seeds (Pollard 2018).

Some fungi and bacteria appear to suppress seed germination, and it has been shown that cell-free filtrates are sufficient and metabolites from microbes are responsible for this effect (Gealy et al. 1996; Widmer and Guermache 2004; Banowetz et al. 2008). Soil microorganisms can cause fatal germination by degradation of seed covering or by release of germination stimulating substances (e.g. stimulation of *Striga hermonthica* by ethylene-producing bacteria; Berner et al. 1999), under conditions where seedlings are unable to establish because of unfavorable conditions (van Leeuwen 1981).

Soil moisture is an important abiotic factor influencing mortality of weed seeds by soil-borne pathogens (Wagner and Mitschunas 2008). It influences seed infection as well as fungal pathogen growth after infection. Understanding the influence of interactions between abiotic and biotic factors is important to characterize seed–pathogen interactions and to develop effective weed control applications.

Biotic interactions also influence seed pathogen dynamics and the seed infection process, which could regulate seed–pathogen interaction in soil. Contamination of uninfected seeds by infected seeds increase infection because of the damage caused by seed-feeding insects (Kremer and Spencer 1989), and protection offered by nonpathogenic microorganisms against pathogens (Kremer 1986, 1987; Links et al. 2014) are some examples of biotic interactions that may influence damage of weed seed by pathogens.

Weed seeds use mechanical (e.g. intact, thick seed coat), chemical (e.g. the presence of phenolics, tannins, and flavonoids), or biochemical (e.g. the presence of chitinase, polyphenol oxidase, and peroxidase enzymes) defenses, mainly in the seed coat, to prevent infection by seed-decaying microorganisms. It has been suggested that mechanical defense is more effective than the chemical defense in persistence of weeds (Dalling et al. 2011; Davis et al. 2016) The ability to germinate rapidly is also a mechanism to avoid mortality due to pathogen infection (Dalling et al. 2011; Davis et al. 2016). Weed seeds also employ microbial defense, where seed-borne bacteria, for example, may be antagonistic to external fungi (Kremer et al. 1984; Kremer 1987). The various defense mechanisms previously described could operate in a complementary way in protecting seeds, i.e. when one mechanism (e.g. mechanical defense) is breached, the other (e.g. chemical and/or microbial) can serve as a secondary line of defense. It has been suggested that species with seeds persisting in the soil for a shorter duration may invest more in chemical defense compared with those species with a long-lived seedbank, which may invest more in a mechanical defense.

Depleting seedbanks by manipulation of microbe–seed interaction is an interesting idea. Three approaches: (i) inoculation of seeds with a host-specific pathogenic organism or microbial metabolite, (ii) manipulating the soil environment to promote seed decay, and (iii) weakening of seed defenses (e.g. damage to seed coat, using seed-feeding insects and germinating stimulating chemicals) have been suggested in this regard (Kennedy and Kremer 1996). Recent advances have increased our understanding of seed–microbe interaction, influences of abiotic and abiotic factors on these interactions, and seed defenses and defense syndrome. While the use of manipulation of microbe–seed interaction appears to have the potential to deplete weed seedbanks, more research including the use of omics technologies is needed before it becomes an effective and practical possibility.

16.5.3 Allelopathy

The following description of the role of allelopathy in weed persistence is based upon Chapter 14 (Latif et al.). Biochemical interactions among plants, microorganisms, and insects mediated through allelochemicals released by these organisms (allelopathy) (Rice 2012) can influence invasion and persistence of weeds (Hierro and Callaway 2003). Techniques allowing detection and quantification have advanced our understanding of these interactions significantly. Allelopathic influence could involve a direct effect of one species on another or an indirect effect, e.g. by affecting root microbiome (Bakker et al. 2018). Secondary metabolites produced by plants can facilitate specific microbial colonization, and the signal molecules produced by these microbes can regulate plant genes over several generations. These interactions could be important for sustainable crop production and agroecosystem health. More research is needed in this area, particularly on how allelopathic interactions influence plant community structure, biodiversity, and crop productivity.

The secondary metabolites responsible for allelopathic interactions are traditionally chemicals involved in plant defense, and do not play a significant role in the primary metabolism of plants. While some of these compounds are widely present in plants, some are produced only in response to stressors (Bartwal et al. 2013). These low molecular weight metabolites include phenolics, terpenoids, and alkaloids (Ahuja et al. 2012). Phenolics followed by flavonoids are the largest group of recently reported allelochemicals (Weston and Mathesius 2013; Macías et al. 2019).

Several mechanisms of action have been reported for plant growth inhibition by allelochemicals. These include effects on respiration, photosynthesis, membrane permeability, nutrient and water uptake, protein and nucleic acid synthesis, and plant growth regulation (Einhellig 2004; Rice 2012). The inhibitory activity of allelochemicals can be influenced by several factors, including their concentration and flux rate, soil activity, stage of plant, environment, and root–microbe interactions (Einhellig 2004; Rice 2012).

While allelochemicals can be present in all plant parts (Rice 2012), their diversity and concentration are higher in shoots compared with roots (Latif et al. 2020). Their high concentration in seeds could offer defense against decomposing microorganisms and enhance their longevity and persistence (Qasem and Foy 2001). Allelochemicals can be released by shoots (by stomates or glandular trichomes) or roots (by root hairs, epidermis, periderm, or border cells). Root exudation is the most common route for the release of allelochemicals in soil. While volatile allelochemicals diffuse from the plant surface, release of nonvolatile allelochemicals involves protein transporters (Weston et al. 2012). Release of allelochemicals is influenced by both intrinsic (e.g. species, age physiology, and tissue type) as well as extrinsic (physical injury, nutrient stress, temperature, irradiation, pathogen infection, soil properties, and plant–plant interactions) factors. Once released, the fate of allelochemicals may be influenced by biotic and abiotic properties of the soil. Metabolic profiling of allelochemicals using metabolomics and associated bioinformatics (Weston et al. 2015) and the impact of bio-activation of secondary metabolites by microbes are interesting and important areas of research in allelopathy. Knowledge obtained from this line of research could allow separation of the role of allelopathic influence from competition for resources in establishment and persistence of weeds. This research could also lead to the development of novel herbicides and may provide better management of species that produce allelochemicals that contribute to their invasion and persistence strategy.

While allelopathy has the potential for use in the development of an environmentally friendly weed management strategy, more research is needed on (i) cross-kingdom interaction using a multidisciplinary approach; (ii) exploration of key pathways and coevolution of invasive species with microbial metagenomes for defense and nutrient uptake; (iii) chemical ecology identifying active allelochemicals and physiology of invasive plants; (iv) release and transformation of allelochemicals in invaded areas and comparison of their biosynthesis and release in invaded versus native areas; and (v) how information on weed allelopathy could be used to manage persistent weeds without harming biodiversity.

16.6 Climate Change and Environmental Influences

16.6.1 Climate Change and Persistence

Weeds have traits that give them advantage over other plants. How these advantages are or will be influenced by climate change is an interesting question. Impacts of climate change on weed distribution, relative response of C_3 versus C_4 species, weed–crop interaction, and associated evolutionary changes and their impact on persistence of weeds are interesting areas for future research, as detailed in Chapter 12 (by Clements and DiTommaso) and summarized here.

Evolutionary processes can make invasive species even more unpredictable by enhancing invasive traits (Gillies et al. 2016; Sezen et al. 2016). There is considerable evidence suggesting that weedy species are able to benefit from climate change (Clements and DiTommaso 2011; DiTommaso et al. 2014; Ziska et al. 2019). Clements and DiTommaso (2011) predicted scenarios under which the traits considered key for success of invasive weeds (Whitney and Gabler 2008) could be selected for via climate change. Climate change can affect plant growth and development directly and positively because of increases in temperature and atmospheric CO_2 concentration. Increased frequency and intensity of droughts, fire, floods, and storms associated with climate change can impact plants adversely (Brown et al. 2018; Maxwell et al. 2019), but the same kinds of extreme events can promote persistence of weeds that colonize after disturbance or that benefit from increased dispersal from such events (e.g. Colleran and Goodall 2015).

Climate change can influence persistence of weeds in a variety of ways (e.g. changes in weed ecophysiology, distribution, and interaction with crops), which may bring about evolutionary changes. Ecophysiological responses may involve effects on plant biochemistry, plant structure (e.g. leaf morphology and anatomy, plant architecture, etc.), community structure, and plant phenology and fitness. Increases in temperature and water stress are two important factors of climate change. Since herbicide actions are influenced by many aspects of plant morphology, physiology, and biochemistry, climate change could also impact herbicide efficacy (Ziska 2016; Matzrafi 2019). Changes in light, ambient CO_2 concentration, temperature, precipitation, relative humidity, soil moisture stress, and wind are also important. While assessment of climate change on herbicide efficacy is a complex issue, weed managers will have to consider effects of climate change on herbicide volatility, photoinactivation, drift, breakdown in soil, absorption, and other management considerations.

With global warming, plant growth range is increasing poleward in North America, Europe, and Oceania. This is opening up new areas for crop production and weed infestations in these lands. Increase in wind velocity and frequency, and intensity of storms can facilitate dispersal of weeds. For example, a tropical storm in 2011 spread Japanese knotweed (*Reynoutria japonica*) by dispersing its fragments to new sites in Vermont, USA (Colleran and Goodall 2015).

The effects of climate change on plant phenology (i.e. life history schedules) could reduce fitness of a species that cannot adjust the timing of flowering and seed production with changes in their environment. Species that are able to modify their phenology by phenotypic plasticity or adaptation may thus be more successful with climate change. Non-native plants have been shown to have a greater ability, compared with native plants, to adjust phenologically to seasonal changes (Zettlemoyer et al. 2019). This suggests that non-native species may become more persistent, compared with native species, with the current scenario of climate change.

Warming associated with climate change could cause a shift from snow to rain, which could favor some weed species and help them outcompete other species (Concilio et al. 2013). Activities of weed pests and pathogens could also be affected, which may influence different species in weed flora differently. This could cause species-specific shifts in relative abundance of different weed species.

Novel environments created by climatic change are selection pressures on weed flora. These pressures could cause changes in traits that help in persistence of weeds (e.g. propagule production and dispersal, seed dormancy and germination behavior, seed longevity, assimilate partitioning, plant architecture, productivity, competitive ability, tolerance to herbivore, pathogen and abiotic environmental factors, etc.). They can also influence crops as well as cropping practices, which would also impact weeds in a variety of ways. Evolutionary changes could allow weeds to adapt to changes in climate, and weed species may differ in this regard. Further research in this area is urgently needed to assess where we are heading to in terms of weed problems throughout the globe and how we are going to deal with them.

16.6.2 Environment and Persistence

Seed dormancy, longevity, and germination requirements are important factors in determining how a soil seedbank will function as a source of seedling emergence and spread over time. In addition to seed predation, pathogen attack, inhibitory effect of allelochemicals, physiological deterioration, and decay, factors such as seed water content, light, O_2, composition of soil atmosphere, and soil chemistry also influence seedbank dynamics significantly. An in-depth discussion of the environmental effects on seedbank persistence is provided in Chapter 5 (by Benech-Arnold and Batlla), and here we summarize the major points.

Persistence of seedbanks and distribution of seed germination and seedling emergence over time enables weeds to infest an area and spread over a longer duration. Seedbanks of weeds with short-term persistence are expected to be exhausted sooner with their maintenance dependent on continual seed input. Seedlings of weeds with transient seedbanks can also emerge at different times, which aids weed persistence in variable and diverse habitats such as crop production systems.

In summer annuals, the level of dormancy decreases during winter, reaching a low level by the beginning of spring. This enables weeds to avoid seedling damage by low winter temperatures and ensures maximum seedling emergence in spring, providing seedlings better conditions to complete their life cycle and produce seeds. Winter annuals exhibit different dormancy dynamics. Their dormancy decreases during the summer and increases during the winter.

Environmental factors that influence seed dormancy can be divided into (i) those that regulate dormancy and (ii) those that terminate dormancy (Benech-Arnold et al. 2000). Factors that regulate dormancy level are responsible for seasonal dormancy patterns and the sensitivity of seeds to environmental signals, ensuring seedling emergence at a time when their survival chances are good (Finch-Savage and Leubner-Metzger 2006). Temperature is an important factor in influencing seed dormancy level in soil banks (Batlla and Benech-Arnold 2010). In summer annuals, low winter temperatures reduce dormancy levels, and high temperatures may induce thermo-dormancy. On the other hand, high summer temperatures reduce dormancy level, and low winter temperatures increase seed dormancy in winter annuals (Karssen 1982). This effect of temperature on seed dormancy can be influenced by seed moisture content. Release of dormancy by low temperatures in summer annuals requires seeds to be imbibed, whereas lowering of dormancy level in winter annuals generally requires drier "after-ripening."

Light and alternating temperatures (thermal amplitude of the cycle and the number of cycles) are two important factors in terminating seed dormancy. Other factors such as nitrate, CO_2, and C_2H_4 can also terminate seed dormancy (Benech-Arnold et al. 2000). A requirement of light and temperature fluctuations for termination of dormancy is ecologically important in determining the effects of state of seed burial, soil disturbance, and loss of vegetation cover (Benech-Arnold et al. 2000) on seed germination.

Seed dormancy is a quantitative character, and fluctuations in its level are associated with a change in the range of environmental factors under which a seed can germinate (Benech-Arnold et al. 2000; Batlla and Benech-Arnold 2010). Highly dormant seeds have a narrower range of environmental conditions required to germinate, and this range broadens as the seed dormancy level decreases. Interactions with other factors (e.g. temperature fluctuations and light) are important in this regard. It has therefore been suggested that dormancy level of seed populations can be assessed via the range of environmental factors (e.g. temperature and water potential) under which seeds can germinate and the sensitivity of seeds to dormancy terminating factors.

Temperature influences the rate of germination, within the range that allows germination, and therefore the timing of seedling emergence. Similarly, soil moisture level can also influence germination percentage and velocity. Various models (e.g. thermal time, hydrotime, and hydrothermal) have been suggested to describe effects of temperature and water potential on seed germination.

Seedbank dynamics are influenced by a complex interaction among seed characteristics and the various components of the biotic and abiotic environment. A sound understanding of how these factors and their interactions influence dormancy cycling in soil seedbanks and seedling emergence is important to predict dynamics of weed populations and to manage seedbanks to maintain populations at an acceptable level. While a persistent seedbank is important for persistence of weeds, it is not a requirement. Some weed species with a transient seedbank can be very successful because of their prolific seed production, effective dispersal, and seed germination behavior.

16.7 Weed Adaptation and Evolution and Persistence of Herbicide-Resistant Weeds

16.7.1 Weed Adaptation and Evolution

According to Harper (1956) "arable weeds constitute an ecological group selected and maintained in association by their fitness for existence under conditions of crop cultivation. They comprise species that have been selected by the very cultural practices that were originally designed to suppress them." While coevolution of weed management practices and weed flora have been occurring for a long time, it has become more intense in recent years because of decreased crop diversity and increased reliance on herbicides for weed control (Harker 2013; Storkey and Neve 2018). Human-mediated selection pressures (e.g. crop selection, rotation, weed management [e.g. herbicide use, tillage, and associated impacts]) have resulted in the origin and further evolution of weeds. Better understanding of weed evolution is necessary to know how plants adapt to changes in climate and crop management practices (Neve et al. 2018). Adaptation of weeds to these changes is important for persistence of weeds and is the subject of Chapter 15 (Neve and Caicedo) of this book, and a brief synopsis is provided here.

Weed adaptation involves several aspects of weed biology, ecology, and management including seed production and dispersal, seed dormancy and longevity, genetics, allelopathy, climate change, and effects of various biotic and abiotic influences and their interactions. Weed adaptation could involve morphological (e.g. plant and seed characters), phenological (life history schedule), and physiological (e.g. resistance to herbicide and environmental stressors) changes. Weeds are capable of mimicking crop seeds or plant morphology in a variety of ways (Barrett 1983), enabling them to persist well in cropping systems.

Changes in agricultural practices over a long period and adaptation of plants to these practices have resulted in the origin of weeds. Three possible origins of weeds have been suggested (De Wet and Harlan 1975; Chapter 15 by Neve and Caicedo): (i) from wild species; (ii) from hybridization between cultivated and wild plants; and (iii) from cultivated species (de-domestication). This area of research could provide insights on the nature and persistence of weeds, however this has received relatively little attention.

Weeds from wild plants may either be preadapted to agricultural practices or may be able to tolerate a wide range of environmental conditions. In either case, they could undergo further adaptation to selection pressures that occur in agroecosystems. Weeds may evolve at multiple locations, or conversely weedy traits may evolve once and then spread by seeds or pollen. Evidence suggesting multiple origins of weedy rice has been reported (Huang et al. 2018). Resistance or tolerance to herbicides (Vigueira et al. 2013) and crop mimicry (Barrett 1983) are good examples of adaptive evolution. Weeds originating from hybridization between crop and plants considered as weeds receive traits to adapt to agricultural environments from crop parents and weedy traits from the weed parents (Vigueira et al. 2013). Weeds can also evolve from crop plants by de-domestication, and a large number of crops, current and domesticated but not grown at present, could develop into weeds although low genetic variation in crops could stifle this somewhat (Vigueira et al. 2013). More research is needed to know if crops possessing certain traits or genotypic variations are more likely to evolve as weeds. Further research is needed on the ability of weeds to evolve from crop volunteers and the adaptive potential of weeds originating from crops.

Genetic variation in weediness and fitness traits is essential for weed evolution. Neve and Caicado in Chapter 15 have proposed the following four questions to help enhance our understanding of genetic variation in weeds and to help develop strategies to decrease weed adaptation: "(i) what is the role of phenotypic plasticity versus genetic variation in facilitating rapid weed adaptation to new and changing environments? (ii) does the majority of phenotypic variation in fitness-related traits arise from variation at single major genes, or are most traits under poly- or oligogenic control? (iii) is there standing genetic variation within populations upon which selection can act, or is variation created *de novo* via novel mutations, gene flow from surrounding populations, and/or hybridization and introgression with related species? and (iv) do weedy traits evolve once or a few times and spread between populations or are there multiple origins of weedy traits?"

If a species is seen as inherently weedy, the natural tendency is for us to focus on its control, but if weediness evolves because of selection, should the focus not be to reduce the selection pressure? The ability of weeds to adapt is probably the most important factor in their persistence, and this ability should be our focus while developing sound weed management strategies. Reduction in crop diversity and diversity of weed control procedures (e.g. overreliance on herbicides) has resulted in rapid adaptation of weeds. Increasing diversity of the crops grown and weed management options could limit weed adaptation. We need to apply evolutionary principles to weed management and foster more research collaborations between weed scientists and evolutionary biologists.

16.7.2 Persistence of Herbicide-Resistant Weeds

Genetic diversity and the number of individuals under selection are important in determining evolutionary response of weeds to selection pressures that occur in agroecosystems (Jasieniuk et al. 1996). Selection pressure exerted by herbicides is a powerful evolutionary force in agroecosystems as discussed in Chapter 7 (Beckie et al.) and encapsulated here. Frequent use of herbicides increases survival of herbicide-resistant weeds over several generations, signifying increased evolutionary fitness, thereby increasing their persistence. Efficacy of herbicide, frequency of application, and duration of its residual activity influence herbicide selection pressure on weed populations (Beckie 2006). Although frequent recurrent herbicide applications tend to represent strong selection pressure resulting in herbicide resistance, weed management with less frequent herbicide applications may result in adaptations to escape control or simply result in reduced control.

Spread of herbicide resistance traits by seeds and pollens, which is generally underestimated (Beckie et al. 2019), can affect persistence of weeds. The majority of flow of herbicide resistance traits is probably due to seed dispersal rather than pollen (Jasieniuk et al. 1996). Seed dispersal can spread herbicide-resistance traits over long distances. Higher rates of gene flow could reduce the time required to achieve high level of herbicide resistance in weed populations (Jasieniuk et al. 1996). Equipment, inadequately composted manure, and contaminated crop seeds and animal feed can potentially carry seeds over long distances (Mt. Pleasant and Schlather 1994; Michael et al. 2010). It is therefore important to control seed dispersal. Controlling spread of herbicide-resistant weeds by pollen, however, is difficult, particularly when resistance is due to a single dominant or semi-dominant nuclear gene.

Seed dormancy, tillage practices, and synchrony of crop and weed maturation can impact development of herbicide resistance in weed populations. In weeds with long-term dormancy, development of herbicide resistance is expected to be slower compared with seeds with a shorter-term dormancy because a smaller number of seeds are exposed to recurrent use of herbicide, and this may influence their persistence. Weeds can also adapt to escape residual effects of preemergent herbicides and persist by selecting for long-term dormancy or by delaying seed germination (Owen et al. 2011). Seeds are present on or near the soil surface in reduced or no-tillage system, in contrast to intensive tillage systems. This exposes them to microhabitats that promote their germination and "turnover" and influences evolution of herbicide resistance (Beckie et al. 2008).

When weeds and crop mature together, weed seeds can be harvested with crop seeds, if they are present at a height equal or greater than the height of crop seeds. Sowing of contaminated crop seeds can facilitate spread and persistence of herbicide-resistant weeds. Harvest weed seed control is becoming increasingly popular in some parts of the world (e.g. Australia). When weeds retain seeds on plants at a height similar or higher than the cutting height of the harvester, they can be collected by the combine. In the long-run, weeds may be able to adapt by developing prostrate growth habit, early seed shattering, and other similar strategies. In wild oats, 13% to 76% seeds fall on the ground before harvest depending on local conditions, which can reduce the effectiveness of harvest weed seed control (Beckie et al. 2018a).

Pleiotropic effects could affect fitness of herbicide-resistant weeds or their response to selection pressure by changing their phenology. This could influence dynamics of seedbanks and weed persistence. There could be a fitness penalty associated with herbicide resistance. For example, delayed germination, emergence and flowering, and reduction of competitiveness and fecundity have been reported in glyphosate-resistant kochia (*Kochia scoparia*) (Martin et al. 2017; Beckie et al. 2018b).

For sustainable agriculture, herbicide-resistant weeds must be controlled, and the widespread incidence of novel resistant weed populations must be curbed. Multidisciplinary research collaborations are required to better understand factors affecting the increase, spread, abundance, and fitness of herbicide-resistant weeds. Weed management systems emphasizing diverse weed control procedures and crop diversification, reducing addition of weed seeds to soil banks, and careful monitoring of herbicide resistance development are necessary.

16.8 Modeling the Persistence of Weed Populations

Because weeds are very persistent, and their seeds can survive in soil seedbanks for extended periods, there is generally a significant lag between a change in weed management and the response of weed flora. Historical crop management practices have changed significantly with time, and these changes are expected to impact weed flora. Traits that helped weeds persist with past management practices may be selected against under the current management practices. Understanding of the effects of past management on persistence traits could help in the development of approaches to predict response of weeds to future management practices.

While short-term experiments can provide information about the trend of a response to a change, they cannot precisely predict the long-term response. This is particularly true regarding environmental changes that occur on long time scales. However, responses to such changes can be assessed by models that simulate changes over a long period. A model consists of equations that predict these relationships using some species-specific parameters. Different approaches to modeling of weed persistence have been described. Models can be used to study relationship between weed management and persistence traits. Traits to be targeted in such studies can be determined based on the reasonably robust literature available. However, because information on weed density and frequency of a trait in an agroecosystem is not sufficient to determine whether a weed is harmful or beneficial, modeling approaches are needed. This was emphasized and described in detail in Chapter 9 (Storkey et al.), and we provide a brief overview of their discussion here.

Maintenance of soil seedbanks is an important factor in persistence of weeds in agroecosystems. Seeds and seedlings undergo many processes (e.g. maintaining viability, predation, microbial attacks, dormancy cycling, germination, seedling emergence, and growth and interplant competition). Crop management operations, the environment, and their interactions can influence these processes. This makes modeling of weed persistence very challenging. Because weeds are very persistent, weed management in one year can have an impact for several years. Since dynamics of seedbanks involve many processes mentioned previously, a process-based approach, integrating the various processes into a multi-annual weed dynamics model, is necessary.

Weed population modeling is important for sustainable weed management. It can help advance the understanding of effects of complex interaction among fluctuating environmental and biotic factors in influencing persistence of weeds. While models can provide useful information on effects of changes in management and the environment on weed communities, they have limitations and thus should be presented and interpreted carefully.

16.9 Conclusions

Taken together, the various aspects that contribute to the persistence of weeds represent a formidable combination of challenges for humanity. Ever since the advent of the agricultural revolution, humankind has had to toil to manage weeds infesting crop fields. The invention of synthetic herbicides has diminished this labor component of agriculture and have made a considerable contribution to weed management. Yet widespread herbicide resistance has highlighted how weeds can persist despite this technology. With globalization and global climate change, invasive weeds are now plaguing many other environmentally and economically important ecosystems and exhibiting persistence despite our best efforts to control them. As highlighted in the discussion on weed evolution in this book, ultimately the best strategy to employ against weed persistence is to understand the root cause of plants evolving weedy traits, to try to create conditions where plants with these traits will not evolve or flourish easily. However, as detailed throughout this book, this is much easier said than done, and further research on the many factors contributing to weed persistence is greatly needed.

References

Adkins, S.W., Simpson, G.M., and Naylor, J.M. (1984). The physiological basis of seed dormancy in *Avena fatua*. III. Action of nitrogenous compounds. *Physiologia Plantarum* 60: 227–233.

Ahuja, I., Kissen, R., and Bones, A.M. (2012). Phytoalexins in defense against pathogens. *Trends in Plant Science* 17: 73–90.

Bakker, P.A., Pieterse, C.M., de Jonge, R., and Berendsen, R.L. (2018). The soil-borne legacy. *Cell* 172: 1178–1180.

Banowetz, G.M., Azevedo, M.D., Armstrong, D.J. et al. (2008). Germination-arrest factor (GAF): biological properties of a novel, naturally-occurring herbicide produced by selected isolates of rhizosphere bacteria. *Biological Control* 46: 380–390.

Baraibar, B., Westerman, P.R., and Recasens, J. (2009). Effects of tillage and irrigation in cereal fields on weed seed removal by seed predators. *Journal of Applied Ecology* 46: 380–387.

Baraibar, B., Canadell, C., Torra, J. et al. (2017). Weed seed fate during summer fallow: the importance of seed predation and seed burial. *Weed Science* 65: 515–524.

Barrett, S.H. (1983). Crop mimicry in weeds. *Economic Botany* 37: 255–282.

Bartušková, A. and Klimešová, J. (2010). Reiteration in the short lived root-sprouting herb *Rorippa palustris*: does the origin of buds matter? *Botany* 88: 630–638.

Bartwal, A., Mall, R., Lohani, P. et al. (2013). Role of secondary metabolites and brassinosteroids in plant defense against environmental stresses. *Journal of Plant Growth Regulation* 32: 216–232.

Batlla, D. and Benech-Arnold, R.L. (2010). Predicting changes in dormancy level in natural seed soil banks. *Plant Molecular Biology* 73: 3–13.

Batlla, D. and Benech-Arnold, R.L. (2014). Weed seed germination and the light environment: implications for weed management. *Weed Biology and Management* 14: 77–87.

Beckie, H.J. (2006). Herbicide-resistant weeds: management tactics and practices. *Weed Technology* 20: 793–814.

Beckie, H.J., Leeson, J.Y., Thomas, A.G. et al. (2008). Risk assessment of weed resistance in the Canadian prairies. *Weed Technology* 22: 741–746.

Beckie, H.J., Francis, A., and Hall, L.M. (2012). The biology of Canadian weeds. 27. *Avena fatua* L. (Updated). *Canadian Journal of Plant Sciences* 92: 1329–1357.

Beckie, H., Blackshaw, R., Harker, K.N., and Tidemann, B. (2018a). Weed seed shatter in spring wheat in Alberta. *Canadian Journal of Plant Science* 98: 107–114.

Beckie, H.J., Blackshaw, R.E., Leeson, J.Y. et al. (2018b). Seedbank persistence, germination and early growth of glyphosate-resistant *Kochia scoparia*. *Weed Research* 58: 177–187.

Beckie, H.J., Busi, R., Bagavathiannan, M.V., and Martin, S.L. (2019). Herbicide resistance gene flow in weeds: under-estimated and under-appreciated. *Agriculture Ecosystems and Environment* 283: e106566. https://doi.org/10.1016/j.agee.2019.06.005.

Begonia, M.F.T. and Kremer, R.J. (1994). Chemotaxis of deleterious rhizobacteria to velvetleaf (*Abutilon theophrasti* Medik.) seeds and seedlings. *FEMS Microbiology Ecology* 15: 227–236.

Benech-Arnold, R., Ghersa, C., Sanchez, R., and Fernandez, A.G. (1988). The role of fluctuating temperatures in the germination and establishment of *Sorghum halepense* (L.) Pers. regulation of germination under leaf canopies. *Functional Ecology* 2: 311–318.

Benech-Arnold, R.L., Sánchez, R.A., Forcella, F. et al. (2000). Environmental control of dormancy in weed seed soil banks. *Field Crops Research* 67: 105–122.

Bentsink, L. and Koornneef, M. (2008). Seed dormancy and germination. In: *The Arabidopsis Book*, vol. 2008, 6, 18. The American Society of Plant Biologists https://doi.org/10.1199/tab.0119.

Berner, D.K., Schaad, N.W., and Völksch, B. (1999). Use of ethylene-producing bacteria for stimulation of *Striga* spp. seed germination. *Biological Control* 15: 274–282.

Bohan, D.A., Boursault, A., Brooks, D.R., and Petit, S. (2011). National-scale regulation of the weed seedbank by carabid predators. *Journal of Applied Ecology* 48: 888–898.

Brown, B.J., McLaughlin, B.C., Blakey, R.V., and Morueta-Holme, N. (2018). Future vulnerability mapping based on response to extreme climate events: dieback thresholds in an endemic California oak. *Diversity and Distributions* 24: 1186–1198.

Burton, N.R., Beckie, H.J., Willenborg, C.J. et al. (2016). Evaluating seed shatter of economically important weed species. *Weed Science* 64: 673–682.

Cardina, J., Regnier, E., and Harrison, K. (1991). Long-term tillage effects on seed banks in three Ohio soils. *Weed Science* 39: 186–194.

Cavers, P.B. (1995). Seed banks: memory in soil. *Canadian Journal of Soil Science* 75: 11–13.

Chahtane, H., Kim, W., and Lopez-Molina, L. (2017). Primary seed dormancy: a temporally multilayered riddle waiting to be unlocked. *Journal of Experimental Botany* 68: 857–869.

Clause, J., Forey, E., Eisenhauer, N. et al. (2017). Seed selection by earthworms: chemical seed properties matter more than morphological traits. *Plant and Soil* 413: 97–110.

Clements, D.R. and DiTommaso, A. (2011). Climate change and weed adaptation: can evolution of invasive plants lead to greater range expansion than forecasted? *Weed Research* 51: 227–240.

Clements, D.R., Benoit, D.L., Murphy, S.D., and Swanton, C.J. (1996). Tillage effects on weed seed return and seedbank composition. *Weed Science* 44: 314–322.

Colbach, N., Bertrand, M., Busset, H. et al. (2016). Uncertainty analysis and evaluation of a complex, multi-specific weed dynamics model with diverse and incomplete data sets. *Environmental Modelling and Software* 86: 184–203.

Colleran, B.P. and Goodall, K.E. (2015). Extending the timeframe for rapid response and best management practices of flood-dispersed Japanese knotweed (*Fallopia japonica*). *Invasive Plant Science and Management* 8: 250–254.

Concilio, A., Loik, M.E., and Belnap, J. (2013). Global change effects on *Bromus tectorum* L. (Poaceae) at its high-elevation range margin. *Global Change Biology* 19: 161–172.

Cuthbertson, E.G. (1972). *Chondrilla juncea* in Australia. 4. Root morphology and regeneration from root fragments. *Australian Journal of Experimental Agriculture and Animal Husbandry* 12: 528–534.

Dalling, J.W., Davis, A.S., Schutte, B.J., and Arnold, A.E. (2011). Seed survival in soil: interacting effects of predation, dormancy and the soil microbial community. *Journal of Ecology* 99: 89–95.

Davis, A.S., Fu, X., Schutte, B.J. et al. (2016). Interspecific variation in persistence of buried weed seeds follows trade-offs among physiological, chemical, and physical seed defenses. *Ecology and Evolution* 6: 6836–6845.

De Wet, J.M. and Harlan, J.R. (1975). Weeds and domesticates: evolution in the man-made habitat. *Economic Botany* 29: 99–108.

DiTommaso, A., Qin, Z., and Clements, D.R. (2014). Identifying climate change as a factor in the establishment and persistence of invasive weeds in agricultural crops. In: *Invasive Species and Global Climate Change* (eds. L. Ziska and J.S. Dukes), 353–370. CAB International.

Eghball, B. and Lesoing, G.W. (2000). Viability of weed seeds following manure windrow composting. *Compost Science and Utilization* 8: 46–53.

Einhellig, F.A. (2004). Mode of allelochemical action of phenolic compounds. In: *Allelopathy: Chemistry and Mode of Action of Allelochemicals* (eds. F.A. Macías, J.C.G. Galindo, J.M.G. Molinillo and H.G. Cutler), 217–238. Boca Raton, FL: CRC Press.

Fierer, N., Schimel, J.P., and Holden, P.A. (2003). Variations in microbial community composition through two soil depth profiles. *Soil Biology and Biochemistry* 35: 167–176.

Finch-Savage, W.E. and Footitt, S. (2017). Seed dormancy cycling and the regulation of dormancy mechanisms to time germination in variable field environments. *Journal of Experimental Botany* 68: 843–856.

Finch-Savage, W.E. and Leubner-Metzger, G. (2006). Seed dormancy and the control of germination. *New Phytologist* 171: 505–523.

Firbank, L.G. and Watkinson, A.R. (1986). Modeling the population-dynamics of an arable weed and its effects upon crop yield. *Journal of Applied Ecology* 23: 147–159.

Gallandt, E.R., Molloy, T., Lynch, R.P., and Drummond, F.A. (2005). Effect of cover-cropping systems on invertebrate seed predation. *Weed Science* 53: 69–76.

Gardarin, A., Dürr, C., Mannino, M.R. et al. (2010). Seed mortality in the soil is related to seed coat thickness. *Seed Science Research* 20: 243–256.

Gealy, D.R., Gurusiddaiah, S., and Ogg, A.G. Jr. (1996). Isolation and characterization of metabolites from *Pseudomonas syringae*-strain 3366 and their phytotoxicity against certain weed and crop species. *Weed Science* 44: 383–392.

Gillies, S., Clements, D.R., and Grenz, J. (2016). Knotweed (*Fallopia* sp.) invasion of North America utilizes hybridization, epigenetics, seed dispersal (unexpectedly) and an arsenal of physiological tactics. *Invasive Plant Science and Management* 9: 71–80.

Gu, X.Y., Chen, Z.X., and Foley, M.E. (2003). Inheritance of seed dormancy in weedy rice. *Crop Science* 43: 835–843.

Harker, K.N. (2013). Slowing weed evolution with integrated weed management. *Canadian Journal of Plant Science* 93: 759–764.

Harper, J.L. (1956). The evolution of weeds in relation to resistance to herbicides. In: *Proceedings of the Third British Weed Control Conference*, 179–188. Farnham, United Kingdom: The British Crop Protection Council.

Hay, F.R., Klin, J., and Probert, R.J. (2006). Can a post-harvest ripening treatment extend the longevity of *Rhododendron* L. seeds? *Scientia Horticulturae* 111: 80–83.

Heggenstaller, A.H., Menalled, F.D., Liebman, M., and Westerman, P.R. (2006). Seasonal patterns in post-dispersal seed predation of *Abutilon theophrasti* and *Setaria faberi* in three cropping systems. *Journal of Applied Ecology* 43: 999–1010.

Hierro, J.L. and Callaway, R.M. (2003). Allelopathy and exotic plant invasion. *Plant and Soil* 256: 29–39.

Hilhorst, H.W.M. and Toorop, P.E. (1997). Review on dormancy, germinability, and germination in crop and weed seeds. *Advances in Agronomy* 61: 111–165.

Huang, Z., Kelly, S., Matsuo, R. et al. (2018). The role of standing variation in the evolution of weediness traits in South Asian weedy rice (*Oryza* spp.). *G3: Genes, Genomes, Genetics* 8: 3679–3690.

Hulme, P.E., Benkman, C.W., 2002. Granivory. In: Herrera,C.M., Pellmyr, O. (Eds.), Plant–Animal Interactions: An Evolutionary Approach. Blackwell, Oxford, pp. 132–154.

Jasieniuk, M., Brûlé-Babel, A.L., and Morrison, I.N. (1996). The evolution and genetics of herbicide resistance in weeds. *Weed Science* 44: 176–193.

Karssen, C.M. (1982). Seasonal patterns of dormancy in weed seeds. In: *The Physiology and Biochemistry of Seed Development, Dormancy and Germination* (ed. A.A. Khan), 243–270. Amsterdam, The Netherlands: Elsevier.

Kennedy, A.C. and Kremer, R.J. (1996). Microorganisms in weed control strategies. *Journal of Production Agriculture* 9: 480–485.

Korres, N.E. and Norsworthy, J.K. (2017). Palmer amaranth (*Amaranthus palmeri*) demographic and biological characteristics in wide-row soybean. *Weed Science* 65: 491–503.

Kremer, R.J. (1986). Microorganisms associated with velvetleaf (*Abutilon theophrasti*) seeds on the soil surface. *Weed Science* 34: 233–236.

Kremer, R.J. (1987). Identity and properties of bacteria inhabiting seeds of selected broadleaf weed species. *Microbial Ecology* 14: 29–37.

Kremer, R.J. and Spencer, N.R. (1989). Impact of a seed-feeding insect and microorganisms on velvetleaf (*Abutilon theophrasti*) seed viability. *Weed Science* 37: 211–216.

Kremer, R.J., Hughes, L.B. Jr., and Aldrich, R.J. (1984). Examination of microorganisms and deterioration resistance mechanisms associated with velvetleaf seed. *Agronomy Journal* 76: 745–749.

Kulkarni, S.S., Dosdall, L.M., Spence, J.R., and Willenborg, C.J. (2015). Depth of seed burial and gender influence weed seed predation by three species of ground beetle (Coleoptera: Carabidae). *Weed Science* 63: 910–915.

Latif, S., Gurusinghe, S., Weston, P.A. et al. (2020). Metabolomic approaches for the identification of flavonoids associated with weed suppression in selected Hardseeded annual pasture legumes. *Plant and Soil* 447: 199–218.

van Leeuwen, B.H. (1981). Influence of micro-organisms on the germination of the monocarpic *Cirsium vulgare* in relation to disturbance. *Oecologia* 48: 112–115.

Leprince, O., Pellizzaro, A., Berriri, S., and Buitink, J. (2017). Late seed maturation: drying without dying. *Journal of Experimental Botany* 68: 827–841.

Links, M.G., Demeke, T., Gräfenhan, T. et al. (2014). Simultaneous profiling of seed-associated bacteria and fungi reveals antagonistic interactions between microorganisms within a shared microbiome on *Triticum* and *Brassica* seeds. *New Phytologist* 202: 542–553.

Long, R.L., Gorecki, M.J., Renton, M. et al. (2015). The ecophysiology of seed persistence: a mechanistic view of the journey to germination or demise. *Biological Reviews* 90: 31–59.

de Luna, L.Z., Kennedy, A.C., Hansen, J.C. et al. (2011). Mycobiota on wild oat (*Avena fatua* L.) seed and their caryopsis decay potential. *Plant Health Progress* 12 https://doi.org/10.1094/PHP-2011-0210-01-RS.

Macías, F.A., Mejías, F.J., and Molinillo, J.M. (2019). Recent advances in allelopathy for weed control: from knowledge to applications. *Pest Management Science* 75: 2413–2436.

Martel, C., Blair, L.K., and Donohue, K. (2018). PHYD prevents secondary dormancy establishment of seeds exposed to high temperature and is associated with lower PIL5 accumulation. *Journal of Experimental Botany* 69: 3157–3169.

Martin, S.L., Benedict, L., Sauder, C.A. et al. (2017). Glyphosate resistance reduces kochia fitness: comparison of segregating resistant and susceptible F2 populations. *Plant Science* 261: 69–79.

Martínková, J., Kočvarová, M., and Klimešová, J. (2004). Resprouting after disturbance in the short-lived herb *Rorippa palustris* (Brassicaceae): an experiment with juveniles. *Acta Oecologica* 25: 143–150.

Martinková, Z., Saska, P., and Honek, A. (2006). Consumption of fresh and buried seed by ground beetles (Coleoptera: Carabidae). *European Journal of Entomology* 103: 361–364.

Matzrafi, M. (2019). Climate change exacerbates pest damage through reduced pesticide efficacy. *Pest Management Science* 75: 9–13.

Maxwell, S.L., Butt, N., Maron, M. et al. (2019). Conservation implications of ecological responses to extreme weather and climate events. *Diversity and Distribution* 25: 613–625.

Merritt, D.J., Martyn, A.J., Ainsley, P. et al. (2014). A continental-scale study of seed lifespan in experimental storage examining seed, plant, and environmental traits associated with longevity. *Biodiversity Conservation Biology* 23: 1081–1104.

Michael, P.J., Owen, M.J., and Powles, S.B. (2010). Herbicide-resistant weed seeds contaminate grain sown in the Western Australian grainbelt. *Weed Science* 58: 466–472.

Mohler, C. and Galford, A. (1997). Weed seedling emergence and seed survival: separating the effects of seed position and soil modification by tillage. *Weed Research* 37: 147–155.

Mohler, C.L., Taylor, A.G., DiTommaso, A. et al. (2018). Effects of incorporated rye and hairy vetch cover crop residue on the persistence of weed seeds in the soil. *Weed Science* 66: 379–385.

Mt. Pleasant, J.M. and Schlather, K.J. (1994). Incidence of weed seed in cow (*Bos* sp.) manure and its importance as a weed source for cropland. *Weed Technology* 8: 304–310.

Myers, J.H. and Sarfraz, R.M. (2017). Impacts of insect herbivores on plant populations. *Annual Review of Entomology* 62: 207–230.

Neve, P., Barney, J.N., Buckley, Y. et al. (2018). Reviewing research priorities in weed ecology, evolution and management: a horizon scan. *Weed Research* 58: 250–258.

Nichols, V., Verhulst, N., Cox, R., and Govaerts, B. (2015). Weed dynamics and conservation agriculture principles: a review. *Field Crops Research* 183: 56–68.

Oerke, E.C. (2006). Crop losses to pests. *Journal of Agricultural Science* 144: 31–43.

Oreja, F.H., de la Fuente, E.B., and Batlla, D. (2017). Role of seed environment and covering structures on large crabgrass germination. *South African Journal of Botany* 111: 170–175.

Osipitan, O.A., Dille, J.A., Bagavathiannan, M.V., and Knezevic, S.Z. (2019). Modeling population dynamics of kochia (*Bassia scoparia*) in response to diverse weed control option. *Weed Science* 67: 57–67.

Ott, J.P., Klimešová, J., and Hartnett, D.C. (2019). The ecology and significance of below-ground bud banks in plants. *Annals of Botany* 123: 1099–1118.

Owen, M.J., Michael, P.J., Renton, M. et al. (2011). Towards large-scale prediction of *Lolium rigidum* emergence: II. Correlation between dormancy and herbicide resistance levels suggests an impact of cropping systems. *Weed Research* 51: 133–141.

Pakeman, R.J., Small, J.L., and Torvell, L. (2012). Edaphic factors influence the longevity of seeds in the soil. *Plant Ecology* 213: 57–65.

Paulsen, T.R., Colville, L., Kranner, I. et al. (2013). Physical dormancy in seeds: a game of hide and seek? *New Phytologist* 198: 496–503.

Petit, S., Trichard, A., Biju-Duval, L. et al. (2017). Interactions between conservation agricultural practice and landscape composition promote weed seed predation by invertebrates. *Agriculture Ecosystems and Environment* 240: 45–53.

Pipatpongpinyo, W., Korkmaz, U., Wu, H. et al. (2020). Assembling seed dormancy genes into a system identified their effects on seedbank longevity in weedy rice. *Heredity* 124: 135–145.

Pitty, A., Staniforth, D.W., and Tiffany, L.H. (1987). Fungi associated with caryopses of *Setaria* species from field-harvested seeds and from soil under two tillage systems. *Weed Science* 35: 319–323.

Pollard, A.T. (2018). Seeds vs fungi: an enzymatic battle in the soil seedbank. *Seed Science Research* 28: 197–214.

Presotto, A., Hernández, F., Casquero, M. et al. (2020). Seed bank dynamics of an invasive alien species. *Helianthus annuus* L. *Journal of Plant Ecology* 13: 313–322.

Probert, R.J., Daws, M.I., and Hay, F.R. (2009). Ecological correlates of ex situ seed longevity: a comparative study on 195 species. *Annals of Botany* 104: 57–69.

Pullaro, T.C., Marino, P.C., Jackson, D.M. et al. (2006). Effects of killed cover crop mulch on weeds, weed seeds, and herbivores. *Agriculture Ecosystems and Environment* 115: 97–104.

Putz, N. and Sukkau, I. (2002). Seedling establishment, bud movement, and subterranean diversity of geophilous systems in Apiaceae. *Flora* 197: 385–393.

Qasem, J. and Foy, C. (2001). Weed allelopathy, its ecological impacts and future prospects: a review. *Journal of Crop Production* 4: 43–119.

Rice, E.L. (2012). *Allelopathy*. Academic Press, 368 pp.

Sarrantonio, M. and Gallandt, E. (2008). The role of cover crops in North American cropping systems. *Journal of Crop Production* 8: 53–74.

Saska, P., Honěk, A., and Martinková, Z. (2019). Preferences of carabid beetles (Coleoptera: Carabidae) for herbaceous seeds. *Acta Zoologica Academiae Scientiarum Hungaricae* 65 (Suppl. 1): 55–74.

Saska, P., Foffová, H., Martinková, Z., and Honěk, A. (2020). Persistence and changes in morphological traits of herbaceous seeds due to burial in soil. *Agronomy* 10: 448. https://doi.org/10.3390/agronomy10030448.

Schuster, M.Z., Pelissari, A., De Moraes, A. et al. (2016). Grazing intensities affect weed seedling emergence and the seed bank in an integrated crop–livestock system. *Agriculture Ecosystems and Environment* 232: 232–239.

Schuster, M., Harrison, S., De Moraes, A. et al. (2018). Effects of crop rotation and sheep grazing management on the seedbank and emerged weed flora under a no-tillage integrated crop-livestock system. *The Journal of Agricultural Science* 156: 810–820.

Schwartz, L.M., Norsworthy, J.K., Young, B.G. et al. (2016). Tall waterhemp (*Amaranthus tuberculatus*) and palmer amaranth (*Amaranthus palmeri*) seed production and retention at soybean maturity. *Weed Technology* 30: 284–290.

Scopel, A., Ballare, C., and Radosevich, S. (1994). Photostimulation of seed germination during soil tillage. *New Phytologist* 126: 145–152.

Sezen, U.U., Barney, J.N., Atwater, D.Z. et al. (2016). Multi-phase US spread and habitat switching of a post-Columbian invasive, *Sorghum halepense*. *PLoS One* 11 (10): e0164584.

Shefferson, R.P., Kull, T., Hutchings, M.J. et al. (2018). Drivers of vegetative dormancy across herbaceous perennial plant species. *Ecology Letters* 21: 724–733.

Shergill, L.S., Bejleri, K., Davis, A., and Mirsky, S.B. (2020). Fate of weed seeds after impact mill processing in midwestern and mid-Atlantic United States. *Weed Science* 68: 92–97.

Soltani, N., Dille, J.A., Burke, I.C. et al. (2016). Potential corn yield losses from weeds in North America. *Weed Technology* 30: 979–984.

Somody, C.N., Nalewaja, J.D., and Miller, S.D. (1985). Self-burial of wild oat florets. *Agronomy Journal* 77: 359–362.

Sosnová, M., Herben, T., Martínková, J. et al. (2014). To resprout or not to resprout? Modeling population dynamics of a root-sprouting monocarpic plant under various disturbance regimes. *Plant Ecology* 215: 1245–1254.

Storkey, J. and Neve, P. (2018). What good is weed diversity? *Weed Research* 58: 239–243.

Swanton, C.J., Shrestha, A., Knezevic, S.Z. et al. (2000). Influence of tillage type on vertical weed seedbank distribution in a sandy soil. *Canadian Journal of Plant Science* 80: 455–457.

Taab, A., Andersson, L., and Bostrøm, U. (2018). Modelling the sprouting capacity from underground buds of the perennial weed *Sonchus arvensis*. *Weed Research* 58: 348–356.

Thompson, K., Bakker, J.P., Bekker, R.M., and Hodgson, J.G. (1998). Ecological correlates of seed persistence in soil in the north-west European flora. *Journal of Ecology* 86: 163–169.

Tidemann, B.D., Hall, L.M., Harker, K.N. et al. (2017). Suitability of wild oat (*Avena fatua*), false cleavers (*Galium spurium*), and volunteer canola (*Brassica napus*) for harvest weed seed control in western Canada. *Weed Science* 65: 769–777.

Verwijst, T., Tavaziva, V.J., and Lunkvist, A. (2018). Assessment of the compensation point of *Cirsium arvense* and effects of competition, root weight and burial depth on below-ground dry weight – leaf stage trajectories. *Weed Research* 58: 292–303.

Vigueira, C.C., Olsen, K.M., and Caicedo, A.L. (2013). The red queen in the corn: agricultural weeds as models of rapid adaptive evolution. *Heredity* 110: 303–311.

Wagner, M. and Mitschunas, N. (2008). Fungal effects on seed bank persistence and potential applications in weed biocontrol: a review. *Basic and Applied Ecology* 9: 191–203.

Westerman, P.M., Liebman, F.D., Menalled, A.H. et al. (2005). Are many little hammers effective? – Velvetleaf (*Abutilon theophrasti*) population dynamics in two- and four-year crop rotation systems. *Weed Science* 53: 382–392.

Weston, L.A. and Mathesius, U. (2013). Flavonoids: their structure, biosynthesis and role in the rhizosphere, including allelopathy. *Journal of Chemical Ecology* 39: 283–297.

Weston, L.A., Ryan, P.R., and Watt, M. (2012). Mechanisms for cellular transport and release of allelochemicals from plant roots into the rhizosphere. *Journal of Experimental Botany* 63: 3445–3454.

Weston, L.A., Skoneczny, D., Weston, P.A., and Weidenhamer, J.D. (2015). Metabolic profiling: an overview-new approaches for the detection and functional analysis of biologically active secondary plant products. *Journal of Allelochemical Interactions* 1: 15–27.

Whitney, K.D. and Gabler, C.A. (2008). Rapid evolution in introduced species, 'invasive traits' and recipient communities: challenges for predicting invasive potential. *Diversity and Distributions* 14: 569–580.

Widmer, T.L. and Guermache, F. (2004). Filtrates of rhizosphere bacteria suppressive to *Centaurea solstitialis* seed germination. *Journal of Plant Diseases and Protection Sonderheft* 19: 497–502.

Zettlemoyer, M.A., Schultheis, E.H., and Lau, J.A. (2019). Phenology in a warming world: differences between native and non-native plant species. *Ecology Letters* 22: 1253–1263.

Zimdahl, R.L. (2018). *Fundamentals of Weed Science*, 5e, 735. Fort Collins, USA: Academic Press.

Ziska, L.H. (2016). The role of climate change and increasing atmospheric carbon dioxide on weed management: herbicide efficacy. *Agriculture Ecosystems and Environment* 231: 304–309.

Ziska, L.H., Blumenthal, D.M., and Franks, S.J. (2019). Understanding the nexus of rising CO_2, climate change, and evolution in weed biology. *Invasive Plant Science and Management* 13: 79–88.

Index

Persistence Strategies of Weeds, First Edition. Edited by Mahesh K. Upadhyaya, David R. Clements, and Anil Shrestha.

b

C

d

e

g

h

i

s